Synthesis Lectures on Communications

This series of short books cover a wide array of topics, current issues, and advances in key areas of wireless, optical, and wired communications. The series also focuses on fundamentals and tutorial surveys to enhance an understanding of communication theory and applications for engineers.

Dhiman Deb Chowdhury

Future of Networks

Modern Communication Infrastructure

 Springer

Dhiman Deb Chowdhury
Hewlett Packard Enterprise
San Jose, CA, USA

ISSN 1932-1244   ISSN 1932-1708   (electronic)
Synthesis Lectures on Communications
ISBN 978-3-031-71439-9   ISBN 978-3-031-71440-5   (eBook)
https://doi.org/10.1007/978-3-031-71440-5

To my beloved mother, whose unwavering love, support, and wisdom have been my guiding light. Your strength and encouragement have inspired every step of this journey. This book is a reflection of your endless belief in me and a tribute to the values you have instilled. Thank you for being my greatest teacher, my strongest advocate, and my forever inspiration.

Preface

The world of networking is undergoing a profound transformation, reshaping how we connect, communicate, and conduct business. As we stand on the brink of the next great leap in technology, the forces driving this change are diverse, dynamic, and deeply interconnected. This book, *Future of Networks: Modern Communication Infrastructure*, is crafted to guide you through the intricate landscape of modern networking, where cutting-edge technologies converge to build the intelligent, resilient, and adaptive networks of tomorrow.

Future of Networks: Modern Communication Infrastructure brings together a comprehensive exploration of the latest networking technologies and trends, offering a detailed yet accessible roadmap for professionals, engineers, and decision-makers who aspire to lead in this ever-evolving domain. From the foundational concepts to the most advanced innovations, each chapter builds on the next, weaving a rich tapestry of insights designed to equip you with the knowledge and strategies needed to design and deploy next-generation networks.

In Chap. 1, we set the stage with a deep dive into the trends that are shaping the networking landscape. By examining the evolution of programmable networks, SDN, and the emergence of open standards, we highlight the shift from rigid, hardware-centric models to flexible, software-defined infrastructures that offer unprecedented control and scalability.

We then transition into Chap. 2, where we delve into the advances in networking SoCs (System on Chips). Here, you'll discover how the latest innovations in silicon design are revolutionizing network performance, enabling faster, smarter, and more energy-efficient devices that form the backbone of modern data centers and enterprise environments.

Chapter 3 introduces you to the realm of cloud networking, exploring how virtualization, SDN, and automation are redefining how we connect in a multi-cloud world. The discussion covers public, private, and hybrid cloud architectures, emphasizing the importance of scalable, secure, and highly available network solutions that cater to the demands of today's distributed applications.

As we move into Chap. 4, the focus shifts to container networking—a key component of the cloud-native ecosystem. This chapter explores the tools, technologies, and best practices that empower seamless communication between microservices, enhancing application agility and resilience in a world where speed and efficiency are paramount.

Chapter 5 takes a deep dive into network automation, showcasing the transformative potential of automating network operations. Through the use of orchestration tools, scripting, and AI-driven insights, this chapter outlines how automation is not just a convenience but a necessity for managing complex, large-scale networks with precision and efficiency.

Chapter 6 expands on the concept of network insights, discussing telemetry, observability, and the critical role of data-driven decision-making in modern networking. This chapter emphasizes the need for comprehensive visibility into network health, enabling proactive management and rapid response to emerging issues.

In Chap. 7, we confront the evolving landscape of network security. With threats growing in sophistication, this chapter highlights the integration of AI, Zero Trust models, and advanced threat detection techniques to create secure, adaptive networks that can defend against today's complex cyberthreats.

Chapter 8 explores the infrastructure required to support AI-driven workloads, delving into AI-native networking and the integration of high-performance computing, advanced interconnects, and scalable storage solutions. This chapter underscores how AI is not only transforming applications but also the networks that support them.

Chapter 9 brings the concept of self-healing networks to the forefront, detailing how AI and automation are converging to create networks that can autonomously detect, diagnose, and resolve issues. This vision of self-managing, resilient networks represents the pinnacle of innovation in network operations.

Finally, Chap. 10 serves as a culmination of the book, providing a strategic framework for building the networks of the future. It integrates the diverse technologies and methodologies discussed throughout the book, offering actionable insights on how to align networking strategies with business goals, enhance scalability, and maintain security in an ever-changing digital landscape.

This book is designed to be more than just a technical guide; it is a call to action for those who will build the networks that shape our future. We hope that by presenting these diverse forces of technology—from SDN and SoCs to AI-driven insights and self-healing capabilities—you will be inspired to rethink traditional approaches, embrace innovation, and craft strategies that propel your organization into the next generation of networking.

Whether you are a seasoned network architect, a business leader, or a student eager to learn, *Future of Networks: Modern Communication Infrastructure* offers a comprehensive outlook on the technological forces at play. It is our hope that this book will empower you to build the resilient, intelligent, and adaptive networks of tomorrow, and help you chart a path forward in this exciting new era of connectivity.

Thank you for embarking on this journey with us.

San Jose, CA, USA   Dhiman Deb Chowdhury

Contents

1.1 Introduction

Since 2008, academia and numerous industry forums have embarked on the journey of developing programmable networks and advancing the decoupling of hardware and software in networking gears. One notable milestone in this journey was the development of OpenFlow, a protocol specification initially crafted at Stanford University in 2008.[1] OpenFlow paved the way for the concept of Software-Defined Networking (SDN), which revolutionizes network infrastructure by enabling centralized management and programmability. While OpenFlow initially faced challenges in gaining widespread adoption, its introduction catalyzed a shift toward programmability in networking gears. Various vendors responded by developing their solutions to achieve SDN objectives, driving innovation and flexibility in network orchestration and management. In 2011, Stanford and Berkeley University established the Open Networking Foundation (ONF) to foster industry collaboration in SDN.[2] The creation of ONF marked a significant milestone, as it provided a platform for industry stakeholders to contribute to the advancement of SDN technologies. This initiative birthed the Open Networking concept, enabling greater industry participation and collaboration in the development of open networking standards.

Another important milestone was Open vSwitch (OvS), which created possibilities for network virtualization and helped in the cloudification of networks. OvS initially started as a project in Nicira and later became open source in 2009. Both OpenFlow and OvS emerged almost simultaneously and ushered in a new era in networking.

Simultaneously, Facebook launched the Open Compute Project (OCP) in 2011 to drive open standardization in computing and networking infrastructure.[3] OCP aimed to promote

---

[1] SDXCentral [1].

[2] ONF [2].

[3] OCP [3].

© The Author(s), under exclusive license to Springer Nature Switzerland AG 2025   1
D. D. Chowdhury, *Future of Networks*, Synthesis Lectures on Communications,
https://doi.org/10.1007/978-3-031-71440-5_1

the development and adoption of open hardware designs, fostering innovation and interoperability in data center technologies. Together, the establishment of ONF and the inception of OCP represented significant milestones in the evolution of networking and computing infrastructure. These initiatives empowered industry stakeholders to collaborate on the development of open standards, driving innovation, and accelerating the adoption of programmable and interoperable networking solutions.

To advance the objectives of network programmability, it became crucial to integrate programmability into the System on Chip (SoC), the fundamental engine driving networking gears. In 2013, Prof. Nick McKeown of Stanford University played a pivotal role in this effort by introducing the concept of the P4 programming language.[4] P4 was designed to empower dynamic SoCs, enabling them to deliver programmable services on demand. This innovation aimed to revolutionize the way networking infrastructure operates, allowing for greater flexibility and adaptability.

Furthermore, a workshop held at Stanford University in 2015 served as a platform to further explore and promote the principles of programmability in networking hardware. This event facilitated collaboration among industry experts and researchers, fostering the exchange of ideas and driving innovation in programmable networking technologies. One notable outcome of these efforts was the commercialization of the concept by Barefoot Networks, a startup founded by Prof. McKeown and Pat Bosshart. Barefoot Networks leveraged P4 and other programmable networking technologies to develop innovative solutions that enable unprecedented levels of flexibility and programmability in networking hardware.

These collective efforts ushered in a new era of industry collaboration and contributed to the emergence of open networks and network programmability. Industry leaders like Google, Amazon, and Facebook led the charge in shifting toward vendor-agnostic networking systems. This movement aimed to depart from traditional vendor-locked systems and embrace more flexible and interoperable solutions. This initiative gained momentum with support from Original Design Manufacturers (ODM) such as Delta, Celestica, Accton, and Quanta Cloud Technology (QCT), among others. As a result, the concept of bare metal switches emerged, aligning with the advancements seen in the computing world.

Bare metal switches provide a foundation for open networking, enabling organizations to leverage their expertise in respective fields. For instance, open-source software forums and software startups focused on developing software solutions for networking gears, while hardware vendors concentrated on producing newer generations of hardware platforms with increased speeds and feeds.

Thie figure below depicts a brief timeline for the development of Open Networking and network programmability. The timeline provided here does not reflect a comprehensive list of contributions toward open networking and network programmability rather a snapshot of some important milestones (Fig. 1.1).

---

[4] Thenextplaftorm [4].

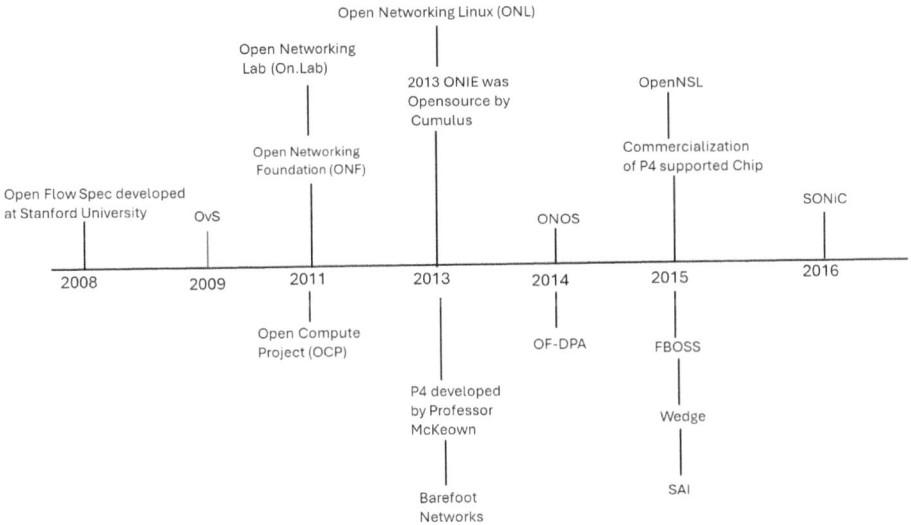

Fig. 1.1 Brief timeline of academia and industry contribution toward the open networks and network programmability

In the proceeding sections of this chapter, we will delve into some of the key milestones and contributions in the development of Open Networking, Network Virtualization, and network programmability. By exploring these milestones and contributions, you will gain a deeper understanding of how the interplay of technologies has led to a paradigm shift in the networking industry.

We will examine the emergence of foundational technologies such as OpenFlow, Open vSwitch (OvS), ONIE, and the P4 programming language, along with other related technologies, which have enabled programmability and flexibility in network hardware and software.

Finally, the chapter will conclude with a glimpse of what the future holds for networking infrastructure, highlighting emerging trends and technologies that are shaping the next generation of networks. Subsequent chapters will provide detailed information on those trends on technologies.

1.2   OpenFlow

In the late 1990s to early 2000s, my colleagues and I worked on a network solution known as "Policy Enable Networking" (PEN) at Nortel. The goal was to provide dynamism to the otherwise static network infrastructure. Our first step toward that goal was to use IP filtering techniques, and later, we made use of COPS (Common Open Policy Service). Until the fall of Nortel, mainly attributed to wrongdoing by its business leaders, PEN was gaining industry attention, and work was in progress for network programmability. However, with Nortel's collapse, we dispersed to different companies. I ended up at Netgear, which had a different focus at the time. Hence, the work on network programmability did not gain traction among the few of us who had contributed significantly to it. I first became aware of OpenFlow in 2010 when the first version of the protocol was introduced. At the time, I was managing R&D at Allied Telesis Inc (ATI) and wanted to bring OpenFlow to ATI's products, but we ran into various flow table issues with OpenFlow specification 1.0. It was not until OpenFlow spec 1.3 was released that some progress was possible, but I had moved out of ATI by then. However, many groups in the industry continued to work on OpenFlow. While at Agema Systems (A Delta Electronics Company), I collaborated with two vendors: NoviFlow and Broadcom. Broadcom contributed OF-DPA (OpenFlow Data Plane Abstraction) to the industry to allow easier interworking between OpenFlow software and Broadcom SoC-based switches. Though OpenFlow continues to exist, it lost its luster of the early days due to lacking scalability and the notion of treating edge switches as dumb. In this section, we will explore OpenFlow in detail, exploring specifications and implementations.

1.2.1   Historical Perspective

The works of OpenFlow can be traced back to 2006, when Martin Casado, a PhD student at Stanford University, developed a new network architecture called Ethane.[5] Martin also published a scholarly article that suggested Ethane allows managers to define a single network-wide fine-grain policy, and then enforces it directly. Ethane couples extremely simple flow-based Ethernet switches with a centralized controller that manages the admittance and routing of flows. While radical, this design is backward-compatible with existing hosts and switches.[6] The article goes on to claim that researchers implemented Ethane in both hardware and software, supporting both wired and wireless hosts supporting over 300 hosts in a large university network.

In the concept of Ethane, two components of a network design are proposed. The first is a central controller that manages global policy, and the second is a simple and "dumb" Ethane switch. These switches only have a flow table and a secure channel to the controller. They forward packets according to the directives of the controller. That means when a packet arrives at the flow table of an Ethane switch, it matches with instructions in the flow table and forwards accordingly. For packets that are not in the flow table, the Ethane switch forwards them to the controller, along with information about which port the packet has arrived on. In an Ethane network architecture, not all switches need to be Ethane switches. The network can be built as a hybrid, allowing Ethane switches to

---

[5] Compterweekly.com [5].
[6] ACM [6].

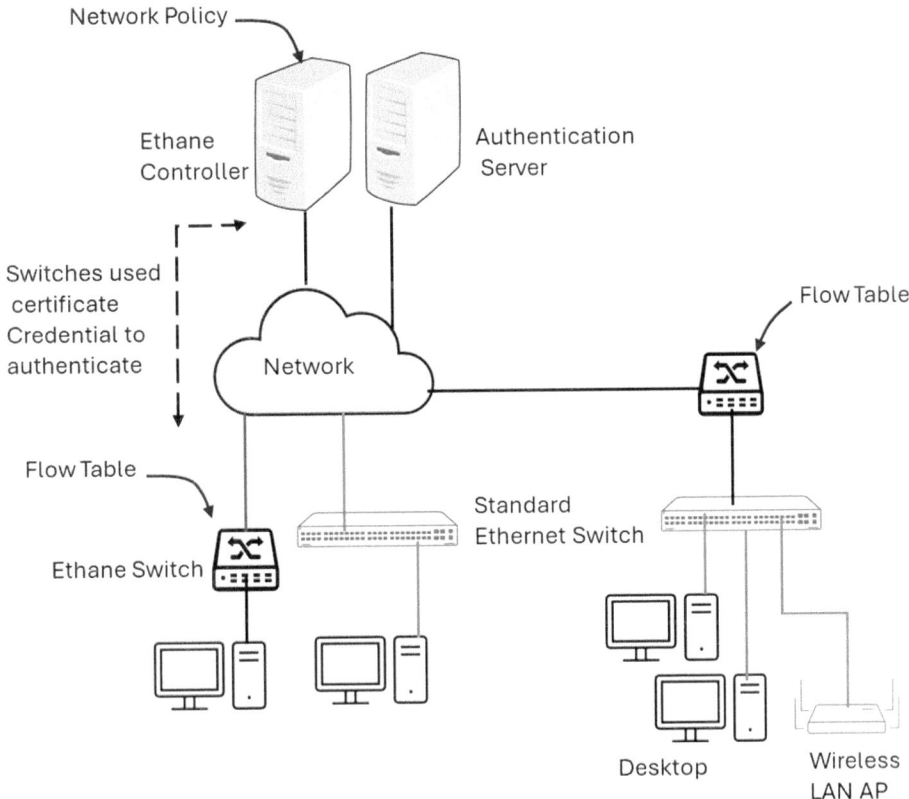

Fig. 1.2 Ethane network design depicting Ethane controller and Ethane switches

coexist with standard Ethernet switches. The following diagram depicts examples of how the flow table works, and the Ethane network architecture is realized (Fig. 1.2).

All switches and host in the Ethane network must register with the controller using different methods. For example, hosts use MAC addresses, users via username, and switches via secure certificate process.

Ethane provided a learning curve for researchers at Stanford University and UC Berkeley[7] to explore how to improve control plane abstraction and create open interfaces for both control plane and forwarding plane.[8] In 2011, Cisco Systems, Facebook, Google, and Microsoft join Stanford and UC Berkeley to form the Open Networking Foundation (ONF) to promote OpenFlow technology and the OpenFlow Switching Protocol.

---

[7] Vaughan-Nichols [7]
[8] Casado et al. [8].

1.2.2 What is OpenFlow?

The quest for network programmability gave birth to Ethane. It was the first step toward Software-Defined Networking (SDN), which aims to make network infrastructure more programmable, flexible, and agile by separating the control plane from the data plane and providing centralized control and management of network devices through software. However, Ethane lacked the open interfaces for operators and vendors to participate in defining their own flows in the control and forwarding plane. The lessons learned with Ethane helped develop the OpenFlow specification with the promise that operators could define their own flows to program the forwarding plane.

The OpenFlow specification defines three main components[9] of the technology as follows:

- OpenFlow Switch: A OpenFlow Switch implements a flow table and performs packet lookup and forwarding according to one or more flow tables and a group table. Both Broadcom and Marvell provided support for OpenFlow in their switching SoCs. For Example, Broadcom contributed OF-DPA to the industry, allowing OpenFlow support in Broadcom switching SoCs.
- OpenFlow Controller: A centralized management technique that implements software applications to provide centralized control of the network. It provides instructions to OpenFlow Switches for packet forwarding in a given network.
- Secure Channel: It is a secure communication channel between the OpenFlow Switch and the OpenFlow Controller.

Central to OpenFlow Network is OpenFlow Protocol that defines communication between OpenFlow Controller and OpenFlow Switch. It uses secure channel for communication as depicted in the Fig. 1.3.

The protocol consists of a set of messages that are sent from the controller to the switch and corresponding messages from the switch to the controller. The message exchanges are done through a secure channel implemented via a TLS connection over TCP. Figure 1.3 shows typical packet processing as you would expect in a switch. The core function takes the packets that arrive at one port and forwards them to another port while making necessary packet modifications along the way. The difference in an OpenFlow switch is that it implements packet matching functions and a flow table. A typical OpenFlow table (please refer to Fig. 1.4) includes flow entries as specified in the OpenFlow specification, which encompass flow entry header fields, counters, and actions associated with the flow.

The header fields are used as match criteria to determine if an incoming packet matches the entry. If a match is found, then the packet belongs to that flow. The counters are used to track statistics related to the flow, such as how many packets have forwarded or dropped

---

[9] Waziri [9].

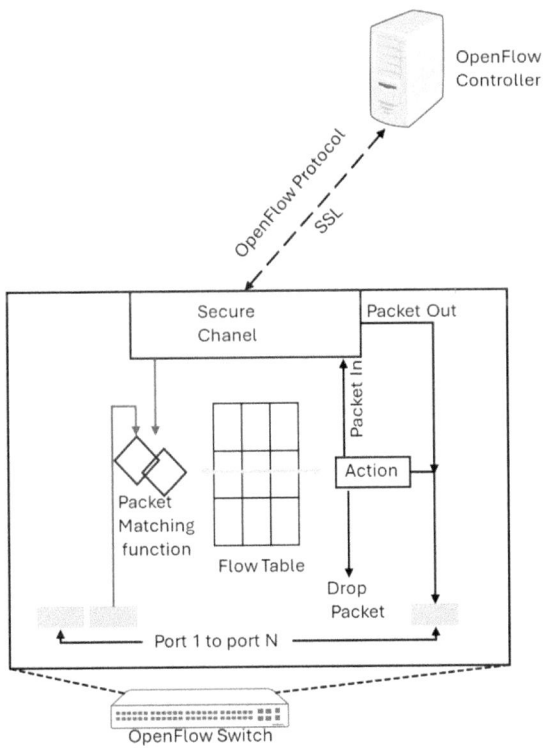

Fig. 1.3 Typical OpenFlow communications between OpenFlow switch and the controller

Flow Entry 0		Flow Entry 1			Flow Entry M		Flow Entry N	
Header fields	Inport 5 192.1.10.1, Port 1012	Header fields	Inport* 209.*.*.* Port*	– – – –	Header fields	Inport 5 192.1.20.1 , Port 995	Header fields	Inport 5 192.1.30.1 , Port 995
Counters	Value	Counters	Value		Counters	Value	Counters	Value
Actions	Value	Actions	Value		Actions	Value	Actions	Value

Fig. 1.4 OpenFlow flow table defined in OpenFlow version 1.0[10]

for the flow. The action fields prescribe what the switch should do with a packet that matches the entry.

When a packet arrives at a given port in OpenFlow switch (please refer Fig. 1.3), it matched against the entries in the flow table for possible match. There are twelve match fields that collectively referred to as "twelve-tuple" and listed herein below [10]:

---

[10] Göransson et al. [10].

- Switch input port.
- VLAN ID.
- VLAN priority.
- Ethernet source address.
- Ethernet destination address.
- Ethernet frame type.
- IP source address.
- IP destination address.
- IP Protocol.
- IP TOS (type of service) bits.
- TCP/UDP source port.
- TCP/UDP destination port.

If no packet match is found, the "table miss" entry in the flow table may define how to process the packet. Options include dropping the packets, passing them to another table, or sending them to the controller. The double-arrowed line depicted in Fig. 1.3 between the packet matching function and action is presented to highlight this operation.

1.2.2.1 Pipeline Processing

The flow table entry and packet processing in an OpenFlow switch are collectively referred to as "pipeline processing." The flow table within the pipeline is sequentially numbered. When a packet enters the ingress port of the switch, the pipeline initiates packet processing, starting with Table 1.1, as depicted in the figure below, and may continue to additional flow tables in the pipeline based on the result in Table 1.1. During this process, a flow entry can direct the packet only to a flow table whose ID is greater than its own table ID. The use of multiple flow tables and group table enhances the efficiency of data packet processing by reducing the length of a single flow table, thereby facilitating efficient entry lookup (Fig. 1.5).

Table 1.1 The OpenFlow 1.3.4 features required by OF-DPA version 2.0[11]

Feature	Description
Pipeline match fields	These metadata fields accompany packets during pipeline processing but are not parsed from packet headers
Experimenter protocol extensions	It provides a standard way to extend the OpenFlow protocol to support additional functionality. OF-DPA version 2.0 defines new Experimenter symmetric messages, multipart messages, flow table match fields, actions, and Meter bands
Select and fast failover group types	These are optional fields in OpenFlow 1.3.4 but required by OF-DPA version 2.0
Local reserved port	

---

[11] Broadcom [11].

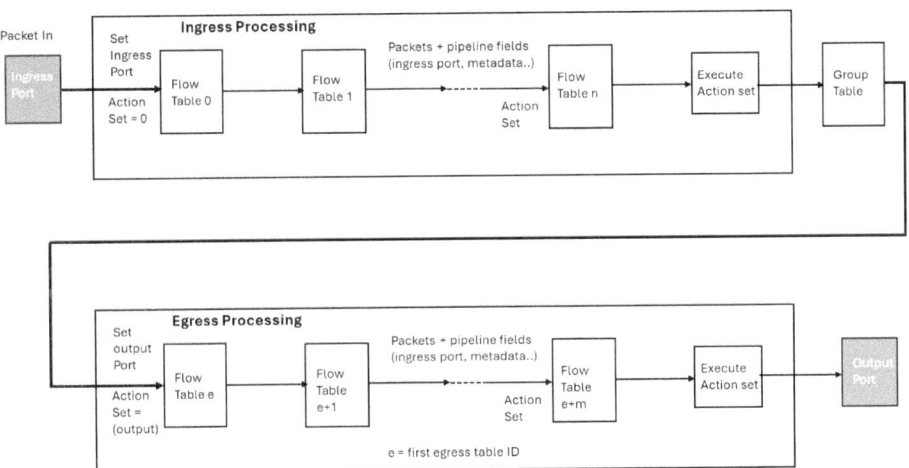

Fig. 1.5 The OpenFlow pipeline of OpenFlow switch as defined in OpenFlow specification 1.5

The group table comprises group entries, each containing one or more action buckets. Before forwarding packets to a port, actions defined in the buckets are applied. The main components of a group entry include a group identifier, group type, counters, and action buckets. The group identifier, a 32-bit unsigned integer, uniquely identifies the group on the OpenFlow switch. Group types determine group semantics, and not all group types are mandatory for implementation. The required group types are "indirect" and "all." For the "indirect" group type, one bucket is typically implemented, allowing centralized actions referenced by multiple flow entries, such as defining the next hop for IP forwarding. The "all" group type includes multiple buckets for handling multicast and broadcast packets, with each incoming packet being replicated and processed by every bucket in the group.

1.2.3 OpenFlow Implementation and Ecosystem

The OpenFlow implementation follows the SDN (Software-Defined Networking) architectural model, in which the control and data planes are decoupled, network intelligence and states are centralized. Furthermore, the network infrastructure is abstracted from applications, as depicted in the Fig. 1.6.

The SDN architecture comprises three primary layers: Infrastructure, Control, and Application. OpenFlow switches form part of the infrastructure layer, where they implement OpenFlow agents and utilize a northbound API to establish connections with the OpenFlow controller at the Control layer of the SDN architecture. Subsequently, the controller employs a southbound interface or API to communicate with OpenFlow switches. OpenFlow protocols facilitate secure communication, ensuring the passage of

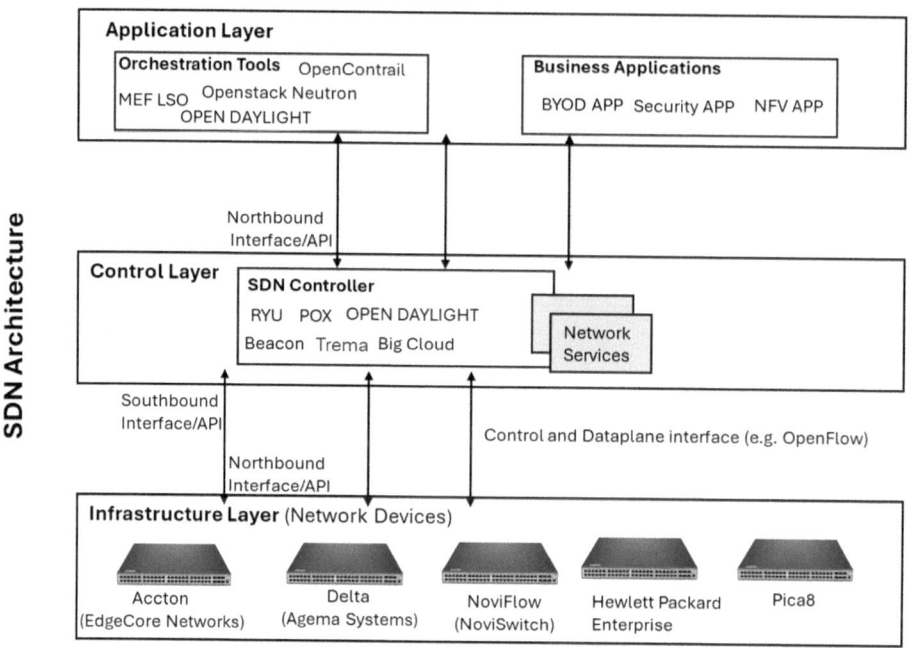

Fig. 1.6 SDN architecture and OpenFlow

flows between these interfaces. Typically, the OpenFlow or SDN controller connects with the Northbound API to the application layer within an SDN architecture. The figure above lists some of the ecosystem devices, controller, and application available in the marketplace to provide complete solutions for OpenFlow-based SDN architecture. The figure above lists some of the ecosystem devices, controllers, and applications available in the marketplace to provide complete solutions for OpenFlow-based SDN architecture. The list provided is a representative sample and is not intended to be exhaustive. Many of the software listed under the control and application layers are available as open source and can be found on GitHub.

1.2.4 OpenFlow Timeline

As discussed, OpenFlow began as an experiment in 2008 at Stanford University and gained industry support by early 2011. Since then, various initiatives have resulted in several specifications that define and enhance the protocol and OpenFlow-based SDN architecture. The diagram below illustrates a timeline of the standardization of OpenFlow (Fig. 1.7).

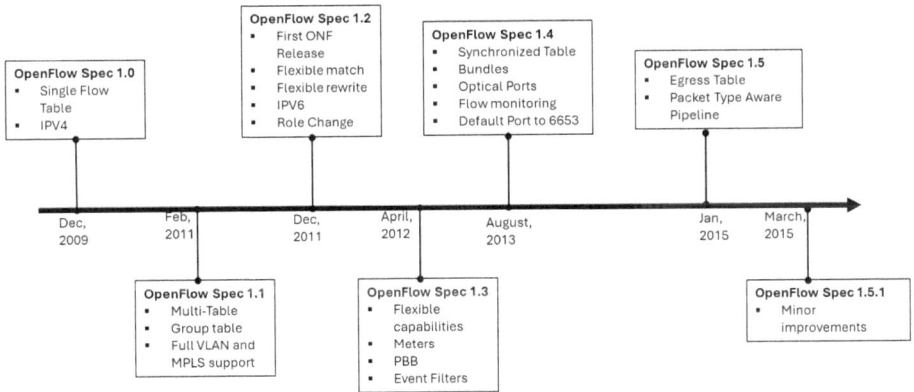

Fig. 1.7 The timeline of OpenFlow specifications

The initial version of OpenFlow featured a single table and support for IPv4. This specification was later updated to include IPv6 in OpenFlow version 1.2. Subsequently, multiple flow tables and group tables were added in version 1.3. Later versions of the protocol introduced improvements and additional features such as meters, flow monitoring, and support for optical ports in subsequent releases. The latest available specification is 1.5.1, which was released in 2015. Industry momentum appears to have waned since that time. A multitude of issues plagued OpenFlow, ranging from vendor support to operational challenges. Additionally, market challenges arose from networking OEM suppliers who began offering their own forms of SDN capabilities. These vendors include, but are not limited to, HPE, Juniper, Cisco, and Arista, who offered SDN solutions that are simpler to implement and require less rigor to manage network infrastructure. We will discuss an example of such vendor-specific SDN solutions in Sect. 1.5.

1.3 OpenFlow Data Plane Abstraction (OF-DPA)

While working at Agema Systems Inc., a technology venture of Delta Electronics, I first became acquainted with OpenFlow data plane abstraction (OF-DPA). As one of the largest suppliers of networking hardware, Delta Electronics is frequently approached by components and SoC vendors seeking to integrate their solutions in Delta networking products. They often shared industry trends and business justifications to integrate their components. Accordingly, Broadcom shared their guidance regarding the future of open networking and the importance of network programmability as an industry trend. At the time, I was leading "Whitebox" product development and exploring ways to create programmable network offerings from different Network Operating System (NOS) vendors. However, the implementation of OpenFlow agents on these NOS, including Broadcom's ICOS, was not robust enough for processing multiple tables as specified in OpenFlow version 1.3. This limitation constrained real-world deployments of OpenFlow. We sought ways to enhance OpenFlow implementation to provide the scalability it required. Therefore, when Broadcom introduced OF-DPA, it came as a relief to us and the industry as a whole. Broadcom, being an industry leader in networking SoCs, opening up their SoC packet pipeline to OpenFlow was considered a significant advancement for OpenFlow scalability at that time.

The OF-DPA provided a standardized way to implement OpenFlow on Broadcom SoC-based networking hardware. It served as a bridge between the OpenFlow control plane and the underlying hardware data plane, abstracting the complexities of hardware programming and providing a consistent interface for OpenFlow controllers to interact with Broadcom SoC-based different vendors' network devices. A summary list of OF-DPA features is given below:

- **Hardware Abstraction:** OF-DPA abstracts the underlying hardware complexities of Broadcom's StrataXGS® SoCs, providing a uniform interface for OpenFlow controllers to interact with the data plane.
- **Support for OpenFlow Versions:** OF-DPA supports multiple versions of the OpenFlow protocol, including OpenFlow 1.0, 1.3, and later versions, allowing for flexibility and compatibility with different controller implementations.
- **Pipeline Configuration:** It defines a pipeline configuration that maps OpenFlow match-action rules to specific hardware forwarding behaviors. This configuration includes support for matching packet headers, performing actions such as forwarding, modifying headers, and executing other packet processing functions.
- **Table Management:** OF-DPA defines a set of logical tables within the hardware pipeline, each responsible for specific packet processing functions. These tables can be programmed by the OpenFlow controller to implement various network policies and forwarding behaviors.
- **Flow Table Capacity:** OF-DPA specifies the capacity and capabilities of the hardware flow tables, including the number of flow entries supported, actions available per-flow entry, and match fields supported for flow classification.
- **Packet Processing Features:** It supports a range of packet processing features commonly used in SDN environments, such as VLAN tagging, MPLS (Multiprotocol Label Switching) processing, QoS (Quality of Service) markings, ACLs (Access Control Lists), and multicast forwarding.
- **Vendor Extensions:** OF-DPA allows for vendor-specific extensions and customizations to accommodate unique hardware capabilities or requirements, enabling vendors to differentiate their offerings while maintaining compatibility with the OF-DPA framework.
- **Documentation and APIs:** Broadcom provides documentation, APIs (Application Programming Interfaces), and software development kits (SDKs) to facilitate the

development, deployment, and management of OF-DPA-based solutions by network operators, software developers, and equipment vendors.

In this section, we will delve into the architecture and use-case scenarios of OF-DPA.

1.3.1   OF-DPA Architecture

As discussed, the software components of OF-DPA provide a hardware abstraction layer between OpenFlow and Broadcom SoC-based networking switches. This layer sits above the Broadcom SDK (Software Development Kit), as depicted in the diagram below (Fig. 1.8).

Networking equipment vendors purchase Broadcom Ethernet switching ASICs along with SDKs for the respective silicon, such as Firebolts, Tridents, Tomahawks, Qumrans, and the Jerichos. The Broadcom SDK enables networking equipment vendors to quickly integrate their software stacks across all the networking silicon that Broadcom supplies. It provides the following benefits[12]:

- A common application programming interface (API) allowing customers to focus on their control plane development while taking care of complex operations such as warm boot, serdes-control and timing or telemetry applications, etc.
- Additionally, functional APIs provide an abstraction from lower level register read or writes, while also providing drivers to the devices.

OF-DPA interacts with the Broadcom SDK by utilizing its APIs for tasks such as table manipulation, flow entry installation, and packet forwarding. This integration ensures that OF-DPA can fully leverage the capabilities and optimizations provided by the Broadcom SDK for managing the underlying hardware. Additionally, it provides an API known as the "OF-DPA API"[13] that connects to the OpenFlow agent deployed in the network operating system (NOS) of the switch. It interworks with a specialized hardware abstraction layer (HAL) that allows programming of Broadcom ASICs using OpenFlow abstractions. OF-DPA does not process OpenFlow messages, for that purpose, OpenFlow agent must be installed in the switch. OpenFlow agent communicates with OpenFlow controller through a secure channel. User applications connects to the Open Flow controller through northbound API and uses it to communicate with the switching hardware.

Broadcom OF-DPA version 2.0 supports any controller that adheres to features specified in Openflow 1.3.4 or higher. These features are depicted in the Table 1.1.

OF-DPA presents the representation of hardware components and functionalities within the Broadcom ASIC (Application-Specific Integrated Circuit) that it controls.

---

[12] Broadcom [12].
[13] Broadcom [13].

Fig. 1.8 OF-DPA layered
architecture

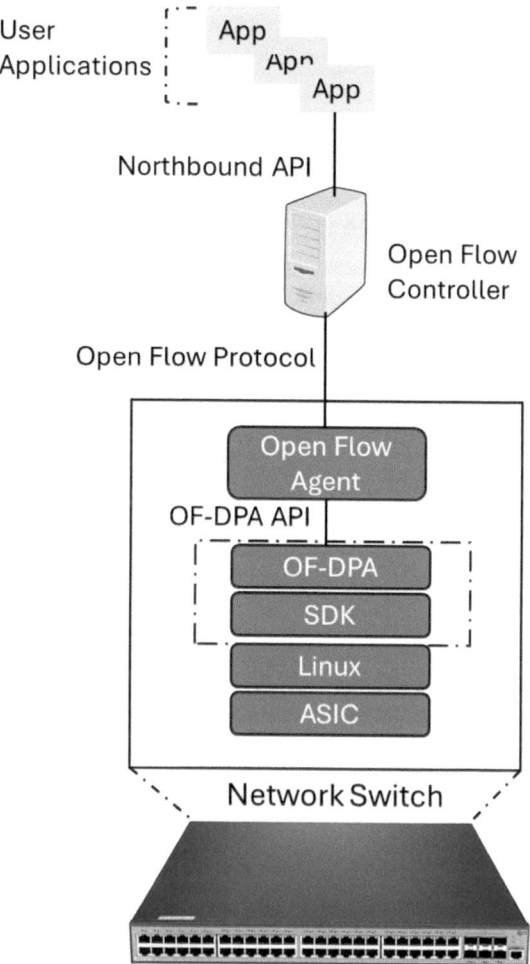

These hardware objects include standard hardware objects and ancillary table objects. The standard hardware objects include, but are not limited to, flow tables, group table entries, queues, and ports that can be programmed using the protocol described in the OpenFlow 1.3.4 specification. The ancillary table objects are needed to support use cases such as MPLS-TP, OAM, and QoS. These objects include per-flow loss management counters and per-flow packet dropping.

Additionally, to support OAM functions, OF-DPA makes use of the "Egress Tables" defined in OpenFlow version 1.5. These tables permit match-action processing after groups are applied and the output port is decided.

1.3.1.1 OF-DPA Abstract Switch

In our discussion, we explored how OF-DPA abstracts the hardware objects within Broad-com ASICs for interaction with the OpenFlow agent. This abstraction is referred to as the "Abstract Switch," which serves as a conceptual model representing network switch to the OpenFlow agent. The OF-DPA abstract switch essentially hides the underlying com-plexity of the Broadcom ASIC (Application-Specific Integrated Circuit) hardware and presents a simplified view of the switch to OpenFlow controllers. It allows controllers to interact with Broadcom-based switches using standard OpenFlow protocols without needing to understand the specific details of the hardware implementation. The OF-DPA model of the "abstract switch" is constructed using the Table Type Pattern (TTP) frame-work defined by the Open Networking Foundation (ONF). TTP defines the forwarding behaviors that controllers can program into a switch. Specifically, it enables application developers to articulate switch requirements, facilitating runtime agreement between the controller and the switch regarding supported features. This implementation allows appli-cations to orchestrate and execute packet processing functions by adding flow entries to OpenFlow flow tables. These flow entries include action lists and/or action sets for packet editing and forwarding. Packet forwarding predominantly utilizes OpenFlow group entries. Moreover, the application can query the status of OpenFlow ports and queues and receive events, such as port state changes or flow expiration, through services provided by the Controller via the OpenFlow agent on the switch.

To better understand how these flow entries are processed in OF-DPA, let us explore some of the use cases as follows:

- **Bridging and Routing:** These are fundamental functions of a switch; in this section, we will explore how OF-DPA abstract switch programs hardware objects to bridging and routing functions.
- **Data Center Overlay Tunnel:** VxLAN is used as an overlay tunnel in data center setup. This section will explore OF-DPA objects and flow entries associated with overlay tunnel.
- **MPLS-TP Customer Edge Device:** Similarly, we will explore how MPLS-TP flow entries are processed in OF-DPA for customer edge (CE) network devices. The MPLS-TP (Multiprotocol Label Switching-Transport Profile) is an extension of MPLS technology, designed to meet the stringent requirements of transport networks.
- **VPWS:** Virtual Private Wire Service is a type of Layer 2 virtual private network (VPN) service that allows for the creation of point-to-point connections between two customer sites over a service provider's MPLS (Multiprotocol Label Switching) network. VPWS is commonly used by enterprises and businesses to establish secure, reliable, and high-performance connections between geographically dispersed sites, such as branch offices, data centers, or remote sites, without the need for dedicated physical circuits. It offers flexibility, scalability, and cost-effectiveness compared to traditional leased

line or dedicated circuit solutions. In this section, we will explore OF-DPA objects for VPWS.

- **MPLS Label Edge Router (LER):** An MPLS Label Edge Router (LER) is a router within a Multiprotocol Label Switching (MPLS) network that resides at the edge of the MPLS domain. This section will discuss OF-DPA objects for LER.
- **MPLS Label Switch Router (LSR):** An MPLS Label Switch Router (LSR) is a key component within a Multiprotocol Label Switching (MPLS) network. LSRs are responsible for forwarding packets based on MPLS labels. This section will explore OF-DPA objects for LSR.

1.3.1.2 Bridging and Routing

In OF-DPA (OpenFlow Data Plane Abstraction), bridging and routing are fundamental functions that can be implemented using specific OF-DPA objects. Here's a brief overview of bridging and routing OF-DPA objects:

Bridging Objects:

(a) **Forwarding Table:** OF-DPA defines forwarding tables that store flow entries for forwarding packets based on specific criteria. Bridging flow entries in these tables typically use MAC addresses for packet forwarding decisions.
(b) **Flow Entries:** Flow entries define forwarding behavior for packets in OF-DPA. Bridging flow entries match on MAC addresses (source and/or destination) and specify actions such as forwarding packets out of specific ports or flooding packets to multiple ports.
(c) **Group Entries:** Group entries in OF-DPA are used to implement multicast or broadcast behavior. Bridging group entries can be used to replicate packets to multiple ports for broadcast or multicast traffic.
(d) **Port Configuration:** OF-DPA allows configuration of port behaviors, including VLAN membership and port mode (access or trunk), which are essential for bridging functionality.

Routing Objects:

(a) **Forwarding Table:** Routing flow entries in the forwarding tables typically match on IP addresses (source and/or destination) and specify actions for routing packets to their next hop or egress port.
(b) **Flow Entries:** Routing flow entries define forwarding behavior for IP packets in OF-DPA. They typically match on IP addresses and specify actions such as forwarding packets to a next-hop MAC address or egress port.
(c) **ECMP (Equal-Cost Multipath) Group Entries:** OF-DPA supports ECMP for load balancing traffic across multiple equal-cost paths. ECMP group entries define multiple next hops for a given destination and distribute traffic among them.

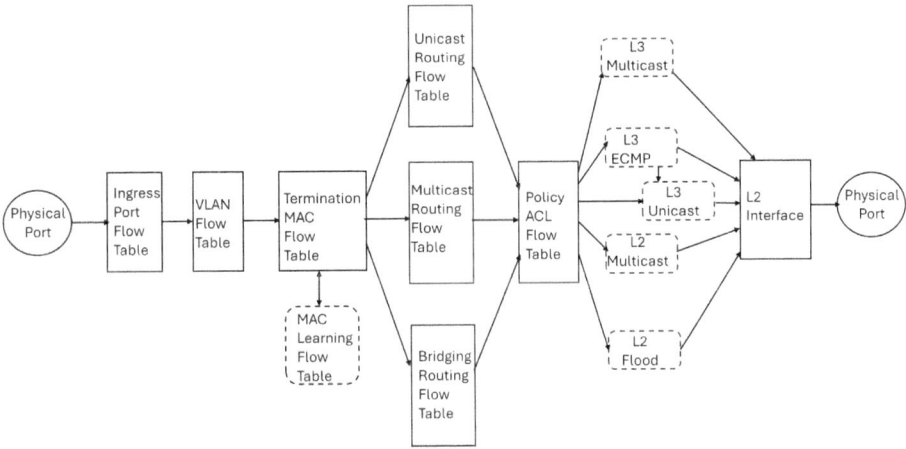

Fig. 1.9 OF-DPA abstract switch object used for bridging and routing [11]

(d) **VLAN Routing:** OF-DPA allows for VLAN routing, where packets are routed between VLANs using VLAN interfaces (SVIs—Switched Virtual Interfaces).

(e) **IPv6 Support:** Some versions of OF-DPA also support IPv6 routing, enabling routing functionality for IPv6 packets in addition to IPv4.

These OF-DPA objects provide the building blocks for implementing bridging and routing functionalities in software-defined networking (SDN) environments using Broadcom ASICs. They allow for flexible and programmable packet forwarding behaviors, enabling the deployment of diverse network architectures and services.

The following diagram depicts OF-DPA objects and packet flows for bridging and routing functions (Fig. 1.9).

Once a packet arrives at the switch, it enters through ingress port and always processed at first ingress port table, also referred to as table 0). The flow entries in this table can sort out traffic from different types of input ports by looking at something called "Tunnel ID" metadata. When packets come from regular ports like physical ones, they have a Tunnel ID of 0. To make things easier to set up, this table has a basic rule that lets through packets with a Tunnel ID of 0 if they don't match any other rules with higher priority.

Every packet in the Bridging and Routing flow needs to have a VLAN tag. The VLAN Flow Table can do two things: it can check if packets have a VLAN tag and decide what to do with them, or it can assign a VLAN to packets that don't have a tag. If a packet has multiple VLAN tags, the table uses the outermost one to figure out where to send the packet. The Termination MAC Flow Table checks the destination MAC address of a packet to decide whether to treat it like a bridge or a router. If it's routing, it figures out if the packet is meant for just one device (unicast) or multiple devices (multicast). It also supports a way to learn MAC addresses using a "virtual" flow table that works together with the Bridging Flow Table.

When MAC learning is enabled, OF-DPA checks the Bridging Flow Table using the source MAC address, the outermost VLAN ID, and the port where the packet came from (IN_PORT). If it doesn't find what it's looking for, it tells the controller about it using a Packet In message. This happens before OF-DPA checks the Termination MAC Flow Table. The controller can't directly see or change the MAC Learning Flow Table.

The ACL Policy Flow Table depicted in the diagram can perform matches with wild-cards on multiple aspects of a packet, similar to how an ACL (Access Control List) works in a regular switch.

OF-DPA uses OpenFlow Group entries a lot. These entries are like containers that hold instructions for what to do with packets. They're handy for things that are tricky to set up in OpenFlow 1.0, like sending packets along multiple paths or broadcasting them to many destinations. Plus, they make good use of the hardware's built-in features.

1.3.1.3 Data Center Overlay Tunnel

The figure shows OF-DPA objects for Data Center Overlay tunnels. These objects haven't changed since OFDPA 1.0.

Data Center Overlay Tunnel processing is about sending traffic for different tenants in separate forwarding areas. The system decides where to send packets based on a special number called "Tunnel-ID," which marks each tenant's forwarding area. To let in Data Center Overlay tunnel packets for a certain tenant, a rule has to be set up in the Ingress Port Flow table. Also, in the Bridging Flow Table and ACL Policy Flow Table, entries look for tunnel traffic using Tunnel ID, not VLAN Id. In OF-DPA, there's a way of naming Tunnel-ID metadata. The first 16 bits of this ID indicate the type of tunnel being used. This helps flow entries tell the difference between different types of tunnel traffic.

Data Center Overlay tunneling is set up using a mix of configuration tools, logical ports, and flow tables. Here's how it works in OF-DPA 1.0.

The Abstract Switch pipeline gets inner packets from "overlay tunnel logical ports" along with some extra info such as Tunnel ID after removing the encapsulation headers. Then, it sends these packets to other logical ports to add the Tunnel ID info back on. But the actual tunnel ends are managed separately from OpenFlow (Fig. 1.10).

OF-DPA 1.0 introduced a way to set up tunnels that worked with OF-Config. This setup was different from what OF-Config 1.2 offered because it could connect Tunnel ID with VxLAN VNI. It also let the system make decisions about where to send traffic based on different VxLAN tunnel endpoints (VTEP). Instead of relying on a vague routing process, it directly sets up how traffic in overlay frames should be forwarded. The way OF-DPA 1.0 set up this tunnel information was similar to how OF-Config did it, making it easy to fit into its YANG model. But it could also be used with ovsdb. Setting up tunnels like this relies on having the right tools for configuration. Typically, it means using a setup where you have a configuration protocol and a setup tool alongside the OpenFlow system.

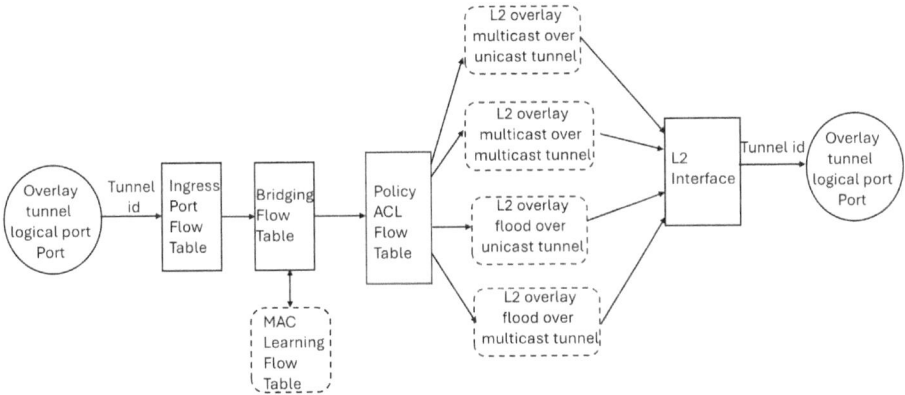

Fig. 1.10 OF-DPA abstract switch object used for overlay tunnel [11]

1.3.1.4 MPLS-TP Customer Edge Device

MPLS-TP functionality was not available in OF-DPA 1.0; it was introduced in OF-DPA 2.0 due to limitations in the OpenFlow 1.3.4 standard. OF-DPA 2.0 added extra features to support MPLS-TP, such as enhanced packet handling capabilities in the data pipeline. Unlike the setup for Data Center Overlay tunnels, MPLS-TP tunnels are directly established within the pipeline itself. To enable this, OF-DPA 2.0 introduced new actions for inserting and removing L2 headers, addressing the lack of support in OpenFlow 1.3.4 for these specific operations.

1.3.1.5 VPWS

Figure 1.11 displays only the OF-DPA Abstract Switch objects involved in VPWS (Virtual Private Wire Service) initiation packet flows. The MPLS-TP L2 VPN groups enclosed within the dotted lines are utilized for 1:1 linear and ring protection switching, as well as label processing.

VPWS is a direct point-to-point service. This means that the pseudo-wire, a virtual connection, decides where packets go without needing to search for a destination through a bridge. VPWS doesn't need to learn, broadcast, or support multicasting.

To start VPWS, packets are put into a specific customer's pseudo-wire based on a mix of where they come from and the VLAN tags they have. Sometimes, packets have multiple VLAN tags, which OpenFlow 1.3.4 can't handle directly. So, OF-DPA uses a special method: it first looks for the outermost VLAN tag and uses that to match packets. Then, it sets a new tag called OVID to the matched value and changes the order of the tags so that the inner one becomes the new outer tag. Finally, it looks for both tags together in the VLAN Flow Table.

For new features like QoS classification or VPLS, OF-DPA introduced a new pipeline match field called MPLS L2 Port. This field represents a logical entry point for the

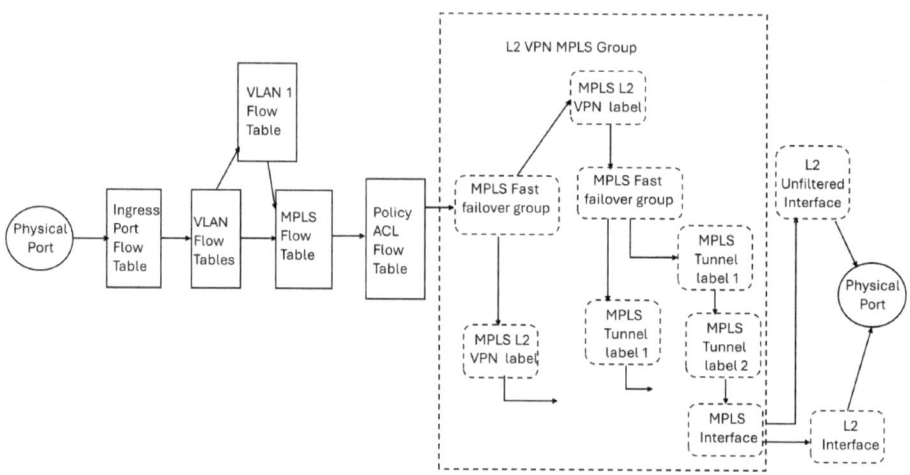

Fig. 1.11 OF-DPA object for MPLS-TP initiation (VPWS) [11]

pseudo-wire, which could be either a local attachment (UNI) or a network (NNI) entry point. OF-DPA sets up a naming system for MPLS L2 Port identifiers, dividing them into UNI and NNI categories. This helps distinguish the source type in flow entries.

Additionally, each flow must have a Tunnel ID metadata value assigned to it. Similar to Data Center Overlay Tunnel logical ports, the MPLS forwarding pipeline creates a separate forwarding domain for each customer pseudo-wire, with packets identified by their Tunnel-ID. A distinct range of tunnel-IDs is defined to differentiate MPLS-TP packets from data center overlay tunnel packets.

For MPLS-TP flows, both MPLS L2 Port and Tunnel-ID values need to be assigned. These values work together to represent packets and their direction for a specific customer flow. Figure 1.12 displays only the OF-DPA Abstract Switch objects used for VPWS termination packet flows. To simplify, the two VLAN flow tables are shown stacked on top of each other. Typically, VPWS termination flows only need to match one VLAN tag.

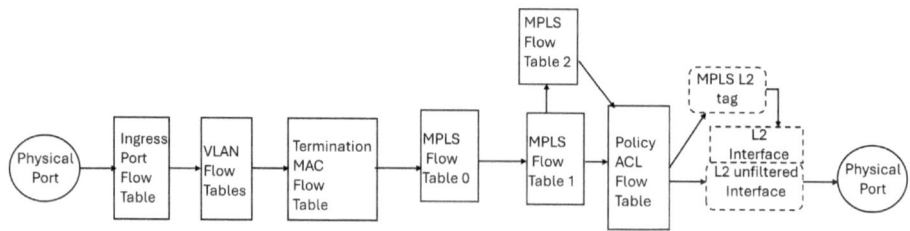

Fig. 1.12 OF-DPA objects for MPLS-TP termination (VPWS) [11]

Table 1.2 P4 versions and timeline [16][14]

P4$_{14}$	
Version 1.0.1	January 2015
Version 1.0.2	March 2015
Version 1.0.3	November 2016
Version 1.0.4	May 2017
Version 1.0.5	November 2018
P4$_{16}$	
Version 1.0.0	May 2017
Version 1.1.0	November 2018
Version 1.2.0	October 2019
Version 1.2.1	June 2020
Version 1.2.3	July 2022

The Termination MAC Flow Table spots MPLS frames that need MPLS tunnel termination processing. It does this by checking flow entries that match the destination MAC address and Ethertype. These entries have instructions to go to the MPLS Flow Tables. Just like how two VLAN tables were used to match two VLAN tags, multiple MPLS Flow Tables are used to match up to three MPLS labels. Each table is used to match an outermost MPLS shim header. The first table, MPLS 0, can be used to match and remove an outermost LSP label. MPLS Flow Tables 1.1 and 1.2 can match another LSP label or a pseudo-wire bottom of stack label. In the latter case, OF-DPA adds new ways to spot and remove a control word if there's one, and to take off the outermost L2 header. The pseudo-wire label also sets the Tunnel-ID and points to a group entry for sending the packet along.

MPLS Flow Table 0 can do fewer things than MPLS Flow Tables 1.1 and 1.2. So, any rules set up in MPLS Flow Table 0 should also be set up in MPLS Flow Tables 1.1 and 1.2. MPLS Flow Tables 1.1 and 1.2 are the same and always have the same rules. So, if you change something in MPLS Flow Table 1.1, it automatically changes in MPLS Flow Table 1.2 too.

1.3.1.6 MPLS Label Edge Router (LER)

The OF-DPA objects for MPLS L3 VPN are displayed in Fig. 1.13 for setting up and Fig. 1.14 for finishing. The packet flow for the MPLS Label Edge Router allows routing into and out of MPLS L3 VPN tunnels. An LER acts as both an IP router and an endpoint for MPLS tunnels. It can handle several VPNs for different customers. To start a tunnel, IP packets are directed to an MPLS L3 VPN. To keep them separate, the system supports

---

[14] P4.org [17].

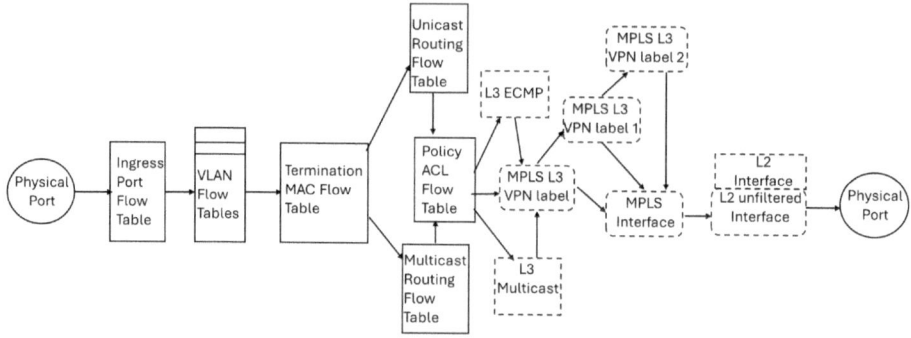

Fig. 1.13 OF-DPA objects for MPLS L3 VPN initiation [11]

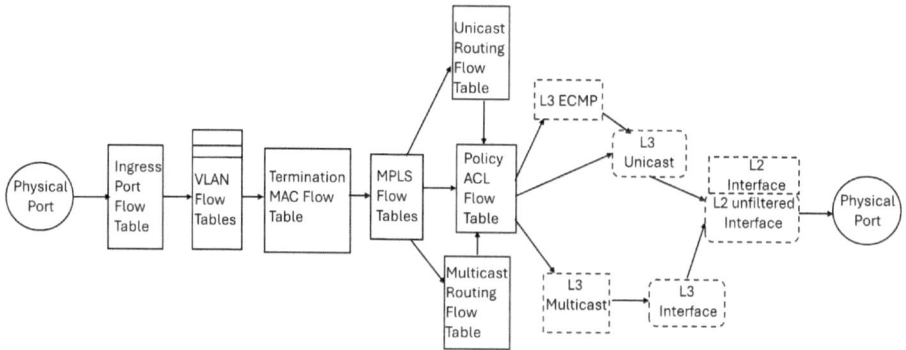

Fig. 1.14 OF-DPA objects for MPLS L3 VPN termination [11]

multiple virtual routing tables (VRFs), which are chosen using VRF pipeline settings. OF-DPA sets up a way to look for VRFs in the data pipeline.

Incoming traffic is sorted by their VLAN, which is like a tag. Then, the OF-DPA abstract switch decides if it wants to send that traffic to a VRF. The data, herein "IP packets," can be forwarded either directly or through an L3 Multicast or L3 ECMP group entry. Eventually, the data goes to MPLS Label Group Entries, which add extra labels to the data and update fields in the Ethernet header.

To end a tunnel, the OF-DPA abstraction switch examines MPLS frames using their destination MAC and VLAN in the Termination MAC Flow Table. Then, it processes the MPLS shim headers based on the information in the MPLS Flow Tables (Fig. 1.14).

For ending MPLS L3 VPN, removing the extra headers turns these frames into IP packets. These IP packets can be guided using the Routing Flow Tables or sent directly to specific group entries like L3 Unicast, L3 Multicast, or L3 ECMP. If the packets are meant for a multicast group, they are directed using the Multicast Routing Flow Table instead of the Unicast Routing Flow Table.

The bottom label in the MPLS stack can set up special information for the VPN process. VPN data is kept separate by sending it to the specific VLAN linked with that VPN.

1.3.1.7 MPLS Label Switch Router (LSR)

The OF-DPA components for an MPLS Label Switch Router (MPLS-TP P node) are displayed in the figure below. The process uses similar components as in the case of a Label Edge Router (LER) (Fig. 1.15).

A Label Switch Router (LSR) moves MPLS frames by sometimes removing one or more labels and then exchanging a label. In OF-DPA, these actions are set up in the MPLS Flow Tables. Label exchange can be done either for a tunnel (LSP) label that is not at the bottom of the stack, or for a PW (pseudo-wire) label that is at the bottom of the stack, especially in PW stitching situations.

The figure above depicts an option to send data to an MPLS ECMP group to help with balancing loads across multiple paths. This is specifically for MPLS L3 VPN and isn't used for MPLS-TP. The decision on which path to take is handled by the hardware platform.

1.3.2 Summary

We discussed the OF-DPA framework developed and open-sourced by Broadcom to facilitate the implementation of OpenFlow on Broadcom SoC-based networking hardware. OF-DPA acts as a bridge between the OpenFlow control plane and the underlying hardware data plane, simplifying hardware programming complexities and providing a consistent interface for OpenFlow controllers. It offers various features, including hardware abstraction for Broadcom's SoCs, support for multiple versions of the OpenFlow protocol, configuration of pipeline behaviors, management of logical tables for packet processing, definition of flow table capacities, and support for packet processing features such as VLAN tagging, MPLS processing, QoS markings, ACLs, and multicast forwarding.

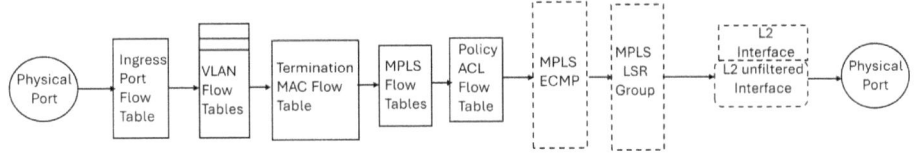

Fig. 1.15 OF-DPA object used for an MPLS LSR [11]

Additionally, OF-DPA allows for vendor-specific extensions and provides documentation, APIs, and SDKs to facilitate development, deployment, and management.

The architecture of OF-DPA includes an abstract switch model, which hides the hardware complexities of Broadcom ASICs and presents a simplified view of the switch to OpenFlow controllers. This abstract switch interacts with the Broadcom SDK and OpenFlow agent deployed in the network operating system, enabling programming of Broadcom ASICs using OpenFlow abstractions.

OF-DPA supports various use-case scenarios, including bridging and routing, Data Center Overlay tunnels, MPLS-TP, VPWS, MPLS Label Edge Router (LER), MPLS Label Switch Router (LSR), and more. We discussed a few use cases and suggest that readers interested in learning more should refer to the OF-DPA specification for further details.

1.4   OpenNSL

In addition to OF-DPA, Broadcom also contributed OpenNSL to the industry as open-source software. OpenNSL aimed to facilitate easier adoption of Broadcom's hardware and provide some level of programmability for Ethernet switching SoCs. However, OpenNSL is limited to advanced customization and optimization of Broadcom SoC-based hardware. For more advanced features, a paid SDK like the "BCM API" is required. Both OpenNSL and BCM API are SDKs (Software Development Kits). The difference lies in their abstraction levels and functionalities. OpenNSL serves as a higher level abstraction layer, offering a common interface for controlling Broadcom's network switch ASICs. It provides a set of APIs (Application Programming Interfaces) enabling developers to manage various networking functionalities such as forwarding tables, VLAN configuration, and packet processing. OpenNSL is designed to simplify the development of network management applications and promote interoperability across different Broadcom switch platforms.

On the other hand, BCM API functions as a lower level programming interface, granting direct access to Broadcom's networking hardware. It offers a more granular level of control over the switch ASICs, allowing developers to optimize performance and implement custom features. The BCM API exposes detailed hardware registers and configuration options, providing greater flexibility but requiring more expertise in hardware programming.

In recent years, Broadcom has also contributed to OpenBCM (an open-source version of the BCM API) and OpenNSA (Open Network Switch API). The latter replaces OpenNSL. For simplicity in this discussion, we will focus on OpenNSL in this section. Readers are advised to refer to the following GitHub repositories:

1. OpenBCM: https://github.com/Broadcom-Network-Switching-Software/OpenBCM.
2. OpenNSA: https://github.com/Broadcom-Network-Switching-Software/OpenNSA.

1.4.1   Software Architecture

As discussed, OpenNSL is a library of network switch APIs that is available as open-source code for programming of Broadcom SOC-based networking platforms. These open APIs allow the creation of networking application software for Broadcom network switch architecture platforms. Due to the diverse range of switch devices compatible with OpenNSL APIs, not all functions are universally available across all devices. The API prioritizes exporting silicon features independently, but it does not implement functions in software if the underlying devices do not support them. From the perspective of OpenNSL, the system comprises interconnected devices, each controlling ports. Operations are executed on specific devices, with ports on these devices often specified as a bitmap. Ports are among the most crucial objects in this system, as evidenced by the extensive set of functions provided by the OpenNSL Port API. The OpenNSL API uses small integers to reference ports, typically corresponding to the physical port numbers on the device. Hence, familiarity with the device's specifications is essential for effectively addressing its ports.

The OpenNSL software package includes the necessary files for developing switch applications: C header files, libraries, and documentation. The following diagram depicts layered architecture for OpenNSL (Fig. 1.16).

The OpenNSL (Open Network Switch Library) layer sits above the Broadcom SDK (Software Development Kit) and provides a higher level abstraction for developers to interact with Broadcom networking hardware. Its function can be summarized as follows:

- **Abstraction of Hardware Complexity:** The OpenNSL layer abstracts the intricate details of Broadcom networking hardware, making it easier for developers to write networking applications without needing in-depth knowledge of the underlying hardware architecture.
- **Standardized API Interface:** OpenNSL provides a standardized API (Application Programming Interface) interface that allows developers to access various networking functionalities such as VLAN management, Quality-of-Service (QoS) configuration, packet forwarding, and more, regardless of the specific Broadcom hardware being used.
- **Facilitation of Cross-Platform Development:** By providing a uniform API interface, OpenNSL enables developers to write networking applications that can run on different Broadcom networking devices without significant modifications. This facilitates cross-platform development and enhances portability.

Fig. 1.16 OpenNSL layered
architecture

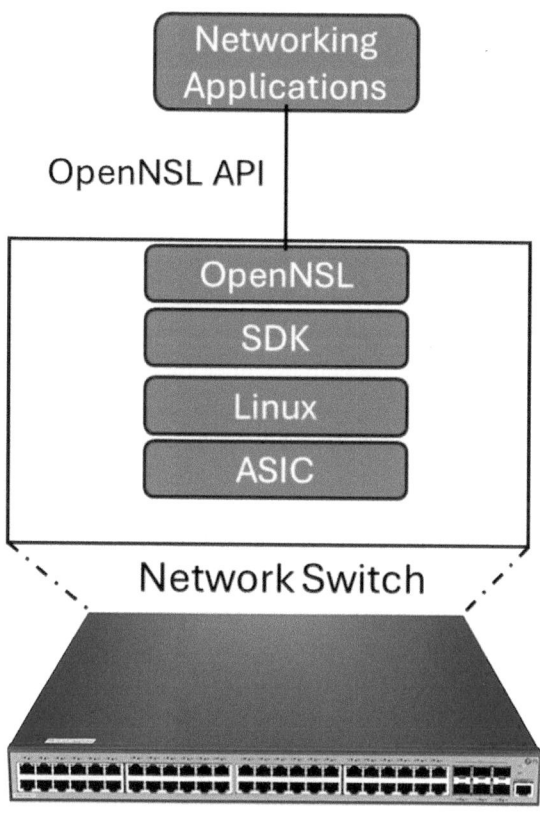

- **Enhanced Productivity:** OpenNSL simplifies the development process by offering
  higher level abstractions and utilities, reducing the amount of code required to implement networking functionalities. This leads to increased developer productivity and
  faster time-to-market for networking applications.
- **Community Support and Resources:** OpenNSL was supported by a community
  of developers and Broadcom itself, providing documentation, tutorials, and forums
  for assistance. This ecosystem helped developers learn and leverage the capabilities of Broadcom networking hardware more effectively. Recently, Broadcom
  introduced OpenNSA as a replacement for OpenNSL. Interested users and readers
  can visit https://www.broadcom.com/products/ethernet-connectivity/software/opennsa
  to learn more about OpenNSA.

The OpenNSL API interacts with the OpenNSL abstraction layer by providing a standardized interface for developers to access the functionalities abstracted by OpenNSL. The
API calls are typically done through software development using programming languages

such as C or C++. The process involves including the appropriate OpenNSL header files
in the source code and linking against the OpenNSL libraries during compilation.

Here's a general overview of how OpenNSL API calls are made:

- **Include OpenNSL Header Files:** At the beginning of the source code file, include
  the necessary OpenNSL header files using "#include" directives. These header files
  contain the function prototypes and definitions required to interact with the OpenNSL
  APIs.

  #include <opennsl/opennsl.h>
- **Initialize OpenNSL:** Before making any API calls, initialize the OpenNSL library by
  calling "opennsl_driver_init()" or "opennsl_init()" function. This function initializes
  the OpenNSL library and sets up necessary data structures.

  opennsl_driver_init(NULL);
- **Configure Devices and Ports:** Configure the network devices and ports as needed for
  your application. This may involve setting up VLANs, configuring port parameters, or
  enabling specific features.
- **Make API Calls:** Use the provided OpenNSL API functions to perform various networking tasks such as forwarding packets, managing VLANs, configuring Quality of
  Service (QoS), etc. These functions typically have names prefixed with "opennsl_."

  opennsl_port_config_t port_config;

  opennsl_port_config_get(0, &port_config); // Get port configuration for unit 0
- Handle Errors: Check the return values of API calls for errors and handle them
  appropriately. OpenNSL API functions usually return an integer indicating success
  ("OPENNSL_E_NONE") or failure ("OPENNSL_E_*" error codes).

  opennsl_error_t rv;

  rv = opennsl_port_config_get(0, &port_config);

  if (rv ! = OPENNSL_E_NONE) {

  printf("Error: Failed to get port configuration\n");

  // Handle error

  }
- **Clean Up:** After completing the required tasks, clean up resources and shut down the
  OpenNSL library if necessary.

  opennsl_driver_exit();

This is a simplified overview of how OpenNSL API calls are made in C or
C++ programs. The specific usage may vary depending on the requirements of your networking application and the functionalities provided by the OpenNSL library. Additionally, the OpenNSL documentation and programming guides provide detailed information
on using the API calls for different tasks. Please visit https://broadcom-switch.github.io/
OpenNSL/doc/html/OPENNSL_OVERVIEW.html for further details.

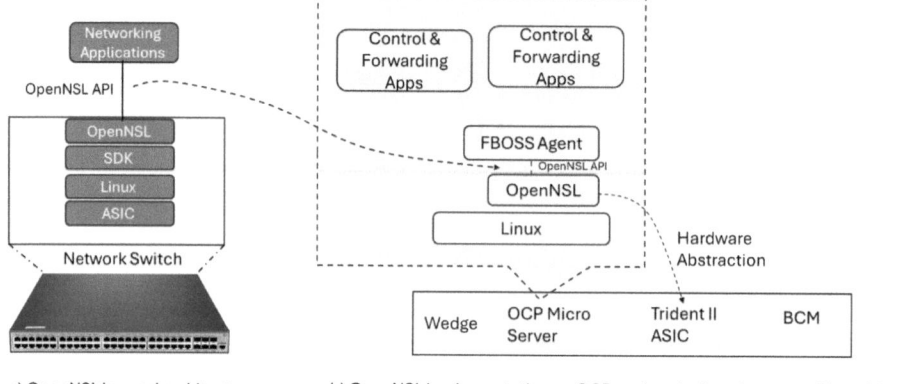

a) OpenNSL layered architecture   b) OpenNSL implementation on OCP wedge platform (courtesy: Meta, 2015)

Fig. 1.17   Example of OpenNSL implementation[15]

1.4.2   OpenNSL Implementation

OpenNSL (Open Network Switch Library) is typically integrated into network switches as a software component that interacts with the Network Operating System (NOS). The following diagram depicts how Meta implemented OpenNSL in the OCP (Open Compute Project) open networking "wedge" platform.

The wedge platform was developed by OCP, and the product specification was made available to the industry in 2015. It utilizes the FBOSS (Facebook Open Switching System) agent to program and control the Broadcom Trident II ASIC. This process operates on each switch and manages the hardware forwarding ASIC [please refer to Fig. 1.17b]. The diagram above illustrates the layered architecture of OpenNSL [Fig. 1.17a] and its correlation with the implementation of OpenNSL in OCP's wedge platform [Fig. 1.17b]. The FBOSS agent, as shown in the diagram [Fig. 1.17b], retrieves information from setup files and thrift APIs. It then configures the chip for proper forwarding and routing leveraging OpenNSL. Additionally, FBOSS agent handles packets from the chip intended for the switch itself, such as control protocol traffic, and other packets that cannot be managed solely in hardware.

Until 2018, OpenNSL was supported in many whitebox platforms as well OEM products. However, its interest has waned since then particularly due to SAI and the introduction of OpenNSA by Broadcom in 2019.

---

[15] Meta [14].

1.5   Switch Abstraction Interface (SAI)

While Broadcom's open-source contributions supporting hardware abstractions for its silicon were well-received by the industry, it became increasingly important to explore the option of vendor-neutral hardware abstractions for various vendors' network switching SoCs. In response to this need, in 2015, the Open Compute Project (OCP) introduced the Switch Abstraction Interface (SAI). SAI is an open-source initiative aimed at standardizing the way network switches are programmed and managed in data center environments. SAI provides a vendor-agnostic API for interacting with network switch ASICs (Application-Specific Integrated Circuits), enabling greater interoperability and flexibility in network control and management software. By abstracting the underlying hardware details of network switches, SAI fosters greater interoperability, flexibility, and innovation in network operating systems (NOS). One of SAI's primary goals is to address challenges such as vendor lock-in, complexity, and fragmentation in the networking industry by providing a common interface that can be used across different switch hardware platforms. Additionally, SAI facilitates the development of vendor-agnostic network operating systems (NOS), SDN (Software-Defined Networking) controllers, and other network management applications, thereby enabling seamless integration and interoperability in modern data center networks.

The diagram below depicts the layered architecture of SAI in comparison to the traditional approach. In the traditional networking stack, vendor-specific SDKs interact with the NOS, similar to what we have discussed thus far.

Please note that NOS often implements CLI and other management interfaces, such as Netconf and REST, while supporting network management protocols like SNMP to facilitate comprehensive management and orchestration of networking platforms. However, modern network orchestration and management platforms also implement aspects of management interfaces that provide a southbound interface to NOS. Therefore, dotted lines in both Fig. 1.18a and b represent parts of the management interface associated with the networking stack residing in the switch. In the networking stack depicted in Fig. 1.18b, the inclusion of SAI enables open NOS such as FRR, SONiC, OpenSwitch, and vendor-specific NOS to interact easily with the underlying ASIC. This concept, known as Open Networking or network disaggregation, involves presenting hardware abstraction for seamless integration with NOS. Further exploration of Open networking will be conducted in a later section.

The Switch Abstraction Interface (SAI) comprises standardized APIs based on the C language for programming network hardware tables. Users are relieved from the necessity of understanding the underlying silicon's switching behavior; they only need to utilize the SAI APIs to configure specific network features of the silicon. The figure below provides an analogy illustrating what the Switch Abstraction Interface offers (Fig. 1.19).

The concept of an adapter, as depicted in Fig. 1.18, is akin to a driver that enables the Switch Abstraction Interface (SAI) to function with a particular forwarding element.

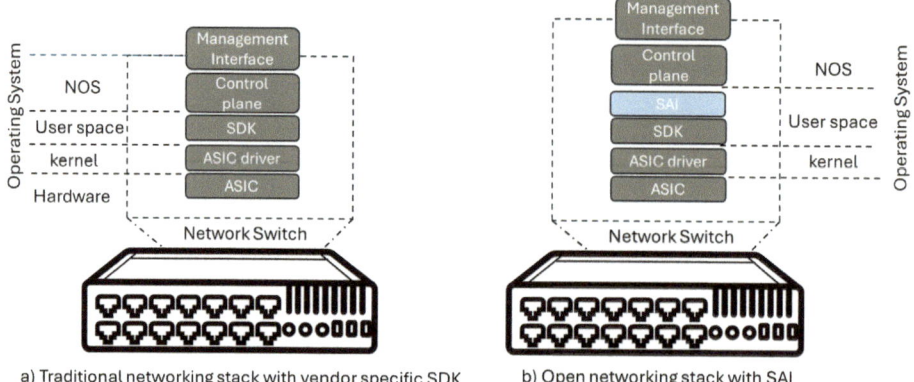

a) Traditional networking stack with vendor specific SDK   b) Open networking stack with SAI

Fig. 1.18   Traditional networking stack and open networking stack with SAI

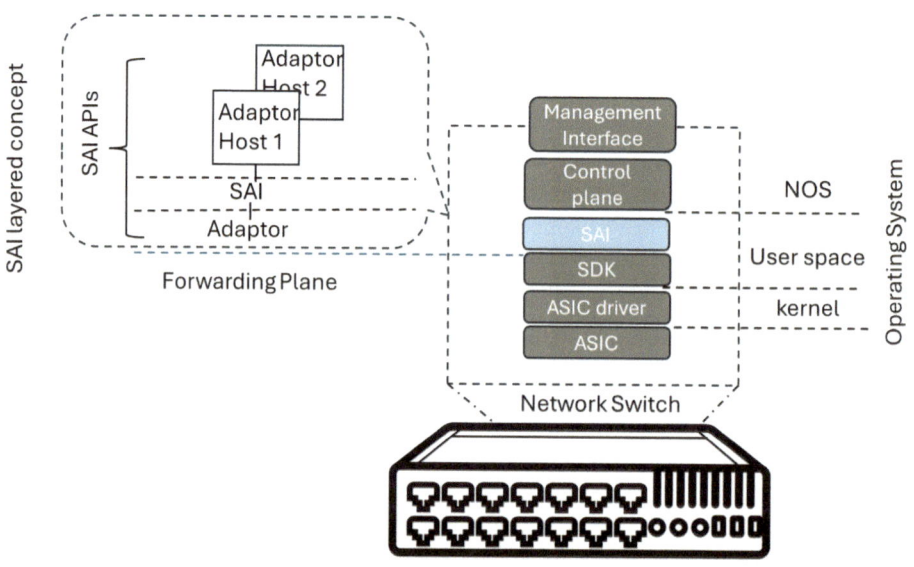

Open networking stack with SAI

Fig. 1.19   SAI APIs in an open networking stack

The forwarding element represents the implementation of a network forwarding plane to which a specific instance of an SAI Adapter adapts. This could be a switch chip and NPU with SDK, or software switch. While the adapter typically accompanies the element, it is not always the case. Its role is to locate and establish a connection with the underlying forwarding element, and to load or attach to any additional components it requires.

Adapters are supposed to be simple, without storing any permanent settings, and only keeping the information it gets from the control stack and the forwarding element. The idea is to make the Adapter as straightforward as possible, shifting the complicated stuff to the control stack whenever we can. To use an Adapter, it gets loaded into a process called an "Adapter Host" and then gets started. When it starts, the Adapter begins looking for the right instance of the forwarding element.

In SAI layered concept, multiple "Adapter Hosts" can run simultaneously. Each "Adapter Host" might handle different functions within the forwarding element. For instance, one Adapter Host could be in charge of updating IP routes on the forwarding element, while another might be responsible for collecting counters from it. For further details about SAI and its implementation guidance, please visit https://github.com/openco mputeproject/SAI/blob/master/doc/spec.md.

1.6   P4 Programming Language

I first learned about P4 in June 2015, shortly after attending a workshop at Stanford University. Later, I had the opportunity to work with P4 at Agema Systems, where we built P4 switches to enable connectivity with the ONOS SDN controller. Developed by Onlabs and later integrated into ONF, the ONOS (Open Network Operating System) controller solutions enabled direct setting of packet processing instructions through P4 runtime to configure the packet pipeline within switch ASICs or SoCs. The following diagram depicts typical SDN solution using ONOS controller and P4 switch (Fig. 1.20).

During our initial experimentation with the ONOS controller, we encountered many issues related to the programmability of the pipeline and achieving scalability with the solution. Another drawback was the lack of support for P4 in mainstream network ASICs to configure the pipeline, which impeded the progress of P4 adoption at the time. However, startups like barefoot introduced P4 programmability in their flagship product "Tofino." The solution allowed network operators to define custom packet processing logic tailored to their specific needs, enabling greater flexibility and innovation in network design.

Despite promising technology and significant industry interest, Barefoot faced challenges in gaining widespread adoption of its products. One key obstacle was the entrenched dominance of established networking vendors, who already had strong footholds in the market with their proprietary switch ASICs and networking solutions. Additionally, the complexity of integrating programmable switch chips into existing network architectures and workflows presented a barrier for many organizations. Adopting Barefoot's technology required significant investments in training, development, and infrastructure changes, which proved challenging for some companies.

In 2019, Barefoot Networks was acquired by Intel Corporation, which saw potential in Barefoot's innovative technology to enhance its own offerings in the data center and

Fig. 1.20 P4 network switch
and ONOS SDN controller[16]

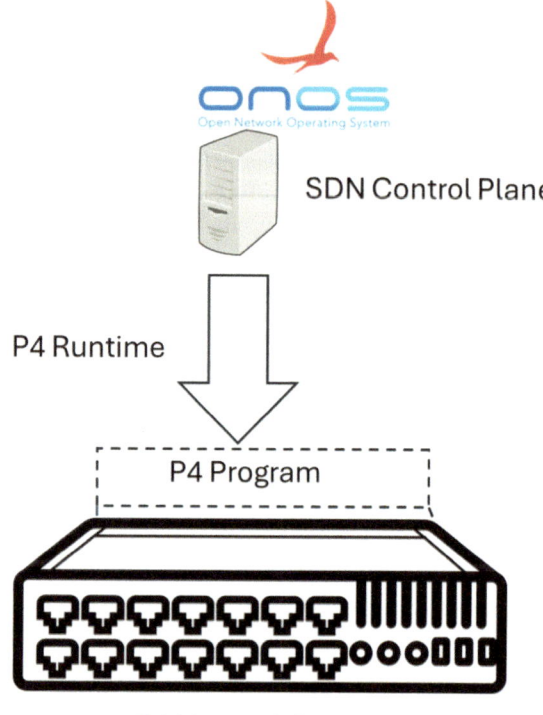

P4 Network Switch

networking markets. However, despite the acquisition, Barefoot's products faced continued challenges in gaining traction in the market. Ultimately, the failure of Barefoot Networks to achieve widespread adoption of its products can be attributed to a combination of factors, including market dynamics, competition from established vendors, and the complexity of integrating new technology into existing networks.

In recent years, P4 has intersected with a new technology known as the Data Processing Unit (DPU), specialized processors designed to accelerate data-intensive workloads such as networking, storage, and security functions. This technological advancement enabled HPE to develop a state-of-the-art and industry-leading product in partnership with Pensando. Through this collaboration, service chaining capabilities were integrated into HPE's flagship product, the "CX10000." Presently, the CX10000 has been deployed in numerous "mission-critical" networks, providing L4 firewall, DDOS protection, NAT, IPSEC, and other services at the edge of networks and data center TOR and DCI solutions. Further details about DPUs will be discussed in Chap. 2.

---

[16] Onosproject.org [15].

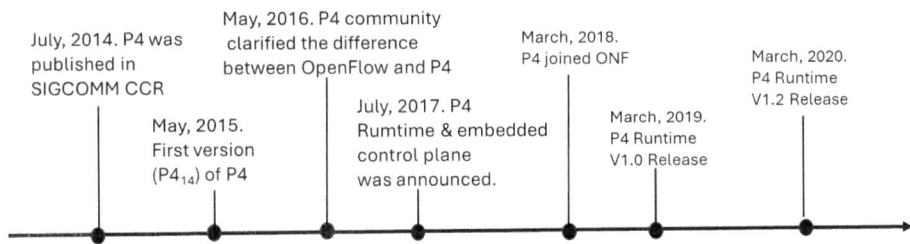

Fig. 1.21 P4 timeline[17]

1.6.1 Rise of P4

P4, short for Programming Protocol-independent Packet Processors, stands as a high-level programming language dedicated to specifying how packets are processed by network devices like switches, routers, and network interface cards (NICs). This open, domain-specific language empowers network engineers and developers to define packet processing pipeline behavior flexibly and programmatically. It made its debut in a 2014 SIGCOMM CCR paper titled "Programming Protocol-Independent Packet Processors," where its alternative name, "P4", was adopted. The inaugural workshop took place in June 2015 at Stanford University, as indicated in Fig. 1.21. In 2016, the P4 community clarified the distinction between the OpenFlow and P4 languages, resolving many misunderstandings. The introduction of the second version, $P4_{16}$, in May 2016 marked a significant milestone, with the previous version being known as $P4_{14}$.

P4Runtime was introduced and integrated into the control plane in July 2017 to execute P4 programs. Its adoption in numerous large-scale networks stems from its scalability, feasibility, and reconfigurability. The release of P4Runtime v1.0 occurred in 2019, followed by the release of P4Runtime v1.2 in 2020 [16]. The table below displays the different versions of P4 from its inception to 2020.

P4Runtime was introduced in July 2017. Following its introduction, P4Runtime v1.0 was released in 2019, followed by the release of P4Runtime v1.2 in 2020. Although $P4_{14}$ is the simpler language, $P4_{16}$ offers additional features that may not be compatible with $P4_{14}$. $P4_{16}$ is considered the latest version in the P4 programming language series.

1.6.2 P4 Architecture

The P4 architecture, also known as the "P4 Abstract Forwarding Model," includes two versions: $P4_{14}$ and $P4_{16}$. The $P4_{14}$ used the concept of PISA (Protocol-Independent Switch

---

[17] Goswami et al. [16]

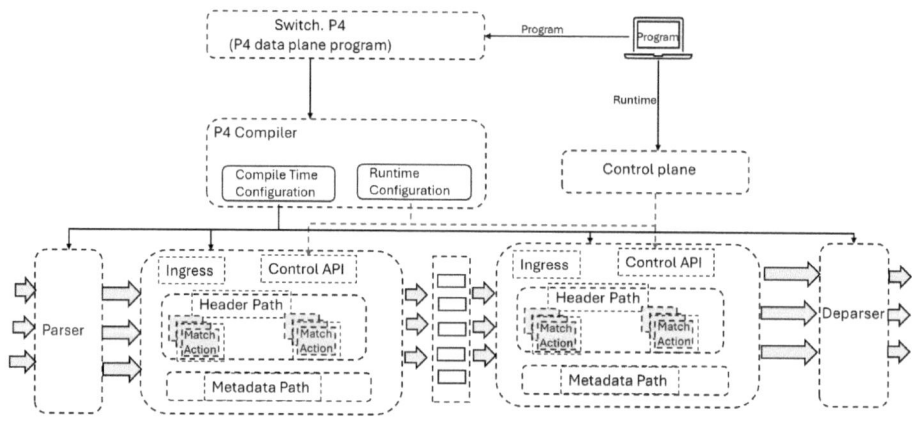

Fig. 1.22 P4 protocol-independent switch architecture (PISA) [18]

Architecture), a single pipeline forwarding architecture [16].[18] A typical PISA switch, as shown in Fig. 1.22, has several parts: a programmable parser, an ingress section, a queue, an egress section, and a deparser. The parser and deparser can be adjusted to support different packet header formats chosen by the user. The ingress and egress pipelines handle packets by going through match-action tables organized in stages. These tables look at the headers of packets based on preset rules and take actions accordingly. Actions can change certain parts of each packet, like headers or metadata, using basic commands.

The black arrows show compiling time configuration, and the red arrows show runtime configurations. The P4 language talks about how the data part of programmable switches works and how it communicates with the control part. The P4 compiler makes specific setups for the data part and P4Runtime APIs from P4 programs. P4Runtime APIs are outputs from the compiler that work with any target. They define how to control the data parts so the control part can do its job. The control part uses a script called PTF (Packet Test Framework) to use APIs. These APIs are available through Thrift-RPC (Thrift Remote Procedure Call). They help with managing ports, changing table entries, and using registers [18].

The newer version of the P4 language, $P4_{16}$, has shifted away from using the term PISA. Instead, it introduced a new architecture called the "V1Model," which helps to define how pipelines are set up for $P4_{14}$-programmable devices. Now, some devices are adopting $P4_{16}$ and defining their pipeline architectures similarly, such as with the V1Model or PSA. For instance, the Netronome NIC initially used $P4_{14}$, but later transitioned to using the V1Model architecture alongside the $P4_{16}$ language.[19] The following diagram shows $P4_{16}$ V1Model architecture (Fig. 1.23).

---

[18] Hang et al. [18].
[19] P4 forum [19].

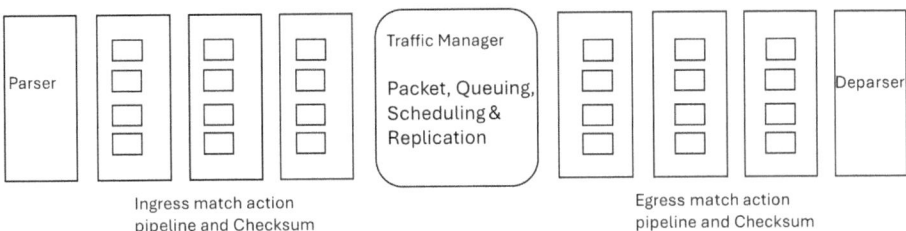

Fig. 1.23 P4 switch architecture: V1Model[20]

In the V1Model, incoming packets are first dealt with by the P4 programmable Parser Stage. The packet parser works like a state machine. Each state in this machine pulls out specific header information (like Ethernet, IPv4, TCP/UDP) as stated in the P4 program. Depending on what's found in these fields, the state changes to another one. Once the packet is parsed, it goes through the P4 programmable Ingress pipeline where Match & Action tables are used. These tables compare the fields in flow entries with packet headers or metadata to figure out what actions to take. These actions could involve adding, removing, or changing fields, or simply passing the packet through the table without any action.

They can also include modifying the header fields of the current packet, which can be either stateless or transient, or performing stateful operations that store information persistently for more than just one packet. P4 allows the programmer to maintain this stateful or persistent information using counters (for counting events associated with entries in tables), meters (for measuring data rates), and registers (counters that can be operated from actions in a general way). Between the Ingress pipeline and the Egress pipeline stages, the V1Model has a block that can't be programmed with P4. This block represents the traffic manager, which has fixed functions provided by the device's maker. After that, the packets move through the Egress pipeline, which also has Match & Action tables for more changes. Finally, the packets go through the Deparser to get put in order and then leave the switch [20].

1.7   OvS (Open vSwitch)

The OvS is a software-based, open-source virtual switch that operates within virtualized environments like data centers and cloud computing systems. It was initially developed by Nicira Networks and later became a collaborative project within the open-source community. Its first release dates back to 2009 and since then OVS has evolved through contributions from open-source community developers.[21]

---

[20] Manzanares-Lopez et al. [20].

[21] Kasal [21].

It is designed to provide switching functions similar to physical switches, enabling network connectivity, traffic management, and security features in virtualized environments. In a compute environment, OVS connects virtual machines (VMs) within a host and across different hosts in a virtualized environment by creating virtual bridges and ports to interconnect the virtual network interfaces. Here's how OVS accomplishes this:

Within a Host:

- When a virtual machine is created on a host system, it typically has a virtual network interface (vNIC) associated with it.
- OVS creates virtual bridges, which act as software-based switches within the host.
- Each virtual machine's vNIC is connected to a port on the virtual bridge.
- OVS dynamically learns the MAC addresses of the VMs by inspecting the traffic passing through its ports (MAC learning).
- When a VM sends a packet, OVS examines the destination MAC address and forwards the packet to the appropriate port connected to the destination VM.

Between Hosts:

- In scenarios where VMs on different hosts need to communicate, OVS can create network overlays using tunneling protocols like VXLAN, GRE, or Geneve.
- OVS instances on different hosts encapsulate packets within tunnel headers and transmit them over the underlying physical network.
- When a packet arrives at the destination host, the OVS instance decapsulates the packet and forwards it to the appropriate virtual bridge and port connected to the destination VM.
- This enables communication between VMs as if they were on the same physical network, even if they are hosted on different physical servers.

Additionally, OVS instances can also be managed and controlled by SDN (Software-Defined Networking) controllers such as OpenDaylight or ONOS. These controllers communicate with OVS instances using the OpenFlow protocol to dynamically configure flow rules and manage network traffic. This allows centralized control and programmability of network forwarding behavior, simplifying the management and automation of network configurations across multiple hosts.

Figure 1.24 illustrates how OVS facilitates VM connectivity and utilizes OpenFlow to connect with the OpenDaylight SDN controller. The OpenDaylight SDN controller employs northbound APIs to interface with OpenStack Neutron for cloudification of the network. Both OpenDaylight (ODL) and OpenStack are open-source software solutions for SDN controllers and cloud computing infrastructure, respectively.

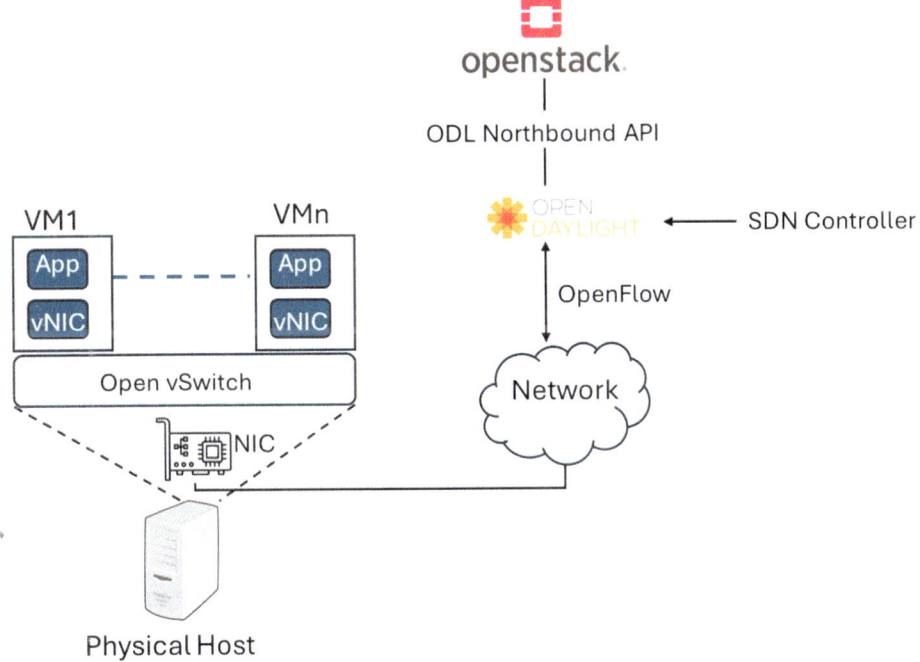

Fig. 1.24 OvS and Opendaylight SDN controller connectivity

1.7.1 Architecture

The architecture of Open vSwitch (OVS) consists of several components working together to provide virtualized network switching capabilities. These are

- Userspace Daemon (ovs-vswitchd).
- Kernel Module (datapath).
- OVSDB (Open vSwitch Database).
- Management Tools.
- Flow Tables.
- Integration Interfaces.

1.7.1.1 Userspace Daemon (Ovs-vswitchd)

The userspace daemon, referred to as "ovs-vswitchd," serves as the central component within the OVS architecture. It is responsible for managing the virtual switch configuration, which encompasses bridges, ports, and flow tables. Furthermore, it establishes communication with the kernel module to configure datapath forwarding rules and oversee packet processing.

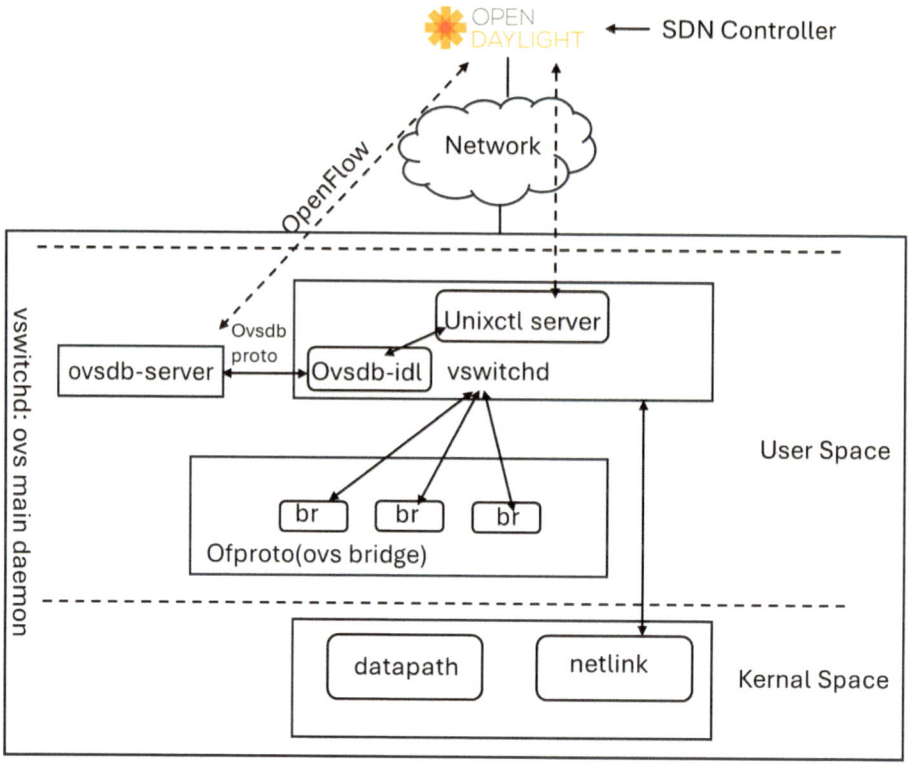

Fig. 1.25 vswitchd: OVS main daemon[22]

The userspace daemon "ovs-vswitchd" retrieves the desired Open vSwitch configuration from ovsdb-server via an IPC channel and subsequently conveys this configuration to the OVS bridges. These bridges, implemented as a library termed "ofproto," are responsible for realizing the desired network topology (please refer to Fig. 1.25). Moreover, ovs-vswitchd facilitates the transmission of specific status and statistical information from the OVS bridges back into the database [22].

1.7.1.2 Kernel Module (Datapath)

The OVS kernel module, also known as the datapath module, interfaces directly with the Linux kernel networking stack. It provides the low-level packet processing and forwarding capabilities necessary for efficient network operation. The datapath module is responsible for implementing packet switching, flow table lookup, and other essential networking functions.

---

[22] Chiao [22].

1.7.1.3 OVSDB (Open vSwitch Database)

OVSDB is a management protocol and database schema used by OVS for configuration and management purposes. It provides a structured way to represent and manipulate the configuration of OVS instances, including bridges, ports, interfaces, and flow tables. OVSDB maintains a centralized database that stores the configuration state of the virtual switch. External management tools and SDN controllers interact with OVSDB to configure and query the state of OVS instances. OVSDB is often accessed using JSON-RPC or other remote procedure call mechanisms. It enables dynamic configuration changes, synchronization of configuration across multiple OVS instances, and integration with higher level management systems.

1.7.1.4 Management Tools

OVS provides a set of command-line tools for managing and configuring the virtual switch. The "ovs-vsctl" tool is used to create, modify, and delete virtual bridges, ports, and other switch components. ovs-ofctl is another tool that allows administrators to interact with the OpenFlow-based flow tables, enabling manual control over packet forwarding behavior. These management tools provide administrators with the flexibility to configure and monitor OVS instances from the command line.

1.7.1.5 Flow Tables

OVS uses flow tables to process incoming packets and make forwarding decisions. Flow entries in the table match specific packet header fields, such as source/destination MAC addresses, VLAN tags, IP addresses, or TCP/UDP ports. When a packet arrives at the switch, it is matched against the flow table entries, and the corresponding action (e.g., forward, drop, modify) is applied based on the matching rule.

1.7.1.6 Integration Interfaces

OVS supports integration with various virtualization platforms, container orchestration systems, and SDN controllers. For example, OVS integrates with OpenStack Neutron, VMware NSX, and Docker networking to provide virtual networking capabilities. It also supports integration with SDN controllers such as OpenDaylight, ONOS, and Ryu using the OpenFlow protocol for centralized network management and control.

1.7.2   Features

As of writing this chapter, the following features are supported in OVS version 3.3.0[23]:

- Visibility into inter-VM communication via NetFlow, sFlow(R), IPFIX, SPAN, RSPAN, and GRE-tunneled mirrors.
- LACP (IEEE 802.1AX-2008).
- Standard 802.1Q VLAN model with trunking.
- Multicast snooping.
- IETF Auto-Attach SPBM and rudimentary required LLDP support.
- BFD and 802.1ag link monitoring.
- STP (IEEE 802.1D-1998) and RSTP (IEEE 802.1D-2004).
- Fine-grained QoS control.
- Support for HFSC qdisc.
- Per VM interface traffic policing.
- NIC bonding with source-MAC load balancing, active backup, and L4 hashing.
- OpenFlow protocol support (including many extensions for virtualization).
- IPv6 support.
- Multiple tunneling protocols (GRE, VXLAN, STT, and Geneve, with IPsec support).
- Remote configuration protocol with C and Python bindings.
- Kernel and userspace forwarding engine options.
- Multi-table forwarding pipeline with flow-caching engine.
- Forwarding layer abstraction to ease porting to new software and hardware platforms.

Readers interested in learning more about OvS should visit https://www.openvswitch.org/ for further details.

1.8   Open Networking

Embarking on the journey into Open Networking proved to be a rollercoaster ride. In early 2015, I joined Delta Electronics, one of the largest Original Design Manufacturers (ODMs), with the mission to define, develop, and deliver open networking solutions under the "Agema" brand. At Delta's subsidiary, Agema Systems, Inc., I led the charge in defining product strategy, creating roadmaps, and overseeing the global rollout of open networking products and solutions. The landscape of open networking was still evolving at the time, with incomplete software support posing a significant challenge. The industry was grappling with uncertainty, torn between the allure of OpenFlow-based programmable switches and the appeal of vendor-agnostic systems akin to those offered by established

---

[23] Openvswitch.org [23].

players like HPE, Cisco, and Juniper. Clients, predominantly large enterprises, hyper-scalers, and global service providers, sought solutions that could trim both CAPEX and OPEX, hence the growing interest in vendor-agnostic switching products, often referred to as "Whitebox."

Some hyperscalers had already taken the plunge, deploying bare metal switches with their proprietary Network Operating Systems (NOS) or embracing open-source alternatives like FRR. The introduction of Cumulus® Linux by Cumulus Networks, along with the contribution of the Open Network Install Environment (ONIE) to the open-source community, marked a turning point for the industry. ONIE's arrival paved the way for the installation of partially decoupled NOS, revolutionizing the landscape. Yet, challenges loomed large. Qualifying various NOS for a diverse portfolio of switching products proved to be a daunting task. Establishing robust global market and logistical support networks was equally demanding. Additionally, creating services and support infrastructure to offer comprehensive global coverage posed a significant challenge. The available NOS, whether open source or developed by startups, often fell short of meeting customer demands for feature richness. Moreover, the lack of automation, orchestration, and management capabilities hindered effective support for complex enterprise, data center, and service provider networks.

As an organization, we were forced to scramble to realign our operations with market demands. This involved developing strategies and solutions for decoupling hardware and software, meticulously qualifying various Network Operating Systems (NOS), conducting rigorous testing with open-source orchestration platforms, and building organizational capabilities to manage global market expansions. These challenges were not unique to Agema Systems but plagued the entire open networking ecosystem.

Despite these initial hurdles, the open networking industry has made significant strides. NOS offerings have matured, encompassing both open-source and vendor-led products. In the following section, we will delve deeper into the Open Networking concept, exploring its ecosystem and the strides made since those early days.

1.8.1 The Concept

Throughout this chapter, we have explored various technologies that have emerged to enable programmability in networking hardware. However, for these technologies to gain traction, open networking has been central. Open networking allows for the decoupling of hardware and software, paving the way for open standards and fostering the utilization of open, standardized, and flexible platforms for building and managing enterprise networks. Through disaggregation, open networking offers enhanced adaptability, lower costs, and more customization options.

While the concept has been around since the 2010s, the formation of the Open Networking Foundation (ONF) by Stanford and Berkeley Universities to pursue Software-Defined Networking (SDN) marked the beginning of a new chapter for Open Networking.

While virtualization and the separation of hardware and software in computing systems have been around for quite some time, networking gear remained tightly coupled. The standardization of OpenFlow by the ONF created momentum, bringing networking technologies closer to how cloud computing operated at the time.

During this period, hyperscalers and service providers felt the need to utilize ODM networking gear to accommodate the growing demand for bandwidth while keeping costs down. They also desired to move away from vendor-locked infrastructure to a multi-vendor network environment. This momentum encouraged ODM vendors to participate in experiments involving the decoupling of hardware and software. In 2013, Cumulus® Networks created ONIE (Open Network Install Environment), an open installer bootloader facilitating the Linux environment and contributed to the open-source community through the "Open Compute Project" (OCP). This "install environment" helped many ODM vendors to offer switches without NOS and allowed customers to choose a qualified NOS from a list of vendors. The networking gears that were supplied without a NOS were called "Bare Metal Switches." The system that was prepared with preinstalled NOS is called "Whitebox." In the following sections, we will further explore these concepts in detail.

1.8.2   ONIE

Open Network Install Environment (ONIE) combines a small operating system and bootloader designed for pre-installation as firmware on bare metal network switches. It serves as a platform for automated operating system provisioning, enabling users to effortlessly install various network operating systems (NOS) on their hardware. ONIE empowers users to select and install their preferred NOS without being bound to any specific vendor or proprietary solution. This flexibility is indispensable for network administrators and organizations seeking to deploy networking hardware with operating systems that precisely meet their requirements. With ONIE, users can automate the provisioning process for network switches, reducing manual intervention and streamlining the deployment of networking infrastructure. This not only saves time and effort but also enhances consistency and reliability across network deployments.

ONIE utilizes the CPU complex of the switch for its operation without interfering with the forwarding plane. During the initial bootup sequence, ONIE locates and executes an NOS vendor's installation program, as illustrated in the figure below. Upon the first power-up, the bootup sequence configures the CPU complex, loads, and boots ONIE from flash memory. A small Linux footprint, known as busybox, configures the management

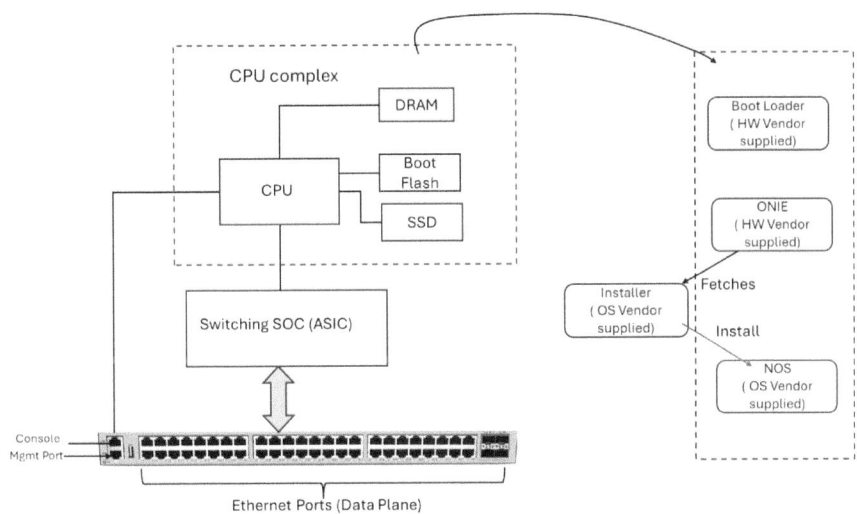

Fig. 1.26 Typical network switch architecture with CPU complex which is utilized by ONIE. The picture shows both CPU complex and first time boot up sequence

(mgmt.) interface, subsequently locating and executing the installer from the network. Additionally, busybox provides tools and an environment for the installer (Fig. 1.26).

This Linux executable installs the vendor-supplied NOS onto the mass storage (e.g., SSD). ONIE is only installed during the initial boot sequence; all subsequent boots directly access the NOS, bypassing ONIE. For further details about ONIE, please refer to the project site at https://opencomputeproject.github.io/onie/.

1.8.3   Bare Metal Switch

As discussed earlier, a bare metal switch comes with all drivers supplied by Hardware (HW) vendors and ONIE. It is ready for installation of third-party NOS. The diagram below depicts typical components of a bare metal switch (Fig. 1.27).

A bare metal networking switch may present a GRUB menu that includes ONIE, diag, and other software components such as switchdev. The device will be ready to connect with network through management port to install NOS. Please note, NOS must be qualified by hardware vendor or software vendor for specific networking switch hardware.

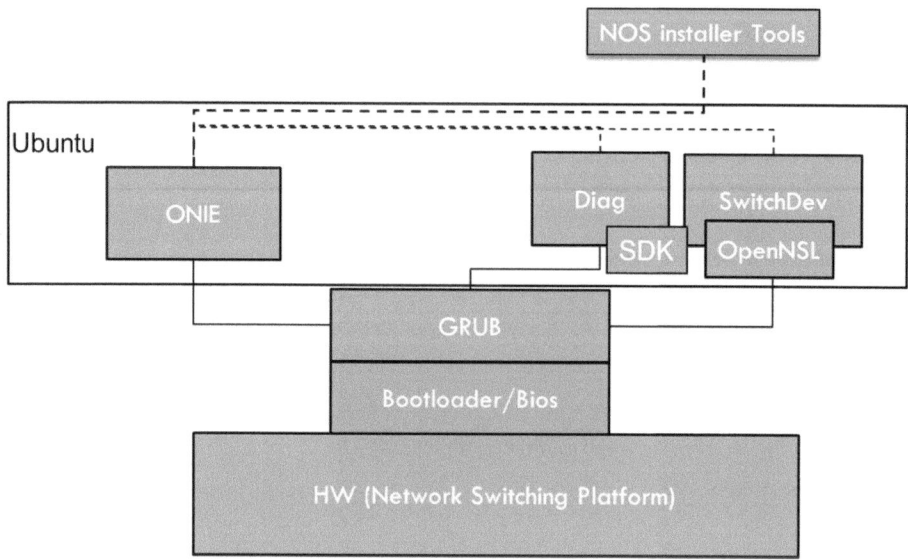

Fig. 1.27 Bare metal switch with its software components

1.8.4 Whitebox

In contrast to a bare metal switch, a whitebox switch will install the appropriate NOS on the networking hardware. It undergoes thorough testing with the NOS and also verifies all transceivers connected to the ports. Generally, whitebox vendors qualify transceivers and provide a list of approved transceivers to the customer. The diagram below should depict the hardware and software components of both whitebox and bare metal switches (Fig. 1.28).

Today, there are few NOS options available on whitebox switches, notably two open-source NOS: SONiC and FRR. Many original design manufacturers (ODM) supply their network hardware products as whitebox switches that are qualified with some of the open-sourced NOS and vendor-specific NOS (please refer to the figure above). The list provided in the figure above is not exhaustive but rather a representative sample.

1.8.5 Brite Box

The "Brite box" is similar to a whitebox switch but is generally offered by leading net-working OEMs such as HPE and Dell. Essentially, we can consider the brite box switch as a middle ground between a traditional switch and a whitebox switch. Brite box switches typically come pre-loaded with an operating system, which means they enable simple plug-and-play deployments and can be used directly. Additionally, professional technical

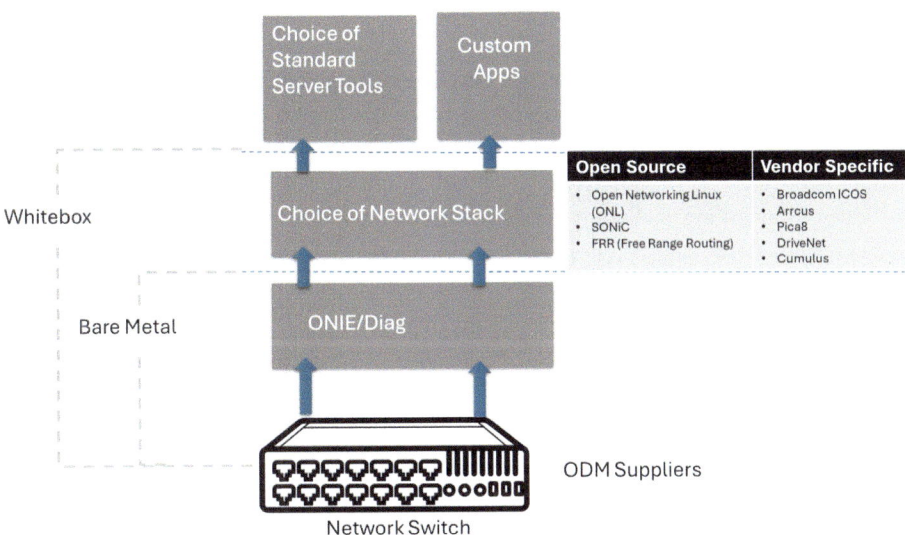

Fig. 1.28 Whitebox and bare metal switch hardware and software components

services from the suppliers are more convenient, as it's a one-stop solution. In terms of cost, brite box switches are more expensive than whitebox switches but much cheaper than traditional switches.

The following table compares Bare Metal, Whitebox, Brite box, and traditional network switches (Table 1.3).

Table 1.3 Comparison among bare metal, whitebox, brite box, and traditional network switches

	Bare metal	White box	Brite box	Traditional OEM switch
Definition	Network switch with ONIE and Diag software installed	Network switch with ONIE/Diag and NOS installed	OEM network switch with open-source software, e.g., SONiC	Network switch and software supplied by OEM vendors
HW suppliers	ODM	ODM	OEM vendors	OEM vendors
Software	N/A	Open source or NOS vendors	Open-source NOS	OEM NOS
Support	ODM for HW only	ODM for hardware and NOS vendor for software	OEM supplier for both hardware and software	OEM supplier for both hardware and software

1.9   The Trend

Throughout this chapter, I have explored various technological advancements. While there are numerous other technologies beyond the scope of those discussed in this chapter, many will be addressed in subsequent chapters. It's important to recognize that programmability and the decoupling of hardware and software, exemplified by OpenFlow and Open Networking technologies, complement each other. This concept arises from the acknowledgment that network infrastructure has been historically rigid and requires dynamism, allowing for programming and service provisioning on demand. Moreover, collaboration within the industry is crucial, necessitating the decoupling of hardware and software. Although OpenFlow initially showcased considerable promise in terms of programmability, it has ceased to be the dominant protocol in traditional networking equipment. Instead, facets of its technology have enabled traditional networking vendors to integrate programmability into their offerings, alongside automation and centralized network management through a unified platform. Consequently, there has been a growing demand for service intent, which tailors network performance based on application requirements, ultimately enhancing user experience. This paradigm is referred to as intent-based networking.

However, modern infrastructure must extend beyond service intent alone. Simplifying and streamlining network planning, configuration, management, optimization, and fault recovery are imperative. Here, self-organizing networks present a promising solution. Although primarily utilized in mobile networks, it's potential in mainstream data networks remains untapped. The current imperative is for networks to autonomously maintain operational integrity, preventing catastrophic failures, mitigating anomalies, and ensuring end-to-end visibility and operational simplicity.

As illustrated in Fig. 1.29, the trajectory of network evolution indicates a trend toward greater intelligence. The future entails intelligent networks capable of self-provisioning, self-healing, and possessing cognitive abilities to address various issues, including security threats and anomalies. The progression toward self-healing and autonomous networks marks a significant stride toward this vision.

In recent months, HPE has catalyzed momentum toward "AI networking." This concept involves networks specifically optimized for AI and machine learning workloads, in response to the growing adoption of such workloads in enterprises. Elements of AI integrated into self-healing and autonomous networks contribute significantly to this paradigm shift toward intelligent networks.

The term "AI networking" encompasses two distinct aspects. Firstly, there's the evolution from High-Performance Computing (HPC) necessitating network infrastructures tailored to the requirements of AI/ML workloads, termed as "Networking for AI." Secondly, there are AI elements inherently embedded within self-healing and autonomous networks. Together, these facets collectively form what is referred to as "AI Networking."

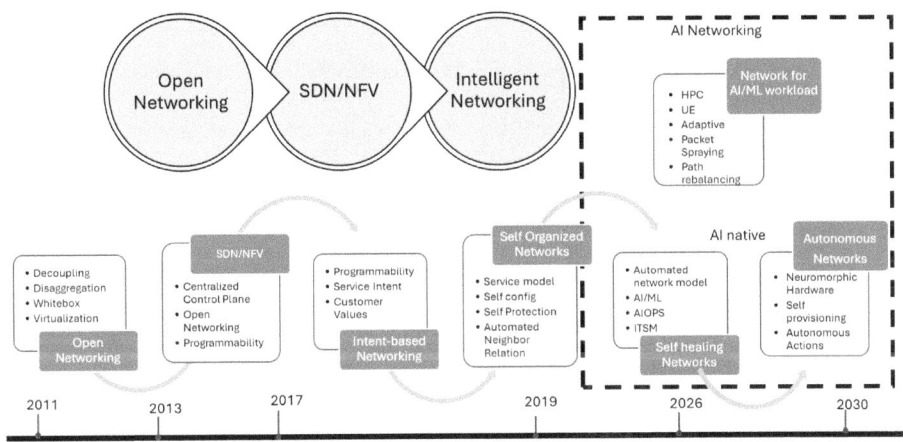

Fig. 1.29 Network technology trends

Understanding this technological prelude is pivotal for grasping the evolving landscape of network technology and its underlying trends. Subsequent chapters will delve deeper into these emerging technologies, equipping readers with comprehensive insights. Prepare yourselves for the journey ahead.

References

1. SDXCentral (2024) What is openflow? Definition and how it relates to SDN. https://www.sdx central.com/networking/sdn/definitions/what-the-definition-of-software-defined-networking-sdn/what-is-openflow/
2. ONF (2024) ONF Journey. https://opennetworking.org/mission/
3. OCP (2024) Open compute project (OCP). https://www.opencompute.org/
4. Thenextplaftorm (2022) Programming the network with intel Nex chief nick Mckeown. https://www.nextplatform.com/2022/08/12/programming-the-network-with-intel-nex-chief-nick-mckeown/
5. Compterweekly.com (2013) The history of OpenFlow. https://www.computerweekly.com/feature/The-history-of-OpenFlow
6. ACM (2007) Ethane: taking control of the enterprise. ACM SIGCOMM Comput Commun Rev 37(4). https://doi.org/10.1145/1282427.1282382
7. Vaughan-Nichols JS (2011) OpenFlow: the next generation of the network? IEEE Computer Society
8. Casado M, McKeown N, Shenker S (2019) From ethane to SDN and beyond. ACM SIGCOMM Comput Commun Rev 49(5)
9. Waziri UJ (2023) Software defined networking (SDN) controller for dynamic network traffic isolation through openflow. ResearchGate. https://www.researchgate.net/publication/374697708

10. Göransson P, Black C, Timothy C (2017) Software defined networks: a comprehensive approach. Elsevier Inc.
11. Broadcom (2014) OpenFlow™ data plane abstraction (OF-DPA): abstract switch specification. Version 2.0
12. Broadcom (2022) Broadcom switch SDK software enables rapid development and deployment. https://www.broadcom.com/blog/broadcom-switch-sdk-software
13. Broadcom (2016) OpenFlow data plane abstraction (OF-DPA) API guide and reference manual. https://broadcom-switch.github.io/of-dpa/doc/html/group__GLOFDPAAPI.html
14. Meta (2015) Facebook open switching system ("FBOSS") and wedge in the open. https://engineering.fb.com/2015/03/10/data-center-engineering/facebook-open-switching-system-fboss-and-wedge-in-the-open/
15. Onosproject.org (2019) ONOS+P4 tutorial for beginners. https://wiki.onosproject.org/?pageId=16122675
16. Goswami B, Kulkarni M, Paulose J (2017) A survey on P4 challenges in software defined networks: P4 programming. IEEE Access
17. P4.org (2022) P416 language specification. https://p4.org/p4-spec/docs/P4-16-v-1.2.3.html
18. Hang Z, Wen M, Shi Y, Zhang C (2019) Programming protocol-independent packet processors high-level programming (P4HLP): towards unified high-level programming for a commodity programmable switch. Electronics 2019(8):958
19. P4 forum (2022) P4 architecture. https://forum.p4.org/t/p4-architecture/246
20. Manzanares-Lopez P, Monoz-Gea J, Malgosa-Sanahuja J (2012) Passive in-band network telemetry systems: the potential of programmable data plane on network-wide telemetry. IEEE Access:3055462
21. Kasal ö (2023) Understanding open vSwitch: part 1. https://medium.com/@ozcankasal/understanding-open-vswitch-part-1-fd75e32794e4
22. Chiao A (2016) OVS deep dive 1: vswitchd. https://arthurchiao.art/blog/ovs-deep-dive-1-vswitchd/
23. Openvswitch.org (2024) Features. https://www.openvswitch.org/features/

2.1   Introduction

In the early 1990s, a plethora of networking technologies vied for dominance within networking devices, encompassing diverse options such as ATM (Asynchronous Transfer Mode), Token Ring, FDDI (Fiber Distributed Data Interface), SONET (Synchronous Optical Network), ISDN (Integrated Services Digital Network), and Ethernet. While each of these technologies held varying degrees of significance in LAN (Local Area Network) and WAN (Wide Area Network) deployments, their roles were not uniform. Some found prominence primarily in WAN and long-distance interconnectivity, while others, like ATM, straddled the realms of both LAN and WAN.

However, as the late 1990s approached, several of these technologies began to wane in popularity, largely due to their inherent complexities and a lack of widespread industry acceptance. In contrast, Ethernet emerged as a steadfast technology of choice, drawing support from industry giants. It's important to note that this widespread adoption of Ethernet does not necessarily imply that it was the unequivocally superior technology for all interconnects. Rather, Ethernet's appeal lay in its simplicity and adaptability across various connectivity domains, including LAN, MAN (Metropolitan Area Network), and to some extent, WAN. As industry leaders rallied behind Ethernet, it solidified its position as a foundational technology underpinning the evolving landscape of network infrastructure.

In the 1990s and into the early 2000s, Ethernet's evolution was chiefly steered by networking vendors, each crafting proprietary ASICs tailored to their specific product lines. This strategy created a landscape where innovations in Ethernet products were closely guarded within vendor portfolios. Throughout this era, Ethernet experienced incremental advancements primarily propelled by these proprietary ASIC designs, with vendors striving to differentiate their offerings through unique features and capabilities.

© The Author(s), under exclusive license to Springer Nature Switzerland AG 2025   49
D. D. Chowdhury, *Future of Networks*, Synthesis Lectures on Communications,
https://doi.org/10.1007/978-3-031-71440-5_2

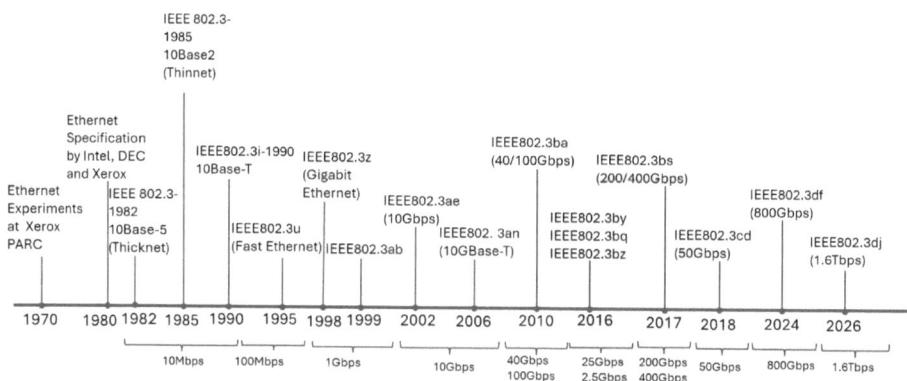

Fig. 2.1 Ethernet timeline

Ethernet's journey began in the early 1970s at Xerox's Palo Alto Research Center (PARC), where Robert Metcalfe and his team developed the original Ethernet protocol to connect computers within local networks. In 1980, Digital Equipment Corporation, Intel, and Xerox (collectively known as DIX) standardized Ethernet's specifications, paving the way for its widespread adoption. However, by the early 2000s, a notable shift emerged within the Ethernet ecosystem. A new wave of PHY (Physical Layer) and Ethernet System-on-Chip (SoC) vendors entered the fray, introducing standardized, higher speed components tailored to meet the escalating demands of network infrastructure. This transition marked a pivotal moment in Ethernet's evolution, facilitating greater interoperability and fostering a more diverse ecosystem of Ethernet devices. The following diagram depicts the timeline of Ethernet (Fig. 2.1).

Over the last four and a half decades, Ethernet has evolved from a modest 10 Mbps per port speed to today's blazing-fast 800 Gbps, with projections reaching 1.6 Tbps by 2026. This remarkable growth has seen Ethernet overshadowing competing technologies from the 1990s, garnering robust industry support.

Initially developed at Xerox PARC in the 1970s, Ethernet saw its first specifications crafted by Intel, DEC, and Xerox[1] in 1980. Subsequently, the standardization process was taken over by the IEEE (Institute of Electrical and Electronics Engineers), a leading technical professional association and standardization body. Under the IEEE's 802.3 task group, the groundbreaking IEEE 802.3–1982 specification was released, paving the way for the rollout of 10Base5 Ethernet in 1982. This inaugural Ethernet deployment, dubbed "Thicknet," utilized thicker COAX cable for connectivity, with "10" denoting a 10 Mbps transport speed and "5" indicating a 500-m distance limit over COAX cable. The system could seamlessly connect up to 100 stations on a single coaxial cable using vampire taps.[2]

---

[1] Goodrich [1].
[2] Wikipedia [2].

Shortly thereafter, IEEE introduced the 802.3–1985 specification for 10 Mbps Ethernet, known as "10Base2." This variant utilized thinner coaxial cable and earned the moniker "Thinnet," with "2" signifying a 200-m distance limit (although 10Base2 could function on 185 m, the number was rounded up for naming convention purposes).

Later in 1990, IEEE released the 802.3i standard for 10Base-T, marking Ethernet's initial foray into copper wire. The word "base" signifies baseband signaling, while "T" denotes "twisted pair." The cabling distance for 10Base-T was up to 100 m. The IEEE specification for Fast Ethernet (100 Mbps) was released in 1995, followed by Gigabit Ethernet (1 Gbps) in 1998. Subsequently, IEEE released the specifications for 10 Gbps Ethernet in 2002, 40 and 100 Gbps in 2010, 2.5 and 25 Gbps in 2016, 200 and 400 Gbps in 2017, 50 Gbps in 2018, and 800 Gbps in 2024. The IEEE standard committee is currently working on the 802.3bj specification for a staggering 1.6 Tbps per port speed, scheduled for release in 2026.

This prelude of Ethernet evolution is crucial for understanding the trajectory of Ethernet technologies and networking infrastructure. As Ethernet speeds increased, the sophistication of Ethernet switching ASICs also grew, evolving into System-on-Chips (SoCs) capable of supporting higher speeds and meeting the escalating demands for security, improved performance, and enhanced user experience. However, the simultaneous demand from end users for improved capabilities to manage complex network environments posed challenges for systems designed with proprietary ASICs. This challenge opened up opportunities for third-party ASIC vendors to enter the marketplace, offering common Commercial Off-The-Shelf (COTS) ASIC/SoCs that could be utilized by different vendors in their Ethernet system designs. Furthermore, these third-party vendors provided superior performance and processing capabilities compared to the in-house ASICs of networking equipment vendors, thereby fostering greater interoperability and diversity in the Ethernet ecosystem.

In this chapter, we will delve into the latest advancements in Ethernet Switching SoCs, examining how these chips contribute to overall system performance and the design of Ethernet-based network infrastructure. We'll explore the pivotal role SoCs play in addressing the evolving needs of modern networks, from enhancing throughput to fortifying security protocols. By analyzing these advancements, we can gain insights into how SoCs are shaping the future trajectory of Ethernet technology and reshaping the landscape of network infrastructure.

2.2 SerDes

The performance of Ethernet Switching SoCs is closely tied to the capabilities of the SerDes integrated within these chips. Higher speed SerDes enable Ethernet Switching SoCs to support faster data rates, accommodate more ports, and deliver improved overall throughput in network switches. Ethernet Switching SoCs often integrate multiple SerDes channels to support various Ethernet standards, including 1 Gigabit Ethernet (GbE), 10 GbE, 25 GbE, 40 GbE, 50GbE, 100 GbE, 200 GbE, 400 GbE, 800 GbE, and 1.6 Tbps per port speed. The speed and performance of SerDes directly impact the maximum achievable port speeds, port density, and overall switching capacity of Ethernet Switching SoCs.

A serializer/deserializer (SerDes) is a crucial component in modern high-speed communication systems, including Ethernet Switching System-on-Chips (SoCs). SerDes functions as an interface between the physical layer (PHY) and the data link layer, facilitating the transmission and reception of serialized data over high-speed serial links. SerDes technology converts parallel data from the data link layer into serial data streams for transmission over high-speed serial links, such as optical fibers or copper interconnects. On the receiving end, it deserializes incoming serial data streams back into parallel data for processing by the data link layer.

A list of functional benefits provided by the SerDes is given below:

- **Data Rate:** SerDes operates at high data rates, typically ranging from several Gbps (gigabits per second) to tens of Gbps, depending on the specific application and implementation.
- **Parallel-to-Serial Conversion:** SerDes converts parallel data streams from the data link layer into serial data streams for transmission over high-speed serial links. This conversion enables efficient data transmission over longer distances with reduced interconnect complexity.
- **Clock and Data Recovery (CDR):** SerDes includes circuitry for recovering the clock signal from the incoming serial data stream, ensuring proper synchronization and accurate data reception. CDR is essential for maintaining signal integrity and reliability, especially in high-speed communication systems.
- **Equalization:** SerDes often incorporates equalization techniques to compensate for signal degradation and distortions introduced by transmission media and interconnects. Equalization helps to mitigate channel impairments and improve overall signal quality, enabling reliable data transmission over challenging communication channels.
- **Jitter Tolerance:** SerDes designs typically include mechanisms for mitigating jitter, which refers to variations in the timing of the received signal. Jitter tolerance ensures that SerDes can reliably recover data from signals with timing variations, enhancing the robustness of communication links.
- **Power Efficiency:** Modern SerDes implementations prioritize power efficiency to minimize energy consumption and heat dissipation, particularly in energy-sensitive applications such as data centers and high-performance computing systems.

2.2.1 Origin and Evolution of SerDes

SerDes technology emerged in response to the need for efficient data transmission over high-speed serial links. In the early days of computing and telecommunications, parallel interfaces were commonly used for data transfer between devices. However, as data rates increased and the complexity of systems grew, parallel interfaces became less practical due to limitations in signal integrity, interconnect complexity, and power consumption. The technology has its roots in communication over fiber-optic and coaxial links. The idea is pretty straight forward—sending data one bit at a time instead of all at once in parallel means fewer cables are needed! This was crucial because having fewer cables allowed for maximizing the data sent over them. The focus was mainly on getting as much data through those cables as possible, while concerns like the size and power usage of SerDes were less important.

In the mid-1980s, the speed of data sent over serial links was mostly driven by telecom needs, especially for things like SONET. Back then, the requirements for data rates like OC-1 and OC-3 were quite modest compared to today's standards. It wasn't until around 1990 that circuits capable of supporting higher speeds, like OC-24 with rates above 1 Gb/s, became available using advanced technologies like bipolar and gallium-arsenide (GaAs) processes. By the early 2000s, 10 Gigabit Ethernet using a 10Gb/s line rate became a reality, marking a significant milestone in high-speed communication. Another significant shift was happening too—SerDes were increasingly used for communication between chips on printed circuit boards (PCBs) and backplanes instead of using parallel links. This change transformed SerDes from being mainly used for long-distance communication to becoming a vital component in System-on-Chip (SoC) designs. One notable example of this shift is PCIe, which debuted around 2002 at 2.5 Gbps and gained popularity in the mid-2000s, demonstrating the versatility and importance of SerDes in modern communication systems.[3]

While it's challenging to pinpoint the exact timeline of SerDes evolution, a representative timeline illustrates the progression of SerDes from 10 to 224G. This timeline underscores the dynamic nature of SerDes technology, which has continually adapted to meet the evolving demands of high-speed communication systems (Fig. 2.2).

The evolution of SerDes, as depicted in the figure above, has significant implications for both Ethernet ASIC design and the per port speed of Ethernet switches. For instance, a port speed of 100 GbE can be achieved using 4 lanes of 25 G SerDes, while a port speed of 400 GbE can be attained through either 8 lanes of 25 G SerDes or 4 lanes of 112 G SerDes. The diagram below illustrates various Ethernet port speeds and their corresponding SerDes lanes (Fig. 2.3).

As of writing this book, 224G SerDes technology is under development and IP block for this is available. IEEE 803dj task group is also working on the draft specification for

---

[3] Galloway [3].

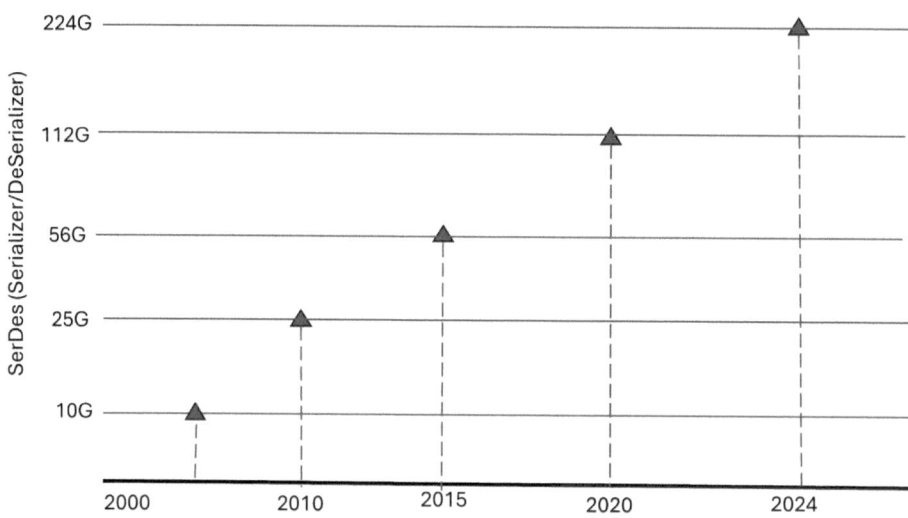

Fig. 2.2 Timeline depicts evolution of SerDes

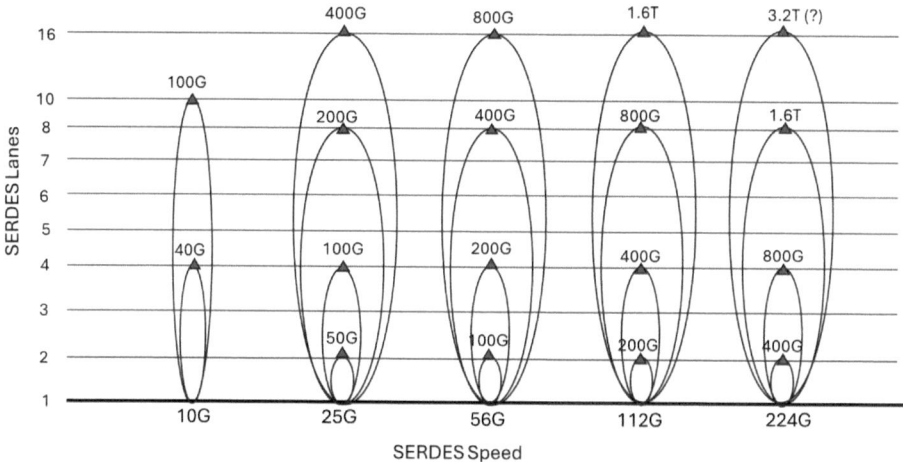

Fig. 2.3 SerDes lanes and corresponding Ethernet speed

1.6Tbps Ethernet. The specification should be available by 2026. With SerDes speed of 224G, it will make it easier to achieve 1.6Tbps per port Ethernet speed with a smaller number of traces in the PCB.

2.2.2 Ethernet ASIC Bandwidth and SerDes

To meet the increasing demand for bandwidth, Ethernet Switching Application-Specific Integrated Circuits (ASICs) are equipped with integrated SerDes technology. This integration enables ASICs to achieve the significant bandwidth necessary to fulfill the requirements of modern networks. By incorporating multiple integrated SerDes within the ASIC design, Ethernet ASICs can attain the substantial data throughput needed to support high-speed data transmission in networking applications. SerDes technology plays a crucial role in converting parallel data streams into serial data streams, allowing for efficient communication over high-speed serial links. The following table depicts a timeline of increases in integrated SerDes in Ethernet switching ASICs. In 2010, the integrated 10Gbps SerDes count was 64 in an Ethernet ASIC, giving a total ASIC bandwidth of 640Gbps. Subsequently, in 2012, that count was increased to 128, allowing the ASIC bandwidth to reach 1.2 Tbps. In 2014, the SerDes count remained the same, but the SerDes speed increased to 25G+, giving the ASIC a 3.2 Tbps capacity. The SerDes count for 25G SerDes increased in 2016, giving the ASIC a 6.4 Tbps capacity.

Until 2016, SerDes speed was below the threshold of the NRZ (Non-Return-to-Zero) method of signaling. The NRZ which is also known as Pulse Amplitude Modulation 2-level (PAM2) utilizes two voltage levels to represent logic 0 and 1. Simply put, NRZ signaling transmits data bits serially, one at a time, with each signal representing either a 1 or a 0 based on the voltage level. In NRZ signaling, the baud rate, or the speed at which a symbol can change, is equal to the bit rate, meaning the NRZ bitrate is equal to its symbol rate. For instance, 1 Gbps is equivalent to 1 Gbaud. In contrast, Pulse Amplitude Modulation 4-level (PAM4) signaling employs four distinct amplitude levels per bit, with two bits grouped and mapped to one symbol. With two bits per symbol, the baud rate is halved compared to the bitrate. For example, 28G baud PAM4 is equal to 56G NRZ. Consequently, PAM-4 achieves twice the throughput using half the bandwidth compared to NRZ [6].

At higher data speeds, the signal can lose strength as it travels through the channel, making it harder for devices to communicate effectively. Sometimes, the channel technology used may not work well for faster data rates. But, with PAM-4, the signal doesn't lose as much strength because it changes less often than with NRZ. This is because in NRZ, the Nyquist frequency, which measures how quickly the signal changes, is half the data speed, while in PAM-4, it's only a quarter. For example, if we look at Fig. 2.4, the signal loss for 56G NRZ at 28 GHz is more than 60 dB, but for 56G PAM-4 at 14 GHz, it's only about 30 dB. This is a big benefit of PAM-4 because it lets us use the same channels and connections for faster data rates without needing to make them work twice as hard or lose as much signal strength.

This is why, starting in 2018, 56G SerDes began using PAM4 signaling, as illustrated in Table 2.1. With 256 SerDes integrated into a single ASIC, the bandwidth capacity of Ethernet ASICs increased to 12.8 Tbps. Subsequently, 112G SerDes with 256 and

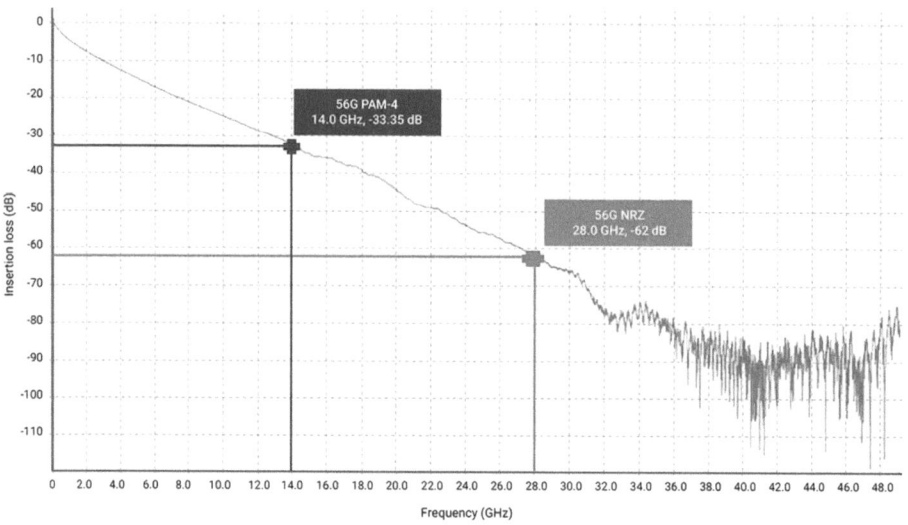

Fig. 2.4 NRZ versus PAM4 signal loss (Courtesy: Synopsys®)[4]

Table 2.1 Timeline of integrated SerDes in Ethernet ASICs[5,6]

Year	2010	2012	2014	2016	2018	2020	2022	2023	2026
SerDes count	64	128	128	256	256	256	512	512	1024 (?)
SerDes speed (Gbps)	10	10	25	25	50	100	100	112	224
Modulation	NRZ	NRZ	NRZ	NRZ	PAM4	PAM4	PAM4	PAM4	PAM4
ASIC bandwidth (Tbps)	0.64	1.28	3.2	6.4	12.8	25.6	51.2	51.2	102.4 (?)
CMOS node (nm)	40	40	28	16	16	7	5	5	?

512 counts elevated ASIC capacity to 25.6 Tbps and 51.2 Tbps, respectively, by 2022. Additionally, as of writing this book, the industry already has an IP block available for 224G SerDes, which will redefine the capacity of future generations of Ethernet ASICs.

---

[4] Horner [6].

[5] Minkenberg et al. [4].

[6] Cady [5].

2.3 Ethernet Systems Architecture

Ethernet switching ASICs today are more like Systems on Chip (SOCs) than simple ASICs. Originally, the term "Ethernet switching ASIC" referred to custom-designed integrated circuits tailored specifically for Ethernet networking functions. This term arose from the predominantly vendor-led development of ASICs serving Ethernet switching functions. However, over the decades, advances in silicon technology have allowed for the integration of more functions within a single ASIC, including Ethernet networking functions, CPU cores, memory controllers, I/O interfaces, and other peripherals. Consequently, most Ethernet Commercial Off-The-Shelf (COTS) devices are now considered SoCs. For the purpose of this discussion, we will use the terms "Ethernet switching ASIC" and "SoC" interchangeably.

An Ethernet SoC serves as the core element of Ethernet switching architecture, as illustrated in Fig. 2.5. A typical high-speed Ethernet system includes dual Power Supply Units (PSUs) for redundancy and a fan module. Both the PSUs and the fan module are connected to the CPU through a CPLD (Complex Programmable Logic Device) to facilitate I2C bus connectivity, power management, reset signals, and interrupt handling. A CPLD is a type of programmable logic device that integrates a large number of programmable logic blocks, interconnects, and input/output pins onto a single chip. In Ethernet systems, CPLDs can provide effective power management, fault detection and protection, temperature monitoring, system control and interfacing, redundancy, and failover mechanisms, among other functions. They serve as interfaces between the CPU and external devices such as PSUs and fans, manage communication protocols, handle interrupts, and coordinate data transfer between the CPU and peripheral devices.

The CPU connects management interface (Mgmt port), console, and USBs of the Ethernet system. In a typical, Ethernet system, CPU connects to Ethernet switching SOC with PCIe which in turn connects to ports with SFI and CPPI-4 depending upon port speeds. In this example, SFI connects to 10/15 GbE ports while CPPI-4 connects to 100GbE Ports. Both SFI and CPPI-4 are electrical interface for SFP+/SFP28 and QSFP28, respectively. It is obvious from this design that all packets are processed through Ethernet switching SoCs. Only certain type of traffic may go to CPU, e.g., BGP. Much of the packet pipeline is processed within the Ethernet switching SoCs. This architectural purview is important to understand how modern network equipment works. With this understand we can move forward to explore more about Ethernet SoCs in the proceeding sections.

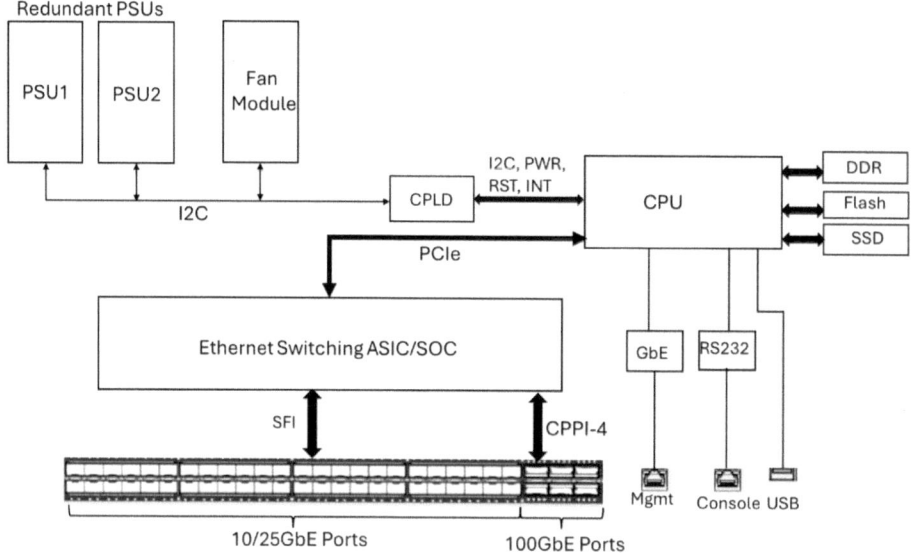

Fig. 2.5 Typical Ethernet switch architecture

2.3.1 Ethernet SoC

Ethernet switching System on Chips (SoCs) consist of several functional blocks that work together to enable the routing, switching, and management of network traffic. Let's consider the Broadcom Hurricane3-MG (BCM56170) Ethernet Switching ASIC,[7,8] as an example. It is a high-performance ASIC capable of delivering up to 380 Gb/s of throughput. This multilayer Ethernet switch supports 24 to 48 ports of MultiGigabit (MGig) connectivity, offering speeds ranging from 1 Gbps, 2.5 Gbps, 5 Gbps, to 10 Gbps, depending on the configuration. Please refer to the diagram below (Fig. 2.6).

The functional blocks depicted in the diagram above are most commonly available in Ethernet switching SoCs. The following sections will elaborate further on those functional blocks.

2.3.1.1 Ethernet Port Interface

This functional block manages the physical interfaces responsible for connecting to Ethernet ports. It encompasses various components, including SerDes (Serializer/Deserializer) interfaces for converting between serial and parallel data, PHY (Physical Layer) interfaces for managing the physical layer signaling, and MAC (Media Access Control) interfaces for handling data link layer functions. As illustrated in the provided figure,

---

[7] STH [7].

[8] Broadcom [8]

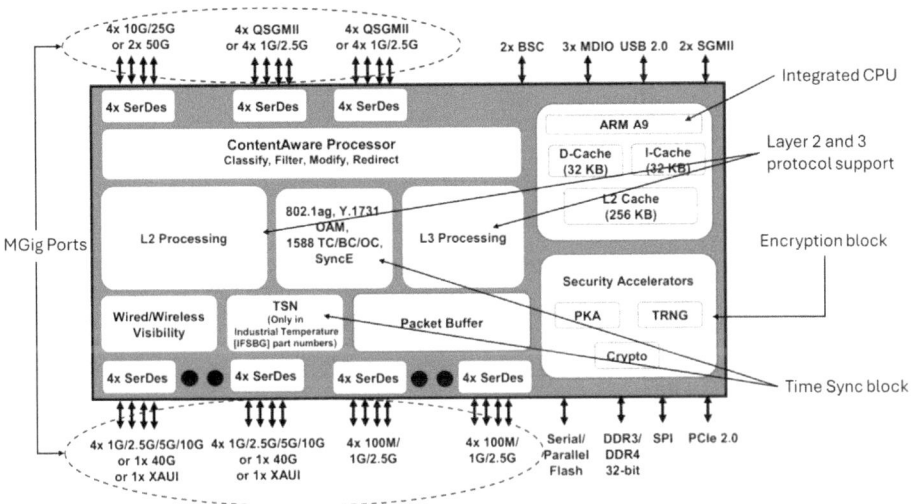

Fig. 2.6 Block diagram of Broadcom® BCM56170 (courtesy: Broadcom Inc) [8]

multiple SerDes are integrated into the Ethernet switching SoC diagram. In the case of the BCM56170, these SerDes facilitate connectivity to MultiGigabit (MGig) front panel ports of a switch. For a better understanding of the connectivity, please refer to the accompanying diagram. Typically, an Ethernet switching ASIC/SOC is connected to an external PHY chip for copper ports or SFP + cages to accommodate optical transceivers. Clock buffers and crystals may also be included in the design based on clocking requirements. It's important to note that the diagram presented here offers a simplified overview of an Ethernet switch's port connectivity and does not represent a reference design. Additionally, the switching chip may connect temperature sensors and DDR4 memory chips as necessary. Advanced Ethernet SoCs often feature a variety of high-speed ports, such as SFP28 for 25 GbE, QSFP28 for 100 GbE, QSFP56 for 200 GbE, QSFPDD for 400G, and so forth (Fig. 2.7).

Optical transceiver modules are available in various configurations tailored for both short- and long-distance connectivity requirements. Additionally, these optical modules may incorporate a PHY chip capable of providing link-level encryption functionalities, such as MACSEC (Media Access Control Security). Similarly, for copper ports, an onboard PHY chip will offer similar capabilities, ensuring secure and reliable data transmission over Ethernet connections.

2.3.1.2 Packet Buffering and Queuing

Ethernet switching SoCs use packet buffering and queuing mechanisms to store incoming and outgoing packets temporarily. This block manages packet storage, prioritization, and

Fig. 2.7 Ethernet interface connectivity to Ethernet switching SoC

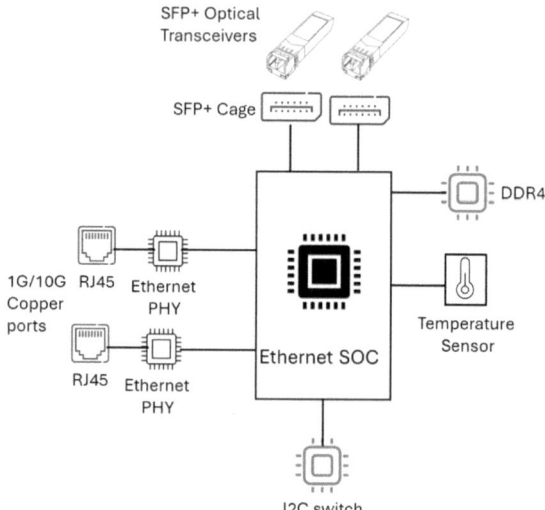

scheduling to ensure efficient handling of network traffic and to prevent packet loss or congestion. These functions can be divided into the following categories:

a. **Buffer Management Policies:** Ethernet switching SoCs implement various buffer management policies to efficiently utilize available memory resources and optimize packet processing. These policies may include techniques such as dynamic buffer allocation, buffer pooling, and buffer recycling to accommodate different traffic patterns and priorities.

b. **Buffer Scheduling Algorithms:** Packet buffering and queuing mechanisms employ sophisticated scheduling algorithms to determine the order in which packets are transmitted from the buffers to the output ports. These algorithms may prioritize certain types of traffic, such as real-time or latency-sensitive packets, over others to meet Quality-of-Service (QoS) requirements.

c. **Congestion Management:** Ethernet switching SoCs incorporate congestion management mechanisms to prevent packet loss and congestion in the network. These mechanisms may include techniques such as congestion avoidance, congestion notification, and flow control to regulate the rate of incoming traffic and maintain optimal network performance.

d. **Traffic Shaping and Policing:** Packet buffering and queuing functionalities may also include traffic shaping and policing mechanisms to control the flow of traffic through the network. Traffic shaping adjusts the rate of outgoing traffic to conform to predefined traffic profiles, while traffic policing enforces traffic rate limits and quality-of-service policies.

e. **Tail Drop and Congestion Avoidance:** In cases where packet buffers become full due to congestion, Ethernet switching SoCs may employ tail drop and congestion avoidance mechanisms to manage incoming packets. Tail drop discards packets when buffer space is exhausted, while congestion avoidance techniques such as Random Early Detection (RED) selectively drop packets based on predefined thresholds to prevent congestion collapse.

f. **Priority Queuing and Class-Based Queuing:** Ethernet switching SoCs support priority queuing and class-based queuing mechanisms to prioritize certain types of traffic over others. Priority queuing assigns different priority levels to packets based on their classification criteria, ensuring that high-priority traffic is transmitted ahead of lower priority traffic. Class-based queuing groups packets into classes based on predefined criteria and applies different queuing policies to each class.

g. **Dynamic Queue Management:** Ethernet switching SoCs may dynamically adjust queue parameters and thresholds based on network conditions and traffic patterns. Dynamic queue management techniques enable the system to adapt to changing traffic loads and optimize packet processing efficiency in real-time.

2.3.1.3 Forwarding and Switching Engine

This block is responsible for making forwarding decisions based on the destination addresses of incoming packets. It incorporates hardware-based lookup tables, such as MAC address tables and VLAN tables, along with forwarding algorithms and logic to identify the suitable output port for each packet. Advanced Ethernet switching chips feature several functional blocks within the forwarding engine, including hashing and load balancing, L2/L3 MAC Tunnel, L2/L3 Processing, L3 routes and Tunnels, and L2/L3 Hosts/ARP. Additionally, a parallel lookup engine may also be included to enhance packet processing efficiency and throughput.

As packets enter an Ethernet switch, they undergo processing within the Ethernet SoC. This processing follows an architectural approach known as pipeline processing, which is implemented by the forwarding engine. Pipeline processing involves breaking down a task or process into smaller stages or steps. Each stage performs a specific function before passing the data to the next stage. This approach allows for concurrent execution of multiple tasks, significantly improving overall system performance and throughput. For the purpose of this discussion, we will explore three distinct pipeline processes: the ingress pipeline, the Memory Management Unit (MMU), and the egress pipeline. These components are responsible for handling the processing of incoming and outgoing packets, respectively. The following diagram illustrates how pipeline processing works (Fig. 2.8).

As depicted in the figure above, packets enter the ingress pipeline, proceed through the Memory Management Unit (MMU), and then traverse the egress pipeline to reach egress ports.

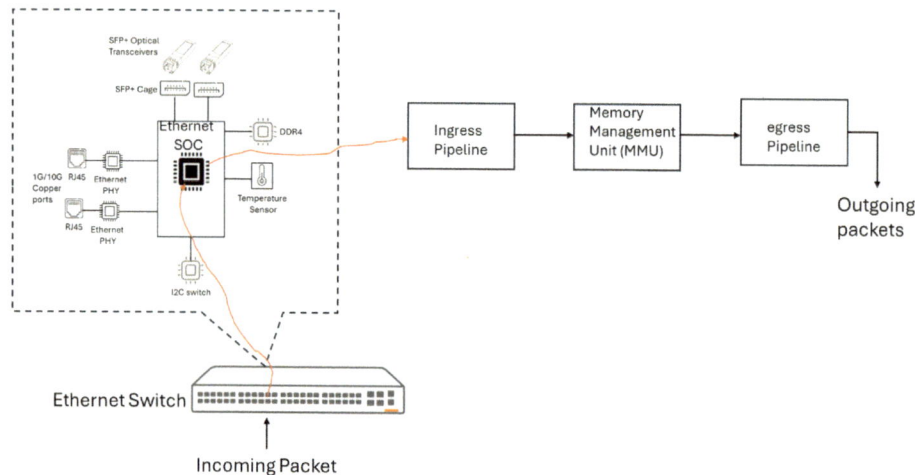

Fig. 2.8 The pipeline processing in an Ethernet SoC

2.3.1.4 Ingress Pipeline Process

The first block encountered by a packet during this process is the parser, a component responsible for extracting relevant information from incoming packets. This information may include details such as source and destination addresses, packet type, protocol headers, payload data, and any other metadata necessary for further processing. The following diagram depicts a representative illustration of pipeline processing within the forwarding engine of an advanced Ethernet SoC. The parser initiates its process by locating the beginning of the packet and proceeds to recognize the headers belonging to various network layers, such as L2, L3, L4, and any additional headers like VLAN or MPLS. Following this, it extracts pertinent information from the Layer 2 (L2) header, typically encompassing details like the source and destination MAC addresses, VLAN tags, and EtherType or Length fields.

This data holds significance in determining the appropriate handling of the packet within the network device. Subsequently, the parser retrieves information from the Layer 3 (L3) header, which usually includes the source and destination IP addresses, IP protocol type (e.g., TCP, UDP, ICMP), and any IP options or flags. This information serves as the foundation for routing and forwarding decisions within the network device. Moving forward, the parser proceeds to extract data from the Layer 4 (L4) header, encompassing details like the source and destination port numbers, sequence numbers, acknowledgment numbers, and flags (e.g., TCP flags). These details are vital for managing higher level protocol operations and session establishment (Fig. 2.9).

In addition to these standard headers, the parser may also extract User-Defined Fields (UDFs) if present in the packet. These UDFs contain application-specific or user-defined information that is not part of the standard protocol headers. The parser identifies and

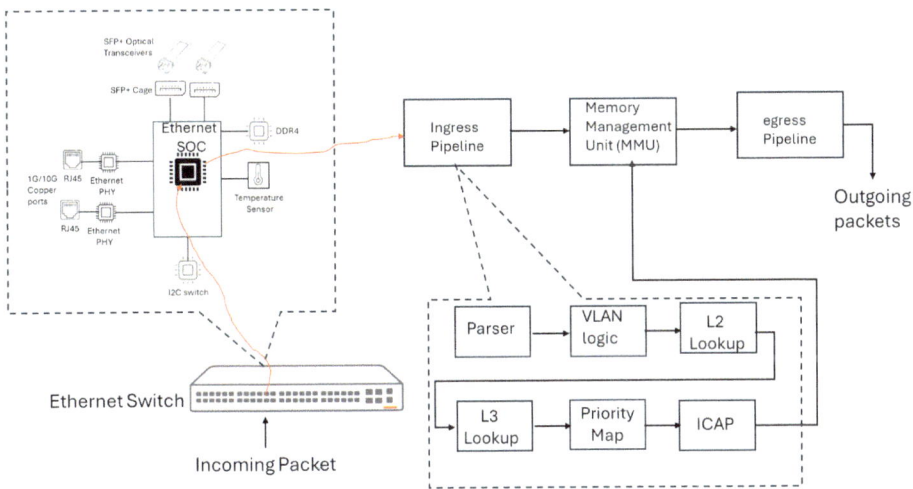

Fig. 2.9 Ingress pipeline processing

extracts these fields based on predefined rules or patterns specified by the user or application. Once the relevant headers and fields have been extracted, the parsed information
is typically forwarded to other modules within the network device for further processing.
It is noteworthy that the Parser utilizes Ingress Control and Parsing (ICAP) qualifiers to
establish specific criteria or conditions for parsing packets and extracting relevant information from them. These qualifiers serve as rules or filters applied by the parser to incoming
packets to determine the appropriate processing steps.

After the packet passes through the parser, it undergoes processing in the VLAN logic
module, where various operations related to VLAN tagging and handling are performed.
Here's an overview of how packets are processed for VLAN tag detection, assignment,
action, and lookup:

- **VLAN Tag Detection:** The VLAN logic module examines the packet headers to detect
  the presence of VLAN tags. VLAN tags are additional header fields added to Ethernet
  frames to identify the VLAN to which the packet belongs.
- **VLAN Assignment:** If the packet is untagged, meaning it does not have a VLAN tag,
  the VLAN logic module may assign a VLAN tag to the packet based on predefined
  rules or configuration settings. This assignment typically involves selecting a VLAN
  ID from a VLAN database or using default VLAN settings.
- **VLAN Tag Action:** Based on the VLAN tag detection and assignment, the VLAN
  logic module determines the appropriate action to take regarding the VLAN tag. This
  action may include adding, removing, or modifying VLAN tags as necessary to ensure
  proper VLAN membership and traffic segregation.

- **VLAN Lookup:** Once the VLAN tags are processed, the VLAN logic module performs a VLAN lookup to determine the forwarding behavior for the packet. This
  lookup involves referencing a VLAN table or database to match the VLAN ID
  extracted from the packet header with the corresponding VLAN configuration.
- **VLAN-based Forwarding:** After the VLAN lookup, the packet is forwarded according
  to the VLAN-specific forwarding rules configured for the matched VLAN. These rules
  may include forwarding the packet to specific ports or VLAN trunks, applying VLAN-
  based access control policies, or routing the packet to other network devices based on
  VLAN membership.

After completing VLAN-related operations, the VLAN logic module forwards the
packet to the Layer 2 (L2) lookup process for further handling. Within the L2 lookup
engine, the packet undergoes identification and processing based on Layer 2 characteristics. Diagrams below illustrate how L2 lookup identifies different types of Layer 2 packets
(Fig. 2.10).

During the packet-type identification process, the L2 lookup engine examines whether
the packet type is a broadcast, unicast, IPV4, or IPV6 multicast. If none of these criteria
match, it then checks if the packet is a L2 multicast and processes it accordingly. For
broadcast handling, the packet is flooded to all members of the VLAN. In the case of a
unicast packet, the L2 lookup engine performs a lookup in the L2 forwarding database
based on the Source Address (SA) and Destination Address (DA), and subsequently
updates the database.

After L2 lookup processing, the packet proceeds to the Layer 3 (L3) lookup engine.
Here, the packet is scrutinized to ascertain if routing is permissible. If not, it will be

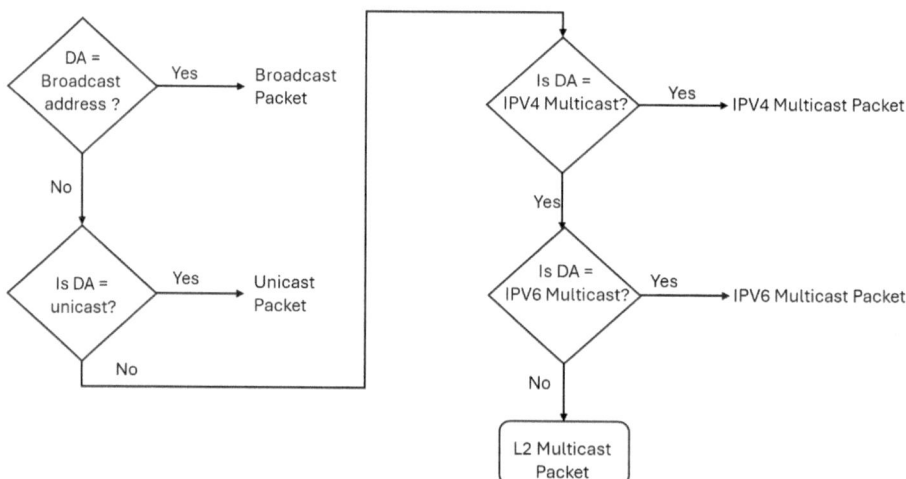

Fig. 2.10 L2 packet-type identification

directed to the L2 logic module for further assessment. In cases where routing is permissible, L3 packets undergo a series of checks, including examination of interface properties in the L3 interface table for Virtual Routing and Forwarding (VRF) determination, Reverse Path Forwarding (RPF), next-hop determination, and L3 unicast destination lookup.

After L3 lookup, packets may be classified and tagged with different priority levels or Quality-of-Service (QoS) markings based on their destination, source, or type of service. These priority markings are often used to determine the order in which packets are processed or forwarded within the network, ensuring that higher priority traffic receives preferential treatment over lower priority traffic.

Priority mapping in the pipeline processing stage, following L3 lookup, enables network administrators to implement traffic prioritization policies, optimize resource allocation, and ensure critical applications or services receive the necessary bandwidth and quality of service to meet performance requirements. Subsequently, the packet is passed to ICAP for further qualification before being handed off to the Memory Management Unit (MMU).

2.3.1.5 Memory Management Unit (MMU)

The figure below illustrates the various functions performed by the MMU to facilitate the pipeline processing of packets. Initially, packets from the ingress pipeline are directed to the MMU for processing (Fig. 2.11).

Within the MMU, several key functions are carried out to manage packet flow effectively. These include buffering, where packets are temporarily stored to accommodate

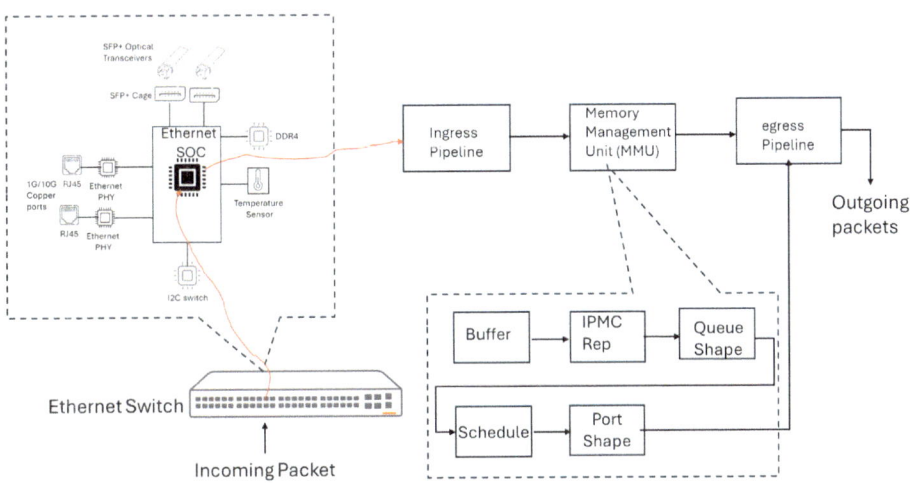

Fig. 2.11 The MMU functional block of pipeline processing

varying traffic loads and prevent congestion. Additionally, the MMU handles packet replication (please refer to IPMC Rep block in the figure above), allowing multicast packets
to be duplicated and forwarded to multiple destinations as required by the network topology. Queue shaping mechanisms are also implemented to regulate the flow of packets into
and out of the MMU, ensuring that each queue receives fair treatment and that Quality-
of-Service (QoS) requirements are met. Furthermore, the MMU incorporates scheduling
algorithms to prioritize packet transmission based on predefined criteria, optimizing network performance and resource utilization. Lastly, port shaping functions are employed to
control the rate of packet transmission from individual ports, aligning with network policies and traffic management strategies. Together, these functions within the MMU play
a crucial role in maintaining efficient packet processing and ensuring reliable network
operation.

2.3.1.6   Egress Pipeline

After processing at the MMU, the packet enters the egress pipeline, which is responsible
for preparing the packet for transmission out of the network device. The egress pipeline
consists of several critical stages, each performing specific tasks to ensure the packet is
properly handled. The following diagram depicts Egress pipeline processing (Fig. 2.12).

   The first stage in the egress pipeline is the parser, which extracts relevant information
from the packet headers and prepares it for further processing. Following the parser, the
packet moves through the Egress L3 stage, where any necessary Layer 3 (L3) processing is
performed. This may include tasks such as route lookups, TTL (Time-to-Live) decremen-
tation, and IP address manipulation. After Egress L3 processing, the packet undergoes a
VLAN check stage, where the VLAN tag is examined to ensure compliance with network

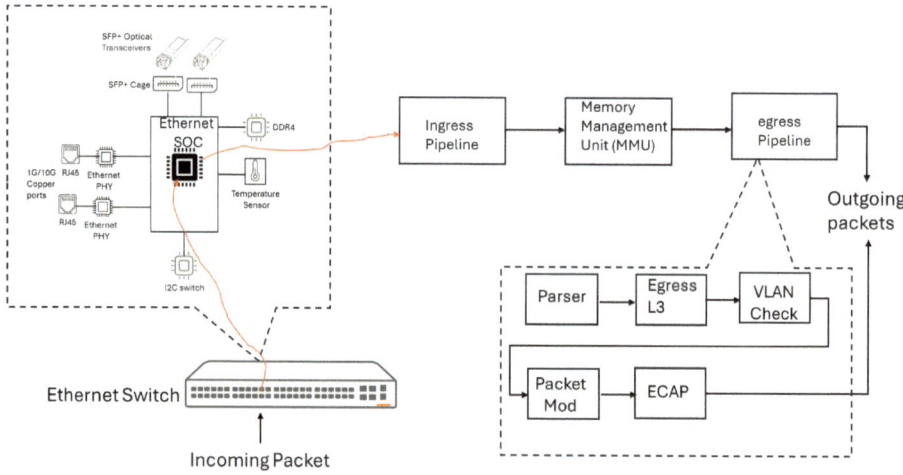

Fig. 2.12 The egress pipeline processing

policies and configurations. Any necessary VLAN modifications or actions are performed at this stage.

Next, the packet enters the packet modification stage, where additional modifications or transformations may be applied based on specific requirements. This could include tasks such as header adjustments, payload encryption, or Quality-of-Service (QoS) marking. Finally, the packet reaches the ECAP (Egress Control and Parsing) stage, where final checks and parsing operations are performed before the packet is transmitted out of the network device. ECAP ensures that the packet meets all necessary criteria and is properly formatted for transmission according to network standards and protocols. The egress pipeline plays a crucial role in preparing packets for outbound transmission, ensuring efficiency, compliance, and adherence to network policies.

2.3.2   Co-op Packaged SoC

Now that you have a good understanding of the advancements in Ethernet switching SoCs and their potentials for offloading L2 and L3 processing, let's delve into a different type of Ethernet switching ASIC known as CO-OP packaged SOC or silicon photonics. In the previous sections, we focused on explaining simple pipeline processing and did not delve into advanced processing techniques such as ECMP, per hop behavior, encryption, and neural-network inference engines for AI/ML workload traffic analysis, etc. However, the information presented thus far should provide a foundation for understanding the potential of modern Ethernet SoCs.

Returning to our discussion on SerDes and the substantial bandwidth offered by modern Ethernet SoCs, several deployment challenges have arisen. These include increased power consumption, the need for larger switch enclosures to accommodate plugin optical modules, and considerations regarding thermal management. To address these challenges and extend the roadmap, a combination of higher lane rates, increased lane counts per module, and alternative deployment models is required. These trends have driven a push toward higher degrees of integration and manufacturing automation for optical assemblies within transceivers, such as those based on silicon photonics. Consequently, the co-integration of switch ASICs and optical engines is seen as the next step in terms of integration density, cost-effectiveness, and energy efficiency [4].

Referring back to Sect. 2.2.2, we presented integrated SerDes and the respective bandwidth of Ethernet SoCs ranging from 640 Gb to 102 tbps in Table 2.1. The table also discussed the development of a 51.2 Tbs Ethernet switch using 5 nm CMOS technology, which integrates 512 lanes of 100G and/or 112G SerDes. The current assumption is that 112G SerDes will be more efficient in energy consumption and presumably capable of transmitting and receiving 112G per lane in a 5 nm CMOS with 1pJ (pico-Joules)/bit.

However, achieving this level of efficiency is not yet feasible. According to publicly available information, Broadcom® Tomahawk5 51.2 Tbps Ethernet ASIC consumes 500W,[9] utilizing 100G SerDes instead of 112G SerDes. Typically, integrated SerDes consumes around 30% power of total ASIC power consumption and operates at a range of 4 pJ/bit to 6 pJ/bit[4]. For our calculation, let's consider 3pJ/bit transmit and receive rate of a 100G SerDes lane. Referring Table 2.1, we can confirm that a 51.2 tbps Ethernet ASIC uses 512 lanes of 100G SerDes.

With these assumptions, let's calculate the total energy consumption per second (in joules) for the 100G SerDes lanes:

Total Energy Consumption of 100G SerDes (J/s) = Energy per bit (pJ/bit) × Total Bandwidth (bits/s).

The total bandwidth of the 100G SerDes lanes can be calculated as: Total Bandwidth (bits/s) = Number of SerDes lanes × Speed per lane (bits/s).

Given that each 100G SerDes lane has a speed of 100 Gbps:

Total Bandwidth (bits/s) = $512 \times 100 \times 10^9$.

Alternatively, Total Bandwidth (bits/s) = 51.2×10^{12}.

Now, let's calculate the total energy consumption per second for the all 100G SerDes lanes:

Total Energy Consumption of 100G SerDes (J/s) = $3 \times 10^{-12} \times 51.2 \times 10^{12}$.

Alternatively, Total Energy Consumption of 100G SerDes (J/s) = 153.6W.

Now, let's find the percentage of the total power consumption of the chip attributed to the 100G SerDes lanes:

Percentage of Power Consumption = $\frac{\text{Total Energy Consumption of 100G SerDes}}{\text{Total Power Consumption of the Chip}} \times 100$.

Percentage of Power Consumption = $\frac{153.6}{500} \times 100$.

Approximately 30.72% of the power consumption of the 51.2 Tbps chip is attributed to the 512 lanes of 100G SerDes. Unless SerDes power consumption is reduced, the ASIC will remain power hungry. Other factors contributing to the overall power consumption of a switch include gearbox and optical modules, among other things.

Additionally, optical transceiver modules lack a roadmap beyond 800G at the time of writing this book. They also incur high costs and consume significant power, mainly through the electrical channel. Moreover, there is an increasing concern about the rack space consumed due to limited bandwidth density.

The following table presents the rack space needed to accommodate high-speed switches in a data center. It displays the respective SerDes, SerDes lanes, ASIC bandwidth, and rack unit (RU) requirements, assuming 32 optical transceiver modules are plugged into a 1RU switch (Table 2.2).

Please note there was no publicly available document depicting the roadmap for 1.6tbps optical transceiver module as of writing this book.

---

[9] The Register [9].

Table 2.2 Rack space needed per switch depending upon bandwidth capacity and number of optical transceiver module used [4]

Switch			Optics module		Switch faceplate	
SerDes Lanes	SerDes (G)	SoC capacity (Tbps)	Lanes	Capacity per SW port	Modules	Size (RU)
128	25	3.2	4	100G	32	1
256	25	6.4	4	100G	64	2
256	50	12.8	8	400G	32	1
512	50	25.6	8	400G	64	2
256	100	25.6	8	800G	32	1
512	100	51.2	8	800G	64	2
1024	100	102.4	8	800G	128	4
1024	100	102.4	16	1.6Tbps	64	2
512	200	102.4	8	1.6Tbps	64	2

As demand for bandwidth increases challenges such as rack space per switch and power consumption will cause more challenges for data center. These factors are driving a shift away from the traditional faceplate-pluggable deployment model to a new approach known as co-packaged optics (CPO). This model brings the optics into closer proximity with the main switching ASIC, aiming to tackle the challenges described herein above.

The concept of onboard optics is not new. It has been implemented in particular scenarios where higher bandwidth density is needed than what pluggable optics can provide. Instances of this can be seen in high-performance computing (HPC) systems like the IBM Power775 interconnect[10] and the Atos/Bull BXI interconnect.[11] Other commercially available examples include mid-board optical engines for optical PCIe.[12] These implementations have primarily relied on multi-mode optics. In 2019, the COBO consortium announced the completion of an MSA for onboard optics, which included specifications for 8- and 16-lane (electrical) onboard optical modules capable of up to 56 Gb/s per lane. The announcement claimed the solution having 30% better power consumption efficiency than QSFPDD and OSPF operating at 7W.[13] Despite this, the adoption of COBO has not gained momentum. Instead, the industry has shown interest in Coherent Optics and CPO (Co-packaged Optics).

One of the primary challenges of the onboard approach is that, despite relocating the modules from the front panel to the main PCB, the electrical channel from the switch to the module does not show significant improvement compared to Front Panel Pluggable

---

[10] Arimilli et al. [10].

[11] Derradji et al. [11].

[12] Samtec [12].

[13] PRnewswire [13].

(FPP) modules, hindering substantial power reduction. Conversely, COBO enhances the thermal environment by dispersing the heat of the modules away from the front panel. Additionally, without FPPs, there is more space available for ventilation [4].

2.3.2.1 CPO Concept

In June 2022, Intel announced the availability of its CPO (Co-packaged Optics) and high-lighted how CPO would revolutionize data center connectivity.[14] However, by late 2023, Intel further announced its decision to divest its silicon photonics business and transfer its further development to Jabil as part of cost-cutting efforts.[15,16] While this move was unexpected, several implementations of co-packaged optics, including both multi-mode[17,18] and single-mode optics[19], have been demonstrated.

The CPO combines optics and silicon on a single chip to deal with the challenges discussed earlier. It involves using different skills in areas like fiber optics, digital signal processing, switch ASICs, and advanced packaging to create better systems for data centers and cloud infrastructure.

The following diagram illustrates the concept of CPO (Fig. 2.13).

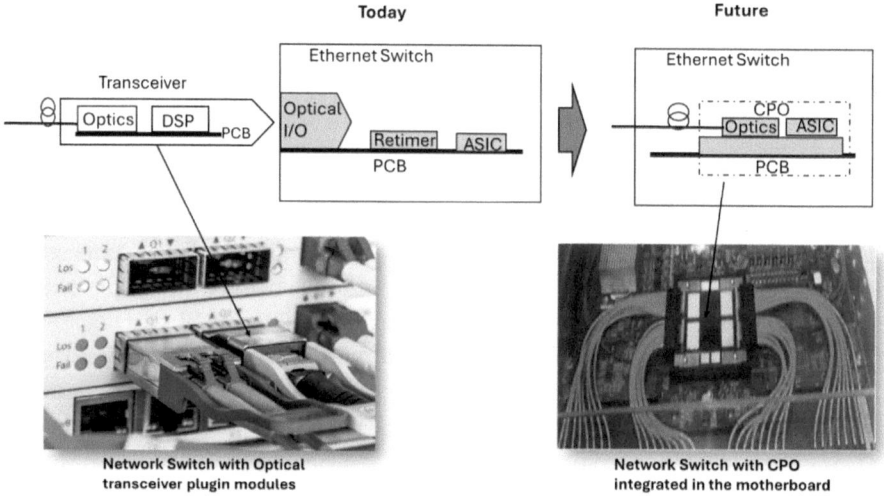

Fig. 2.13 Optical module today and CPO integrated in the motherboard of network switch

---

[14] Intel [14].

[15] The Register [15].

[16] CRN [16].

[17] Cook et al. [17].

[18] Krishnamoorthy et al. [18].

[19] Meade et al. [19].

The figure shows how standard optical modules are connected to Ethernet switch today and how the functional elements of optical module will be moved to Ethernet switch motherboard in future. As depicted in the figure above, co-packaged optics (CPO) integrates photonic devices with high-performance electronics using advanced packaging. This solution significantly shortens the SerDes distance, resulting in greatly reduced power consumption. The concept of CPO relies on advanced photonics integration, which can be achieved through various platforms, including silica (planar lightwave circuits),[20] indium phosphide (InP),[21,22] silicon on insulator (SOI/silicon photonics), silicon nitride (SiN), lithium niobate (LiNbO3), glass, and polymer. While these platforms have complementary strengths and weaknesses, other critical factors beyond photonic density, functionality, and performance come into play. These factors include how well the platform facilitates (a) the integration of different materials systems and (b) high-volume manufacturing across various applications. The first factor ensures maximum versatility in selecting the best material for each functional unit, while the second factor is vital for meeting volume requirements and achieving economies of scale [4]. Considering these concerns, a heterogeneous integration approach that uses different processes for electronics and photonics appears to be more suitable. The current CPO roadmap includes 2.5D CPO, 2.5D chiplet CPO, and 3D CPO approach that denotes level of increased integration of optics and ASIC.

The term "2.5D" heterogeneous integration refers to the method of integrating discrete chips on a shared substrate. In a 2D architecture, two or more active devices are positioned side by side and interconnected through a common substrate (Fig. 2.14).

In the figure above, both 2.5D and 3D heterogeneous approaches of CPO are shown. In the 3D packaging approach, an interposer is shown underneath the ASIC and opto engine. Interposers are utilized in heterogeneous integration to connect different types of chips, such as silicon chips and organic chips, enabling the creation of more powerful and versatile devices. It is worth noting that the 2.5D method also utilizes an interposer. The 2.5D packaging approach is useful for 50 Tbps ASICs, while the 2.5D chiplet + opto engine configuration is employed for 50 to 100 Tbps CPO solution. The 3D approach is utilized for CPO solutions exceeding 100 Tbps.

In conclusion, the integration of chips using 2.5D and 3D heterogeneous approaches plays a pivotal role in enhancing the performance and versatility of Co-Packaged Optics (CPO) solutions, catering to a wide range of bandwidth requirements in modern data center and cloud infrastructures. As we move forward, the next section will delve into another transformative technology: Data Processing Units (DPUs). These advanced processors are revolutionizing data center infrastructure by offloading and accelerating key

---

[20] Doerr and Okamoto [20].

[21] Williams et al. [21].

[22] Smit et al. [22].

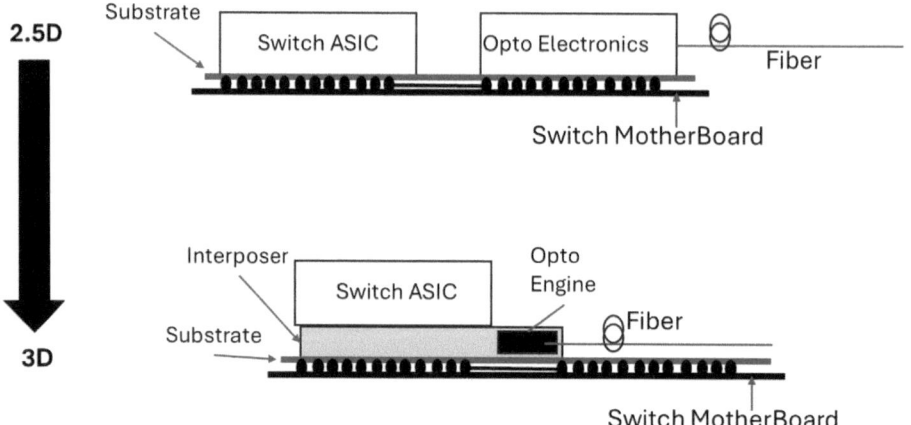

Fig. 2.14 2.5D and 3D heterogenous approach to CPO development [4][23]

networking, security, and storage functions. Let's explore the capabilities and benefits of DPUs in optimizing data center performance and efficiency.

2.4   Data Processing Unit (DPU)

In 2016, my journey into the realm of hardware-level network programmability began with the emergence of the Barefoot Tofino chip, which garnered attention in certain networking circles. The industry buzzed with anticipation over the transformative potential of this innovative technology. However, as the dust settled, it became apparent that while Tofino offered programmability, its software stack was not as mature as expected for building complex network infrastructures. Delving into P4 programming, the language supported by Tofino proved challenging for many network engineers due to its specialized nature and the complexities of manipulating hardware through software. Despite these hurdles, the acquisition of Barefoot by Intel in 2019 marked a significant turning point. However, despite Intel's backing, Tofino's promising technology struggled to achieve widespread adoption in mainstream networking equipment.

This experience underscores the challenges and complexities inherent in integrating cutting-edge hardware solutions into existing network architectures. However, it also serves as a catalyst for exploring alternative avenues for advancing network programmability and performance. Amid this backdrop, a new frontier emerges in the form of Data Processing Units (DPUs), poised to redefine the landscape of networking infrastructure.

---

[23] Nagarajan et al. [23].

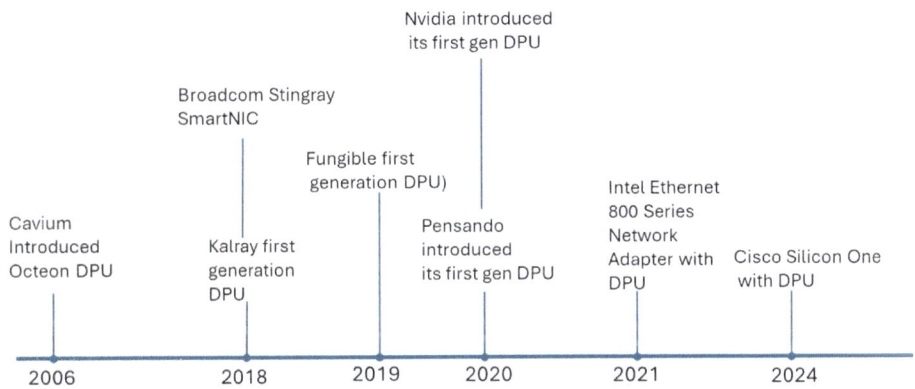

Fig. 2.15 Representative timeline of DPU

In 2006, Cavium Networks (now Marvell Technology) introduced the Octeon family of multi-core processors, which included models specifically designed for networking applications. These early DPUs primarily focused on accelerating packet processing tasks such as routing, forwarding, and security functions. Since 2018, several companies have entered the DPU marketplace. A representative timeline is presented in the diagram below. The momentum around DPU started to pickup in 2018 and evident in the timeline presented below (Fig. 2.15).

In 2018, Kalray unveiled its inaugural DPU offering, dubbed the Kalray Coolidge processor. Engineered to meet the escalating demand for efficient and scalable data processing across diverse applications such as networking, storage, and edge computing, the Coolidge processor marked a significant step forward in DPU technology. Concurrently, Broadcom® introduced its Stingray SmartNIC, incorporating DPU capabilities, in the same year. This was closely followed by Fungible's debut of its first-generation DPU in 2019. Subsequently, both Nvidia and Pensando entered the DPU arena with their inaugural DPU releases in 2020. Intel® followed suit in 2021 with the launch of its Ethernet 800 series featuring integrated DPU functionality. Finally, Cisco introduced its Silicon One platform with integrated DPU in 2024, further expanding the landscape of DPU-enabled solutions.

In recent years, demand for DPUs has surged due to several key factors:

- **Demand for Data Processing:** With the exponential growth of data in various industries, there's a pressing need for efficient data processing solutions. DPUs offer specialized hardware acceleration for data-intensive tasks such as networking, security, storage, and AI inference, meeting the demand for high-performance data processing.
- **Rise of Edge Computing:** The proliferation of edge computing applications, where data is processed closer to the source rather than in centralized data centers, has fueled the need for DPUs. These units enable real-time processing, low latency, and efficient

data transfer at the edge, facilitating applications like IoT, autonomous vehicles, and smart cities.

- **Network Performance Optimization:** DPUs enhance network performance by offloading packet processing tasks from the CPU to dedicated hardware. This offloading frees up CPU resources for other critical tasks, improves throughput, reduces latency, and enhances overall network efficiency.
- **Security Requirements:** With the increasing complexity and frequency of cyberthreats, there's a growing demand for robust security solutions. DPUs offer hardware acceleration for encryption, decryption, and other security protocols, bolstering network security and protecting sensitive data.
- **Cloud-Native Infrastructure:** The shift toward cloud-native architectures and microservices has spurred the adoption of DPUs. These units enable efficient data processing and traffic management within cloud environments, supporting the scalability, agility, and flexibility required by modern cloud-native applications.
- **Emerging Technologies:** DPUs play a crucial role in enabling emerging technologies such as AI/ML inference, 5G networks, and software-defined networking (SDN). Their hardware acceleration capabilities optimize performance and efficiency for these advanced technologies, driving their adoption across industries.

The convergence of these factors has created a fertile ground for the DPU market, leading to increased investments, innovation, and competition among vendors in this space.

2.4.1 DPU Architecture

The DPU has emerged as a pivotal component in modern computing, often recognized as one of the three fundamental pillars alongside the CPU and GPU. It encompasses a programmable processor architecture that integrates several key components, including programmable multi-core CPUs, high-performance network interfaces, and programmable acceleration engines. At its core, the DPU serves as a versatile and programmable processing unit, capable of executing a wide range of tasks efficiently. The programmable multi-core CPU provides the computational horsepower needed for general-purpose processing tasks, while the high-performance network interface enables seamless connectivity and data transfer across networks. Additionally, the programmable acceleration engine enhances performance by offloading specialized tasks such as encryption/decryption, compression/decompression, and AI/ML inference from the CPU or GPU.

This integrated architecture offers unparalleled flexibility and scalability, allowing developers to tailor the DPU's functionality to specific application requirements. Whether it's accelerating network packet processing in data centers, optimizing storage performance in edge computing environments, or enhancing AI/ML workloads, the DPU plays

Fig. 2.16 The Fungible
DPU™ architecture [24]

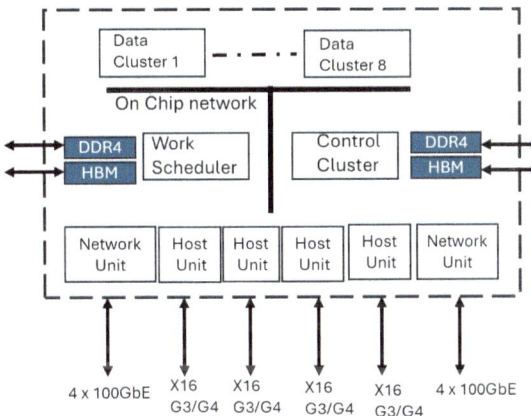

a crucial role in unlocking the full potential of modern computing systems. The following
diagram depicts "Fungible" DPU™ architecture (Fig. 2.16).[24]

The architecture of the Fungible DPU™ comprises eight data clusters, each equipped
with 192 processor threads, featuring full cache coherency and tightly integrated accelerators. In addition to these data clusters, there is a control cluster with eight processor
threads, a work scheduler, memory and I/O units, and dedicated network and host units.
These data clusters serve as the primary processing units, optimized for parallel execution
and capable of handling various tasks simultaneously, making them suitable for compute-
intensive workloads. The control cluster acts as the central management unit, coordinating
tasks and resource allocation across the DPU.

The memory and I/O units manage data storage and handle input/output operations,
ensuring smooth communication with external devices and efficient data transfer. The
work scheduler dynamically assigns computing resources based on task priorities and
workload characteristics to maximize the DPU's processing efficiency. The network unit
manages network-related operations such as packet processing and routing, leveraging
specialized hardware accelerators for fast data transfer and low-latency communication.
On the other hand, the host units facilitate communication with external systems and
devices. Finally, the on-chip network serves as the communication backbone, facilitating
seamless data exchange and task synchronization among different functional units within
the DPU.

2.4.2   DPU in Ethernet Switch

In recent years, DPU is offering unparallel potentials in networking gears. With integration
of Pensando (AMD) DPU, Hewlett Packard Enterprise introduced first distributed services

---

[24] Fungible [24].

switch in the industry. This example is important to depict potential of DPU in networking gears. The distributed services switch (DSS) CX10000 brings unique service chaining capabilities offering microsegmentation, firewall capabilities, DDOS protection, NAT, and IPSEC among other services available in the product. In such product, DPU is added along with standard Ethernet SoC and CPU to offer additional services beyond those available in Ethernet SoC. All these are done on demand through software programming abilities of DPU.

The following diagram illustrates typical Ethernet networking equipment featuring DPUs. The number of DPUs added depends on the intended traffic that will flow through them. In a typical Ethernet switch, the DPU or DPUs connect to the CPU via a PCIe link and to the Ethernet SOC via a 400GbE link (Fig. 2.17).

To direct traffic to the DPUs for additional processing beyond the capabilities of the Ethernet SOCs, a deterministic hashing algorithm is employed. This algorithm guides traffic based on factors such as VLAN or virtual L2 or L3 interfaces. The DPU can serve various purposes, as exemplified by Hewlett Packard Enterprise's DSS product portfolio, such as the "CX10000" series. Moreover, it can handle more advanced services like advanced DDoS protection, load balancing, TCP optimization, and others. The following diagram depicts two distinct use cases in which Hewlett Packard Enterprise (HPE) CX10000 (CX10K) is used (Fig. 2.18).

As depicted in the diagram above, HPE's DSS product CX10K fulfills multiple roles, showcasing its versatility and capability in different network scenarios. In the first use case, it serves as the L4 firewall, offering near-line-rate stateful inspection and DDoS protection at the campus core, where diverse end-user traffic traverses through the CX10K.

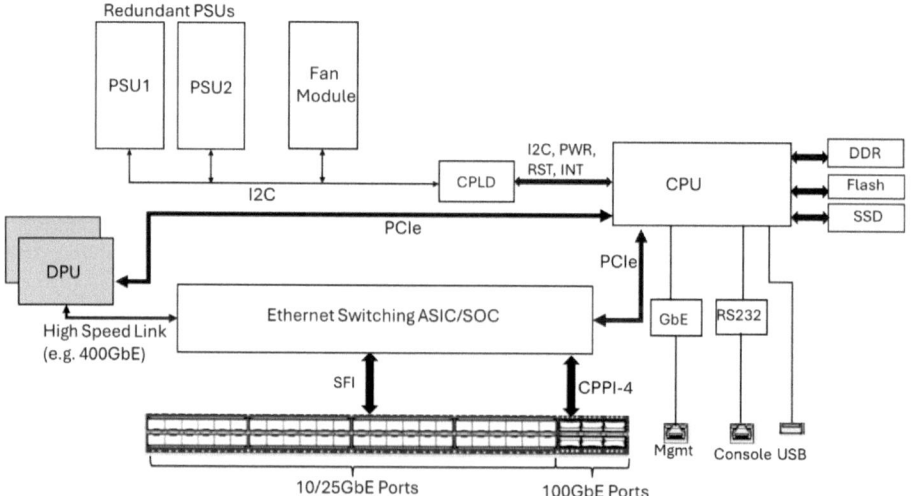

Fig. 2.17 Typical Ethernet switch architecture with DPUs

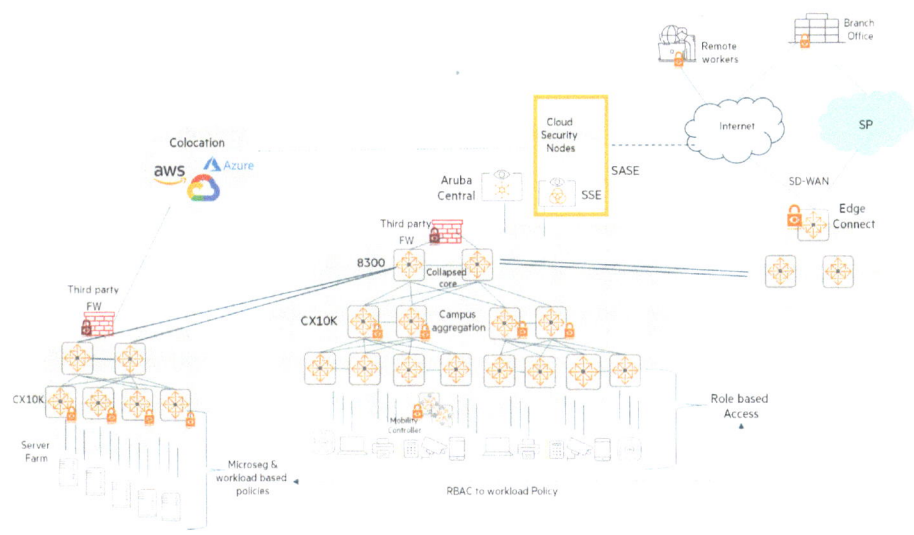

Fig. 2.18 An enterprise network with HPE's DSS product CX10K

Meanwhile, in the second use case, the CX10K operates as the Top-Of-Rack (TOR) switch, interconnecting the server farm. Here, it provides crucial microsegmentation for east–west traffic among servers, enhancing security and isolation within the network infrastructure. Moreover, the CX10K extends user-role-based access to workloads, integrating seamlessly with HPE's dynamic segmentation solution. This collaborative approach ensures granular control over network access, enhancing security posture and compliance adherence.

These use cases underscore the multifaceted capabilities of switches equipped with DPUs, highlighting their pivotal role in modern network architectures. While the CX10K excels in L4 firewalling and TOR switch functionalities, it also integrates seamlessly with third-party routers to handle L7 traffic and IPS/IDS functions, further enhancing network security and performance.

References

1. Goodrich J (2023) Ethernet is still going strong after 50 years: the technology has become standard LAN worldwide. IEEE Spectrum. https://spectrum.ieee.org/ethernet-ieee-milestone
2. Wikipedia (2023) 10BASE5. https://en.wikipedia.org/wiki/10BASE5
3. Alloway J (2020) Why do we need Serdes? ElectronicDesign. https://www.electronicdesign.com/technologies/analog/article/21132088/why-do-we-need-serdes
4. Minkenberg C, Krishnaswamy R, Zilkie A, Nelson D (2020) Co-packaged datacenter optics: opportunities and challenges. The Institute of Engineering and Technology, Willey

5. Cady Ed (2021) Ethernet Alliance holds a successful technical exploration forum. Connector TIPS. WTWH Media LLC. https://www.connectortips.com/ethernet-alliance-holds-a-successful-technical-exploration-forum/

6. Horner R (2024) Shift from NRZ to PAM-4 signaling for 400G ethernet. Synopsys, Inc. https://www.synopsys.com/designware-ip/technical-bulletin/pam4-400g-ethernet-2019q3.html

7. STH (2024) Broadcom BCM56170 block diagram. https://www.servethehome.com/fs-s5860-20sq-switch-review-20x10gbe-4x25gbe-2x40gbe-1u/broadcom-bcm56170-block-diagram/

8. Broadcom (2016) Switching technology product brief—BCM56170 24/48 port MGig campus switch with VxLAN and 25 GbE uplinks. https://docs.broadcom.com/doc/56170-PB

9. The Register (2022) Broadcom challenges Nvidia's Spectrum-4 with 51.2T switch silicon. Situation Publishing. https://www.theregister.com/2022/08/16/broadcom_nvidia_switch/

10. Arimilli B et al (2010) The PERCS high-performance interconnect. In: 18th IEEE symposium on high performance interconnects. IEEE, pp 75–82

11. Derradji S et al (2015) The BXI interconnect architecture. In: IEEE 23rd annual symposium on high-performance interconnects. IEEE, pp 18–25

12. Samtec (2020) FireFly^{TM} micro flyover system. Samtec. https://www.samtec.com/optics/optical-cable/mid-board/firefly

13. PRnewswire (2019) Consortium for on-board optics demonstrates the industry's first packet transmission between on-board and multi-supplier agreement optical modules. https://www.prnewswire.com/news-releases/consortium-for-on-board-optics-demonstrates-the-industrys-first-packet-transmission-between-on-board-and-multi-supplier-agreement-optical-modules-300925165.html

14. Intel (2022) Intel labs announces integrated photonics research advancement. Intel Inc. https://www.intc.com/news-events/press-releases/detail/1555/intel-labs-announces-integrated-photonics-research

15. The register (2023) Intel dumps its silicon photonics bells and whistles into Jabil's lap. Situation Publishing. https://www.theregister.com/2023/10/31/intel_silicon_photonics_jabil/

16. CRN (2023) Intel offloads silicon photonics product line to Jabil. The Channel Co. https://www.crn.com/news/components-peripherals/intel-divests-another-non-strategic-business-this-time-to-jabil

17. Cook C et al (2003) A 36-channel parallel optical interconnect module based on optoelectronics-on-VLSI technology. IEEE J Sel Top Quant Electron 9(2):387–399

18. Krishnamoorthy AV et al (2011) Progress in low-power switched optical interconnects. IEEE J Sel Top Quant Electron 17(2):357–376

19. Meade R et al (2019) Teraphy: a high-density electronic-photonic chiplet for optical I/O from a multi-chip module. In: 2019 Optical fiber communications conference and exhibition (OFC). OSA, p M4D.7

20. Doerr C, Okamoto K (2006) Advances in silica planar lightwave circuits. J Lightwave Technol 24(12):4763–4789. https://www.academia.edu/86353112/Advances_in_Silica_Planar_Lightwave_Circuits

21. Williams K et al (2018) Introduction to the JSTQE issue on indium phosphide integrated photonics. IEEE J Sel Top Quant Electron 24(1):1–4

22. Smit M, Williams K, van der Tol J (2019) Past, present, and future of InP-based photonic integration. APL Photonics 4:050901

23. Nagarajan R, Ding L, Coccioli R, Kato M, Tan R, Tumne P, Patterson M, Liu L (2023) 2.5D heterogeneous integration for silicon photonics engines in optical transceivers. IEEE J Sel Top Quant Electron 29(3)

24. Fungible (2020) The fungible DPU™: a new category of microprocessor for the data-centric era: hot chips 2020

Cloud Networking

3.1 Introduction

Cloud networking has emerged as a critical component in modern IT infrastructure, revolutionizing the way enterprises manage and deploy their network resources. At its core, cloud networking leverages virtualized infrastructure provided by cloud service providers to facilitate efficient and scalable communication between various components and services over a network. This paradigm shift enables organizations to dynamically provision and manage network resources, optimizing performance, scalability, and cost-effectiveness. Enterprises are increasingly adopting cloud networking solutions to address the evolving demands of digital transformation. Public cloud environments offer scalability, flexibility, and ease of access to a wide array of services, making them ideal for hosting globally distributed applications and workloads. By leveraging public cloud services, enterprises can offload the burden of managing physical infrastructure and benefit from on-demand provisioning, automated scaling, and pay-as-you-go pricing models. However, reliance solely on public cloud environments may raise concerns regarding data privacy, compliance, and security, particularly for sensitive workloads and regulated industries.

In contrast, private cloud environments provide organizations with greater control, customization, and security over their network infrastructure and data. By deploying a private cloud within their own data centers or through dedicated hosting environments, enterprises can maintain compliance with industry regulations, ensure data sovereignty, and mitigate security risks associated with multi-tenant architectures. Private cloud deployments are well-suited for mission-critical applications, sensitive data workloads, and industries with stringent regulatory requirements, such as healthcare, finance, and government.

D. D. Chowdhury, *Future of Networks*, Synthesis Lectures on Communications, https://doi.org/10.1007/978-3-031-71440-5_3

Hybrid cloud networking has emerged as a compelling solution for enterprises seeking to leverage the benefits of both public and private cloud environments. By seamlessly integrating on-premises infrastructure with public cloud services, organizations can achieve greater flexibility, scalability, and resilience while addressing diverse workload requirements. Hybrid cloud architectures enable workload portability, allowing applications to run on the most suitable platform based on performance, cost, and compliance considerations. Additionally, hybrid cloud networking facilitates seamless data migration, disaster recovery, and workload orchestration across heterogeneous environments, enabling enterprises to optimize resource utilization and enhance business continuity.

In this chapter, we will explore the intricacies of cloud networking, delving deeper into various types of cloud networking solutions. We will examine the evolution of networking within cloud environments, exploring the key components and technologies that enable efficient communication and connectivity between cloud resources. Additionally, we will discuss the advantages and challenges associated with different types of cloud networking architectures, including public, private, and hybrid clouds. By the end of this chapter, readers will gain a comprehensive understanding of cloud networking principles and be equipped with insights to effectively design, implement, and manage networking infrastructure in the cloud.

3.2   History of Cloud Networking

The inception of cloud networking can be traced back to the early days of computing and network experimentation. In the 1960s, MIT's time-sharing project for mainframes laid the groundwork, allowing simultaneous access to computing resources. This primitive form of system virtualization evolved, culminating in IBM's CP-40 in 1964, which provided a more meaningful solution for resource sharing. By 1999, VMware introduced its own virtualization solution, marking a turning point in mainstream adoption. Concurrently, from 1992 to 1996, the need for secure communication over the Internet drove research into IP-layer encryptions. The US Navy's establishment of Simple Internet Protocol Plus (SIPP) in 1992 paved the way for Virtual Private Networks (VPNs), enabling secure connectivity between sites over the Internet or WAN. These developments laid the foundation for cloud networking, bridging the gap between virtualized resources and secure connectivity (please refer to the timeline below) (Fig. 3.1).

The evolution of cloud computing further propelled cloud networking forward. In 1999, Salesforce pioneered Software as a Service (SaaS), showcasing the potential of cloud computing. This was followed by the OpenNebula project in 2005, which aimed to develop an open-source cloud computing platform. Subsequently, major players like Amazon with AWS, Google with GCP, and Microsoft with Azure introduced Infrastructure as a Service (IaaS) offerings, collectively referred to as public cloud.

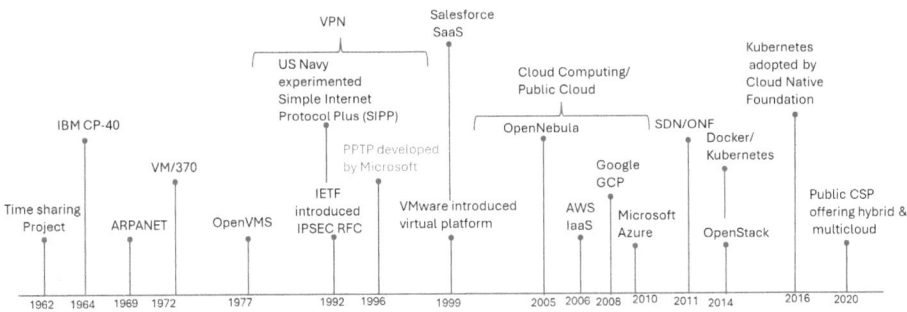

Fig. 3.1 Cloud networking timeline

In 2014, OpenStack emerged as an open-source solution for cloudifying compute, storage, and networks, empowering organizations to develop their own private cloud solutions. The rise of containerization solutions like Docker and Kubernetes further revolutionized cloud computing, with Kubernetes being adopted by the Cloud Native Computing Foundation in 2016. Fast forward to 2020, Cloud Service Providers (CSPs) began offering multi-cloud solutions, enabling organizations to leverage multiple cloud platforms seamlessly. This timeline illustrates the intertwined evolution of system virtualization, secure connectivity, and cloud computing, shaping the landscape of modern cloud networking.

3.3 Virtualization

Virtualization is a foundational technology that revolutionized the way computing resources are provisioned, managed, and utilized. At its core, virtualization enables the creation of virtual instances of computing resources, such as servers, storage devices, and networks, abstracting them from the underlying physical hardware. These virtual instances, also known as virtual machines (VMs) or containers, operate independently of each other and can run multiple operating systems or applications simultaneously on the same physical hardware.

The concept of virtualization predates cloud computing and has its roots in the early days of computing, with mainframes being among the first systems to employ virtualization techniques. However, virtualization gained widespread adoption and prominence in the early 2000s with the emergence of × 86-based server virtualization solutions, such as VMware's ESX Server and Microsoft's Hyper-V. Virtualization brings several key benefits that have made it integral to modern computing environments:

- **Resource Utilization:** By abstracting physical hardware into virtual instances, virtualization allows for more efficient utilization of computing resources. Multiple virtual machines can run on a single physical server, maximizing its capacity and reducing resource wastage.

- **Isolation:** Virtualization provides isolation between virtual instances, ensuring that applications and workloads running on one VM do not interfere with those running on another. This enhances security, stability, and reliability within the computing environment.
- **Flexibility and Scalability:** Virtualization enables rapid provisioning and scaling of resources, allowing organizations to adapt quickly to changing workload demands. New virtual machines or containers can be spun up or down as needed, providing agility and responsiveness to business requirements.
- **Cost Savings:** By consolidating multiple workloads onto fewer physical servers, virtualization helps reduce hardware, power, cooling, and maintenance costs. It also enables organizations to optimize their IT infrastructure and achieve better return on investment.

Cloud computing builds upon the principles of virtualization to deliver computing resources and services over the Internet or leased lines. In a cloud computing environment, virtualization is used to create a pool of shared computing resources, which are then accessed and consumed by users on-demand via the Internet. Cloud service providers leverage virtualization to provision virtual machines, containers, storage, and networking resources, enabling users to deploy and manage applications and workloads in a scalable, flexible, and cost-effective manner. Virtualization is thus an essential component of cloud computing, providing the foundation for the scalability, flexibility, and resource optimization that characterize cloud-based environments. Without virtualization, the dynamic allocation and management of computing resources in the cloud would not be possible, making it a fundamental building block of modern IT infrastructure.

In subsequent sections, we will delve into server/compute virtualization, network virtualization, and storage virtualization to elucidate how cloud computing virtualizes and shares each of these resources.

3.3.1   Server Virtualization

Server virtualization involves creating multiple virtual instances on a single server, abstracting server resources such as individual physical machines, processors, and operating systems. This consolidation of resources is depicted in Fig. 3.2. In traditional computer setups, hardware and software designs were geared toward supporting single applications, leading to underutilized processors, memory capacity, and network bandwidth. This resulted in an escalation of server hardware counts as organizations expanded their application and service deployments, straining data center resources and escalating costs for space, power, cooling, and connectivity.

Fig. 3.2 Virtualization

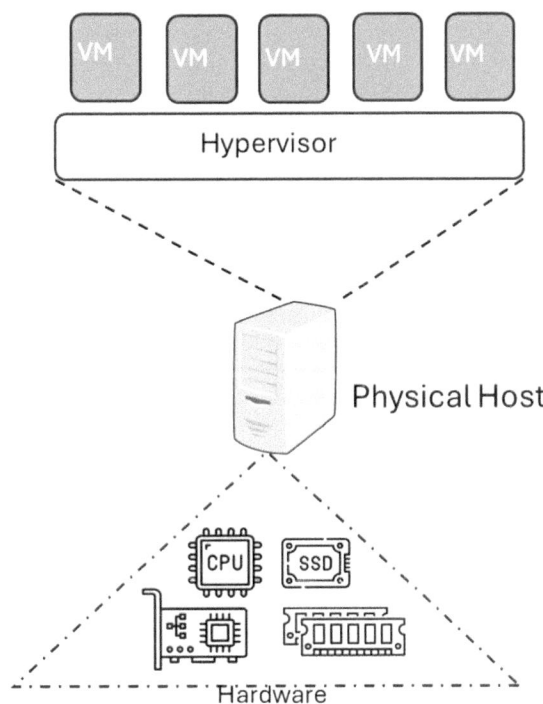

The introduction of server virtualization revolutionized this landscape. A hypervisor software layer abstracts the underlying hardware from the software running above it, translating physical resources into virtual equivalents. The hypervisor then organizes and manages these virtualized resources, creating logical instances known as virtual machines (VMs), each capable of functioning as an independent server. The primary advantage lies in resource utilization. Through hypervisor-managed virtualization, multiple VMs can leverage the available hardware simultaneously, effectively consolidating the workload of multiple servers onto a single machine. This optimization reduces the need for additional servers, alleviates strain on data center resources, enhances IT flexibility, and reduces overall IT costs.

Various server virtualization techniques exist, each offering distinct approaches to creating and managing virtual servers. Here are some of the most common types:

- **Full Virtualization:** In full virtualization, a hypervisor sits between the physical hardware and the virtual machines. The hypervisor abstracts the physical hardware, allowing multiple guest operating systems (OS) to run on the same physical server simultaneously. Each guest OS interacts with the virtual hardware provided by the hypervisor, which translates the instructions to the underlying physical hardware.

- **Para-Virtualization:** Unlike full virtualization, para-virtualization requires modifications to the guest operating systems to be aware of the virtualization layer. This allows for more efficient communication between the guest OS and the hypervisor, resulting in improved performance compared to full virtualization.
- **Hardware-Assisted Virtualization:** Hardware-assisted virtualization leverages specialized CPU features, such as Intel VT-x or AMD-V, to improve virtualization performance. These features allow the hypervisor to run more efficiently and reduce overhead, leading to better performance and scalability.
- **Container-Based Virtualization:** Container-based virtualization, also known as operating system-level virtualization, involves running multiple isolated userspace instances, known as containers, on a single host operating system (OS). Containers share the host OS kernel and resources, making them lightweight and fast to deploy. Popular containerization platforms include Docker and Kubernetes.
- **OS-Level Virtualization:** OS-level virtualization, similar to container-based virtualization, involves partitioning a single OS instance into multiple isolated environments, known as virtual private servers (VPS) or containers. Each VPS operates independently, with its own file system, users, and processes, but shares the same kernel and resources of the host OS.

Organizations select server virtualization techniques based on factors such as performance requirements, resource utilization, and management complexity. Often, a combination of these techniques is employed to optimize IT infrastructure and address specific needs.

3.3.1.1 VM Definition and Attributes

When discussing server virtualization and its integral role in cloud computing, the term "VM" arises frequently. Therefore, it's crucial to grasp its meaning before delving further into cloud networking. The concept of a VM was first defined by Popek and Goldberg[1] in 1974 as "an efficient, isolated duplicate of the real machine." Seawright and MacKinnon[2] in 1979 elaborated that a VM allows for "multiple software copies of real computing systems on one real processor." Essentially, while a physical computer is a tangible and complete machine, a virtual machine resembles a full-fledged machine but exists as a collection of files and programs on a real computer (which may go unnoticed by the user). Various definitions of VM exist; for instance, VMware defines it as a software entity and basic computer components that interface with VMM (such as VMware vSphere hypervisor) to facilitate infrastructure virtualization[3]. In a computer system, a VM helps connect to the underlying VMM, allowing different operating systems to coexist on a single physical system. For instance, a Linux OS and a Windows OS can share the same

---

[1] Popek and Goldberg [1].
[2] Seawright and MacKinnon [2].
[3] VMware [3].

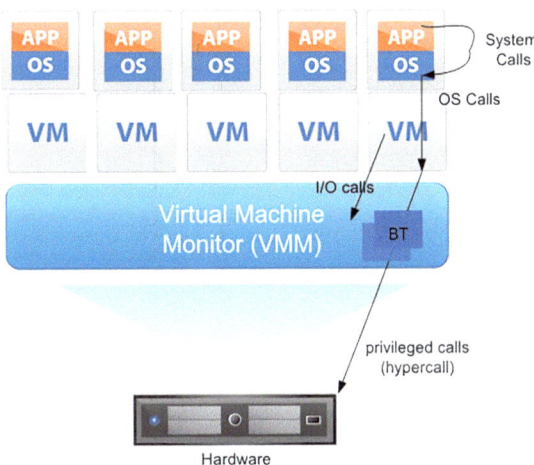

Fig. 3.3 Server virtualization—depicting VMM and systems commands

hardware resources if there's virtualization software in place, which includes VM and VMM support (Fig. 3.3).

Without this software layer, sharing resources like this wouldn't be possible. Think of the VM as a copy of a real machine that provides a layer of abstraction for operating system commands to reach the hardware. The figure above illustrates how system commands directly reach the hardware through the VMM's execution process known as BT (Binary Translation). Meanwhile, input/output commands are handled by drivers within the VMM. VMs are divided into two main types based on their use: those that support an entire system platform (including the operating system) and those that support a single process and its associated program. This distinction is crucial to our study, as we'll be focusing on system virtual machines in the context of cloud computing environments.

SVM and HVM

In server virtualization, the goal is to create multiple virtual instances of servers, each running its own operating system and applications, on a single physical server. This allows for better resource utilization, improved scalability, and easier management of infrastructure. There are two approaches to creating virtual machines in a server virtualization environment, these are SVM (System Virtual Machine) and HVM (Hardware Virtual Machine).

- **SVM (System Virtual Machine):** SVMs are software-based virtual machines that run on top of a host operating system. They use software emulation to provide a virtualized environment for guest operating systems to run. SVMs are often used in environments where compatibility with different guest operating systems is important, as they can support a wide range of operating systems without requiring specific hardware support.

- **HVM (Hardware Virtual Machine):** HVMs, on the other hand, run directly on the underlying hardware without the need for a host operating system. They leverage hardware virtualization features, such as Intel VT-x or AMD-V, to provide a more efficient and performant virtualization environment. HVMs typically offer better performance compared to SVMs, especially for CPU and memory-intensive workloads.

The choice between SVM and HVM depends on factors such as performance requirements, compatibility needs, and the specific use case of the virtualization environment.

VM Attributes

Although the implementation of VMs may vary depending on the VMM environment, all VMs share a set of attributes as follows:

- **Software Compatibility:** VMs provide a compatible abstraction to support different software written for them. In the case of process VMs, like the JVM, the mantra is "write once, run anywhere." Therefore, it's essential that whether it's an SVM (System VM) or process VM, software compatibility is provided for applications intended for it.
- **Partitioning:** In multiprocessor systems, there may be a need to partition a large system into smaller ones, requiring partitioning. In physical partitioning, hardware resources of one VM are disjoint from the others, providing complete physical separation. In logical partitioning, underlying hardware resources are time-multiplexed between different partitions, thereby improving system resource utilization.
- **Isolation:** Isolation is a crucial feature of VMs, ensuring they isolate VMs from each other and the real machine. For example, if a virus infects a particular VM, it will only spread within that VM and won't infect other VMs or the host OS. Fault isolation, software isolation, and performance isolation are essential aspects. Fault isolation ensures that privileges contained within the VM boundaries prevent issues such as a VM with a buggy guest operating system scribbling over physical memory. Software isolation addresses problems like DLL (Dynamic Link Library), OS, and library corruptions, containing them within VM boundaries. Performance isolation at the VM level is vital regardless of ongoing hardware activities, achieved through smart scheduling and resource allocation policies in the monitor.
- **Encapsulation:** Encapsulation provides better execution at the VM level by correcting and optimizing software abstraction. For example, HLL (High-Level Language) VMs support runtime checks to reduce errors in programming, including type-safe, memory-safe, and garbage-collected memory management.
- **Performance:** VM-level performance is crucial, and reducing the layer of software at the VM level reduces overhead and improves performance. Various techniques, such as those discussed in the isolation section, can enhance VM performance.

Now that we have a good understanding of VMs, let's explore Virtual Machine Monitor (VMM), often known as a hypervisor, in the next section. Both VMs and VMMs are integral parts of server virtualization.

3.3.2 Virtual Machin Monitor (VMM)

Central to server virtualization is the Virtual Machine Monitor (VMM), which originated in the 1960s and saw a resurgence in 1998 with the VMware patent "6,397,242," as previously discussed. The VMM, also known as a hypervisor, is a software layer that typically resides beneath the operating system and presents a hardware interface to the OS. It manages computer resources, multiplexes resources between multiple virtual machines (VMs), and provides performance isolation between VMs[4].

There are several benefits of using a VMM or hypervisor:

- **Efficient and Isolated Environments:** The VMM creates efficient and isolated programming environments, giving users the impression of direct access to the real machine environment through duplicates or VMs.
- **Configurable Environment:** Administrators can configure environments for virtual machines, which can differ from the underlying hardware configuration. For example, a VM can have 8 MB of memory while the real machine supports 32 MB.
- **Concurrent Execution:** The VMM allows for the concurrent execution of multiple operating systems on the same hardware.
- **Isolation of Untrusted Applications:** It provides isolation for untrusted applications. For instance, if a user downloads a program containing a virus, its execution will be confined within a single VM and typically won't affect other VMs or the host OS.
- **Upgrade Capabilities:** The VMM enables the upgrade of guest OS without losing legacy operational capabilities[5].

Furthermore, the VMM provides logical separation or partitioning to guest OSes, along with the function set that each OS can call to perform tasks while remaining isolated from other guest OSes. Logically separated or partitioned OS platforms receive non-overlapping subsets of platform resources, such as processor interrupt regions, memory, and I/O access. The partitioned hardware resources are represented by their own firmware device tree to the OS image[6] (Fig. 3.4).

To illustrate this operation, refer to figure above, which depicts logically partitioned OS-based platforms. For each hardware resource, such as an I/O Adaptor (IOA), the hypervisor or VMM provides a secure DMA (Direct Memory Access) window.

---

[4] Fiuczynski [4].

[5] Robin and Irvine [5].

[6] Arndt [6].

Fig. 3.4 The operational
perspective of hypervisor [6]

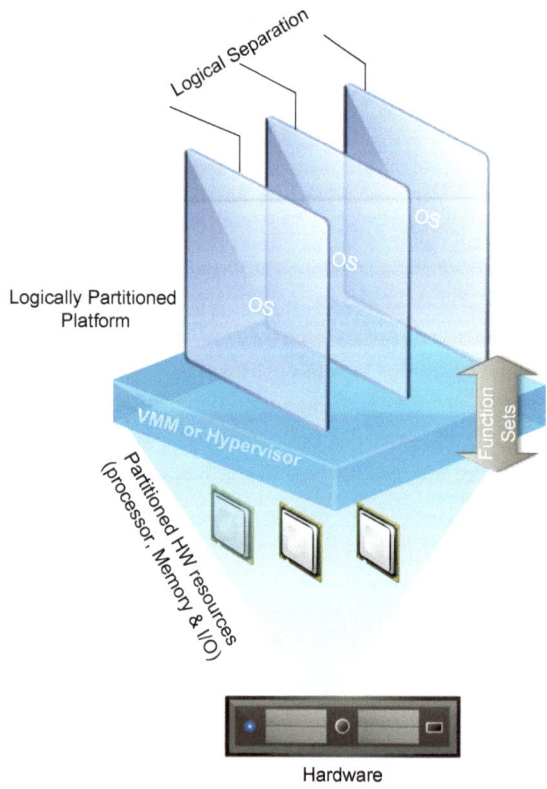

To delve further into this operation, it may be helpful to discuss how an OS virtualizes the CPU or underlying hardware, both of which employ a technique known as "limited direct execution."[7] In a traditional computing system, the operating system (OS) virtualizes and/or shares the CPU among multiple concurrently running jobs by employing time-sharing techniques. This allows one process to access the CPU for a period before switching to another process, and so forth. However, this approach introduces several challenges. Firstly, if a process needs to perform privileged operations, such as accessing I/O devices, the OS cannot simply grant such privileges to any process. To address this, developers designate processes to operate in user mode or kernel mode during platform design, enabling kernel mode programs to receive privileged access.

However, this segregation raises another issue: what if a user mode program requires privileged access to I/O services? How does the OS discern such a need? Presently, most systems support a mechanism called "system call" through which a user program can request services. The program triggers a special trap instruction to initiate a system call,

---

[7] Arpaci-Dusseau and Arpaci-Dusseau [7].

Fig. 3.5 The operational perspective of how OS handles user programs or processes

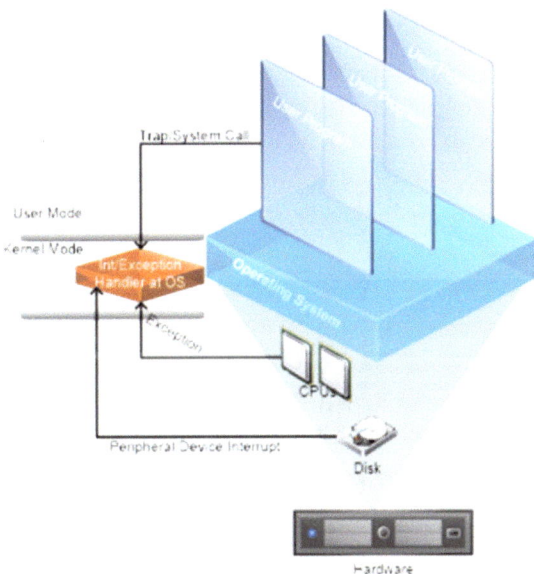

which then reaches the kernel, as depicted in the figure below, and elevates the privilege level for the request (Fig. 3.5).

The operating system (OS) responds to calls through a special "return-from-trap" instruction, which is handled at the hardware level through a per-process Kernel stack. While different hardware may use different conventions, the basic concept remains the same. During boot time, the OS kernel that handles exceptions sets up a trap table, ensuring that it knows which code to execute upon a trap. These procedures are carried out through the Limited Direct Execution (LDE) protocol. The LDE protocol involves two phases: first, the kernel initializes the trap table during boot time, and the CPU remembers its location. Secondly, the kernel sets up various configurations, such as memory allocation, before issuing a "return-from-trap" to initiate the execution process. Another critical task performed by LDE is "Saving and Restoring Context."

Switching between processes to handle their requests is not straightforward. It's possible for a process to run on the CPU without the OS having control over it. This leads to the third issue: how to control the CPU. One approach is the "Cooperative" approach, where the OS trusts the processes to behave reasonably and return CPU control by making system calls. In such cases, an explicit system call is issued to transfer CPU control back to the OS. Once CPU control is regained, the OS can decide whether to continue running the current process or switch to a different one. This decision is made through a scheduler, and a low-level program known as a context switch is executed.

Similar to the context switch process, a hypervisor or Virtual Machine Monitor (VMM) must perform a "machine switch" between running VMs. The VMM saves the entire state of each VM before moving to the next one to be processed. In the case of a bare

Fig. 3.6 The semantic gap bridging mechanism that improves interactions between guest OS and VMM [9]

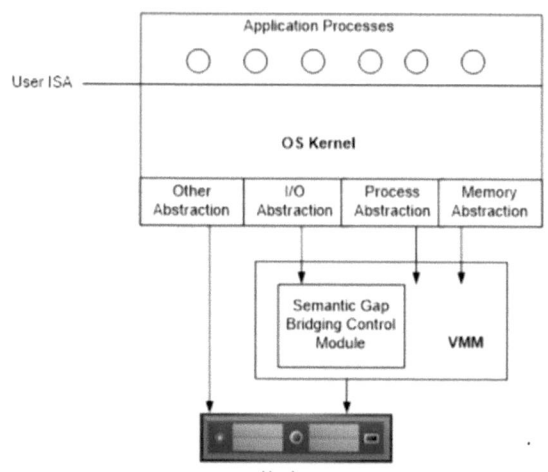

metal VMM, where the guest OS does not provide privileged operations, the VMM must rely on the OS trap handler, recorded during the guest OS bootup sequence, to handle system calls. Once the guest OS completes its task, it forwards a "return-from-trap" back to the VMM, which then provides the actual "return-from-trap" and puts the VM back into user mode. This process involves a lot of operations that can potentially slow down system calls, thereby degrading system performance. Additionally, guest OSs cannot run in kernel mode and must manage their own data structures while preventing access to their data from user processes operating in a seemingly less privileged mode. This issue has been partially addressed in Disco works, where Bugnion et al.[8] were able to run an MIPS processor in supervisory mode, allowing for a bit more memory even in user mode. In systems that do not support supervisory mode in the CPU, the OS must run in user mode and utilize memory protection mechanisms, such as page tables and Translation-Lookaside Buffers (TLBs), to protect guest OS data structures.

Earlier, we discussed how VMMs may not know what guest OS is running in a VM, similarly, this lack of knowledge applies to applications running on top of guest OSes. This information gap creates inefficiencies and potential security loopholes. Several scholars have attempted to address this semantic information gap. For example, Xiong et al.[9] proposed using "libvmi," a library designed to bridge the semantic information gap between guest OSes and VMs (Fig. 3.6).

The proposal by Xiong et al. [9] depicted in the figure above utilizes the libvmi library, which interfaces with the libxc of Xen and the libvirt of KVM. Xen and KVM are two open-source Virtual Machine Monitors (VMMs) or hypervisors. Xen implements a Type I hypervisor, also known as a bare metal hypervisor, while KVM implements a Type II or

---

[8] Bugnion et al. [8].

[9] Xiong et al. [9].

Fig. 3.7 Hypervisor implementation: (1) Type I or bare metal, (2) Type II or hosted, and (3) Hybrid

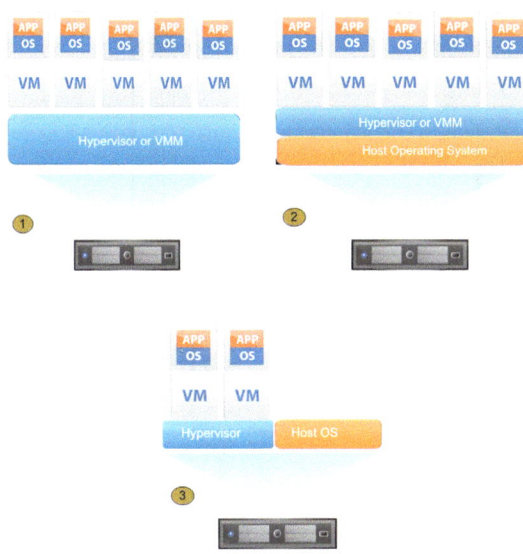

hosted hypervisor. Additionally, there is a third type of hypervisor referred to as a "hybrid hypervisor" (Fig. 3.7).

The concept of a hybrid hypervisor is relatively uncommon, but Microsoft Virtual PC and Server have employed such techniques. However, there appears to be some ambiguity regarding the definition of a hybrid VMM. For instance, Sawazaki, Maeda & Yonezawa[10] consider it to be a fusion of QEMU and KVM architectures, enabling dynamic switching between them. According to their perspective, the hybrid VMM allows security- and reliability-critical software to run on the software-based VMM, while performance-critical software can utilize the hardware-based VMM (although referencing KVM may not be entirely appropriate here, as it is a Type II hypervisor). In this framework, Sawazaki et al. [10] regard QEMU (an open-source emulator) as the software-based hypervisor, responsible for translating all instructions, whereas other VMMs only translate essential instructions for virtualization purposes. The hybrid hypervisor model proposed by Sawazaki et al. [10] incorporates paravirtualized code, such as KVM (Fig. 3.8).

In the hybrid hypervisor model illustrated in figure above, the key feature is its ability to offer both a security mechanism and dynamic switching capability between KVM and QEMU for specific tasks. In contrast, Weng et al.[11] introduced a different hybrid virtualization structure, presenting a combination of high-throughput VM and concurrent VM to enhance performance, addressing the limitations of traditional hypervisors like Xen, which lack efficient scheduling strategies. Interestingly, the implementation of the hypervisor in Microsoft's Virtual PC diverges significantly from the concepts outlined in

---

[10] Sawazaki et al. [10].

[11] Weng et al. [11].

Fig. 3.8 A hybrid hypervisor
model [10]

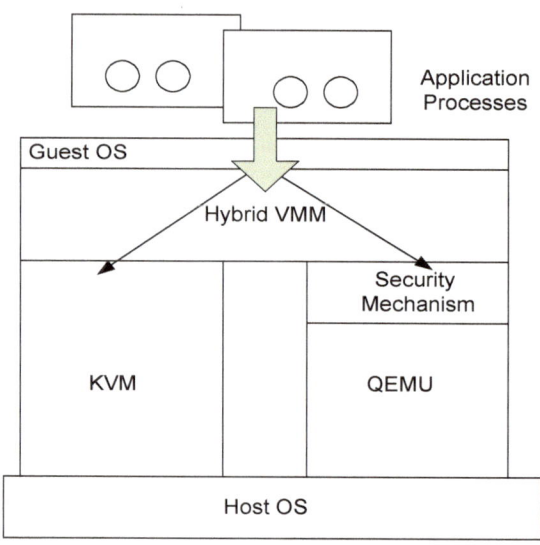

Table 3.1 Type I and type II hypervisors

Category	Name	Remarks
Type I	Xen	Open source
	Citrix Xen server	Citrix; includes full-featured Xen hypervisor
	Microsoft Hyper-V	Microsoft
	VMware VSphere hypervisor	VMware
	POWER5 hypervisor	IBM
Type II	Microsoft virtual PC and virtual server	Microsoft
	KVM	Open source
	Red hat enterprise virtualization (RHEV)	Red hat

scholarly literature. While Microsoft Virtual PC and Virtual Server are hosted solutions, meaning they run on top of Windows, their VMM operates at "Ring 0" level, the most privileged or kernel level in $\times$ 86 architecture. However, access to hardware is routed through the OS.[12]

Type I and Type II hypervisors are the most commonly used, and many vendors currently support such implementations. Instead of providing a comprehensive list, the following table illustrates some of the commonly used hypervisors in both Type I and Type II categories (Table 3.1).

---

[12] Armstrong [12].

The list of Type I and II hypervisors provided in the table above is not exhaustive; rather, it offers a representative sample of available hypervisors. The primary distinction between Type I and Type II hypervisor implementations lies in whether the hypervisor operates directly on the hardware or on top of a Host OS.

3.3.3 Network Virtualization

Network virtualization is a broad term encompassing various technologies and concepts aimed at abstracting the physical network infrastructure to create multiple virtual network instances, each operating independently and securely. Let's break down the components and technologies involved, comparing and contrasting them along the way:

- **Virtual Local Area Network (VLAN):** VLANs are a basic form of network virtualization that logically segment a physical network into multiple isolated broadcast domains. VLANs operate at the data link layer (Layer 2) of the OSI model. They are typically limited in scalability and flexibility, as they are bound to the constraints of the underlying physical network.
- **Virtual Extensible LAN (VXLAN):** VXLAN is an overlay network technology that extends VLANs across different data centers or networks. It encapsulates Layer 2 Ethernet frames within Layer 3 UDP packets, enabling communication between virtual machines (VMs) across Layer 3 boundaries. VXLAN provides greater scalability compared to traditional VLANs and is often used in cloud environments.
- **Network Functions Virtualization (NFV):** NFV involves virtualizing network services traditionally implemented in dedicated hardware appliances, such as firewalls, load balancers, and routers. By virtualizing these functions, they can be instantiated on commodity hardware or virtual machines, offering greater flexibility, scalability, and cost-effectiveness. NFV abstracts network functions from the underlying hardware, enabling them to be dynamically deployed, scaled, and managed as virtualized instances.
- **Virtual Network Functions (VNF):** VNFs are the virtualized instances of network functions enabled by NFV. They can be deployed and chained together to build complex network services and architectures. VNFs offer advantages such as rapid deployment, scalability, and flexibility compared to traditional hardware-based implementations.
- **NFV Infrastructure (NFVI) Framework:** NFVI refers to the hardware and software infrastructure that supports NFV deployments. It includes compute, storage, and networking resources that are virtualization-aware and capable of hosting virtual network functions. NFVI frameworks ensure resource isolation, performance optimization, and efficient management of virtualized network functions.

- **Network Virtualization in Cloud Networking:** In cloud networking, network virtualization plays a crucial role in enabling multi-tenant environments, where multiple users or organizations share the same physical infrastructure. Cloud providers leverage network virtualization technologies like VLANs, VXLAN, NFV, and VNFs to provide isolated virtual networks, security, and network services to their customers. Network virtualization in cloud networking abstracts the underlying physical infrastructure, providing tenants with flexible, scalable, and secure network connectivity tailored to their specific requirements.

In the upcoming sections, we will provide a concise overview of select network virtualization technologies pertinent to cloud networking. However, it's worth noting that certain technologies will not be covered here, with the assumption that readers can explore those independently through other sources. This approach ensures focus on the discussed topics, acknowledging the vast depth of each subject, which could easily fill entire books.

3.3.3.1   Virtual Extensible LAN (VXLAN)

Virtual Extensible LAN (VXLAN) is a network virtualization technology primarily utilized to mitigate the constraints of traditional VLANs, particularly in large-scale cloud computing environments. VXLAN operates by overlaying virtual Layer 2 networks atop existing Layer 3 infrastructure, thereby facilitating the creation of scalable and adaptable multi-tenant network architectures. In the context of data center setups or connections between data centers and colocation facilities or clouds, VXLAN serves as a potent tool for extending VLANs. The illustration below depicts a typical use case of VXLAN in a multi-tenant scenario, wherein VXLAN enables the expansion of VLANs across disparate sites. In multi-tenant scenarios, clients can seamlessly extend Layer 2 domains from their premises to the tenant POD or rack within the service provider's data center. When VXLAN is configured alongside VRF (Virtual Routing and Forwarding) leaking, it permits VXLAN hosts within the fabric to establish communication with external hosts outside the fabric. Consequently, users gain access to resources, services, and applications hosted on external networks, such as the public cloud, colo, or other data centers (Fig. 3.9).

Conceptually, VXLAN operates as a tunnel established on top of existing IP networks. This VXLAN tunneling is commonly referred to as an overlay, with the IP networks forming the underlying infrastructure termed as the underlay. Unlike VLAN, which is limited to 4096 Layer 2 segments, VXLAN offers the remarkable ability to extend this boundary to a staggering 16 million segments. It encapsulates Layer 2 Ethernet frames into a Layer 4 User Datagram Protocol (UDP) packet with a VXLAN header. The process helps create virtualized Layer 2 subnets that span across Layer 3 networks. Each segmented subnet is uniquely identified by a VXLAN Network Identifier (VNI). A device that performs the encapsulation and decapsulations of the packet is known as VTEP. Depending on the needs of the network, a VTEP can be set up in a TOR switch or on the server. Below

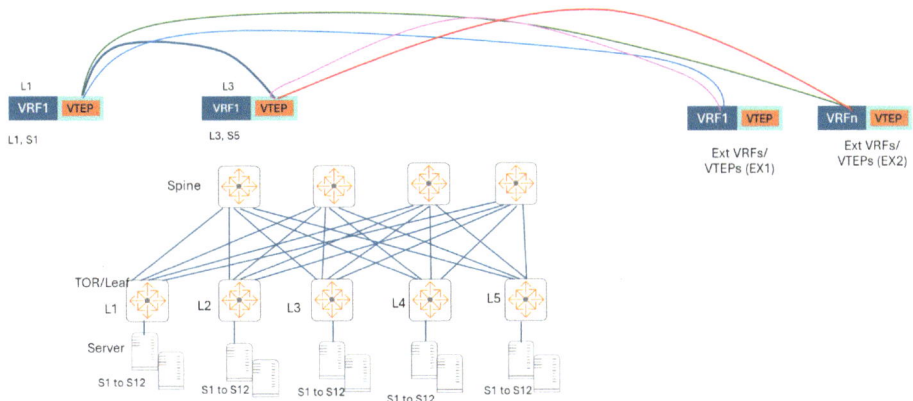

Fig. 3.9 VxLAN—stretching in a typical multi-tenant scenario

is a diagram of a VXLAN packet. The packet consists of 14 bytes for the outer Ethernet header, 20 bytes for IP, and 8 bytes for UDP. The VxLAN header is 8 bytes long. Additionally, the packet contains the original L2 frame along with the FCS.

The outer Ethernet header contains fields for destination address, VLAN, and Ethertype. If the destination VTEP is local, the destination address is set to its MAC address. Otherwise, if the destination VTEP is on a different L3 network, it's set to the MAC address of the next-hop device, usually a router. The VLAN field is optional in VXLAN implementation. If present, it's indicated by an Ethertype of 0×8100 and includes a VLAN ID tag. The Ethertype field is set to 0×0800 for IPv4 packets only. The Outer IP header can be either IPv4 or IPv6. In an IPv4 header, the Protocol field is set to 0×11, indicating that the frame contains a UDP packet. The source IP field shows the IP address of the originating VTEP, while the destination IP field displays the IP address of the target VTEP. However, if this information is unknown, such as in the case of a target virtual machine not previously targeted by the VTEP, a discovery process must be initiated by the originating VTEP (Fig. 3.10).

In networks utilizing an IPv6 underlay, VTEPs encapsulate VXLAN packets within an IPv6 outer header and route them through an IPv6 underlay network. Configurations for the IPv6 underlay closely resemble those for IPv4 underlay, with the distinction that VTEP source addresses are specified as IPv6 addresses. Furthermore, IPv6 addresses are assigned within the underlay, and reachability is established using the IPv6 protocol. The UDP header comprises source port, destination port, UDP length, and UDP checksum fields. The source port is determined by the VTEP, with the recommended range being 49,152–65,535. The UDP destination port corresponds to the VXLAN port, assigned the value 4789 by IANA. Typically, the UDP checksum is set to zero for packet decapsulation.

The VXLAN header includes VXLAN flags, VNI, and reserved fields. If the I bit in the VXLAN Flags field is set to 1, it signifies that the VXLAN ID is valid; conversely, if

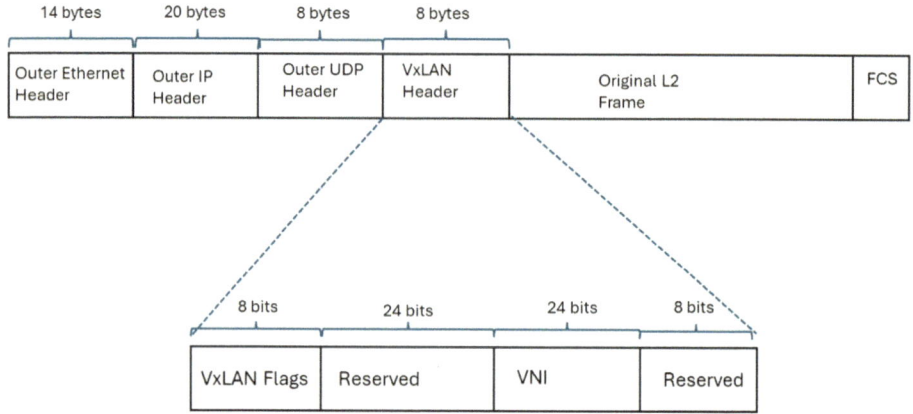

Fig. 3.10 VXLAN packet format

the I bit is 0, the VXLAN ID is considered invalid. All other bits in the VXLAN flags are reserved and set to 0. The VXLAN Network Identifier (VNI) field identifies the VXLAN of the frame, with values ranging from 4096 to 16,777,215.

A VXLAN overlay can be setup in two ways: static (also known as unicast VXLAN) and dynamic. The static VXLAN is the easiest way to connect two VTEPs. The static approach is ideal for smaller network environments generally few racks setup whereas dynamic VXLAN is appropriate for larger network environments.

The following table depicts the difference between static VXLAN and dynamic VXLAN (Table 3.2).

Table 3.2 The difference between static and dynamic VXLAN

Static VXLAN	Dynamic VXLAN
In this configuration, network admin needs to manually set up the VTEP as a member in the virtual network Additionally, flood and learn based on multicast control plane is also considered as static VxLAN	There's no need for manual setup. Every remote VTEP is automatically recognized as part of the virtual network through the EVPN routes received from it. Once the address of a remote VTEP is learned, VXLAN traffic can be both sent to and received from the VTEP
Data packets learn remote hosts after decapsulation of VxLAN header	The control plane protocol assists in the learning of remote hosts MAC addresses through BGP EVPN routes and MAC/IP advertisements

VXLAN Control Plane

In VXLAN setups, two common control plane methods are the flood-and-learn approach using multicast and the MP-BGP EVPN control plane. When using flood and learn multicast, the approach typically involves flooding traffic at the data link layer and dynamically learning MAC addresses. This method is used to transmit broadcast, unknown unicast, and multicast (BUM) traffic, as well as to discover remote VTEPs and learn MAC addresses and mappings for each VXLAN segment. Moreover, IP multicast is employed to limit the scope of flooding to only the hosts participating in the VXLAN segment. Each VXLAN segment, known as a VNI, is associated with an IP multicast group in the transport IP network. Each VTEP device is configured to join this multicast group independently, acting as an IP host through the Internet Group Management Protocol (IGMP). When VTEPs join, it triggers Protocol-Independent Multicast (PIM) signaling across the transport network for the specific multicast group. The transport network then builds a multicast distribution tree based on the locations of the participating VTEPs.

The MP-BGP EVPN control plane is a standardized approach commonly utilized in large-scale cloud data centers and VXLAN deployments spanning across Data Center Interconnects (DCIs) to remote or customer sites. In such configurations, control plane learning is utilized for end hosts located behind remote VTEPs. This method ensures a clear separation between control and data planes and establishes a unified control plane for both Layer 2 and Layer 3 forwarding within a VXLAN overlay network. This control plane operates as an MP-BGP-based VPN technology, where EVPN leverages multiprotocol BGP (MP-BGP) for exchanging information among VXLAN tunnel endpoints (VTEPs). EVPN was initially introduced in RFC 7432, with RFC 8365 outlining VXLAN-based EVPN as a next-generation VPN solution. VXLAN-based EVPN is designed to supersede previous-generation VPNs like Virtual Private LAN Service (VPLS).

3.3.3.2 NFV, VNF, and NFVI

Within the realm of network engineering, Network Functions Virtualization (NFV) and Virtualized Network Functions (VNF) are often subjects of discussion, albeit not without some degree of confusion. The Virtualized Network Functions (VNFs) typically denote the software-based versions of network appliances like routers, firewalls, and load balancers. These VNFs are typically deployed as virtual machines (VMs) on hypervisors such as Linux KVM or VMware vSphere, running on standard commercial hardware. In contrast, Physical Network Functions (PNFs) refer to traditional network appliances built on proprietary hardware. Meanwhile, Cloud-Native Network Functions (CNFs) pertain to containerized versions of VNFs, which may include container networking and service mesh functionalities for microservice architectures.

On the other hand, Network Function Virtualization (NFV) represents a network architecture primarily advocated by traditional telecom service providers aiming to leverage network virtualization technologies to enhance the speed and agility of diverse network nodes. Traditional networking equipment is often costly and lacks the scalability required for large-scale deployment by telecom service providers. This has prompted the search for an alternative network architecture, leading to the emergence of NFV. The European Telecommunications Standards Institute (ETSI) plays a pivotal role in this domain, having introduced the initial NFV standard in October 2013. Subsequently, ETSI has released additional specifications addressing various facets of NFV and its constituent components, solidifying its position as a leading authority in the field. Both the ETSI and the Linux Foundation (LF) play key roles in the development and cultivation of the reference architecture and standards for the NFV framework commonly known as NFV MANO (Management and Orchestration). ETSI's Open Source MANO (OSM) and the Linux Foundation's Open Network Automation Platform (ONAP) stand out as the primary open-source NFV projects, garnering support from service operators and network vendors alike.

According to ETSI's definition, the NFV MANO framework is responsible for critical operational tasks. It encompasses the NFV Orchestrator, VNF Manager, and Virtualized Infrastructure Manager (VI Manager), and their interplay with other operational systems. Additionally, the Linux Foundation's ONAP architecture incorporates all aspects of the MANO layer functionalities outlined in the ETSI NFV Framework. Furthermore, ONAP offers a network service design framework and features for fault management, configuration, accounting, performance, and security (FCAPS). The following diagram shows ETSI NFV framework. Since its first release in 2012, the NFV framework undergone many releases, current NFV release 4 was initiated in 2019. Some of the work is still ongoing but the release specified improvements in the following areas [13]:

- **NFVI (Network Function Virtualization Infrastructure):** The improvements proposed in NFVI will accommodate to accommodate lightweight virtualization technologies like OS containers. These enhancements optimize NFVI abstraction to minimize the dependence of Virtualized Network Functions (VNFs) on infrastructure. Additionally, they streamline networking integration into the infrastructure fabric, simplifying connectivity for VNFs and Network Services (NSes).
- **NFV automation and capabilities:** It encompasses various enhancements, including lifecycle management and orchestration improvements. These advancements introduce more policy-based management, simplify VNF and NS management through virtualization, and address developments in autonomous networking (Fig. 3.11).
- **NFV-MANO (Management and Orchestration) framework:** The works primarily emphasize optimizing the exposure and utilization of internal NFV-MANO capabilities. This involves exploring service-based transformation and enhancing reliability and availability through features such as NFV-MANO upgrades and robustness improvements.
- **Operations and Security:** The area of works involves simplification of NFV to ease development and deployment of sustainable NFV-based solutions, verification (and certification) procedures, and security hardening.

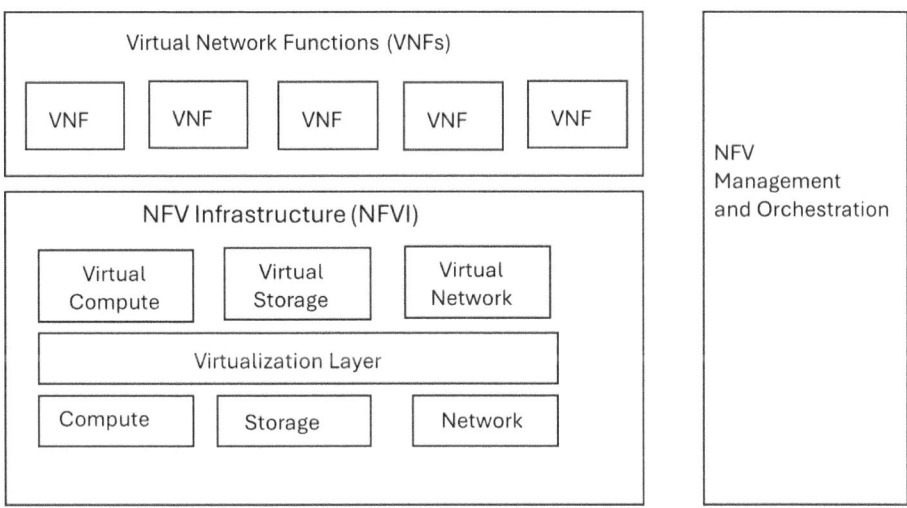

Fig. 3.11 ETSI NFV—framework (Courtesy: ETSI)[13]

The implementation of NFV, VNFs, and NFVI extends beyond traditional telecom service providers to include cloud service providers (CSPs) and Managed Service Providers (MSPs) as well. These entities leverage NFV frameworks and components to enhance their cloud networking capabilities, offering more flexible and scalable services to their customers. As NFV continues to evolve, its adoption by CSPs and MSPs underscores its significance in shaping the landscape of cloud networking.

3.3.4 Storage Virtualization

Storage virtualization is a technology that abstracts physical storage resources and presents them as logical entities to users or applications. It allows multiple physical storage devices, such as hard disk drives (HDDs), solid-state drives (SSDs), or storage area networks (SANs), to be aggregated into a single virtual storage pool.

Key aspects of storage virtualization include the following:

- **Abstraction:** Storage virtualization abstracts the underlying physical storage hardware, providing a layer of abstraction that separates the logical view of storage from its physical implementation. This abstraction simplifies storage management and enables greater flexibility in allocating and managing storage resources.

---

[13] ETSI [13].

- **Aggregation:** Storage virtualization aggregates multiple physical storage devices into a unified storage pool. This pooling of resources allows for efficient utilization of storage capacity and facilitates centralized management of storage resources.
- **Virtual Volumes:** Virtual volumes, also known as logical volumes or virtual disks, are created from the pooled storage resources and presented to users or applications as if they were physical disks. These virtual volumes can be dynamically provisioned, resized, and managed to meet changing storage requirements.
- **Data Migration and Mobility:** Storage virtualization enables seamless data migration and mobility across different storage devices or systems without disrupting access to data. This facilitates tasks such as data replication, data migration, and storage tiering for optimization of performance and cost.
- **Data Protection and Redundancy:** Storage virtualization often includes features for data protection and redundancy, such as RAID (Redundant Array of Independent Disks) configurations, snapshots, and replication. These features help ensure data integrity, availability, and resilience against hardware failures or disasters.
- **Interoperability:** Storage virtualization solutions are typically designed to be vendor-agnostic, allowing them to work with a variety of storage hardware from different vendors. This interoperability enables organizations to leverage existing investments in storage infrastructure and adopt new technologies without vendor lock-in.

The following diagram illustrates a typical storage virtualization scenario in VMware environment. Two commonly used storage virtualization elements are depicted in the diagram: file storage and block storage. The NFS and VMFS are two different file system supported in VMware environment. The NFS is a file-level file system while VMFS is a block-level file system. The VMFS file system is created by the vSphere, while the NFS file system is on storage side and is only mounted as a shared folder on the vSphere (Fig. 3.12).

There are several types of storage virtualization, each with its own approach and characteristics. Here is a list of various storage virtualization types:

- **Block-Level Storage Virtualization:** This type of virtualization operates at the block level, where storage blocks from multiple physical storage devices are aggregated into a single pool of storage. It presents these blocks as logical volumes or disks to the host systems. Technologies such as RAID (Redundant Array of Independent Disks) and Storage Area Networks (SANs) often employ block-level storage virtualization.
- **File-Level Storage Virtualization:** File-level storage virtualization operates at the file level, where files from multiple storage devices are aggregated into a single namespace. It allows users and applications to access files using a unified directory structure, regardless of the underlying storage system. Network-attached storage (NAS) solutions often utilize file-level storage virtualization.

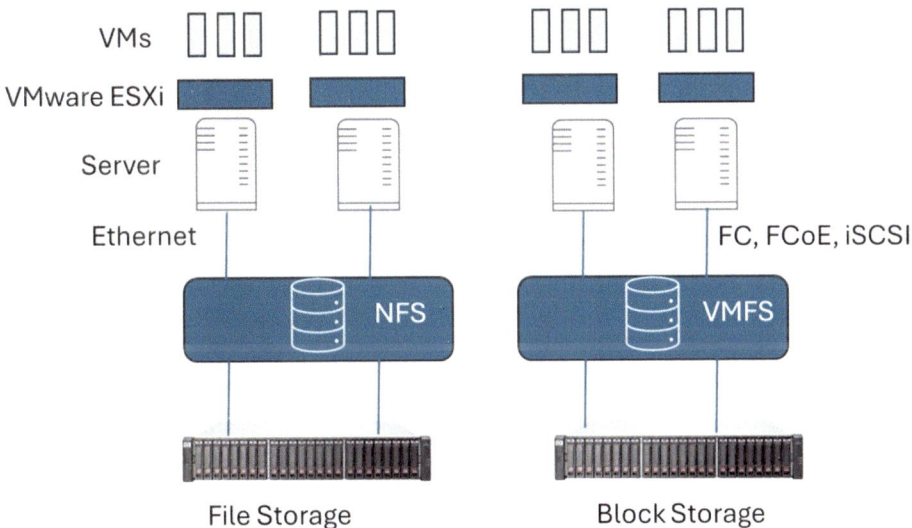

Fig. 3.12 Storage virtualization

- **Object-Level Storage Virtualization:** Object-level storage virtualization abstracts storage resources at the object level, where data is stored as objects with metadata. It provides a scalable and flexible way to store and manage large volumes of unstructured data, such as multimedia files, documents, and backups. Object storage systems like Amazon S3 and OpenStack Swift use object-level storage virtualization.
- **Unified Storage Virtualization:** Unified storage virtualization combines block-level, file-level, and sometimes object-level virtualization into a single storage platform. It offers a unified storage pool that supports multiple access methods, such as block-based access for applications and file-based access for users. Unified storage systems aim to provide greater flexibility and efficiency in managing diverse storage workloads.
- **Storage Virtualization Appliances:** Storage virtualization appliances are standalone devices or software solutions that sit between host systems and storage devices, acting as a virtualization layer. These appliances provide centralized management and control of storage resources, regardless of the underlying storage systems. They often offer features such as data migration, data replication, and storage tiering.
- **Host-Based Virtualization:** Host-based storage virtualization occurs directly on the host system itself, rather than being implemented within dedicated storage hardware or appliances. This type of virtualization typically involves software installed on the host server, which abstracts, aggregates, and manages the underlying storage resources. In host-based storage virtualization, the host system controls and manages the storage resources, presenting them to applications or users as logical volumes or disks.

This approach provides flexibility and scalability without requiring additional hardware or complex configurations. Host-based storage virtualization is often used in environments where centralized storage solutions, such as SAN or NAS, are not feasible or cost-effective. It allows organizations to leverage existing server hardware and storage devices while still benefiting from the advantages of storage virtualization. Common examples of host-based storage virtualization solutions include software-defined storage (SDS) platforms, volume managers, and file system-level virtualization tools. These solutions enable host systems to abstract, manage, and optimize storage resources independently of the underlying storage hardware.

- **Array-Based Storage Virtualization:** In array-based storage virtualization, virtualization capabilities are built into the storage array hardware itself. This type of virtualization occurs at the block level, where storage blocks from multiple physical storage devices within the array are aggregated into a single pool of storage. The storage array then presents these blocks as logical volumes or disks to the host systems. Array-based storage virtualization offers benefits such as centralized management, scalability, and efficient resource utilization. It allows organizations to manage diverse storage workloads and optimize performance and capacity across multiple storage devices within the array. The "block-level storage virtualization" is often associated with array-based storage virtualization.

- **Network-Based Storage Virtualization:** The network-based storage virtualization occurs at the network level, typically within the storage area network (SAN) infrastructure. In this approach, virtualization capabilities are implemented in network devices such as switches, routers, or specialized appliances. This type of storage virtualization abstracts and virtualizes storage resources across multiple storage devices within the SAN, presenting them as a single, unified pool of storage to host systems or applications. Network-based storage virtualization enables centralized management, scalability, and flexibility in allocating and managing storage resources across the SAN. Both block-level and file-level storage virtualization can be implemented within the SAN infrastructure, leveraging network-based virtualization capabilities to provide centralized management, scalability, and flexibility in allocating and managing storage resources across the SAN.

In revisiting our earlier discourse on storage virtualization within a VMware environment, various implementations come to light. One such instance involves the utilization of an external storage array to furnish shared storage across multiple ESXi hosts. This storage can be accessed by ESXi hosts through either file-based or block-based storage protocols. In the case of file storage, the storage array, also known as a filer, creates and manages a file system, presenting it to hosts for utilization. VMware ESXi can access file-based storage using the NFS protocol.

Alternatively, block storage implementation entails the storage array presenting a raw set of hard drive blocks, termed as a Logical Unit (LUN), to the connected hosts. It is then the responsibility of the hosts to format and establish a file system on this allocated space. Block-based storage, in conjunction with storage adapters (HBA) and storage fabric (switches and cabling), constitutes a Storage Area Network (SAN). ESXi supports various SAN protocols including Fiber Channel (FC), Fiber Channel over Ethernet (FCoE), or iSCSI. When connecting to iSCSI SANs, ESXi hosts utilize either 1Gb or 10Gb Ethernet connections. The SAN provides ESXi with a raw LUN, which ESXi formats using the VMware File System (VMFS). Virtual Machines are stored within the VMFS datastore.

The discourse on storage virtualization and this succinct overview of VMware implementation are provided to acquaint you with storage virtualization. This serves as an introductory exposition, aimed at enabling readers to grasp the concept of virtualization within the context of both private and public clouds, as we introduce various cloud networking concepts in the subsequent sections of the chapter.

3.4   Virtual Private Cloud (VPC)

A Virtual Private Cloud (VPC) is a cloud computing environment that provides a logically isolated section of a public cloud infrastructure. It enables users to deploy resources such as virtual machines (VMs), storage, and networking components within a dedicated, customizable virtual network. In essence, a VPC serves as a means to allocate dedicated bandwidth within a public cloud setting. Users leverage VPCs to carve out a portion of the computing resources within a public cloud for their exclusive needs. By offering data isolation akin to private cloud setups, VPCs seamlessly blend the convenience and scalability of public cloud infrastructure. The diagram below illustrates a typical VPC configuration and the services it offers (Fig. 3.13).

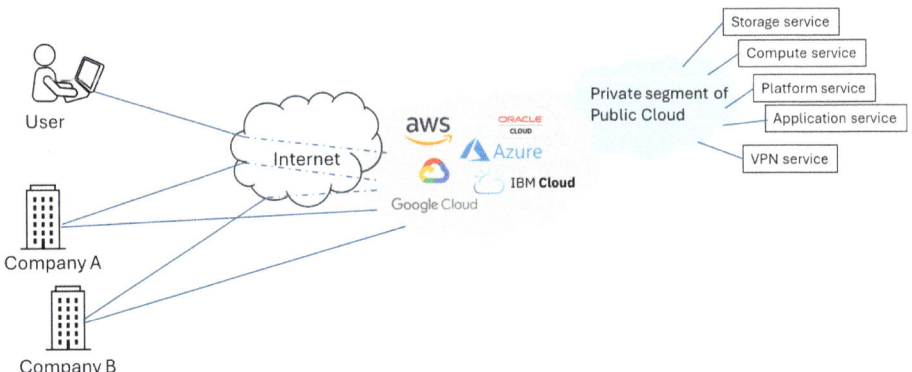

Fig. 3.13 Typical VPC setup

In most cloud service providers (CSPs), the standard VPC setup enables users and tenants to connect via the Internet and access various services such as SD-WAN, cross-connect through colocation facilities, and leased lines. The services provided through VPCs include, but are not limited to, storage, computing, platforms, applications, and VPN services.

Key components of a Virtual Private Cloud include the following:

- **Virtual Network:** A VPC allows users to define and configure their own virtual network, including subnets, IP address ranges, and routing tables. This network is isolated from other VPCs and provides a secure environment for deploying resources.
- **Subnets:** Subnets within a VPC allow users to partition the virtual network into smaller, more manageable segments. Each subnet can be associated with specific availability zones or regions within the cloud provider's infrastructure.
- **Security Groups:** Security groups act as virtual firewalls for resources deployed within a VPC. They allow users to define inbound and outbound traffic rules to control access to their resources based on IP addresses, port numbers, and protocols.
- **Network Access Control Lists (NACLs):** NACLs are stateless packet filters that control traffic at the subnet level. They provide an additional layer of security by allowing users to define rules that govern traffic flow in and out of subnets.
- **Internet Gateways (IGWs):** IGWs enable communication between resources within a VPC and the Internet. They serve as entry and exit points for traffic flowing in and out of the VPC.
- **Virtual Private Gateways (VGWs):** VGWs establish secure connections between a VPC and an external network, such as an on-premises data center or another VPC. They enable hybrid cloud deployments and extend the reach of the VPC.
- **VPN Connections:** VPN connections allow users to establish encrypted tunnels between a VPC and an external network using standard VPN protocols such as IPsec. This enables secure communication between on-premises infrastructure and resources deployed within the VPC.
- **Peering Connections:** Peering connections enable communication between separate VPCs within the same cloud provider's infrastructure. They allow resources in one VPC to communicate directly with resources in another VPC without traversing the public Internet.

A VPC grants enterprises the advantages of a private cloud setup while integrating features of a public cloud. This hybrid model enables companies to harness the benefits of both cloud deployment types concurrently[14]. It's important to note that a VPC is not inherently a private cloud; instead, it offers a curated set of resources within the public cloud infrastructure. The hybrid model mentioned here combines the convenience of the public cloud with the isolation requirements of a private cloud.

---

[14] Google [14].

3.5   Virtual Private Network (VPN)

In simple terms, a VPN establishes a secure and encrypted connection between an end-point device and the Internet via a VPN server operated by a provider. This is particularly vital for end-user access to corporate networks, remote sites, or applications in public clouds like Azure. VPNs mask IP addresses and encrypt Internet traffic, thwarting attempts by hackers, ISPs, or governments to monitor users' online activities. Furthermore, VPNs play a crucial role in facilitating site-to-site connectivity, linking data centers with public cloud environments. Specifically in cloud networking, VPNs provide secure access to cloud resources. Many organizations leverage cloud platforms such as AWS, Azure, or GCP for data storage, application hosting, and infrastructure management.

VPNs ensure secure connectivity between an organization's on-premises network and their public cloud resources. This guarantees that data transmitted between the organization's network and the cloud remains encrypted and safeguarded against unauthorized access or interception by malicious entities. By establishing VPN connections to the cloud, users can securely access cloud-based applications, databases, and other resources as if they were directly connected to the organization's internal network. Moreover, VPNs offer an additional layer of security when accessing cloud services. In instances where network restrictions or firewalls limit access to certain cloud resources, VPNs prove invaluable by encrypting traffic and routing it through a VPN server located outside the restricted network. This enables users to bypass such restrictions, accessing cloud resources that may otherwise be inaccessible.

The diagram below illustrates a typical setup in which an enterprise data center is interconnected with a public cloud via colocation facilities, often referred to as "colos." In this scenario, the enterprise data center (DC) is linked to two colocation facilities to ensure redundancy (Fig. 3.14).

Colocation facilities generally offer connectivity to multiple cloud providers, catering to the diverse needs of customers. Enterprises may opt to connect with a single cloud provider or multiple providers through colocation facilities or direct connections. In the diagram, HPE's CX10000 (CX10K) switch serves as the border leaf. In Chap. 2, we introduced the CX10000 (CX10K) as a unique L2/L3 switch featuring service chaining capabilities through its DPU technology. In this instance, the CX10K serves as both a Top-of-Rack (TOR) switch, providing microsegmentation to servers, and a border leaf, facilitating IPsec site-to-site VPN connectivity between the data center (DC) and colocation facilities. Additionally, the device performs stateful inspection as an L4 firewall, adding an extra layer of security to the traffic path. It's worth noting that the IPsec endpoint could terminate to a gateway within the colo or connect through the colo to gateways at different Cloud Service Providers (CSPs). The diagram below illustrates various CSP gateways that provide IPsec terminations. For instance, AWS utilizes a transit gateway to terminate IPsec tunnels, enabling an enterprise to connect to its Virtual Private Cloud (VPC) at AWS (Fig. 3.15).

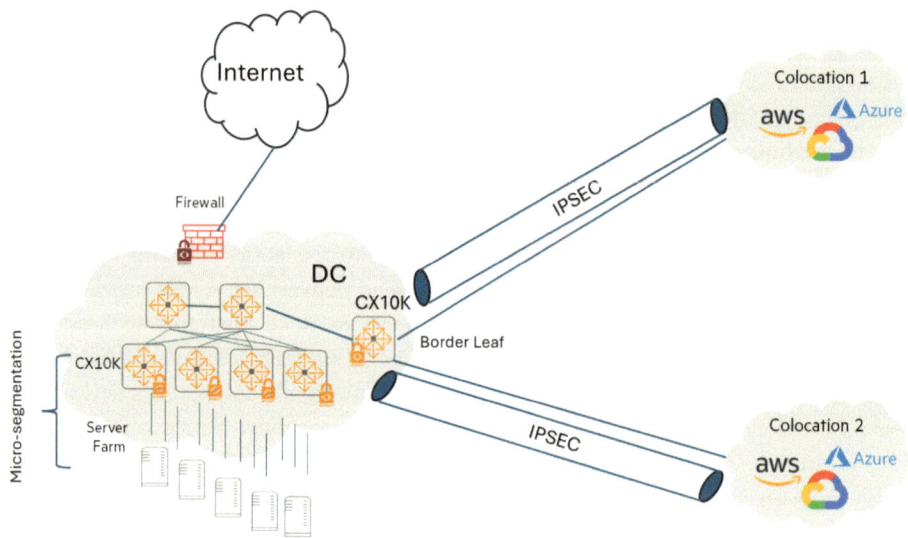

Fig. 3.14 Data center to colo connectivity using IPSEC as VPN

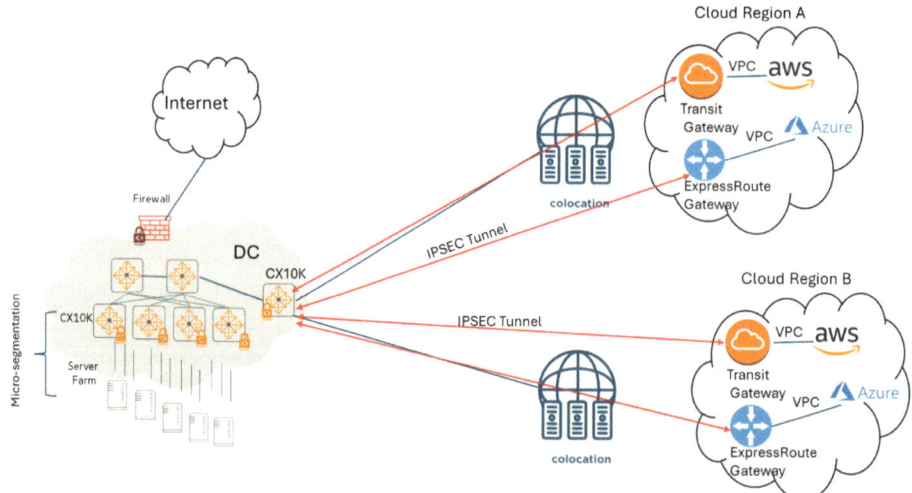

Fig. 3.15 IPSEC tunneling terminating at a CSP gateway

Each CSP may have its own transit gateway to facilitate IPsec tunnel termination and allow enterprises to connect to their respective VPCs. There are numerous CSP gateways that offer IPsec termination and connectivity to their respective VPCs. We will briefly explore some of them below.

3.5.1   AWS Transit Gateway

Amazon AWS transit gateway is a scalable and fully managed service that facilitates connectivity between Virtual Private Clouds (VPCs) and on-premises networks in a hub-and-spoke architecture. With Transit Gateway, customers can seamlessly connect thousands of VPCs. All hybrid connectivity, including VPN and Direct Connect connections, can be attached to a single Transit Gateway instance. This consolidation enables centralized control over the organization's entire AWS routing configuration. Transit Gateway governs traffic routing among the connected spoke networks through route tables, simplifying management and reducing operational costs. In this hub-and-spoke model, VPCs only need to connect to the Transit Gateway instance to gain access to the connected networks, streamlining network management and enhancing scalability (Fig. 3.16).

The diagram above depicts an IPSEC tunnel originating from CX10K, terminating at the AWS Transit Gateway, which is equipped with a BGP route peering router connecting to AWS backbone networks. The GRE tunnel further extends extra layer of security directly to VPCs. An enterprise customer has the option to create their own Virtual Private Cloud (VPC), incorporating EC2 VM instances and various cloud resources. This setup facilitates access from their Data Center (DC) or offices whenever necessary. Should a customer opt for SD-WAN, their choice may hinge upon the available connectivity services. In such cases, the AWS Transit Gateway offers a managed, highly available, and scalable regional network transit hub, effectively interconnecting VPCs and the SD-WAN network.

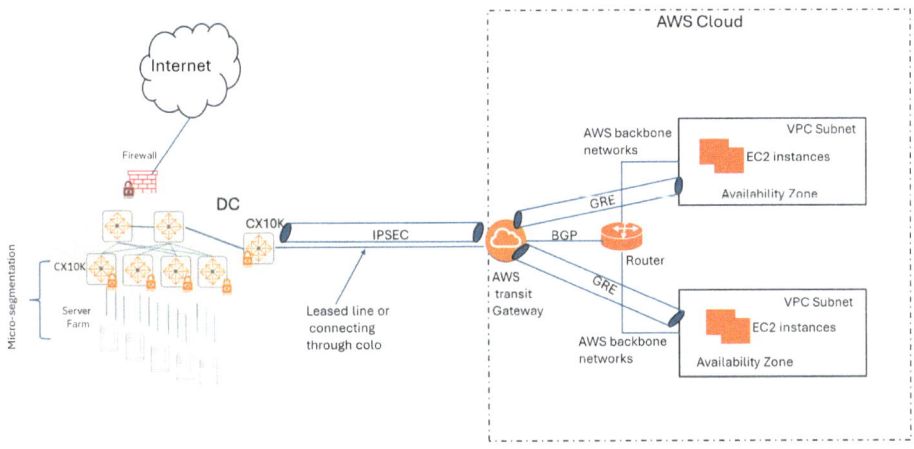

Fig. 3.16 Enterprise data center to AWS cloud network connectivity

3.5.2   Microsoft Azure ExpressRoute Gateway

Microsoft ExpressRoute Gateway is a networking feature provided by Microsoft Azure that facilitates secure and private connectivity between your on-premises network and Azure cloud services. It acts as a bridge, enabling data transfer between your on-premises infrastructure and resources hosted in Azure over a dedicated, private connection. ExpressRoute Gateway comes in different SKUs (Service Level Agreements) to cater to varying performance and feature requirements. These virtual SKUs include the following:

- **Standard SKU:** This is the basic offering, providing a standard set of features and performance capabilities suitable for many scenarios. It offers redundancy and high availability, making it suitable for mission-critical applications.
- **High-Performance SKU:** This SKU is optimized for high bandwidth and low-latency requirements. It offers increased performance compared to the Standard SKU, making it suitable for applications that demand high throughput and low latency, such as large-scale data transfers and real-time analytics.
- **Ultra-Performance SKU:** This is the premium offering, providing the highest level of performance and scalability. It offers ultra-low latency and extremely high throughput, making it ideal for ultra-demanding workloads, such as high-frequency trading and large-scale scientific computing.

The diagram below illustrates the typical process of connecting an on-premises network to virtual networks on Azure through Azure ExpressRoute. This connection utilizes a private, dedicated link facilitated by a third-party connectivity provider, effectively extending your on-premises network into the Azure cloud environment (Fig. 3.17).

Moreover, the virtual instances of ExpressRoute gateway are capable of also providing BGP peering services as needed. This allows for dynamic routing and enhanced control over the traffic flow between your on-premises network and the Azure virtual networks, ensuring optimal performance and efficient management of network resources.

3.5.3   Dynamic Routing Gateway

For Oracle cloud services, the Site-to-Site IPSEC tunnel terminates at Oracle's virtual router, known as a dynamic routing gateway (DRG). The DRG serves as the gateway into the Virtual Cloud Network (VCN) from an enterprise's on-premises network. Whether the customer is utilizing the Site-to-Site VPN or Oracle Cloud Infrastructure FastConnect private virtual circuits to establish the connection between their on-premises network and VCN, all traffic traverses through the DRG. From the perspective of a network engineer, the DRG functions as the VPN headend. Upon creating a DRG, it is essential to attach it to the respective VCN, a process achievable through either the Console or API. Additionally,

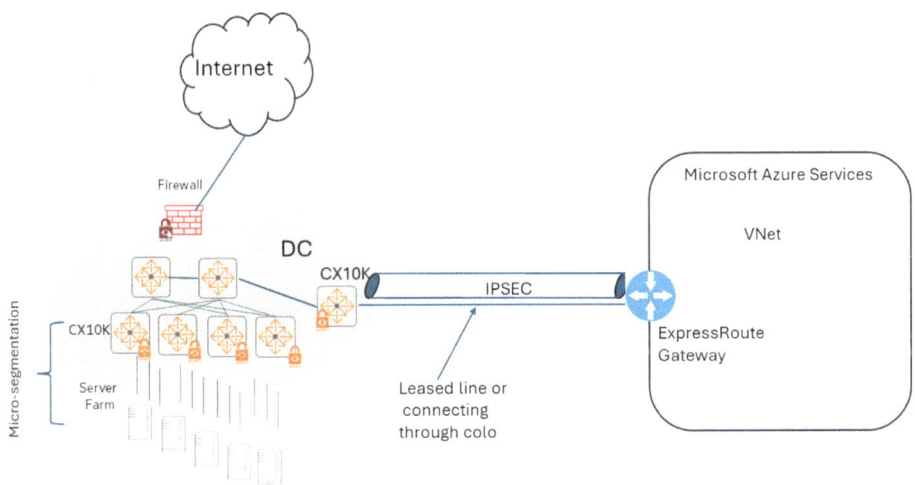

Fig. 3.17   IPSEC connectivity with Microsoft Azure ExpressRoute gateway

the customer must incorporate one or more route rules directing traffic from the VCN to the DRG. Without this attachment and the presence of route rules, traffic cannot flow between a given VCN and the on-premises network. Notably, the DRG can be detached from the VCN at any time while preserving all other VPN components. Subsequently, the customer has the flexibility to reattach the DRG to the same VCN or attach it to another VCN as needed.

The diagram below depicts how IPSEC tunnel terminates at DRG in an Oracle Cloud Infrastructure (OCI). To create an IPSEC connection, customers must first setup a CPE object at OCI. The CPE object is virtual representation of the device used for IPSEC tunnel at the on-premises (Fig. 3.18).

After creating both the CPE object and DRG at OCI, customers can connect them by establishing an IPSec connection, which serves as a parent object representing the site-to-site VPN. There are two commonly used methods in IPSEC connectivity: transport mode and tunnel mode. In transport mode, IPSec encrypts and authenticates the actual payload of the packet while keeping the header information intact. In tunnel mode, however, IPSec encrypts and authenticates the entire packet. After encryption, the packet is encapsulated to create a new IP packet, leading to altered header information. For OCI connectivity, customers should use tunnel mode only.

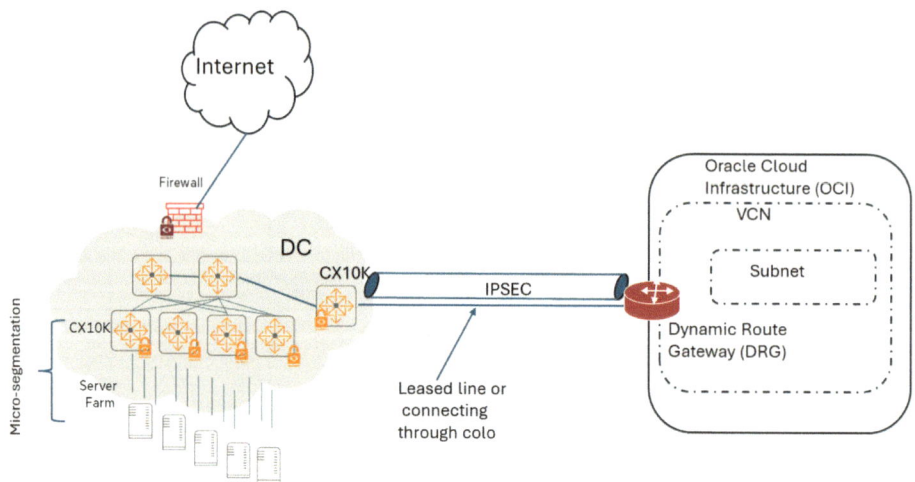

Fig. 3.18 IPSEC connectivity to Oracle dynamic route gateway (DRG)

3.5.4   Google Cloud VPN Connection

Google Cloud allows IPSEC VPN connectivity to VPC, but customers must choose one of the two methods specified: HA VPN and classic VPN. HA VPN represents a high-availability (HA) Cloud VPN solution enabling secure connections between your on-premises network and your VPC network via an IPsec VPN connection. Depending on the topology and configuration, HA VPN can offer an SLA of either 99.99% or 99.9% service availability.

Upon creating an HA VPN gateway, Google Cloud automatically selects two external IPv4 addresses, one for each interface. These IPv4 addresses are chosen from distinct address pools to ensure high availability. Each interface of the HA VPN gateway supports multiple tunnels, and multiple HA VPN gateways can be created. When deleting the HA VPN gateway, Google Cloud releases the IP addresses for reuse. While it's possible to configure an HA VPN gateway with only one active interface and one external IP address, this setup doesn't guarantee an availability SLA. The following diagram depicts typical Google Cloud HA VPN (Fig. 3.19).

Customers have the option to connect to the HA Gateway either through the Internet or via a partner VPN. For the latter, compatibility for the IPSEC tunnel needs to be verified between the customer's VPN end device and the partner VPN gateway. The Classic VPN is an older setup of Google Cloud VPN wherein a single interface with a single external IP address is utilized, and it supports tunnels that use static routing (either policy based or route based). However, customers can only configure BGP routes through a partner VPN gateway that is running Google Cloud VPN instances. Both Classic VPN and HA VPN gateways utilize external (Internet-routable) IPv4 addresses. Only ESP, UDP 500, and

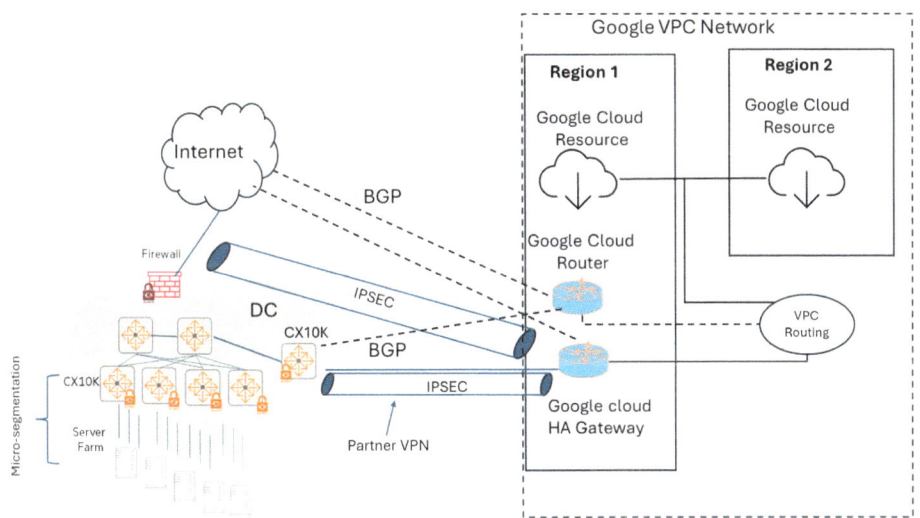

Fig. 3.19 Google Cloud HA VPN

UDP 4500 traffic are allowed to reach these addresses. This rule applies to Cloud VPN addresses configured by you for Classic VPN or to automatically assigned IP addresses for HA VPN. It's important to note that Google Cloud VPN preferably uses tunnel mode for IPSEC VPN connectivity, with a bandwidth limit of up to 3Gbps per tunnel.

3.5.5   IBM Cloud VPN

IBM offers both direct link and connectivity through Internet to its VPC. Its VPN services are supported in either type of connectivity. The VPN services include SSL VPN, IPSEC VPN, and VPN Appliances like Juniper vSRX or AT&T vRouter. The IPSEC VPN is for access to an entire VPC network and generally used for site-to-site connectivity. The SSL Connection is for remote users who would like to access specific resources in IBM Cloud or VPC. For Direct link IPSEC VPN connectivity, customers can connect to IBM VPC network through one of its PoP (Point of Presence) in their respective region. The following diagram depicts a typical VPN connectivity for IBM cloud (Fig. 3.20).

The IKE policy is applied to IPSEC tunnels through Internet and IPSEC policy is applied to tunnels over direct link. The diagram shows VPN gateway is placed in a different subnet than the cloud resources, e.g., VSI (Virtual Server instances) within a landing zone. IBM Cloud VPN for VPC provides static-route-based and policy-based VPN modes. In a policy-based VPN, traffic matching negotiated CIDR ranges is routed through the

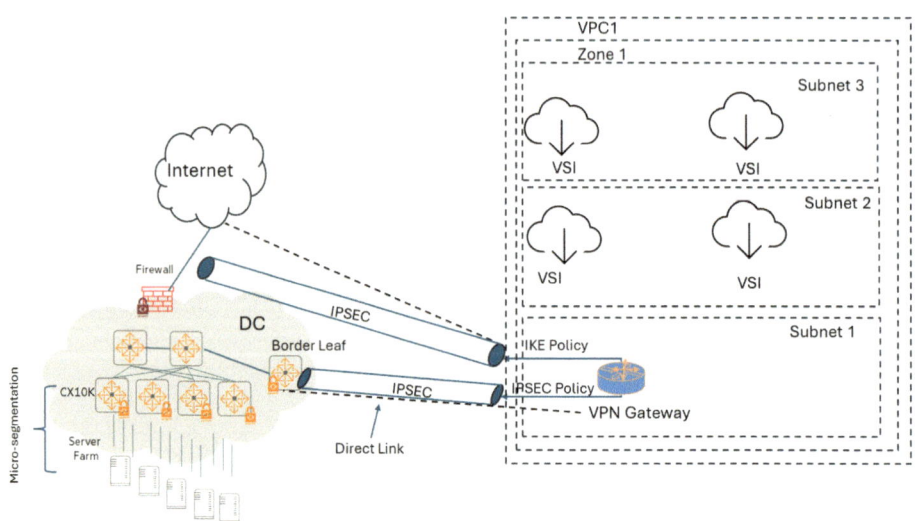

Fig. 3.20   VPN connectivity to IBM cloud

VPN. In a static-route-based VPN, virtual tunnel interfaces are established, and any traffic routed toward these logical interfaces using custom routes is directed through the VPN. Both VPN options offer identical features.

3.6   Private Cloud Networks

A private cloud is a cloud computing environment dedicated solely to one organization, providing exclusive access to computing resources such as servers, storage, and networking infrastructure. Unlike public clouds, where services are shared among multiple users, a private cloud is designed to meet the specific needs of a single organization, offering greater control, customization, and security.

Key characteristics of a private cloud include the following:

- **Dedicated Infrastructure:** Private clouds typically utilize dedicated hardware and software resources deployed within an organization's own data centers or hosted in a third-party data center exclusively for that organization.
- **Resource Pooling:** Private clouds pool together computing resources, including processing power, storage capacity, and network bandwidth, to support multiple workloads and applications within the organization.
- **Elasticity and Scalability:** Private clouds offer the ability to dynamically scale resources up or down based on changing demand, allowing organizations to optimize resource utilization and accommodate fluctuations in workload requirements.

- **Self-service Provisioning:** Private clouds often provide self-service portals or APIs that enable authorized users to provision and manage computing resources on-demand, without requiring manual intervention from IT administrators.
- **Virtualization:** Virtualization technologies such as hypervisors and virtual machines (VMs) are commonly used in private clouds to abstract physical hardware and create virtual instances of computing resources, enabling greater flexibility, efficiency, and resource utilization.
- **Security and Compliance:** Security is a primary focus in private clouds, with organizations implementing robust security measures to protect sensitive data, applications, and infrastructure. Private clouds offer greater control over security policies, access controls, encryption, and compliance requirements compared to public clouds.
- **Customization and Control:** Private clouds provide organizations with greater customization and control over their cloud environments, allowing them to tailor infrastructure configurations, network settings, and application environments to meet specific business requirements and regulatory constraints.
- **Isolation and Privacy:** Private clouds offer enhanced isolation and privacy compared to public clouds, ensuring that computing resources are dedicated exclusively to the organization and data remains within the organization's own infrastructure, reducing the risk of unauthorized access or data exposure.

Private clouds are commonly utilized by organizations with stringent security and compliance requirements, including government agencies, financial institutions, healthcare providers, and enterprises operating in regulated industries. They offer a balance between the benefits of cloud computing, such as agility, scalability, and cost efficiency, and the imperative for control, security, and privacy over sensitive data and workloads. An organization may choose to construct a private cloud independently or enlist companies like Hewlett Packard Enterprise (HPE) to assist in building a private cloud solution. Depending on the organization's needs, either an entire data center or a subset of it, such as a few racks of servers, storage, and networking equipment, can be cloudified. HPE's GreenLake solutions provide a "pay-as-you-go" model for private cloud services, enabling customers to acquire a few rack solutions and deploy them in their data centers or colocation facilities. The following diagram illustrates an example of such private cloud solutions offered by HPE's GreenLake offerings (Fig. 3.21).

The private cloud solution encompasses both hardware and software components. Hardware elements consist of compute, network, and storage racks, along with associated gear. The diagram illustrates compute resources, storage resources, control plane switches, management switches, and the customer's data center leaf switch. HPE facilitates the virtualization of these resources, simplifying customer utilization. This virtualized form of physical infrastructure is referred to as "IaaS" (Infrastructure as a Service). On the right-hand side of the diagram, software components are depicted, including elements for managing physical infrastructure and SDN (Software-Defined Networking). Customers

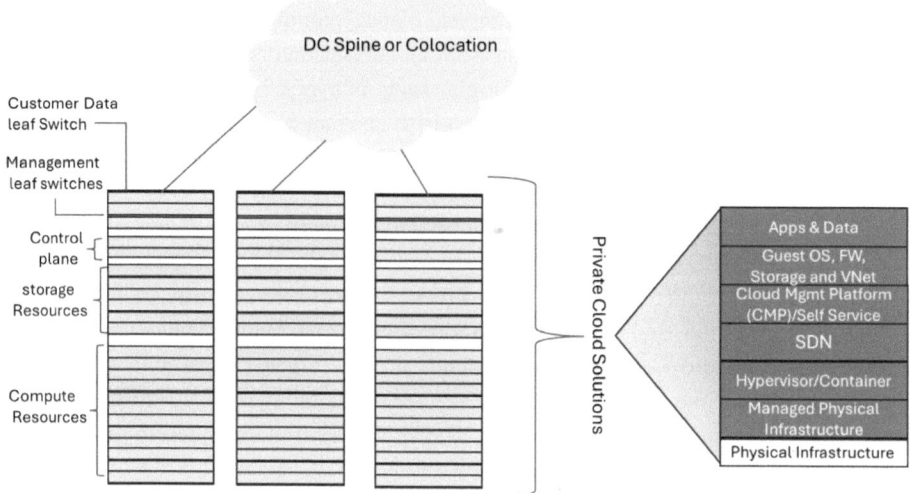

Fig. 3.21 Private cloud solution depicting various components of the solution

have the option to use either a hypervisor or container, both of which are supported by private cloud solutions like HPE's GreenLake. Notably, HPE offers a self-service portal from the public cloud called "GreenLake Central," enabling easier provisioning, orchestration, and control of all virtualized components, including guest OS, firewall, virtual networks (VNet), data, and applications. The third element in private cloud is "As a Service," encompassing the entire package deployed at customer premises and the life cycle management thereof. Customers may opt to manage these aspects themselves or entrust companies like HPE to provide complete solutions for private cloud deployment.

3.6.1   Infrastructure as a Service (IaaS)

Infrastructure as a Service (IaaS) represents a foundational cloud computing model, offering virtualized computing resources accessible via the Internet. Within the realm of private cloud, IaaS extends to furnishing essential infrastructure components like computing, networking, and storage resources as a service within an organization's dedicated data center or private cloud environment.

In the framework of a private cloud setup, IaaS empowers organizations to dynamically allocate and manage their IT infrastructure resources on-demand, circumventing the necessity for physical hardware ownership and upkeep. IaaS encompasses virtualization of resources, self-service provisioning, scalability, isolation, security, customization, control, and the capacity to integrate with existing infrastructure.

Fig. 3.22 IaaS layered concept

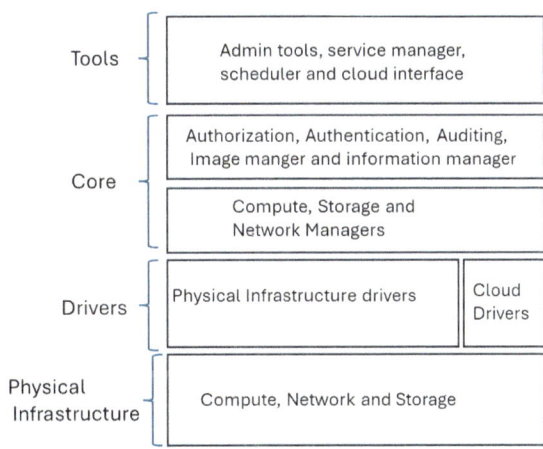

The IaaS layered concept consists of four tiers: physical infrastructure, drivers, core elements, and tools. The ensuing diagram illustrates the IaaS layered concept (Fig. 3.22).

The physical infrastructure incorporates compute, storage, and network devices, which can be virtualized and harnessed through the composable elements of the infrastructure, executed at the driver layer. Furthermore, control and orchestration are provided through the core layer, which includes functions such as compute, storage, and network managers, alongside image managers, authorization, authentication, and auditing functionalities. Additionally, the tool layer furnishes administration tools, service managers, schedulers, and cloud interfaces. Other services offered within the realm of IaaS may include BMaaS (Bare Metal as a Service), VMaaS (VM as a Service), and CaaS (Container as a Service). Unlike traditional IaaS, BMaaS offers direct access to hardware instead of virtualized elements. VMaaS and CaaS, on the other hand, are subsets of IaaS that provide access to virtual machines and containers as resource pools. Additionally, IaaS orchestration platforms like Morpheus provide persona-based access to IaaS for private cloud environments.

3.6.2 Software for Private Cloud

A suite of software tools is essential for private cloud environments, tailored to meet the specific demands and requirements of customers. These tools encompass centralized orchestration and management, operations management, and CI/CD (Continuous Integration/Continuous Deployment/Delivery) capabilities. Certain vendors, such as HPE, offer these tools as part of a comprehensive package solution for private cloud deployment. The diagram below illustrates some of the essential tools required for private cloud environments, which can be customized based on the unique needs and preferences of customers (Fig. 3.23).

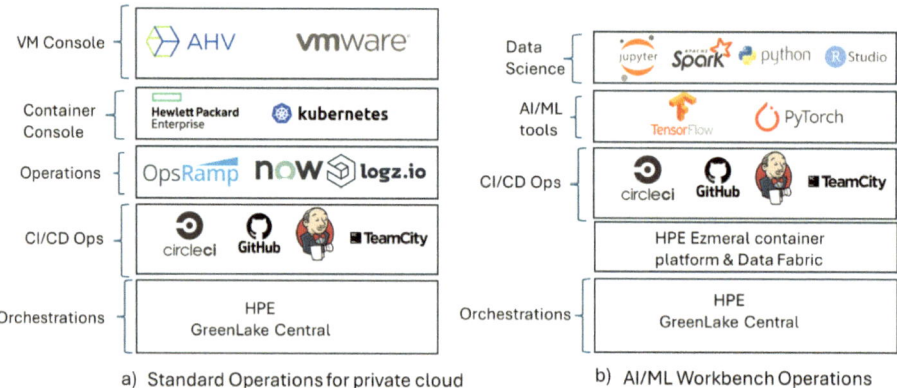

a) Standard Operations for private cloud   b) AI/ML Workbench Operations

Fig. 3.23   Software tools for private cloud operations

Let's consider HPE's GreenLake Private Cloud Enterprise (PCE) solution package as an example to illustrate. The diagram depicts two solution packages: one for standard operations (a) and another for AI/ML workbench operations (b). Without such vendor-led solutions, customers may opt to build their own software tools and packaging for private cloud services. In the diagram, the orchestration layer comprises a vendor-specific centralized management and control platform, such as HPE GreenLake Central. This platform empowers users to determine the placement and management of workloads and data within the private cloud environment. The purchased services offer capabilities for monitoring security, compliance, capacity, resource utilization, and costs.

Moving to the CI/CD layer, various tools like CircleCI, GitHub Actions, Jenkins, and TeamCity are utilized for automating software development processes. Each tool possesses its unique strengths and weaknesses: CircleCI offers simplicity and rapid setup, GitHub Actions provides seamless integration with GitHub, Jenkins allows extensive customization, and TeamCity offers scalability and robustness for large-scale projects. Selecting the appropriate tool depends on the specific requirements and limitations of the project at hand.

The Operations layer encompasses tools such as OpRamp, ServiceNow, and Logz.io, among others. These tools serve distinct purposes within the realm of private cloud operations, providing capabilities that span from infrastructure management to IT service management and log management, respectively.

- **OpRamp:** This software tool is primarily a cloud operations management platform designed to provide visibility and control over hybrid and multi-cloud infrastructure. It offers features such as infrastructure monitoring, event management, automation, and incident response. OpRamp can be used in a private cloud environment to centrally monitor and manage the performance and availability of virtualized infrastructure, ensuring optimal resource utilization and minimizing downtime. Its automation capabilities enable administrators to streamline routine tasks and enforce policies across the private cloud environment.

- **ServiceNow:** It is an IT service management (ITSM) platform that helps organizations streamline and automate various IT processes, including incident management, change management, asset management, and service catalog management. In the context of private cloud operations, ServiceNow can be utilized to manage IT services and resources deployed in the private cloud environment. It facilitates efficient incident resolution, change tracking, and service request fulfillment, thereby improving the overall operational efficiency of the private cloud infrastructure. ServiceNow also offers integration capabilities with other IT management tools, allowing for seamless orchestration and automation of workflows across the private cloud environment.

- **Logz.io:** This is a cloud-based log management and analytics platform that enables organizations to centralize, analyze, and visualize logs generated by various components of their IT infrastructure, including applications, servers, and networking devices. In a private cloud environment, Logz.io can be used to collect, index, and analyze logs from virtualized infrastructure, applications, and services deployed within the private cloud. By gaining insights into log data, organizations can proactively identify and troubleshoot issues, optimize performance, and ensure compliance with regulatory requirements. Logz.io's scalable architecture and advanced analytics capabilities make it well suited for managing log data generated in dynamic and distributed private cloud environments.

For container deployments, customers may choose to use either Kubernetes or HPE Container Platform, depending on their specific requirements and preferences. Kubernetes is a widely adopted open-source container orchestration platform known for its flexibility and scalability, while HPE Container Platform offers additional enterprise features and support services tailored to specific use cases.

Regarding VM console options, customers often have a preference between Nutanix AHV and VMware. Nutanix AHV is a hypervisor developed by Nutanix, known for its simplicity, scalability, and integration with the Nutanix hyper-converged infrastructure platform. VMware, on the other hand, offers a range of virtualization solutions including vSphere, known for its robustness, feature set, and extensive ecosystem of third-party integrations. Customers may evaluate factors such as cost, feature set, and existing infrastructure when selecting between Nutanix AHV and VMware for their virtualization needs.

In a private cloud environment, vendors like HPE GreenLake offer a comprehensive AI/ML workbench solution with a range of tools tailored for data science and machine learning tasks. This includes HPE's proprietary Ezmeral data fabric, along with popular frameworks such as TensorFlow and PyTorch for deep learning applications. Additionally, HPE GreenLake provides tools like Jupyter, Spark, Python, and R Studio to facilitate data exploration, model development, and analysis within the data science workflow. This integrated suite of tools enables organizations to efficiently build, deploy, and manage AI/ML solutions in their private cloud infrastructure.

3.7   Hybrid and Multi-cloud Networks

Hybrid cloud and multi-cloud networks are both strategies for deploying and managing cloud infrastructure, but they differ in their approach and implementation. Hybrid cloud refers to an IT architecture that combines on-premises infrastructure (private cloud) with public cloud services. This approach allows organizations to leverage the scalability and flexibility of public cloud resources while retaining control over sensitive data and applications within their own data centers. A hybrid cloud environment typically involves seamless integration between the private and public clouds, enabling data and workloads to move securely between them as needed. This architecture offers several benefits, including improved agility, cost efficiency, and flexibility, as organizations can dynamically scale their IT resources based on fluctuating demand while maintaining compliance and security requirements. On the other hand, multi-cloud refers to an IT strategy that involves using multiple cloud service providers to host different components of an organization's infrastructure, applications, and services. Unlike hybrid cloud, which combines public and private cloud environments, multi-cloud involves leveraging services from multiple public cloud providers such as AWS, Microsoft Azure, Google Cloud Platform, and others. Organizations adopt a multi-cloud approach for various reasons, including avoiding vendor lock-in, accessing best-of-breed services, optimizing costs, and enhancing redundancy and resilience. However, managing multiple cloud environments can introduce complexity in terms of interoperability, data consistency, security, and governance.

The following list outlines key considerations:

- **Interoperability:** Both hybrid cloud and multi-cloud environments require robust interoperability between different cloud platforms and on-premises infrastructure to ensure seamless data and workload mobility.
- **Data Consistency:** Maintaining data consistency and integrity across multiple cloud environments is crucial for ensuring accurate analytics, compliance, and business continuity.
- **Security and Compliance:** Organizations must implement consistent security measures and compliance controls across hybrid and multi-cloud environments to protect sensitive data and mitigate cyberthreats.
- **Cost Management:** Optimizing costs and resource utilization across hybrid and multi-cloud deployments requires careful monitoring, governance, and automation to avoid unnecessary expenses and ensure cost-effective operations.

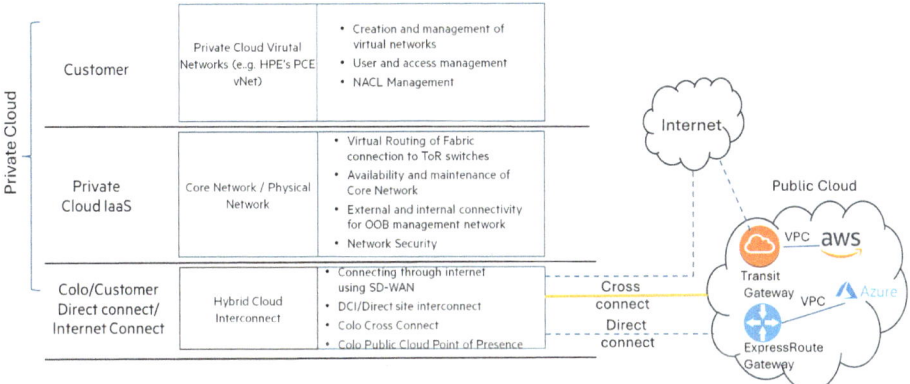

Fig. 3.24 Hybrid cloud solution connecting private cloud directly to public cloud

- **Vendor Management:** Managing relationships with multiple cloud service providers involves evaluating performance, reliability, support, and pricing to ensure alignment with business requirements and objectives.

We discussed about HPE GreenLake PCE solutions in earlier section, this same solution can be extended to both hybrid cloud and multi-cloud solutions. The following diagram depicts hybrid cloud deployment (Fig. 3.24).

For the sake of simplicity in our discussion, we have considered connectivity between a private cloud and a public cloud. However, it's worth noting that a data center can be directly connected to a public cloud, assuming appropriate security measures are in place. Whether a data center extended to the public cloud, a private cloud within a data center, or a standalone deployment connected to the public cloud, all are considered forms of hybrid cloud. Connectivity options may include direct connect, Internet connection, colocation cross-connect, or connecting through public points of presence within a colocation facility. In all cases, VPC connection (as described in Sect. 3.4) is established through a VPN mechanism (refer to Sect. 3.5).

A variation of this deployment model is hybrid multi-cloud, where multiple cloud providers are connected. The figure above illustrates a hybrid multi-cloud scenario, depicting connectivity to both AWS and Azure.

3.7.1   Multi-cloud Networks (MCN)

Multi-cloud deployment differs from hybrid cloud or hybrid multi-cloud setups. In a multi-cloud environment, users have the option to connect directly to various cloud providers from their home, as well as from their organization's data center and private

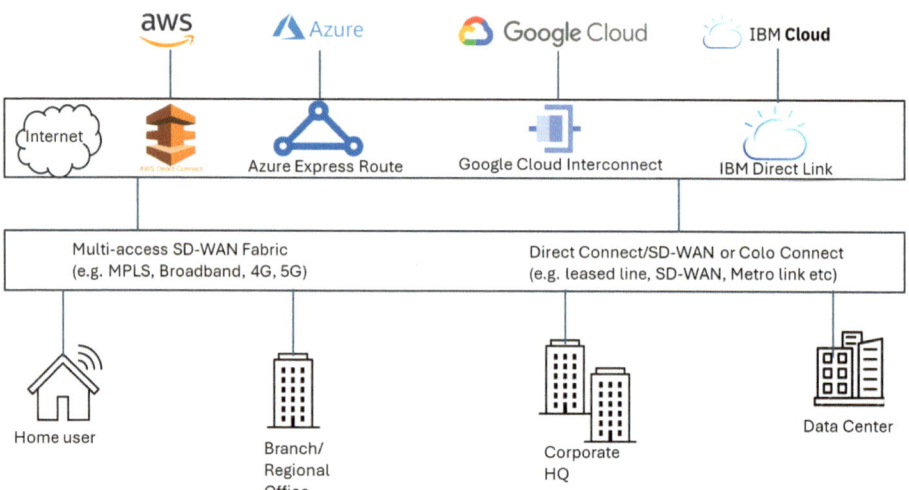

Fig. 3.25 Multi-cloud networking

cloud, extending connectivity to public clouds. However, to ensure seamless operation across multiple cloud environments, a Multi-cloud Network (MCN) solution should provide consistent network policy, security, governance, and visibility through a unified management interface.

SD-WAN is among the technologies employed in conjunction with other VPN technologies, catering to diverse customer connectivity needs for MCN. The diagram below illustrates a typical MCN deployment (Fig. 3.25).

The SD-WAN (Software-Defined Wide Area Network) technology simplifies the management and operation of a wide area network (WAN) by decoupling networking hardware from its control mechanism. Unlike traditional hardware-based network appliances, SD-WAN solutions leverage software to intelligently route traffic across the WAN. They offer centralized management, dynamic path selection, traffic prioritization, WAN optimization, and security, among other features.

For cloud connectivity, SD-WAN facilitates seamless access to cloud-based applications and services by providing direct connectivity to cloud service providers' networks. This eliminates the need to route traffic through the data center, enhancing performance and user experience for cloud-based applications. Additionally, SD-WAN often provides cost-effective solutions for connectivity. As depicted in the diagram above, SD-WAN enables multi-cloud connectivity to remote workforces and branches across various geographical locations. It can utilize a range of physical connectivity technologies, such as service provider MPLS networks, 4G, 5G, and broadband connections. However, SD-WAN is not the only technology used in multi-cloud networks.

For instance, a corporate headquarters and their data center might establish direct connections through leased lines, metro links, or cross-connects within colocation facilities to connect to multiple public cloud providers across different regions. For a deeper understanding of such connectivity options, please refer to the earlier discussions in Sects. 3.4 and 3.5.

3.8 Summary

In this chapter, we explored cloud networking in depth, emphasizing its crucial role in modern IT infrastructure. It begins by highlighting how cloud networking leverages virtualized infrastructure to optimize the management, scaling, and performance of network resources. The chapter discusses public, private, and hybrid cloud environments, each offering distinct advantages in terms of scalability, flexibility, security, and compliance. The historical evolution of cloud networking is traced from early time-sharing systems in the 1960s to pivotal technologies like VMware virtualization and the rise of major cloud service providers such as AWS, Google Cloud, and Microsoft Azure. Virtualization is presented as the backbone of cloud computing, allowing multiple virtual instances of computing, storage, and networking resources to operate efficiently on shared physical infrastructure. This abstraction enhances resource utilization, scalability, and isolation, making it fundamental to modern cloud environments.

Network virtualization technologies such as VLANs, VXLANs, and Network Functions Virtualization (NFV) are explored, showcasing how these innovations enable the creation of isolated, scalable, and secure virtual networks within cloud environments. The chapter delves into the components of NFV, including Virtual Network Functions (VNFs) and the supporting infrastructure (NFVI), underscoring their role in flexible network management.

Storage virtualization is also covered, demonstrating how it abstracts physical storage into logical pools, facilitating efficient data management, mobility, and redundancy. Various forms of storage virtualization—block, file, and object-level—are discussed, illustrating their applications in cloud computing environments.

The concept of Virtual Private Cloud (VPC) is introduced as a secure, customizable virtual network within a public cloud, allowing users to deploy resources with greater control over data security and access. The chapter further examines Virtual Private Networks (VPNs) as essential tools for secure connectivity between on-premises networks and cloud environments, providing safe and encrypted data transmission across different network boundaries.

Private clouds are discussed as exclusive, dedicated environments that offer enhanced control, customization, and security for organizations, particularly those with stringent regulatory requirements. The chapter explores the architecture of private cloud deployments, emphasizing the role of Infrastructure as a Service (IaaS) and the tools necessary for managing private cloud operations.

The chapter concludes with a discussion on hybrid and multi-cloud networks, highlighting how organizations leverage the strengths of multiple cloud environments to optimize performance, reduce costs, and increase resilience. Hybrid clouds combine on-premises and public cloud resources, while multi-cloud strategies involve using multiple cloud providers to avoid vendor lock-in and enhance operational flexibility.

Overall, the chapter provides a comprehensive understanding of cloud networking, equipping readers with the knowledge needed to design, implement, and manage cloud networking infrastructures effectively. The insights gained from this chapter lay the foundation for exploring more advanced cloud networking strategies in subsequent chapters.

References

1. Popek GJ, Goldberg RP (1974) Formal requirements for virtualizable third generation architectures. Association for Computing Machinery, Inc. http://www.logos.ic.i.u-tokyo.ac.jp/~tau/lecture/os/gen/papers/p412-popek.pdf
2. Seawright LH, MacKinnon RA(1979) VM/370-a study of multiplicity and usefulness. IBM Syst J 18(1). http://cseweb.ucsd.edu/classes/wi08/cse221/papers/seawright79.pdf
3. VMware (2013) VSphere: compute. http://www.vmware.com/products/vsphere/features-compute
4. Fiuczynski EM (2009) Virtual machine monitor: CS318. Department of Computer Science, Princeton University. http://www.cs.princeton.edu/courses/archive/fall09/cos318/lectures/VirtualMachine.pdf
5. Robin SJ, Irvine EC (2000) Analysis of the intel pentium's ability to support a secure virtual machine monitor. Center for Information Systems Security Studies and Research Computer Sciences Department Naval Postgraduate School Monterey, CA
6. Arndt LR (2005) Hypervisor function sets: US patent "6892383 B1." International Business Machine Corporation
7. Arpaci-Dusseau R, Arpaci-Dusseau A (2013) Operating systems: three easy pieces. Remzi Arpaci-Dusseau and Andrea Arpaci-Dusseau
8. Bugnion E, Devine S, Govil K, Rosenblum M (1997) Disco: running commodity operating systems on scalable multiprocessors. ACM Trans Comput Syst 15(4):412–447. http://research.cs.wisc.edu/areas/os/Qual/papers/disco.pdf
9. Xiong H, Liu Z, Xu W, Jiao S (2012) Libvmi: a library for bridging the semantic gap between guest OS and VMM. In IEEE 12th conference on computer and information technology
10. Sawazaki J, Maeda T, Yonezawa A (2010) Implementing a hybrid virtual machine monitor for flexible and efficient security mechanisms. In Pacific rim international symposium on dependable computing. IEEE Computer Society
11. Weng C, Wang Z, Li M, Lu X (2009) The hybrid scheduling framework for virtual machinesystems. VEE09; ACM—78-1-60558-375-4/09/03. http://www.cs.columbia.edu/weng-cl/index.files/vee09-paper.pdf

12. Armstrong B (2006) VMMs versus hypervisors. Ben Armstrong's virtualization blog. Microsoft Corporation. http://blogs.msdn.com/b/virtual_pc_guy/archive/2006/07/10/661958.aspx

13. ETSI (2012) Network functions virtualisation (NFV); architectural framework: ETSI GS NFV 002 V1.2.1 (2014–12). https://www.etsi.org/deliver/etsi_gs/NFV/001_099/002/01.02.01_60/gs_NFV002v010201p.pdf

14. Google (2024) Google cloud VPN overview. https://cloud.google.com/network-connectivity/docs/vpn/concepts/overview

Container Networking

4

In the ever-evolving landscape of modern computing, containers have risen as the bedrock of application deployment, offering unparalleled agility, scalability, and efficiency. With organizations increasingly embracing containerized architectures to streamline operations, the demand for tailored networking solutions becomes paramount. Moreover, networking infrastructure must adapt to accommodate microservices architectures, scaling out connectivity for container pods, such as those in Kubernetes environments. Concurrently, addressing network security for container pods and facilitating support for new applications within containerized networking devices are pressing needs in modern network infrastructure.

This chapter embarks on a comprehensive exploration into the realm of container networking, delving deep into its fundamentals, challenges, and innovative solutions. From grasping the rudiments of container networking to navigating the complexities of orchestrating communication among distributed containerized applications, this introduction lays the groundwork for a journey through the intricate web of container networking.

Over the following pages, we will voyage through the multifaceted layers of container networking, unraveling its complexities and unveiling strategies to optimize performance, security, and resilience. Whether you're a seasoned DevOps engineer aiming to bolster your container networking expertise or a newcomer eager to grasp foundational concepts, this chapter aims to furnish invaluable insights and practical guidance for navigating the dynamic landscape of container networking effectively.

Join us as we traverse the digital highways and byways of container networking, forging connections, surmounting challenges, and paving the path for a future where containerized applications seamlessly communicate and collaborate in the digital realm.

D. D. Chowdhury, *Future of Networks*, Synthesis Lectures on Communications, https://doi.org/10.1007/978-3-031-71440-5_4

4.1   Evolution of Containers and Microservices

The evolution of containers and microservices architecture represents a transformative shift in how applications are developed, deployed, and managed. This evolution has been driven by a combination of technological advancements, changing business requirements, and the need for greater agility and scalability in software development. The early roots of containerization can be traced back to "Chroot", short for "change root", which was developed and added in Version 7 Unix in 1979. It functions as both a Unix system call and command, enabling the alteration of the apparent root directory for the current running process and its children. When a process operates within a chroot environment, it perceives the specified directory as the root of the filesystem hierarchy, effectively isolating it from the remainder of the system.

Chroot was developed with the aim of bolstering the security and isolation capabilities of Unix-like operating systems. Its primary objective is to establish a confined environment in which a process can operate with restricted access to the filesystem. By employing chroot, administrators can create a segregated environment that limits a process's access to certain parts of the filesystem, enhancing system security.

The following diagram illustrates the Linux Standard File System and the nested filesystem structure after the execution of the chroot() command (Fig. 4.1).

Chroot's ability to create isolated environments paved the way for subsequent advancements in containerization technology, laying the groundwork for the development of more

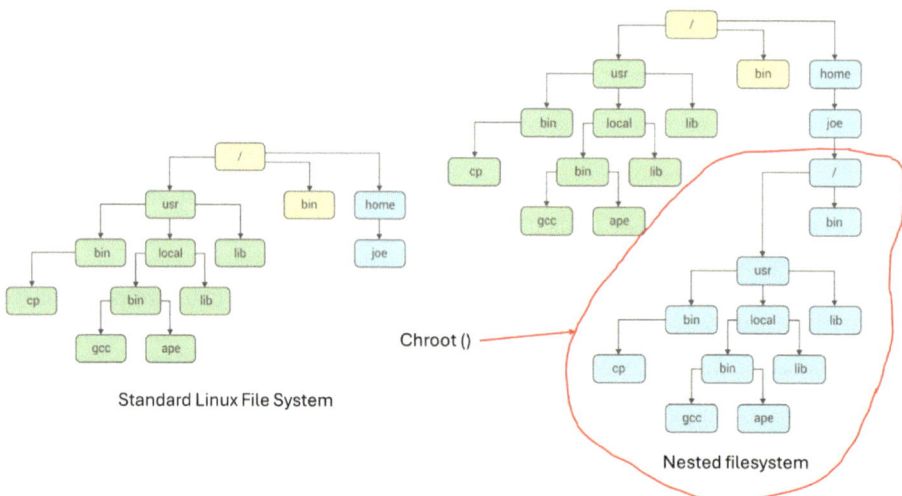

Fig. 4.1 The diagram depicts standard linux file system and nested filesystem using "chroot" command

sophisticated containerization solutions such as Docker and Kubernetes. This isolation offered several benefits:

- **Security:** By limiting the access of a process to a specific directory and its subdirectories, chroot helps prevent unauthorized access to sensitive system files and directories. This confinement reduces the risk of security breaches and protects the integrity of the system.
- **Testing and Development:** Chroot is commonly used in software development and testing environments to create isolated environments for building and testing software. Developers can install dependencies and experiment within the chroot environment without affecting the rest of the system.
- **System Recovery:** In the event of system corruption or compromise, administrators can use chroot to boot into a minimal environment with a clean filesystem. This allows them to diagnose and repair the system without the risk of further damage from malicious software or corrupted files.
- **Software Installation:** Chroot can also be used for installing and running software packages in a controlled environment. This is particularly useful for installing software that may have conflicting dependencies or for running legacy applications on modern systems.

Chroot stands as the precursor to modern containerization, introducing the concept of isolation well before namespaces and cgroups were developed. While namespaces and cgroups have since become the cornerstone of containerization technology, the historical significance of chroot in establishing isolated environments remains undeniable. It's worth noting that the term "chroot" is derived from "change root," underscoring its fundamental role in altering the root of a filesystem to establish a confined environment.

While Chroot serves as the precursor to containerization, virtual machines (VMs) and subsequent system virtualization emerged in 2000, as discussed in Chap. 3, offering a method to run multiple operating systems on a single physical machine. However, VMs, while providing isolation, were resource-intensive and slow to boot, leaving a gap in the development and deployment of applications. Containers emerged as a solution to address these shortcomings. Unlike VMs, containers share the host operating system kernel, resulting in more lightweight and efficient deployments. With containers, developers could package their applications along with their dependencies into portable, self-contained units, allowing for consistent deployment across different environments.

The following diagram illustrates a timeline depicting the evolution of containers and microservices. For simplicity of discussion, we've highlighted major milestones and omitted many works in the field (Fig. 4.2).

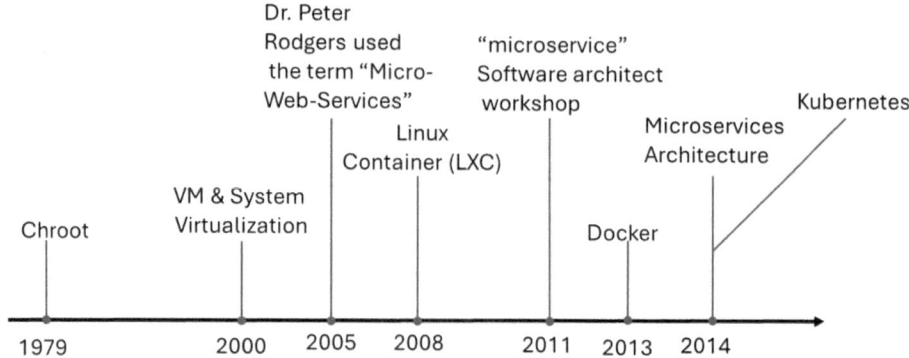

Fig. 4.2 The timeline of container and microservices development

As depicted in the timeline, the origin of the term "microservices" can be attributed to a presentation by Dr. Peter Rodgers at a cloud computing conference. Dr. Rodgers introduced the concept of "micro-web services,"[1] advocating for software components that support micro-web services. His presentation established a functional model of microservices, which eventually became a reality, revolutionizing the way software is developed and deployed in modern computing environments.[2] However, the term "microservices" was officially coined during a software architects meeting in May 2011 near Venice, Italy. This event marked a significant milestone in the evolution of software architecture, as it provided a clear and concise term to describe the architectural style of decomposing applications into smaller, independently deployable services. Dr. Peter Rodgers' earlier presentation on "micro-web services" laid the conceptual groundwork for this architectural approach, but it was during this meeting that the term "microservices" gained widespread recognition and adoption within the software development community.

Meanwhile, in 2008, Linux Containers (LXC) introduced lightweight process isolation using kernel namespaces and control groups (cgroups). This technology revolutionized the way applications were deployed and managed, providing an efficient and lightweight alternative to virtual machines. While VMs offered isolation, they were resource-intensive and slow to boot, leaving a gap in the development and deployment of applications. Linux Containers, or simply containers, addressed these shortcomings by sharing the host operating system kernel, resulting in more lightweight and efficient deployments. With containers, developers could package their applications along with their dependencies into portable, self-contained units, enabling consistent deployment across different environments.

In 2013, Docker emerged as a game-changer in the world of containerization. Docker simplified the process of creating, distributing, and running containers by providing a

---

[1] Vsourz [1].

[2] Foote [2].

user-friendly interface and standardized container images. Docker's rise in popularity accelerated the adoption of containerization, making it accessible to a wider audience of developers and organizations. Following the emergence of containerization, the microservices architecture gained traction as a scalable and flexible approach to building and deploying applications. Microservices architecture decomposes applications into smaller, loosely coupled services that can be developed, deployed, and scaled independently. This architectural style aligns well with containerization, as containers provide the lightweight isolation needed to deploy microservices at scale.

In 2014, Kubernetes, an open-source container orchestration platform, was released by Google. Kubernetes automates the deployment, scaling, and management of containerized applications, providing features such as service discovery, load balancing, and self-healing. Kubernetes quickly became the de facto standard for container orchestration, further accelerating the adoption of containers and microservices architecture in modern software development.

4.1.1   Linux Container

Linux Containers (LXC) introduced a groundbreaking approach to process isolation and resource control within the Linux operating system. Developed in 2008, LXC leverages kernel features such as namespaces and control groups (cgroups) to create lightweight, isolated environments known as containers. The following is a diagrammatical representation of LXC environment (Fig. 4.3).

At its core, LXC utilizes kernel namespaces to provide process isolation, filesystem isolation, network isolation, and hostname isolation. Namespaces enable the illusion of separate namespaces for various system resources, ensuring that processes within a container have a restricted view of the system. For example, a container can have its own

Fig. 4.3 The LXC environment

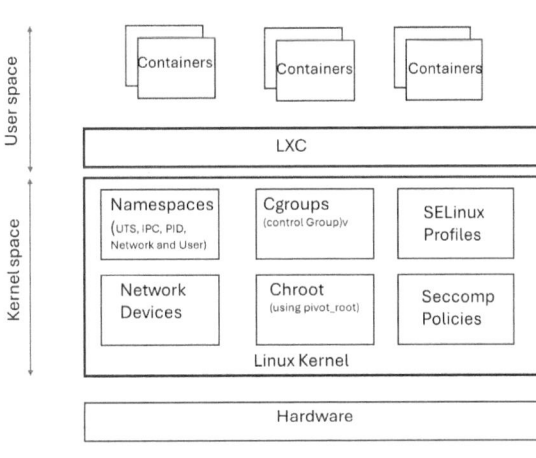

process ID (PID) namespace, which means processes running inside the container are unaware of processes running outside the container. This isolation prevents processes in one container from affecting processes in other containers or the host system.

Additionally, LXC leverages control groups (cgroups) to manage resource allocation and utilization for containers. Cgroups allow administrators to limit the CPU, memory, disk I/O, and network bandwidth usage of containers, ensuring fair allocation of resources among multiple containers running on the same host. With Cgroups, administrators can enforce resource quotas, prioritize critical workloads, and prevent resource contention between containers. Furthermore, LXC provides a user-friendly command-line interface and APIs for managing containers. Administrators can use commands like lxc-create, lxc-start, lxc-stop, and lxc-destroy to create, start, stop, and destroy containers, respectively. LXC also supports configuration files (typically in YAML format) that define container properties such as network configuration, storage configuration, and resource limits. One of the key advantages of LXC is its efficiency and low overhead. Since LXC containers share the same kernel as the host operating system, they incur minimal performance over-head compared to virtual machines. This lightweight approach makes LXC well suited for scenarios where high-density container deployments and low latency are essential, such as cloud computing environments and microservices architectures. Moreover, LXC is highly flexible and extensible, allowing developers and system administrators to cus-tomize container environments to meet specific requirements. Advanced users can create custom templates, define custom storage backends, and integrate LXC with other tools and technologies to build robust container ecosystems.

4.1.1.1 Docker

While LXC provided a neat and powerful interface at the userspace level, it was and still is not easy to use. Moreover, it didn't generate mass interest or garner widespread adoption like later containerization technologies such as Docker. Several factors contributed to this:

- **Complexity:** LXC, while powerful, was relatively complex to use compared to later containerization solutions. It required users to have a deep understanding of Linux kernel features such as namespaces and Cgroups, which posed a barrier to entry for many developers and operators.
- **Usability:** LXC lacked user-friendly tools and interfaces for container management. Working with LXC containers often involved using command-line tools and manual configuration, which made it less approachable for users accustomed to more intuitive interfaces.
- **Limited Ecosystem:** LXC had a smaller ecosystem compared to later containerization platforms like Docker. It lacked a centralized repository for sharing and distributing container images, which hindered collaboration and adoption among developers.
- **Community and Marketing:** Docker, on the other hand, benefited from a strong community and effective marketing efforts. It gained momentum through developer

advocacy, community engagement, and strategic partnerships, which helped drive adoption and awareness.
- **Standardization:** Docker introduced standardized container images and a simple, portable format for packaging applications and their dependencies. This standardization made it easier for developers to create and share containerized applications, further accelerating Docker's adoption.

Docker emerged as a transformative force in the world of containerization, addressing many of the usability issues that existed with earlier technologies like LXC. While LXC provided powerful capabilities for process isolation and resource control within the Linux operating system, it struggled to gain widespread adoption due to its complexity and lack of user-friendly interfaces. Docker changed the game by abstracting away many of the complexities involved in container management, making containerization accessible to a wider audience of developers and operators. By providing a simple and intuitive interface, Docker democratized the process of creating, distributing, and running containers. Developers no longer needed an in-depth understanding of Linux kernel features like namespaces and Cgroups to leverage the benefits of containerization.

One of the key innovations of Docker was its introduction of standardized container images and a portable format for packaging applications and their dependencies. This standardization made it easier for developers to create, share, and distribute containerized applications, fostering collaboration and accelerating the adoption of containerization across industries. Furthermore, Docker came with a rich ecosystem of tools and services, including Docker Hub, a cloud-based registry service for storing and sharing container images. Docker Hub provided a central repository where developers could discover, download, and contribute container images, further enhancing the collaboration and distribution of containerized applications.

With Docker, developers gained support for automatically building, versioning, and reusing containers, streamlining the workflow for packaging and deploying applications. Docker's user-friendly approach and robust ecosystem contributed to its rapid rise in popularity, making it the de facto standard for containerization in modern software development and deployment.

Relation to LXC

The Docker project, initiated by Solomon Hykes as part of dotCloud, a platform-as-a-service company, was later released as an open-source project in 2013. Initially, Docker utilized LXC as its default execution environment. However, this reliance on LXC was short-lived. Close to a year later, LXC was replaced with an in-house execution environment, libcontainer, written in the Go programming language.

Although Docker no longer uses LXC as its default execution environment, it maintains compatibility with LXC, as well as other isolation tools like libvert and systemd-nspawn.

Fig. 4.4 Docker with LXC

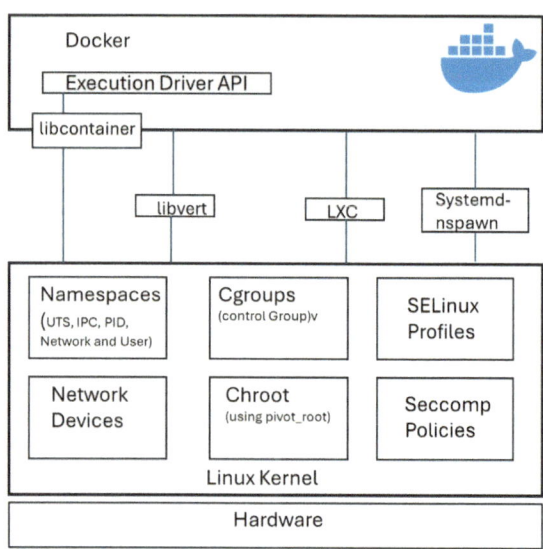

This compatibility is made possible through the use of an execution driver API. Additionally, Docker's adaptability extends to non-Linux systems, thanks to this API, enabling Docker to run seamlessly across various platforms.[3,4] The diagram below illustrates Docker deployed alongside LXC, libvirt, and systemd-nspawn. Through libcontainer, Docker gains the capability to directly manage namespaces, Cgroups, SELinux profiles, network interfaces, and firewall rules in a controlled and predictable manner. Unlike previous dependencies like LXC, libcontainer empowers Docker to handle these functionalities internally, enhancing its autonomy and flexibility (Fig. 4.4).

In contrast to Docker, LXC is lightweight and offers bare metal performance, making it particularly suitable for resource-intensive applications. Docker, on the other hand, excels in rapid deployment and portability across diverse environments. While LXC provides greater performance efficiency, Docker's strength lies in its ease of use, flexibility, and compatibility with different infrastructures. Depending on the specific requirements of an application, developers may choose between LXC for performance-driven workloads and Docker for streamlined deployment and flexibility.

Docker Architecture

While Docker's introduction is relatively recent, its core architecture has undergone multiple evolutions. Notably, we observed the transition from LXC to libcontainer as the default execution environment. Although there have been additional alterations, we will focus solely on Docker's current architecture to maintain clarity. The Docker architecture

---

[3] Swan [3].

[4] Chandrakant [4].

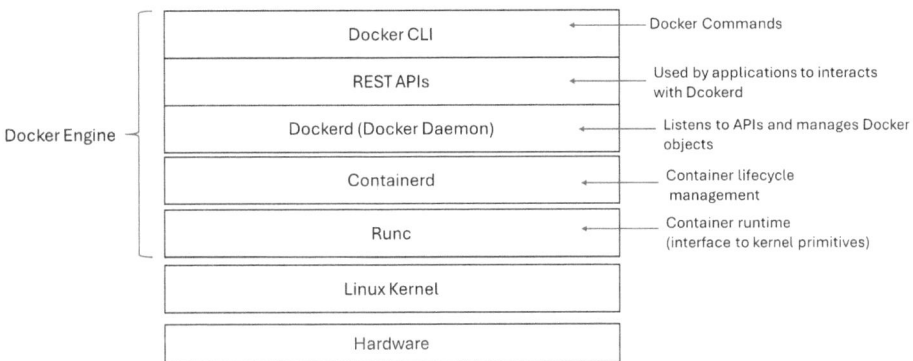

Fig. 4.5   Docker engine architecture

encompasses several essential components: Docker CLI, REST APIs, Dockerd (Docker Daemon), Containerd, and Runc, all illustrated in the accompanying diagram (Fig. 4.5).

Together, these elements constitute the "Docker Engine," functioning collaboratively to deliver a robust containerization solution. Let's delve into the specific roles of each component:

- **Docker CLI (Command-Line Interface):** The Docker CLI is a command-line tool that allows users to interact with the Docker Engine. It provides a user-friendly interface for managing Docker containers, images, networks, volumes, and other resources. With the Docker CLI, users can perform actions such as building Docker images, running containers, managing container lifecycles, and inspecting container logs.
- **REST APIs:** Docker exposes a set of RESTful APIs that enable programmatic interaction with the Docker Engine. These APIs allow developers to automate Docker tasks, integrate Docker with other tools and services, and build custom applications on top of Docker. With the Docker REST APIs, developers can perform actions such as creating and managing containers, querying container status, inspecting Docker objects, and controlling Docker daemon behavior.
- **Dockerd (Docker Daemon):** Dockerd, also known as the Docker Daemon, is a background process that runs on the host system and manages Docker objects. It listens for Docker API requests, handles container lifecycle management, manages container networking and storage, and coordinates interactions with the container runtime. Dockerd is responsible for creating and managing containers, scheduling container tasks, and enforcing security policies defined by Docker.
- **Containerd:** Containerd is an industry-standard container runtime used by Docker to manage container execution. It is responsible for low-level container operations, including container lifecycle management, image distribution, and container snapshotting.

Containerd abstracts away the complexities of interacting directly with container runtimes like runc, providing a standardized interface for container management. Docker leverages Containerd to ensure compatibility with different container runtimes and to provide a stable and reliable container execution environment.

- **Runc:** Runc is a lightweight, portable container runtime used by Docker to execute container processes. It implements the Open Container Initiative (OCI) runtime specification, which defines a standardized interface for container runtimes. Runc is responsible for creating and running container processes, setting up container namespaces and control groups, managing container filesystems, and handling container lifecycle events. Docker uses Runc as its default container runtime, ensuring consistency and interoperability across different containerization platforms and environments.

In summary, the Docker Engine components collaborate synergistically to offer a comprehensive containerization solution, enabling users to efficiently build, deploy, and manage containerized applications.

4.1.2 Microservices

Traditional monolithic architectures, where an application is built as a single, tightly coupled unit, began to show limitations as applications grew in size and complexity. Monolithic applications were difficult to scale, deploy, and maintain, leading to long release cycles and high operational overhead. Microservices architecture emerged as an alternative approach to building and deploying software. In a microservices architecture, applications are decomposed into smaller, loosely coupled services that can be developed, deployed, and scaled independently. This approach enables teams to iterate quickly, scale more efficiently, and deliver value to customers faster (Fig. 4.6).

The microservices architecture breaks down an application into smaller, focused services that communicate through well-defined APIs. These services handle specific tasks, such as user authentication, data storage, or order processing. Each microservice typically encapsulates a single business capability and manages its own data, business logic, and, if applicable, user interface. Collaboration among individual services to handle end-to-end user requests is a primary concern in microservices architectures. To facilitate communication between these services, a communication layer such as RabbitMQ or Kafka is commonly employed. Furthermore, each microservice typically defines an API specification outlining how it can be interacted by other services.

Containers and microservices architecture complement each other, with containers providing lightweight isolation necessary for deploying microservices at scale. Containers package individual microservices along with their dependencies, simplifying the management and scalability of complex, distributed systems. Container orchestration platforms like Kubernetes have emerged to automate the deployment, scaling, and management of

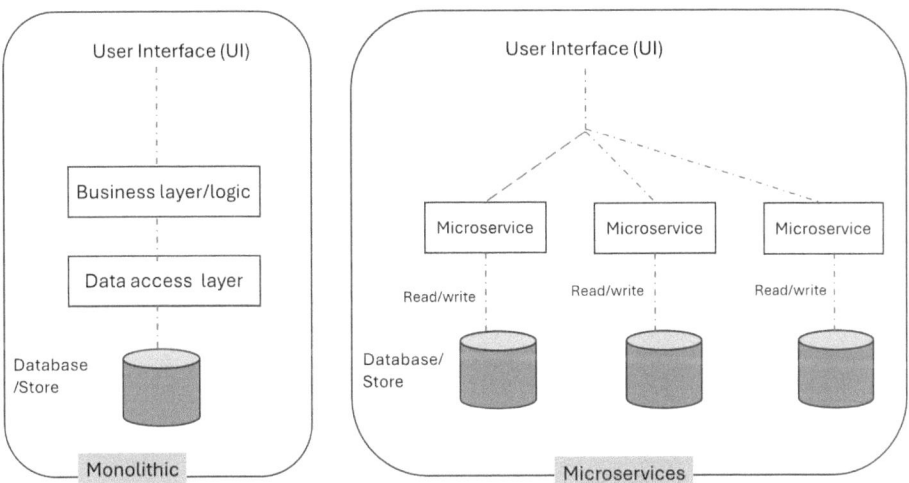

Fig. 4.6 Monolithic versus microservices architecture

containerized microservices. Kubernetes offers features such as service discovery, load balancing, and self-healing, enabling teams to deploy and manage microservices at scale effortlessly.

4.1.3   Kubernetes

Kubernetes, often abbreviated as K8s, is an open-source container orchestration platform designed to automate the deployment, scaling, and management of containerized applications. Originally developed by Google and later donated to the Cloud Native Computing Foundation (CNCF), Kubernetes has become the de facto standard for container orchestration in the industry.

At its core, Kubernetes provides a flexible and extensible platform for deploying and managing containerized workloads across a cluster of machines. It abstracts away the underlying infrastructure and provides a unified API for managing containers, making it easier to deploy and scale applications. One of the key features of Kubernetes is its ability to manage containerized applications in a highly resilient and efficient manner. Kubernetes ensures high availability by automatically restarting containers that fail, replacing containers that do not respond to health checks, and distributing workloads evenly across the cluster to optimize resource utilization. Kubernetes also offers powerful networking and storage capabilities, allowing containers to communicate with each other securely and access persistent storage volumes. It provides built-in support for service discovery, load balancing, and traffic routing, making it easier to build distributed and scalable applications.

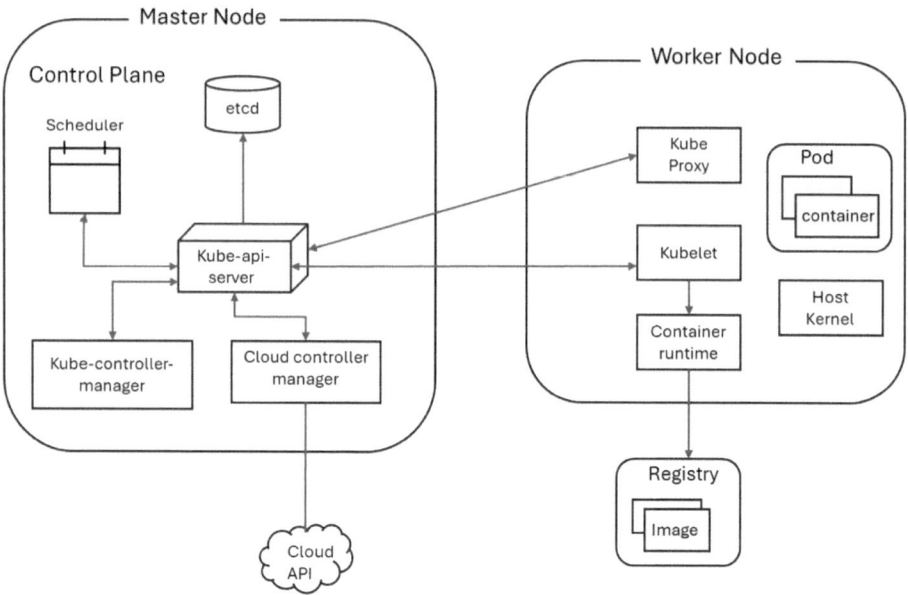

Fig. 4.7 Kubernetes architecture depicting different components of Kubernetes cluster

Another notable feature of Kubernetes is its declarative configuration model, which allows users to define the desired state of their applications using YAML or JSON manifests. Kubernetes then takes care of reconciling the actual state of the cluster with the desired state, ensuring that applications are always running as intended. Furthermore, Kubernetes supports a wide range of deployment patterns, including rolling updates, blue-green deployments, and canary releases, allowing for seamless and non-disruptive updates to applications (Fig. 4.7).

The architecture of Kubernetes, as illustrated in the diagram above, is organized into a cluster comprising different nodes, which include both Master Nodes and Worker Nodes. The Master Node hosts all control plane components, including the scheduler, etcd, kube-api-server, Kube-controller-manager, and Cloud Controller Manager. Conversely, the Worker Node houses components such as Kube Proxy, Kubelet, the container runtime, and Pods.

The Master Node in a Kubernetes cluster plays a central role in orchestrating and managing the deployment and operation of containerized applications across the cluster. The Scheduler, located within the Master Node, is responsible for assigning Pods (groups of containers) to available Worker Nodes based on resource requirements, affinity, and anti-affinity rules. Etcd, a distributed key-value store, stores the configuration data and state of the entire cluster, serving as the cluster's database. The kube-api-server exposes Kubernetes APIs that clients, including the kubectl command-line tool and the

Kubernetes Dashboard, use to interact with the cluster. Additionally, the Kube-controller-manager manages various controllers that regulate the state of the cluster, such as the Replication Controller, Namespace Controller, and Service Account Controller. The Cloud Controller Manager interacts with the underlying cloud infrastructure (if applicable) to manage cloud-specific resources such as load balancers, storage volumes, and virtual machines.

On the Worker Node, Kube Proxy runs on both the Master and Worker Nodes, maintaining network rules to enable communication between Pods across the cluster. The Master Node interacts with Kube Proxy to configure network policies and routes. The Kubelet, running on each Worker Node, manages the lifecycle of Pods on that node. The Master Node communicates with Kubelet to schedule Pods, monitor their health, and enforce the desired state. The container runtime is responsible for running containers within Pods. The Master Node communicates with the container runtime (such as Docker or containerd) on each Worker Node to start, stop, and manage containers as specified by the Kubernetes API. Pods are the smallest deployable units in Kubernetes, consisting of one or more containers that share networking and storage resources. The Master Node instructs the Worker Nodes to create, update, and delete Pods as needed based on scheduling decisions and workload requirements.

In summary, the Master Node acts as the brain of the Kubernetes cluster, overseeing the scheduling, deployment, and management of containerized applications. It interacts with the Worker Nodes to distribute workloads, monitor their status, and maintain the desired state of the cluster, ensuring efficient and reliable operation.

4.1.4   Importance of Container Networking

Container networking plays a pivotal role in modern computing and network infrastructure, serving as a fundamental component in the deployment and operation of containerized applications. As organizations increasingly adopt containerized architectures to streamline their operations, the need for robust networking solutions tailored to the unique demands of containerized environments becomes paramount. One of the key aspects of container networking is its ability to facilitate seamless communication between containers, both within a single host and across distributed systems. Containers typically run as isolated instances, each with its own network namespace, allowing them to have their own network stack and IP address. This isolation enables containers to communicate with each other while remaining independent of the underlying host system.

Container networking also offers significant advantages in terms of scalability and efficiency. By abstracting networking functionality into lightweight, portable units, containers enable developers to deploy and scale applications more easily and efficiently. Containers can be rapidly provisioned and torn down as needed, allowing for dynamic allocation of resources and improved utilization of infrastructure resources. Moreover, container networking supports microservices architectures, where applications are decomposed into smaller, loosely coupled services that communicate with each other over the network. This approach enables teams to develop, deploy, and update individual services independently, leading to greater agility, scalability, and resilience.

In modern network infrastructure, container networking introduces new challenges and considerations. As the number of containers and microservices grows, managing network connectivity and security becomes increasingly complex. Container networking solutions must provide mechanisms for service discovery, load balancing, network segmentation, and traffic routing to ensure optimal performance and reliability. Security is another critical aspect of container networking. Containers share the same kernel and underlying infrastructure, making them potentially vulnerable to security threats such as container escape attacks and network breaches. Container networking solutions must implement robust security measures, including network segmentation, encryption, and access control, to protect sensitive data and applications from unauthorized access.

Several technologies and solutions have emerged to address the challenges of container networking. Container orchestration platforms like Kubernetes provide built-in networking capabilities, including support for overlay networks, network policies, and service discovery. Additionally, software-defined networking (SDN) solutions offer advanced networking features such as virtual network overlays, traffic management, and security policies tailored to containerized environments.

In the upcoming sections, we will explore container networking in depth, focusing on its relevance to modern computing environments. We'll examine network guidelines tailored for such environments, as well as how network functions are virtualized and delivered through containers in modern network infrastructure gear, particularly in L2/L3 network switches. By delving into these topics, we aim to provide a comprehensive understanding of container networking's role in both computing and network infrastructure landscapes.

4.1.5 Role of Container Networking in Microservices Architecture

Container networking plays a pivotal role in microservices architecture, providing the essential connectivity infrastructure that allows individual microservices to communicate and collaborate effectively within a distributed system. In microservices architecture, applications are decomposed into smaller, loosely coupled services, each responsible for a specific business function. These services communicate with each other over the network, often asynchronously, to fulfill complex business processes. The seamless communication between microservices is facilitated by container networking, which enables the creation of networked environments where services can discover, connect to, and interact with each other. Container networking ensures that microservices can communicate securely and efficiently, regardless of their location or underlying infrastructure.

One of the key benefits of container networking in microservices architecture is its ability to support service discovery and dynamic routing. As microservices are deployed and scaled dynamically, container networking solutions automatically discover and register services, making them accessible to other services within the system. This dynamic service discovery enables microservices to locate and communicate with each other without manual intervention, promoting agility and flexibility in application development and deployment. Additionally, container networking provides mechanisms for load balancing and traffic routing, ensuring that requests are distributed evenly across multiple instances of a service. This helps optimize resource utilization and improve application performance by directing traffic to healthy and available instances of a service.

Container networking also plays a crucial role in securing microservices communication. By implementing network policies, encryption, and access control mechanisms, container networking solutions help protect sensitive data and prevent unauthorized access to microservices. Secure communication between microservices is essential for maintaining the integrity and confidentiality of data in distributed systems. Furthermore, container networking enables the implementation of advanced networking patterns such as circuit breaking, retries, and timeouts, which enhance the resilience and fault tolerance of microservices. These patterns allow microservices to gracefully handle failures and recover from errors, ensuring the reliability and availability of the overall system.

In summary, container networking is integral to the success of microservices architecture, providing the connectivity infrastructure that enables seamless communication, dynamic service discovery, load balancing, and security. By leveraging container networking solutions, organizations can build scalable, resilient, and agile microservices-based applications that meet the demands of modern distributed systems.

4.2   Fundamentals of Container Networking

Engaging with containers can evoke a sense of wonderment, akin to magic, for those well versed in their inner workings, yet may provoke apprehension for those less familiar. Fortunately, we've explored containerization technology, revealing that containers are essentially isolated and constrained Linux processes.

Now, let's delve into the intricacies of container networking, beginning with LXC container networking, followed by Docker and Kubernetes. LXC networking is a vital component of managing and connecting containers within a Linux environment. It utilizes various networking features and configurations to enable communication between containers and the external environment. These include network namespaces, virtual Ethernet (veth) pairs, bridge interfaces, IP address management, and network bridge configuration tools.

The concept of network namespaces plays a crucial role in LXC container networking. Network namespaces allow the creation of isolated network environments within a

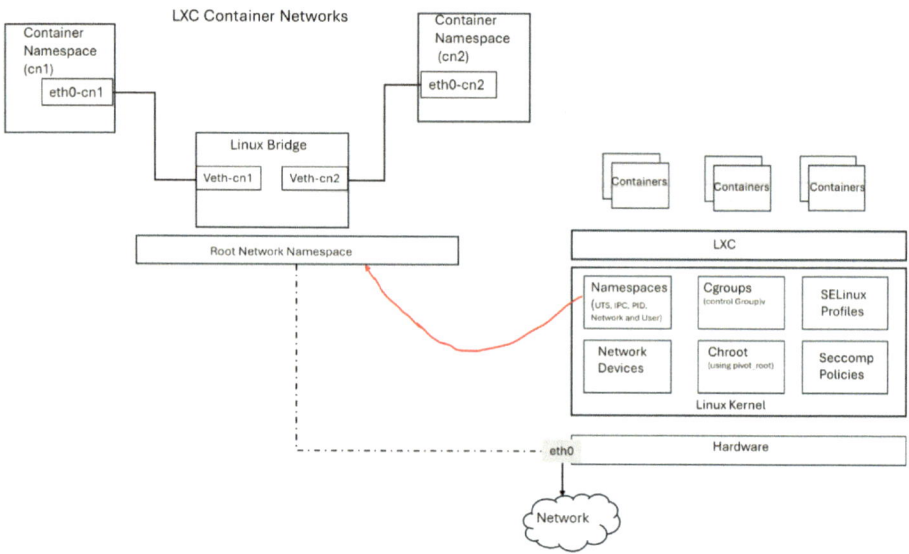

Fig. 4.8 Linux container (LXC) networking using network namespace

single operating system. Each network namespace possesses its own network devices, IP addresses, routing tables, and firewall rules. LXC leverages network namespaces to establish isolated network environments for each container, ensuring that they operate independently of each other and the host system (Fig. 4.8).

The diagram illustrates how two LXC containers are linked through network namespaces, with each container possessing its own dedicated network namespace. This configuration ensures seamless communication while upholding isolation, thereby facilitating efficient container networking within a Linux environment. Such a setup lays the groundwork for scalable and secure containerized applications. Furthermore, LXC employs Linux bridges to enable communication among multiple containers on the same host. A bridge interface functions as a virtual switch, connecting various veth interfaces from different containers to the host's physical network interface. Consequently, containers can interact with each other and external networks via the host's network interface.

LXC also offers diverse IP address management techniques, including static IP assignment, DHCP (Dynamic Host Configuration Protocol), and IP masquerading. Containers can be assigned static IP addresses or configured to acquire IP addresses dynamically from a DHCP server. IP masquerading (NAT) facilitates the access of containers with private IP addresses to external networks via the host's public IP address. To configure network bridges and manage network interfaces on the host system, LXC provides utilities such as "brctl" (bridge control) and "ip" (networking configuration tool). These tools are commonly utilized to create and administer network bridges, add interfaces to bridges, and configure IP addresses and routes, ensuring robust and efficient networking within LXC environments.

4.2.1 LXC Network Types

LXC (Linux Containers) supports various types of networks, each tailored to different use cases and networking requirements. Here are the different types of LXC networks:

- **Bridged Networking:** Bridged networking allows containers to communicate with external networks as if they were directly connected to the host's physical network interface. In this configuration, each container is assigned its own IP address on the external network, enabling seamless communication with other devices on the network. Bridged networking is ideal for scenarios where containers need full network access and visibility.
- **Host-Only Networking:** Host-only networking restricts communication between containers and the external network, isolating them from other devices on the network. In this setup, containers can communicate with each other and with the host system but cannot access external networks directly. Host-only networking is commonly used for testing and development environments where isolation is desired.
- **NAT (Network Address Translation) Networking:** NAT networking allows containers to access external networks through the host's IP address. In this configuration, the host acts as a gateway for outbound traffic from the containers, translating their private IP addresses to the host's public IP address. NAT networking is useful for scenarios where containers need Internet access but do not require direct communication with external devices.
- **Overlay Networking:** Overlay networking enables communication between containers running on different hosts in a distributed environment. It creates a virtual network overlay that spans multiple hosts, allowing containers to communicate with each other as if they were on the same local network. Overlay networking is commonly used in container orchestration platforms like Kubernetes to facilitate inter-container communication in a cluster.
- **Macvlan Networking:** Macvlan networking assigns a unique MAC address to each container, allowing them to appear as individual devices on the host's physical network. This enables containers to communicate directly with other devices on the network without NAT or port forwarding. Macvlan networking is useful for scenarios where containers need direct access to the physical network, such as running network services or appliances.

Each type of LXC network offers distinct advantages and is suited to different use cases, allowing users to tailor their networking configurations to meet specific requirements in their containerized environments.

4.2.2 Docker Container Networks

Docker container networks share similarities with LXC networks, but Docker introduces some unique features such as user-defined bridging, which enables users to create custom bridges. Docker supports both the default Linux bridge and user-defined bridges, offering flexibility in network configuration. User-defined bridges provide automatic DNS resolution between containers, allowing them to resolve each other by name or alias. In contrast, containers on the default bridge network can only access each other by IP addresses, although the legacy link option can be used as an alternative. The diagram below illustrates Docker container networks with both the default Linux bridge and a user-defined bridge (Fig. 4.9).

It's worth noting that Docker's default configuration prohibits communication between containers in both the Linux bridge and user-defined bridge setups. However, Kubernetes circumvents this limitation, allowing communication across different bridges. Additionally, within a Docker environment, each container must have a unique name and ID, even if they are on separate bridges.

Docker offers support for seven network types, encompassing the five types discussed previously, along with "none" and "user-defined bridge" networks mentioned earlier. In a "none" network configuration, containers are detached from any network, rendering them isolated from external networks and other containers, and incapable of communication.

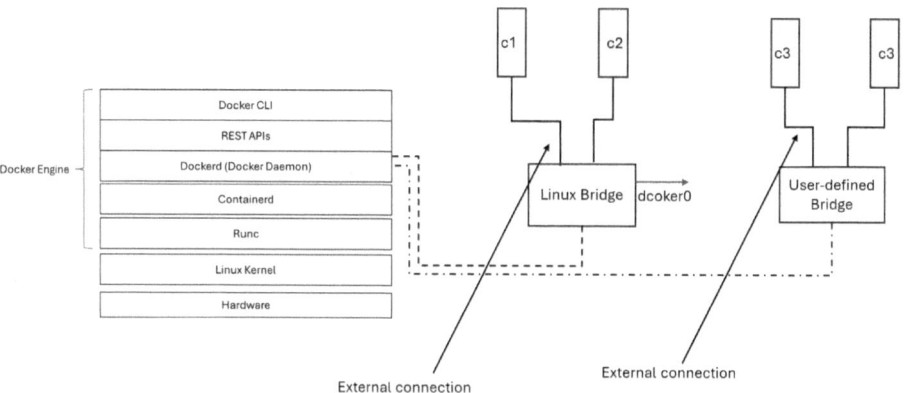

Fig. 4.9 Docker container networks

4.2.3   Kubernetes Container Networks

Kubernetes container networking plays a pivotal role in facilitating communication between containers within a Kubernetes cluster, enabling seamless interaction and collaboration among distributed applications. Kubernetes network model defines following types of configurations:

- **Pod Networking:** In Kubernetes, containers are encapsulated within Pods, which are the smallest deployable units. Pod networking ensures that containers within the same Pod can communicate with each other over the localhost interface, enabling efficient inter-container communication.
- **Cluster Networking:** Kubernetes clusters consist of multiple nodes, each hosting Pods that collectively form applications. Cluster networking establishes communication between Pods running on different nodes within the cluster. This is achieved through various networking solutions, such as overlay networks, that provide virtualized network connectivity across the cluster.
- **Service Networking:** Kubernetes services abstract away the complexity of individual Pods by providing a stable endpoint for accessing a set of Pods. Service networking ensures that applications can discover and communicate with services, regardless of the underlying Pod topology or network configuration. Kubernetes services utilize cluster networking to route traffic to the appropriate Pods based on service selectors.
- **Ingress Networking:** Ingress resources in Kubernetes enable external access to services within the cluster. Ingress networking manages external traffic routing and load balancing to backend services based on defined rules and policies. This allows for the implementation of advanced routing, SSL termination, and HTTP/HTTPS traffic management for applications deployed in Kubernetes.
- **Network Policies:** Kubernetes network policies define rules for controlling inbound and outbound network traffic to Pods based on labels, namespaces, and IP addresses. Network policies provide fine-grained control over communication between Pods, allowing administrators to enforce security policies, isolate workloads, and segment network traffic within the cluster.

Kubernetes container networking allows the use of two distinct networking solutions: Kubenet and Container Networking Interface (CNI). Kubenet is a simple and lightweight networking solution for Kubernetes clusters. It works by configuring the Linux networking stack on each node to provide communication between Pods. This is ideal for smaller organization that needs some basic network configuration for Kubernetes cluster. With Kubenet, each Pod is assigned an IP address from the same subnet as the host node. Communication between Pods within the same node occurs over the localhost interface, while communication between Pods on different nodes is routed through standard Linux networking mechanisms, such as IP routing and NAT (Network Address Translation).

Kubenet typically relies on the kube-proxy component running on each node to handle network proxying and load balancing for services within the cluster.

4.2.3.1 Container Networking Interface (CNI)

The CNI (Container Networking Interface) is a project under the Cloud Native Computing Foundation aimed at establishing a standard interface for configuring network interfaces within Linux containers. It offers a modular framework that enables the integration of various networking solutions into Kubernetes clusters, promoting interoperability and flexibility in container networking configurations. With CNI, networking plugins are responsible for setting up networking for Pods. These plugins can implement various networking configurations, such as overlay networks, bridge networks, and more complex network policies. CNI plugins are executed by the kubelet on each node when Pods are created or destroyed. Each plugin is responsible for configuring the network interface of the Pod and ensuring network connectivity within the cluster. The following diagram provides a basic overview of cross-pod networks facilitated by CNI (Fig. 4.10).

It illustrates how CNI simplifies the networking aspect for cross-Pod communication within a Kubernetes cluster. In such a setup, multiple CNI plugins may operate concurrently, each handling distinct networking functions, such as routing and firewall management, thereby enhancing the overall efficiency and flexibility of the network setup.

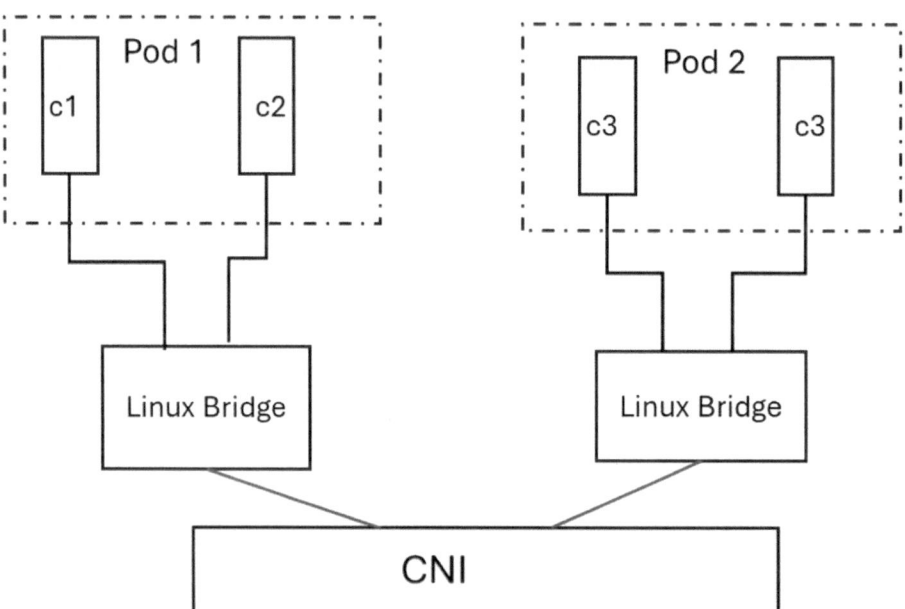

Fig. 4.10 Cross-Pod networks using CNI

Some of the commonly used CNI plugins include NSX-T container plugin (NCP), Calico, Flannel, Cilium, Weave, Canal, Contiv, and Multus. These plugins offer diverse functionalities and features, catering to different networking requirements and preferences within Kubernetes environments. Each plugin contributes to the seamless operation and optimization of container networking, ensuring robust connectivity and security across the cluster.

NSX Container Plugin (NCP)

An effective strategy for facilitating networking and security within containerized applications involves integrating NSX-T with Kubernetes clusters. NSX-T offers a software-defined networking and security platform that enables organizations to establish dynamic, scalable, and secure network infrastructures. By leveraging Kubernetes for the automation of deployment, scaling, and management of containerized applications, organizations can streamline their operations. The integration of NSX-T with Kubernetes clusters provides a unified platform for administering network and security policies. Additionally, it enhances visibility into containerized workloads, allowing organizations to better monitor and manage their infrastructure. This integration enables seamless coordination between networking and security measures, ensuring comprehensive protection for containerized environments.

NSX-T leverages NCP, installed within a container, to enable communication between NSX-T Manager and the Kubernetes API server, especially within k8s/OpenShift environments. NCP is tasked with monitoring changes to containers and other associated resources, as well as managing networking resources like logical ports, switches, routers, and security groups for containers through interactions with the NSX API. As a software component, the NSX Container Plugin (NCP) is packaged as a container image, typically deployed as a Kubernetes pod within the environment.

Calico

Calico is a popular CNI plugins designed specifically for Kubernetes clusters. It provides a scalable and efficient networking solution that enables communication between pods and external networks while also enforcing network policies for security. One of the key features of Calico CNI is its use of a flat L3 IP fabric, which simplifies networking and eliminates the need for overlays. This approach allows for efficient routing and scalability, making it well suited for large-scale Kubernetes deployments. The following diagram depicts how Calico in a Kubernetes cluster works with DCN (Data Center Network) over an IP networks fabric (Fig. 4.11).

Calico CNI utilizes BGP (Border Gateway Protocol) routing to dynamically manage the routing of traffic between pods and external networks herein DCN. This dynamic routing enables efficient load balancing and failover, ensuring high availability for applications running in Kubernetes clusters. In addition to its networking capabilities, Calico

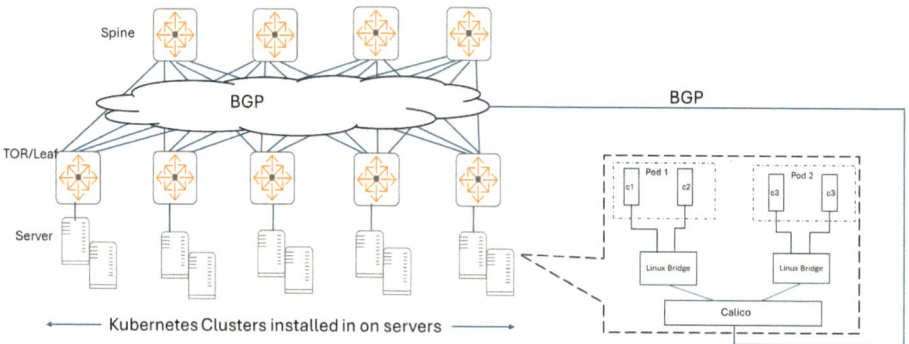

Fig. 4.11 Calico CNI of a Kubernetes cluster connecting to DCN using BGP

CNI also provides powerful network policy enforcement. Administrators can define fine-grained network policies to control traffic between pods based on labels, namespaces, and other attributes. This allows for granular control over network traffic, helping to enforce security policies and compliance requirements.

Flannel

Similar to Calico, Flannel is also a popular networking solution specifically designed for Kubernetes clusters. It provides a simple and reliable way to establish networking connectivity between pods running on different nodes within a Kubernetes cluster. At its core, Flannel creates an overlay network that spans across all nodes in the Kubernetes cluster. This overlay network allows pods to communicate with each other regardless of the underlying network configuration of the physical nodes. Flannel achieves this by assigning each pod a unique IP address from a predefined address range, ensuring that pods can communicate with each other seamlessly (Fig. 4.12).

One of the key features of Flannel is its simplicity and ease of use. It requires minimal configuration and can be easily deployed in Kubernetes clusters of any size. Additionally, Flannel integrates seamlessly with Kubernetes and does not require any changes to the underlying infrastructure. Flannel supports various backends for managing the overlay network, including VXLAN, UDP, and host-gateway mode. Each backend has its own advantages and is suited for different use cases, allowing users to choose the one that best fits their requirements.

Cilium

Cilium is an advanced networking and security solution designed to enhance Kubernetes environments. It provides advanced networking features, including load balancing, network policy enforcement, and transparent encryption, to enable secure communication between pods and services within a Kubernetes cluster. The diagram below illustrates the integration of Cilium with Kubernetes (K8s), showcasing how it serves

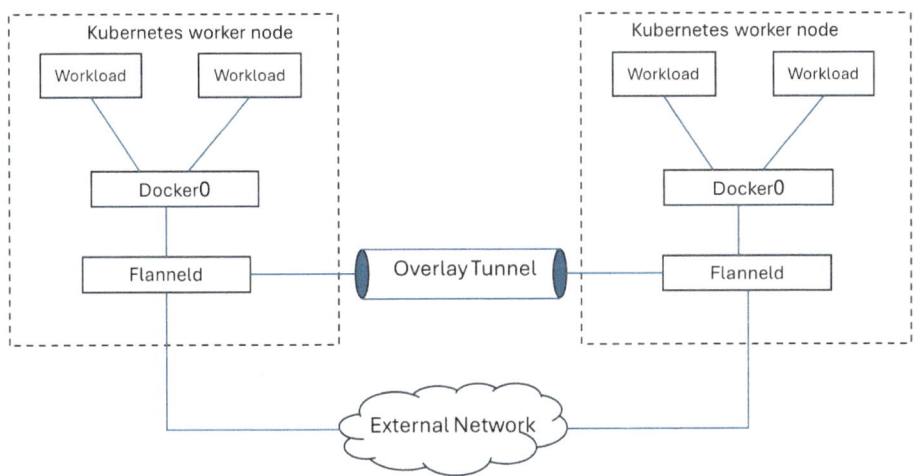

Fig. 4.12 Overlay networks with Flannel

as a CNI (Container Network Interface) plugin for K8 nodes, offering a range of services (Fig. 4.13).

As depicted in the figure above, Cilium leverages eBPF (extended Berkeley Packet Filter) technology to implement efficient and scalable networking and security policies among other services. By utilizing eBPF, Cilium is able to enforce policies at the kernel level, ensuring minimal overhead and maximum performance. One of the key features of Cilium is its support for transparent encryption of pod-to-pod communication. This allows organizations to encrypt traffic between pods without requiring any changes to the application code or configuration, thereby simplifying the process of securing communication within the cluster. Cilium also provides advanced load balancing capabilities, allowing organizations to distribute traffic across pods and services based on various criteria such as round-robin, least connections, or IP affinity. This helps optimize resource utilization and improve application performance within the Kubernetes cluster.

In addition to networking features, Cilium offers powerful network policy enforcement capabilities. Organizations can define fine-grained policies to control traffic between pods and services based on factors such as pod labels, namespaces, and application protocols. This allows for comprehensive security enforcement and isolation within the Kubernetes cluster.

Weave

Weave, also known as Weave Net, is a networking and observability solution specifically designed for Kubernetes clusters. It provides seamless networking between pods and services within a Kubernetes environment, while also offering comprehensive monitoring and troubleshooting capabilities.

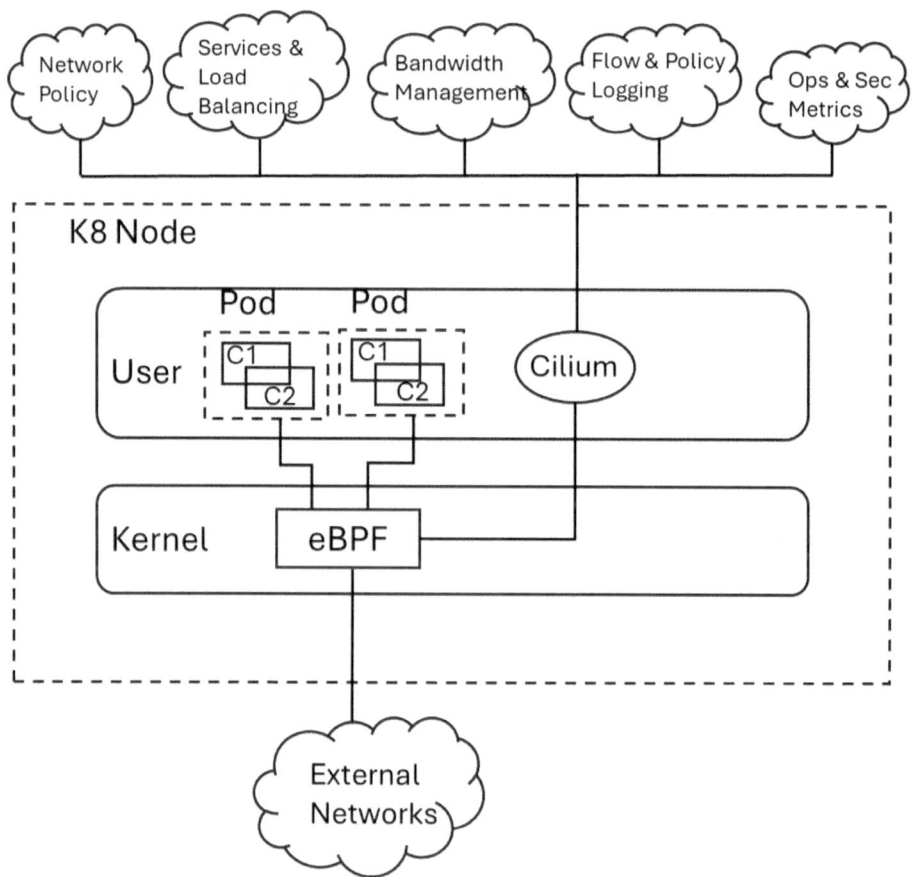

Fig. 4.13 K8 node with Cilium CNI services

At its core, Weave utilizes a software-defined networking (SDN) approach to connect pods and services across the Kubernetes cluster. It establishes a virtual network overlay that spans across nodes, allowing pods to communicate with each other regardless of their physical location within the cluster. This overlay network is designed to be highly resilient and scalable, ensuring reliable communication between pods even as the cluster scales up or down. One of the key features of Weave Net is its simplicity and ease of use. It is designed to be lightweight and easy to deploy, requiring minimal configuration to get started. With Weave, users can quickly set up networking for their Kubernetes clusters without the need for complex manual configurations.

In addition to networking, Weave Kubernetes also offers robust observability features to help users monitor and troubleshoot their Kubernetes environments effectively. It provides real-time visibility into network traffic, allowing users to identify and diagnose

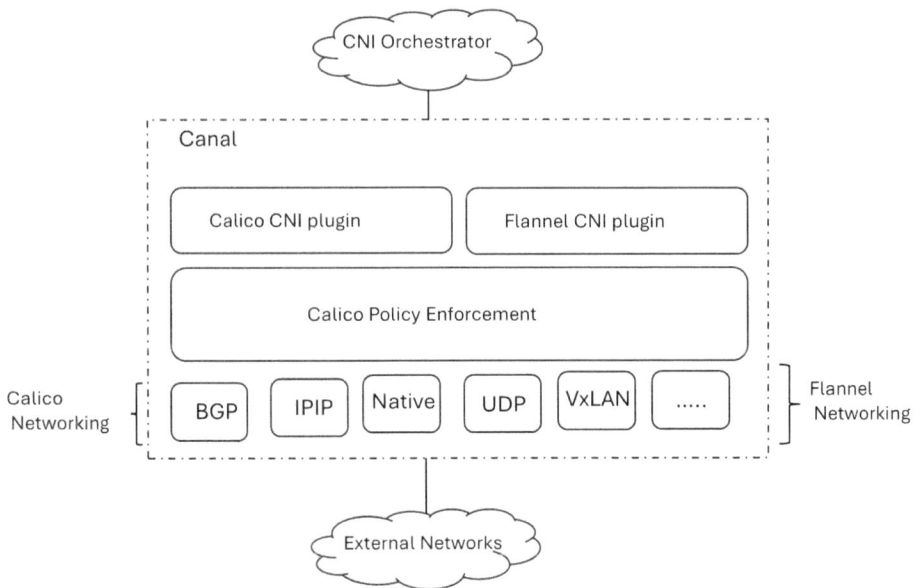

Fig. 4.14 Canal CNI plugin combines Calico and Flannel projects to provide network and policy enforcement services for K8s nodes

issues quickly. Additionally, Weave integrates with popular monitoring tools such as Prometheus and Grafana, enabling users to visualize and analyze network performance metrics easily.

Overall, Weave Kubernetes is a powerful networking and observability solution that simplifies networking in Kubernetes clusters while also providing advanced monitoring capabilities. It is widely used in production environments to enhance the reliability, scalability, and performance of Kubernetes deployments.

Canal

Canal is a CNI network provider for Kubernetes clusters that combines the capabilities of two popular projects: Calico and Flannel. It is designed to provide network policy enforcement and secure communication between pods in a Kubernetes cluster. Canal leverages the strengths of Calico for network policy enforcement and Flannel for pod networking (Fig. 4.14).

The diagram above illustrates the integration of Calico and Flannel plugins within the Canal implementation for Kubernetes clusters. Calico provides a range of network services, including BGP and IPIP, along with segmentation and isolation through policy enforcement mechanisms. Conversely, the Flannel plugin offers network overlay and encapsulated transport capabilities.

By combining the strengths of Calico and Flannel, Canal delivers a holistic networking solution for Kubernetes clusters. It ensures secure and dependable communication between pods while also facilitating fine-grained network policy enforcement to bolster security and compliance measures. Overall, Canal is instrumental in establishing robust networking infrastructure within Kubernetes environments, offering both reliability and flexibility for modern containerized applications.

Contiv

Contiv is a comprehensive networking solution designed to address the complexities of modern containerized environments, particularly within Kubernetes clusters. As a Container Network Interface (CNI) plugin, Contiv offers a range of features and capabilities tailored to enhance networking performance, security, and scalability.

At its core, Contiv provides a unified networking fabric that seamlessly integrates with Kubernetes deployments. It enables seamless communication between containers and services running within Kubernetes clusters while offering advanced networking functionalities to meet diverse workload requirements. One of the key features of Contiv is its support for multiple networking modes, including Layer 2, Layer 3, and Overlay networking. This flexibility allows users to choose the networking mode that best suits their specific use case, whether it involves low-latency communication within a single data center or distributed connectivity across multiple locations.

Contiv also offers robust security features, including network segmentation and isolation through the enforcement of fine-grained network policies. Users can define access controls and traffic rules to restrict communication between different parts of their application stack, enhancing overall security posture. Additionally, Contiv provides built-in support for network observability and troubleshooting. It offers comprehensive monitoring and logging capabilities, allowing operators to gain insights into network traffic, performance metrics, and potential issues affecting containerized workloads.

In terms of deployment, Contiv seamlessly integrates with Kubernetes clusters, leveraging Kubernetes-native APIs and resources for configuration and management. Users can easily deploy and manage Contiv as a CNI plugin within their Kubernetes environments, leveraging familiar tools and workflows.

Contiv-VPP, or Contiv Vector Packet Processing, is an enhanced version of the Contiv networking solution that leverages the Vector Packet Processing (VPP) technology for accelerated networking performance. VPP is an open-source networking stack developed by FD.io (Fast Data project under the Linux Foundation) that provides high-speed packet processing capabilities.

Contiv-VPP extends the capabilities of Contiv by integrating VPP into its networking stack, enabling higher throughput, lower latency, and improved scalability for containerized workloads. By leveraging VPP's advanced packet processing features, Contiv-VPP enhances network performance and efficiency, making it well suited for high-performance use cases and demanding networking environments. In addition to accelerated packet processing, Contiv-VPP retains all the features and functionalities of Contiv, including support for multiple networking modes, network segmentation, security policies, observability, and seamless integration with Kubernetes clusters. Users can leverage Contiv-VPP to achieve high-performance networking while benefiting from the rich set of features provided by Contiv for containerized applications.

Multus

Multus is a Kubernetes network plugin that enables the attachment of multiple network interfaces to pods. Developed by Intel, Multus allows users to define and attach multiple network interfaces to a single pod, providing greater flexibility and versatility in networking configurations for containerized workloads.

One of the key features of Multus is its ability to support multiple network attachment definitions (NADs), which specify how additional network interfaces are configured and attached to pods. These NADs can be implemented using different network plugins, such as CNI (Container Network Interface) plugins, allowing users to mix and match networking solutions based on their specific requirements.

With Multus, users can create custom network configurations tailored to their applications' needs. For example, they can attach separate network interfaces for different purposes, such as data traffic, management traffic, or specialized network services. This capability enables the deployment of complex network topologies and scenarios within Kubernetes clusters. Multus integrates seamlessly with Kubernetes' native networking features and components, such as the kubelet and the CNI interface, ensuring compatibility with existing Kubernetes deployments and workflows. It also provides support for advanced networking features like network policy enforcement and network segmentation, enhancing security and isolation for containerized workloads.

4.2.3.2   eBPF

We previously touched upon eBPF during our discussion on the Cilium CNI plugin. eBPF, or extended Berkeley Packet Filter, stands as a transformative technology enabling programmable packet processing within the Linux kernel. It offers a secure and efficient method for executing custom code within the kernel, thereby unlocking advanced networking capabilities and performance optimizations. Originally introduced as a successor to the traditional packet filter BPF, eBPF's integration into the Linux kernel marked a significant milestone in networking innovation. One year following its initial merge into the kernel, the eBPF backend was incorporated into the LLVM compiler suite, enabling LLVM to emit eBPF bytecode. Concurrently, integration into the kernel's traffic control layer made Linux networking programmable with eBPF (Fig. 4.15).

The diagram above depicts the integration of eBPF within the Linux kernel. When applications interact with the network, they utilize system calls, which are handled by the Linux kernel. Traditionally, userspace networking frameworks must traverse the socket layer of the Linux kernel, even if they aim to remain transparent to the application.

Fig. 4.15 eBPF embedded in
the linux kernel

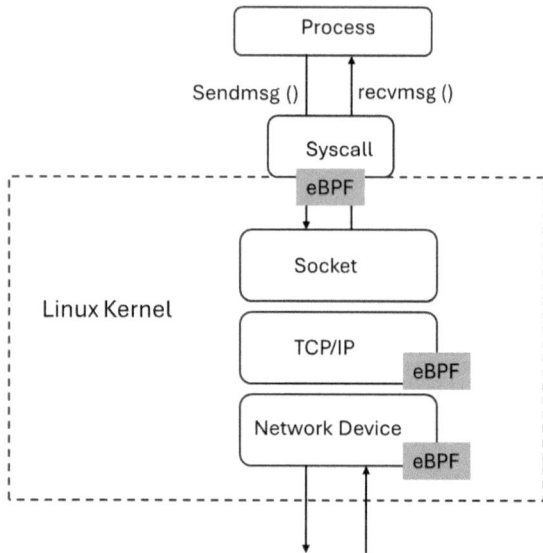

However, eBPF circumvents this requirement by residing entirely within the kernel. This allows eBPF to execute directly within the kernel space, avoiding the need to traverse the socket layer, thereby streamlining packet processing and enhancing performance.

Another pivotal development occurred in 2016 with the merger of XDP into the Linux kernel, ushering in a high-performance data path by enabling eBPF programs to execute directly within the driver of a network device. XDP furnishes an incredibly efficient data path for packet processing, boasting minimal overhead and low latency. Leveraging XDP, eBPF programs can achieve high throughput and low latency, rendering them suitable for demanding networking tasks like DDoS mitigation, load balancing, and packet filtering.

In the context of Cilium, eBPF assumes a critical role in implementing various networking and security features. Cilium harnesses eBPF to execute tasks such as transparent encryption, load balancing, service discovery, and network policy enforcement directly within the Linux kernel. This strategy yields substantial performance gains compared to traditional userspace networking solutions, as packet processing occurs closer to the network interface, circumventing costly context switches between user and kernel space. Moreover, eBPF equips Cilium with profound visibility and control over network traffic, enabling fine-grained policy enforcement based on application-level attributes like HTTP headers, DNS names, and Kubernetes metadata. By amalgamating eBPF with Kubernetes and CNI, Cilium delivers a potent and scalable networking solution tailored to the requirements of modern cloud-native environments.

4.3   Container Networking in Network Infrastructure

Now that we've established a foundational understanding of container networking, let's explore its deployment within network infrastructure. Our discussion will focus on two key areas: firstly, considerations for network infrastructure to support container networking, and secondly, the implementation of container-based Virtual Network Functions (VNFs) within networking equipment.

We'll begin by examining how Kubernetes container networks are deployed in data center environments and the corresponding considerations for data center network (DCN) design. For simplicity, we'll concentrate on the prevalent spine/leaf DCN architecture, where servers housing Kubernetes nodes are connected to leaf or Top-of-Rack (TOR) switches. Within this architecture, a Kubernetes (K8s) cluster may span multiple racks, each with its own Classless Inter-Domain Routing (CIDR) range in a Layer 3 (L3) deployment. Consequently, nodes in different racks require communication capabilities, and the diagram below illustrates how inter-rack traffic traverses the network fabric. However, this deployment, as depicted in the diagram below, has one drawback: the single leaf or TOR switch connectivity to Master or worker nodes creates a single point of failure (Fig. 4.16).

To address this issue, there are two potential solutions: Multi-Chassis Link Aggregation (MLAG) and routing within the host with Equal-Cost Multi-Path (ECMP) routing. The most common approach for a scale-out design is a BGP-based Layer 3 (L3) data

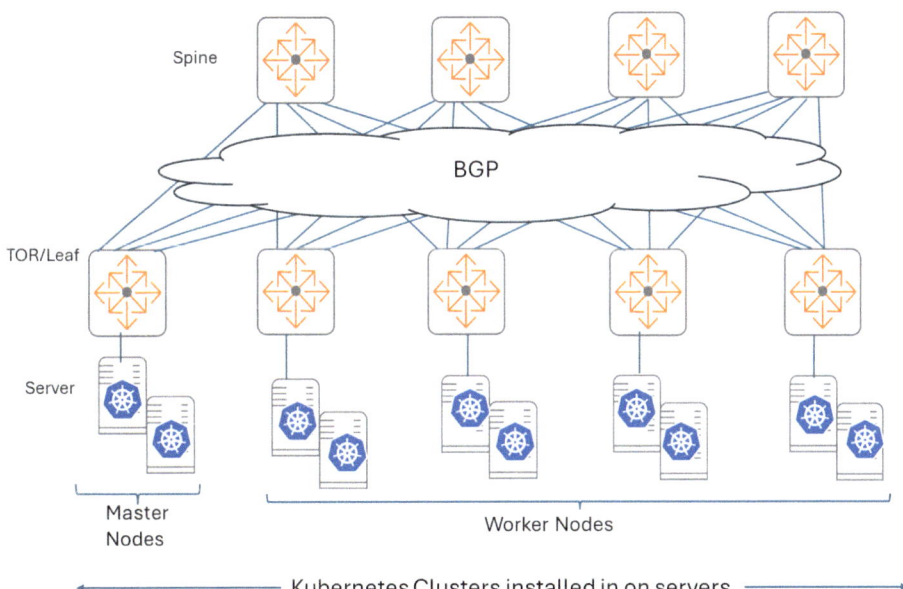

Fig. 4.16   Kubernetes clusters are connected through a spine/leaf data center network

center network (DCN), in which the Kubernetes cluster connects to multiple Top-of-Rack (TOR) switches using ECMP routing. The following diagram shows a simplistic purview of Kubernetes nodes connectivity to leaf/TOR switches. When considering Layer 3 connectivity from each node, it's important to distinguish between two types of networks: one for the node itself and another for the pods. From the node standpoint, every node in a rack connects to two Top-of-Rack (TOR) switches, each belonging to a different Classless Inter-Domain Routing (CIDR) range. Both links are active-active, meaning that even if one link or TOR switch goes down, connectivity remains intact.

Calico CNI plugin offers robust configuration options for achieving high scalability for both nodes and pods by seamlessly integrating with the network fabric. Implementing this solution using the Calico CNI plugin involves disabling peering between Calico nodes and instead establishing peering between the Calico nodes and the TOR switches. In this setup, each Calico node runs a BGP daemon (BIRD) and establishes iBGP peering with the TOR switches.[5] Please note dual TOR is not supported if you are using BGP with encapsulation (VXLAN or IP-in-IP).[6]

In the dual TOR peering configuration, AS numbering and label assignment play crucial roles. These measures enable BGP peering to be applied based on the assigned label, facilitating efficient communication and routing within the network. Within a Calico network, each endpoint represents a route, and the number of routes can reach tens of thousands or even hundreds of thousands for the Kubernetes clusters. Therefore, it's essential to ensure that the network switches deployed to build such a scalable network have a routing table size that can accommodate the scale of Kubernetes compute resources without limitations (Fig. 4.17).

4.3.1 VNF Containers in Network Switch

VNF containers, or Virtual Network Function containers, represent a significant advancement in network infrastructure. Traditionally, network functions such as firewalls, load balancers, and routers were implemented using dedicated hardware appliances. However, with the advent of virtualization technologies, these functions can now be implemented as software running on standard server hardware or even within containers. In the context of network switches, VNF (Virtual Network Function) containers offer several advantages. They allow network operators to deploy and manage network functions in a more flexible and dynamic manner. Instead of deploying separate physical appliances for each network function, operators can run multiple VNF containers on a single switch, reducing hardware costs and simplifying management. The integration of VNF containers into network deployments enhances agility and scalability, offering significant benefits compared to traditional hardware appliances. VNFs, implemented as software, provide the flexibility

---

[5] Gupta [5].

[6] Tigera [6].

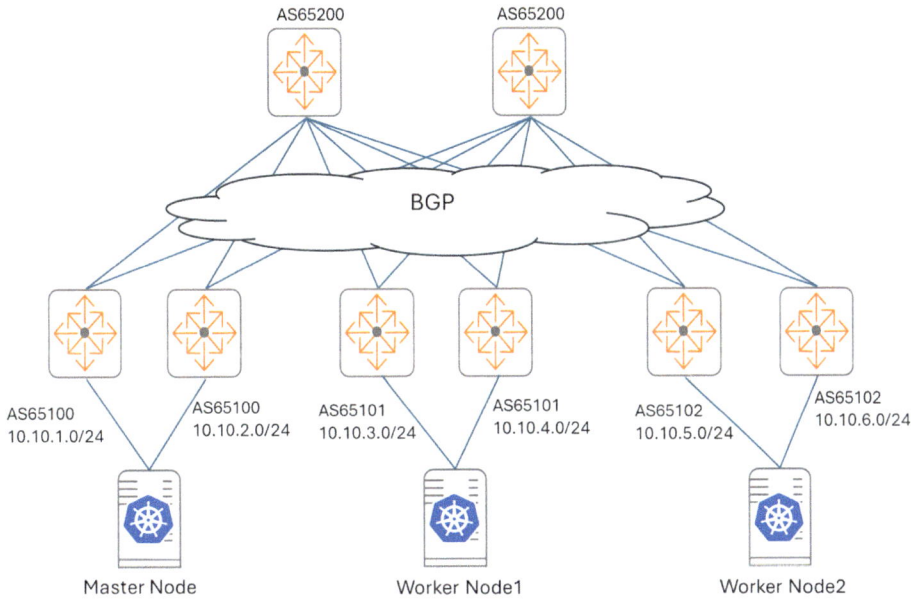

Fig. 4.17 L3 connectivity between Kubernetes nodes and leaf/TOR switches

to scale up or down in response to changes in network traffic or demand. This agility enables operators to quickly adapt to evolving network conditions and optimize resource allocation efficiently.

The diagram illustrates a network switch equipped with multiple containerized VNFs (Fig. 4.18).

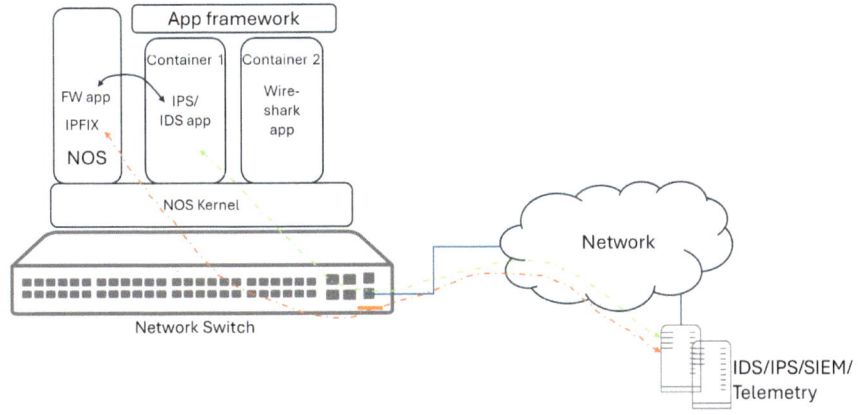

Fig. 4.18 A network switch with multiple container VNFs

Building upon our earlier discussion on the distributed services switch (DSS) in Chap. 2, which integrates Data Processing Units (DPUs) alongside switching ASICs and CPUs, this setup enables service chaining capabilities within the network switch.

One of the essential services that can be achieved through DSS architecture is Layer 4 (L4) firewall (FW) functionality coupled with stateful inspection. As depicted in the diagram, this stateful inspection and firewall (FW) function can be enhanced further by implementing an Intrusion Prevention System (IPS) or Intrusion Detection System (IDS) app agent within a container. The DSS, capable of providing near-line-rate stateful inspections, can leverage the IPS/IDS for detecting and preventing network anomalies that require additional scrutiny.

The IPS/IDS app agent communicates with dedicated IDS/IPS/Security Information and Event Management (SIEM) server(s) to compare against known threats or trigger further actions by network administrators. This service can also benefit from real-time telemetry information provided through IPFIX for enhanced threat detection and response. Moreover, the network switch can host a Wireshark app within a container, enabling the capture of specific packets for in-depth packet analysis. This functionality enhances the network switch's capability to provide detailed insights into network traffic and potential security threats.

In summary, the adoption of VNF containers in network infrastructure offers a cohesive approach to enhancing agility, resource efficiency, and security, empowering operators to meet the dynamic demands of modern networks effectively.

4.3.2   Microsegmentation for Containers

Microsegmentation for containers is a crucial aspect of modern network security, enabling organizations to enforce granular access controls and isolate workloads within their containerized environments. This approach involves dividing the network into smaller segments or zones, where each segment represents a specific set of workloads or applications. By implementing microsegmentation, organizations can strengthen security posture, minimize the impact of potential breaches, and maintain compliance with regulatory requirements.

Network switches play a significant role in providing microsegmentation for containers by leveraging advanced technologies such as eBPF (extended Berkeley Packet Filter) and Cilium. eBPF is a powerful framework within the Linux kernel that allows for programmable packet processing, enabling fine-grained control over network traffic at the kernel level. Cilium, on the other hand, is an open-source project that utilizes eBPF to provide network and security services for containerized environments.

With eBPF and Cilium, network switches can implement microsegmentation by enforcing policies based on application-level attributes such as Kubernetes labels, HTTP headers, or DNS names. This approach allows organizations to define security policies that dictate which containers can communicate with each other, as well as what types of traffic are permitted or denied. For example, organizations can create policies to restrict communication between containers belonging to different departments or teams, thereby reducing the risk of lateral movement in the event of a security breach.

Furthermore, network switches can leverage eBPF and Cilium to provide visibility and control over network traffic within containerized environments. By monitoring and analyzing traffic flows in real-time, organizations can detect and respond to security threats more effectively. For instance, network switches can use eBPF to implement intrusion detection and prevention systems (IDS/IPS), allowing them to identify and block malicious activity before it reaches critical workloads.

Integrating such capabilities into a Distributed Services Switch (DSS) that implements IDS/IPS could tremendously benefit an organization, creating an appropriate security posture for today's network environment. While some of the Container Network Interface (CNI) plugins discussed here can be integrated with a Kubernetes cluster, combining these functionalities within a DSS switch enhances the overall security of the network infrastructure.

4.4   Summary

This chapter explores the critical role of container networking in modern computing and network infrastructure, highlighting its importance as containers become foundational to application deployment. It begins by discussing the evolution of containers and microservices, tracing the roots of containerization from the introduction of "chroot" in Unix systems in 1979 to the development of modern solutions like Docker and Kubernetes. The chapter emphasizes how containers enable lightweight, efficient deployments by sharing the host operating system kernel, contrasting this with the resource-intensive nature of virtual machines.

The chapter details the development of Linux Containers (LXC), which introduced namespaces and control groups (Cgroups) for lightweight process isolation. It explains the rise of Docker, which simplified container management with user-friendly interfaces, standardized container images, and a rich ecosystem, making containerization accessible to a broader audience. Docker's architecture, including its shift from LXC to libcontainer, is thoroughly explored, alongside its integration with other tools like Kubernetes. Kubernetes is presented as a powerful container orchestration platform that automates the deployment, scaling, and management of containerized applications. The chapter discusses Kubernetes' core components, such as Pods, services, and ingress networking, which facilitate communication between containers and external systems. Key features of Kubernetes, including service discovery, load balancing, and declarative configuration models, are highlighted.

The chapter delves into the fundamentals of container networking, covering LXC, Docker, and Kubernetes networking models. It explains various types of container networks, including bridged, host-only, NAT, overlay, and Macvlan networking, describing how each supports specific use cases within containerized environments. Detailed explanations of networking interfaces like Kubenet and Container Networking Interface (CNI) plugins such as Calico, Flannel, Cilium, Weave, Canal, Contiv, and Multus are provided, illustrating their roles in managing network connectivity, security, and performance in Kubernetes clusters.

The integration of container networking in network infrastructure is examined, focusing on its deployment in data centers with spine/leaf architectures and the role of Virtual Network Functions (VNFs) within network switches. The chapter discusses the use of container-based VNFs to enhance agility and scalability in network operations, offering insights into service chaining and advanced functionalities like stateful inspection, IPS/IDS, and real-time telemetry.

The chapter concludes by highlighting the importance of container networking in microservices architecture, where it facilitates seamless communication, service discovery, dynamic routing, load balancing, and security between microservices. Container networking's role in implementing microsegmentation is also covered, showcasing how technologies like eBPF and Cilium enable fine-grained control and isolation within containerized environments.

Overall, this chapter provides a comprehensive understanding of container networking, detailing its evolution, technologies, and impact on modern network infrastructure. The insights gained from this chapter equip readers with the knowledge to effectively navigate and implement container networking solutions in various computing and networking scenarios.

References

1. Vsourz (2022) Microservices explained—all you ever wanted to know about microservices architecture.   https://www.vsourz.com/blog/microservices-explained-all-you-ever-wanted-to-know-about-microservices-architecture/
2. Foote (2021) A brief history of microservices. Dataversity. https://www.dataversity.net/a-brief-history-of-microservices/
3. Swan C (2014) Docker drops LXC as default execution environment. Infoq.com. https://www.infoq.com/news/2014/03/docker_0_9/
4. Chandrakant K (2024) Evolution of Docker from Linux containers. Baeldung. https://www.baeldung.com/linux/docker-containers-evolution
5. Gupta B (2020) Designing on-prem kubernetes networks for high availability. Tigera Inc. https://www.tigera.io/blog/designing-on-prem-kubernetes-networks-for-high-availability/
6. Tigera (2024) Deploy a dual ToR cluster. Tigera Inc. https://docs.tigera.io/calico-cloud/networking/configuring/dual-tor

Network Automation 5

In today's interconnected world, where businesses operate across borders and time zones, the need for seamless digital connectivity has become synonymous with success. However, with this heightened interconnectivity comes a myriad of challenges, chief among them being the management of sprawling network infrastructures. As organizations navigate the intricate webs of devices, applications, and data flows that underpin their operations, the traditional methods of manual network management struggle to keep pace. In response, the paradigm of network automation and telemetry has emerged as a beacon of efficiency and reliability in the face of mounting complexity.

Automation, the cornerstone of modern network management strategies, empowers organizations to streamline their operations, reduce human error, and respond rapidly to changing demands. By automating routine tasks such as device provisioning, configuration management, and troubleshooting, businesses can unlock newfound agility and scalability, enabling them to adapt swiftly to evolving market conditions. Moreover, the integration of telemetry technologies provides invaluable insights into network performance and health, facilitating proactive maintenance and optimization efforts.

Against this backdrop, this chapter embarks on a journey into the heart of network automation, shedding light on its transformative potential and practical applications. Through a comprehensive exploration of its underlying principles, challenges, and future trajectories, we aim to equip readers with the knowledge and insights needed to embark confidently on their automation journey. From understanding the fundamental concepts to navigating the complexities of implementation and integration, this chapter serves as a roadmap for harnessing the full potential of automation and telemetry in modern network management.

© The Author(s), under exclusive license to Springer Nature Switzerland AG 2025   159
D. D. Chowdhury, *Future of Networks*, Synthesis Lectures on Communications,
https://doi.org/10.1007/978-3-031-71440-5_5

5.1   Evolution of Network Automation

The history of network automation represents a captivating journey, characterized by significant technological leaps and the evolution of paradigms in network management. In the nascent days of networking, configuring devices like routers, switches, and firewalls was a manual and arduous process. Network administrators were tasked with individually accessing each device and effecting configuration changes via command-line interfaces (CLIs).

As networks expanded in both scale and intricacy, the imperative for automation became increasingly evident. Throughout the 1990s, network administrators began harnessing scripting languages such as Perl and Expect to streamline repetitive tasks. These scripts ushered in the automation of configuration alterations, backups, and rudimentary monitoring duties (Fig. 5.1).

The timeline of network automation's evolution is depicted above. Notably, in the 1990s, the Simple Network Management Protocol (SNMP) and Remote Monitoring (RMON) emerged as standardized protocols for overseeing and managing network devices. Although not strictly classified as automation technologies, SNMP and RMON laid the foundation for remote management and monitoring, granting administrators the capability to collect data and execute basic management functions programmatically.

The early 2000s witnessed the rise of configuration management tools like RANCID (Really Awesome New Cisco confIg Differ) and CiscoWorks. These tools afforded administrators the ability to automate the backup and version control of device configurations, thereby mitigating the risk of configuration drift and simplifying network failure recovery processes. Network Management Systems (NMS) underwent a profound transformation, evolving into increasingly sophisticated platforms for monitoring and managing network devices. NMS solutions introduced features such as fault detection, performance monitoring, and configuration management, laying the groundwork for more holistic network automation capabilities.

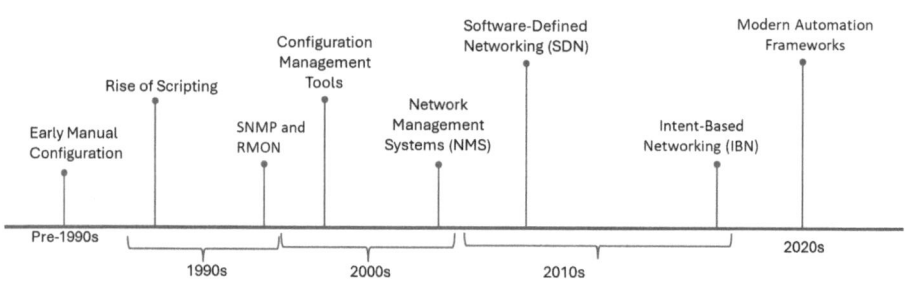

Fig. 5.1 Network automation timeline

The advent of Software-Defined Networking (SDN) in the 2010s marked a seismic shift in network architecture, delineating the control plane from the data plane and consolidating network management via software controllers. SDN promised heightened agility, scalability, and automation prowess, empowering administrators to dynamically provision and configure network resources. Subsequently, Intent-Based Networking (IBN) emerged as a pivotal advancement in network automation, focusing on translating overarching business objectives into automated network policies and configurations. IBN systems leverage machine learning and artificial intelligence to decipher intent, validate network changes, and ensure alignment with business imperatives.

In the contemporary landscape, network automation stands as a cornerstone of network management strategies, propelled by the proliferation of cloud computing, virtualization, and the Internet of Things (IoT). Modern automation frameworks such as Ansible, Puppet, and Chef furnish robust platforms for automating the provisioning, configuration, and management of network infrastructure, spanning from on-premises data centers to cloud environments.

5.2 Defining Network Automation

Network automation refers to the utilization of software-based solutions and technologies to streamline and expedite the management, configuration, provisioning, monitoring, and operation of network infrastructure. At its core, network automation aims to replace manual, repetitive tasks traditionally performed by network administrators with automated processes, thereby enhancing efficiency, reducing errors, and enabling rapid adaptation to changing network requirements.

Through network automation, organizations can achieve several key objectives:

- **Efficiency:** By automating routine tasks such as device provisioning, configuration changes, and troubleshooting, network automation reduces the time and effort required for network management activities, freeing up valuable human resources to focus on strategic initiatives.
- **Reliability:** Automation minimizes the risk of human error, ensuring consistency and accuracy in network configurations and operations. Automated processes follow predefined rules and standards, reducing the likelihood of misconfigurations and enhancing the reliability and stability of the network.
- **Scalability:** Automated workflows enable networks to scale dynamically to accommodate changing demands, whether it be the addition of new devices, services, or users. Network automation facilitates rapid provisioning and configuration of resources, allowing organizations to adapt quickly to evolving requirements without manual intervention.

- **Consistency and Compliance:** Automation ensures consistent application of network policies and configurations across the entire infrastructure, promoting adherence to security standards, compliance regulations, and best practices. Automated monitoring and auditing capabilities help identify deviations from established norms, facilitating prompt remediation actions.
- **Agility and Innovation:** By accelerating the deployment of network services and applications, automation fosters greater agility and responsiveness to business needs. It enables organizations to experiment with new technologies, implement changes more rapidly, and drive innovation within their network environments.

Network automation encompasses a broad range of technologies, tools, and methodologies, including scripting, configuration management, orchestration, software-defined networking (SDN), intent-based networking (IBN), and the integration of artificial intelligence (AI) and machine learning (ML). These automation techniques enable organizations to design, deploy, and manage complex network infrastructures efficiently and effectively in today's dynamic and rapidly evolving digital landscape.

5.2.1 Contrasting Network Automation with Traditional Network Management

Network management has evolved significantly over the years, transitioning from traditional manual methods to more advanced, automated approaches. Below is a comparison of traditional network management and network automation, highlighting the key differences, advantages, and limitations of each.

5.2.1.1 Traditional Network Management

Traditional network management involves manual configuration and oversight of network devices, which is labor-intensive. Network administrators are responsible for configuring devices, managing settings, and troubleshooting issues manually. These tasks are time-consuming, requiring significant effort for provisioning new devices, updating configurations, and performing regular maintenance. Manual processes are prone to human error, which can lead to misconfigurations, network outages, and security vulnerabilities. Additionally, without standardized procedures, configurations and management practices can be inconsistent, leading to variability across the network.

The approach to problem detection in traditional network management is generally reactive. Issues are often identified and addressed only after they have occurred, resulting in network downtime and reduced performance. As networks grow in size and complexity, managing them becomes increasingly difficult and less scalable. Monitoring and troubleshooting in traditional network management rely on basic tools and methods. Simple Network Management Protocol (SNMP) and manual log reviews are commonly used for monitoring, while diagnosing and resolving network issues often require manual intervention. This can involve physical access to devices or remote command-line interface (CLI) access, which is both time-consuming and inefficient.

Effective traditional network management requires specialized knowledge. Network administrators must have a deep understanding of network protocols, device-specific commands, and troubleshooting techniques. This high demand for skilled administrators makes network management resource-intensive, as these tasks necessitate continuous oversight and expertise.

5.2.1.2 Network Automation

Automated configuration and management streamline routine tasks such as device provisioning, configuration updates, and monitoring, significantly reducing the time and effort required. Automated processes adhere to predefined scripts and protocols, minimizing the risk of errors and misconfigurations. This approach ensures uniform application of configurations and policies across all network devices, enhancing consistency.

Network automation also adopts a proactive approach. Predictive maintenance leverages telemetry and monitoring data to foresee and prevent issues before they occur, thereby enhancing network reliability. The scalability of automated systems allows them to easily accommodate growing networks, enabling dynamic provisioning and configuration of resources as needed.

Advanced monitoring and troubleshooting are integral components of network automation. Comprehensive monitoring tools provide real-time insights and analytics, while automated troubleshooting employs scripts and AI/ML algorithms to diagnose and resolve common network issues without human intervention. Skill requirements for network automation encompass a broad skill set, including knowledge of automation tools, scripting languages such as Python, and modern networking concepts like Software-Defined Networking (SDN) and Intent-Based Networking (IBN). Automation frees up network administrators to focus on strategic tasks, such as optimizing network performance and planning for future growth.

Advantages of Network Automation Over Traditional Network Management

Network automation increases efficiency by automating repetitive tasks, freeing up human resources for higher level functions. It enhances reliability by reducing the likelihood of human error, leading to more stable and dependable network operations. Automation offers greater agility, allowing for rapid deployment and configuration changes, quickly adapting to new business requirements. It improves scalability, handling large, complex networks with ease and supporting dynamic growth and changes. Proactive maintenance utilizes predictive analytics to anticipate potential issues and address them before they impact network performance.

5.2.1.3 Limitations and Considerations

Traditional network management is labor-intensive and prone to errors due to the significant manual effort involved. It also struggles with limited scalability, making it challenging to manage large, complex networks effectively.

Network automation requires an initial investment in automation tools and staff training. The transition to automation can be complex, often necessitating the redesign of existing processes. Additionally, automated systems need regular updates and maintenance to ensure they remain effective and secure.

In summary, while traditional network management relies heavily on manual processes and is often reactive, network automation offers a more efficient, reliable, and scalable approach to managing modern network infrastructures. Automation not only enhances operational efficiency but also supports proactive network management, allowing organizations to stay ahead of potential issues and adapt swiftly to changing demands.

5.3 Network Automation Framework

As discussed in Sect. 5.2, network automation encompasses a wide array of software-based tools and technologies that facilitate the configuration, provisioning, management, operation, and testing of network infrastructure. Providing a comprehensive account of all available tools and technologies would be challenging, but this section presents a framework to guide readers in deploying network automation within their environments. This framework is designed to be useful not only for practitioners but also for those learning about network automation for the first time, enabling them to create a network automation solution for their organization.

While this framework does not list every available tool and technology, it provides sufficient guidance to develop a solution tailored to specific network infrastructures. Some tools or technologies mentioned may be integrated into orchestration and management software provided by various vendors. For example, leading network equipment suppliers like HPE offer orchestration and management software that integrates many of these tools or their functions. The framework presented here aims to remain vendor-agnostic, enabling readers to build their own network automation solutions using a variety of tools, including readily available open-source options.

The following diagram illustrates a typical network automation framework, categorizing tools into several key areas, each designed to address specific aspects of network management. Let us delve further into the subject and explore the components of this framework in detail (Fig. 5.2).

The framework presented above depicts several key categories essential for comprehensive network automation: Security, Source of Truth, Software Version Control, Automation Hub and CI/CD, Infrastructure as Code, Configuration Management, Testing

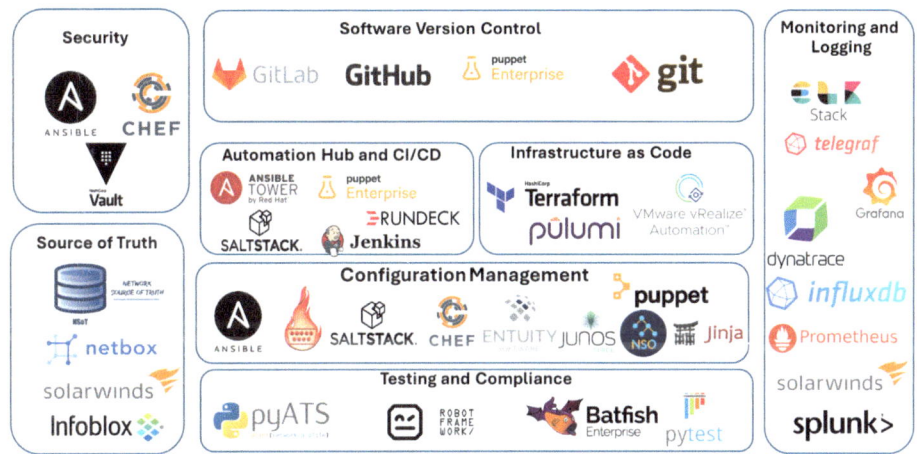

Fig. 5.2 Typical network automation framework

and Compliance, and Monitoring and Logging. By leveraging this network automation framework, organizations can enhance their network management strategies, ensuring greater efficiency, reliability, and scalability in their network operations. Each category represents a crucial aspect of the overall automation landscape, contributing to a more streamlined and secure network environment. In the following sections, we will delve further into each of these categories, exploring the tools depicted in the diagram above and discussing their specific roles and functionalities in network automation.

5.3.1 Security in Network Automation

The security category in the network automation framework is crucial for ensuring that network configurations and operations are conducted securely. In today's network environment, security remains a vital component of network infrastructure, and network automation is no different. This section delves into the tools and practices that provide robust security in network automation, focusing on key tools such as Ansible, Chef, and Vault. Each of these tools plays a significant role in protecting sensitive data, automating secure configurations, and maintaining the integrity of network operations.

5.3.1.1 Ansible
Ansible is a powerful open-source automation tool that can manage configurations, deploy applications, and orchestrate complex IT workflows. The following diagram depicts various components of Ansible and its operation (Fig. 5.3).

Ansible utilizes an agentless architecture to connect with network devices, hosts, cloud services, and Configuration Management Databases (CMDB). Connectivity to network

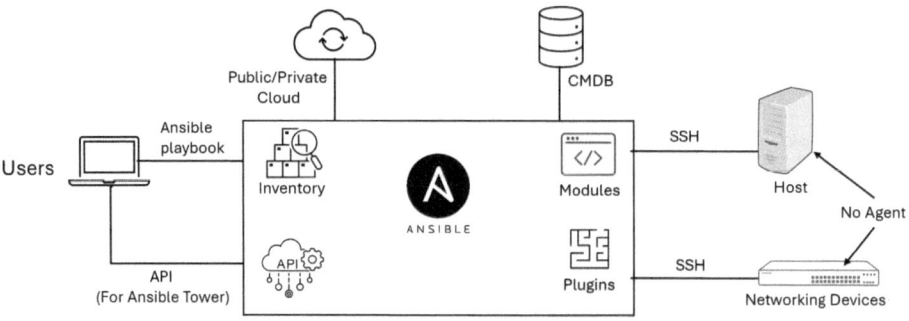

Fig. 5.3 Ansible agentless architecture and secured connectivity

devices is typically established through the SSH protocol or device APIs when available. The Ansible engine comprises an inventory that includes information about IP addresses and databases of hosts and devices, as well as APIs, plugins, and modules. Cloud services and users can connect to the Ansible control node via APIs. Modules are executed directly on remote hosts through playbooks, managing system resources such as services, packages, and files, or executing system commands. These modules operate by interacting with system files, installing packages, or making API calls to network services.

In the context of security, Ansible enhances network operations through several features. Its agentless architecture minimizes the attack surface and potential vulnerabilities by eliminating the need for software agents on managed nodes. Ansible ensures secure communication by using SSH for Unix/Linux systems and WinRM for Windows systems, providing encrypted and secure connections. Ansible Vault allows the encryption of sensitive data such as passwords and API keys, ensuring their protection both at rest and in transit with AES-256 encryption. Ansible Tower's role-based access control (RBAC) manages who can execute specific playbooks, access inventories, and manage projects, thus ensuring that only authorized personnel can make changes. Additionally, Ansible automates the enforcement of security policies and maintains detailed logs of changes, facilitating security audits and ensuring accountability.

5.3.1.2 Chef

Chef is an automation platform that revolutionizes infrastructure management by converting configurations into code, which ensures consistent and repeatable setups. This method greatly reduces the risk of configuration drift and unauthorized changes, thereby maintaining the security integrity of the network. Chef's core principle of Infrastructure as Code (IaC) allows users to define network configurations as code. This practice ensures that configurations can be version-controlled, audited, and reliably reproduced across different environments, making it easier to identify and revert unauthorized changes, thus enhancing overall security.

Chef Automate is an integrated platform that extends Chef's capabilities across infrastructure, applications, and compliance. It enhances security by providing comprehensive visibility, facilitating collaboration, and enforcing compliance with security standards. Through real-time monitoring, teams can oversee infrastructure changes, collaborate on updates, and ensure that all configurations align with organizational security policies, significantly reducing the likelihood of security breaches due to misconfigurations. The diagram below depicts various components of Chef automation framework (Fig. 5.4).

Chef automation components include the Chef server, workstations, and clients. The Chef server is the main infrastructure component that stores data and acts as a hub of information. Cookbooks and policy settings are uploaded to the Chef Infra Server by users from workstations. Chef clients access the Chef server from the nodes on which they reside; please refer to the diagram above. For example, a Chef client could reside on a server, virtual platform, public cloud, storage, or network device. The Chef client accesses the Chef server to get configuration data, perform searches of historical Chef Infra Client run data, and pull down the necessary configuration data.

To protect sensitive information, Chef uses encrypted data bags. These data bags securely store configuration items, including passwords, API keys, and other secrets, ensuring that sensitive data remains protected during the automation process, both at rest and in transit. This encryption is crucial for maintaining the confidentiality and integrity of

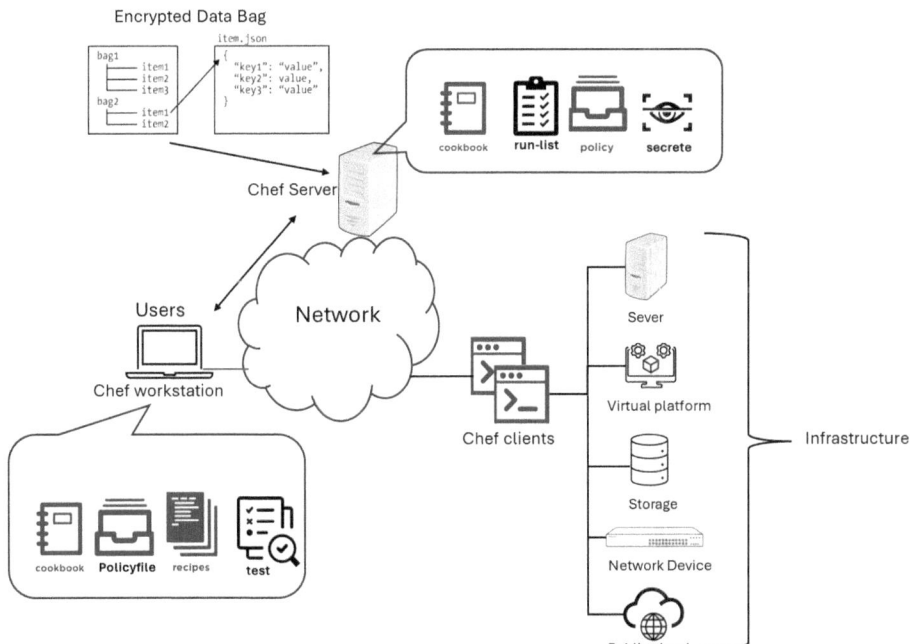

Fig. 5.4 Chef network automation components

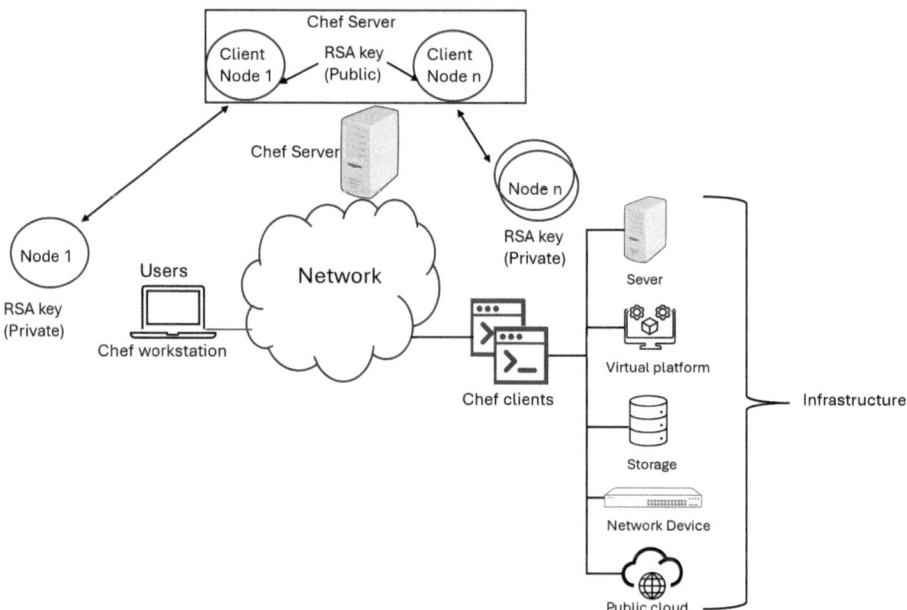

Fig. 5.5 Chef vault solutions for secured data storing in Chef automation framework

sensitive information within automated workflows. An alternative to the shared key mechanism is the Chef Vault, which does not require the distribution of a shared key that can be tampered with. Instead, it uses an RSA key pair, keeping public half in the Chef server and the private half with the infrastructure nodes, such as network devices. This approach enhances security by eliminating the risks associated with shared key distribution. The following diagram illustrates how chef vault works (Fig. 5.5).

As depicted in the diagram, the Chef server holds client objects for each node, which also retain the public half of the RSA key, while the nodes themselves retain the private half of the RSA key pair (usually stored in /etc/chef/client.pem).

Additonally, Chef InSpec is a powerful testing framework focused on compliance and security. It allows users to define security and compliance requirements as code, automating compliance testing across the infrastructure. By integrating InSpec into the automation pipeline, organizations can continuously verify that their systems adhere to defined security policies and compliance standards. This continuous compliance testing quickly identifies and addresses potential security issues, ensuring a secure and compliant network environment.

Overall, Chef's approach to automation, with its emphasis on Infrastructure as Code, the integrated capabilities of Chef Automate, the secure handling of sensitive data through encrypted data bags, chef vault and the compliance automation provided by Chef InSpec, collectively enhances the security of network operations. These features ensure that configurations are consistent, secure, and compliant with organizational policies, thereby maintaining the integrity and security of the network infrastructure.

5.3.1.3 Vault by HashiCorp

HashiCorp Vault is a robust open-source tool designed for securely managing secrets, such as API keys, passwords, and certificates. It offers comprehensive security features for handling sensitive data, which are essential for maintaining a secure network environment. Vault allows for the secure storage and access to secrets and sensitive information, reducing the risk of credential leaks. It can dynamically generate secrets, such as database credentials, which minimizes the potential for misuse by ensuring these credentials are short-lived. The following diagram depicts how Vault encrypts and decrypts data. App sends data to Vault to get encrypted and decrypted as needed (Fig. 5.6).

Vault also provides encryption services to securely store and manage sensitive data, supporting multiple encryption methods to ensure data is protected both at rest and in transit. This feature is crucial for safeguarding information against unauthorized access. Access control policies in Vault define who can access specific secrets and what actions they can perform, offering granular control over sensitive information. This ensures that only authorized users and applications have access to the data they need, adhering to the principle of least privilege.

Detailed audit logs are another critical feature of Vault, recording all access and operations. These logs help monitor access to sensitive data, detect unauthorized access, and

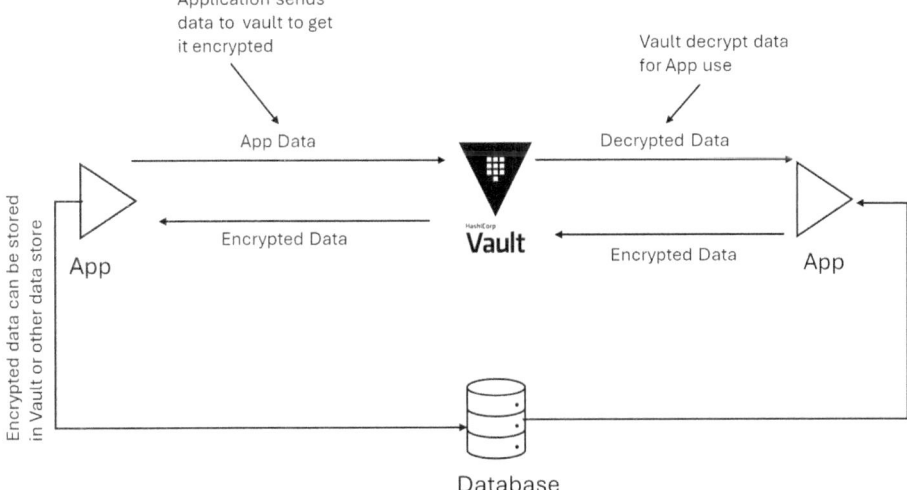

Fig. 5.6 Hashicorp vault used for app data encryption and decryption

ensure compliance with security policies, thus enhancing overall security oversight. Moreover, Vault can generate dynamic secrets on-demand. For instance, it can produce database credentials that are valid only for a specific period or session. This dynamic approach reduces the risk of credential misuse and enhances security by limiting the lifespan of sensitive information.

In summary, Vault by HashiCorp provides essential security capabilities for network automation. Its features for secure secret management, robust encryption, granular access control, detailed auditing, and dynamic secret generation collectively ensure a secure and compliant network environment.

5.3.2 Configuration Management

Configuration management is a critical aspect of network automation, ensuring that all network devices are configured correctly and consistently maintained over time. The diagram illustrates various tools and practices used in configuration management to enhance network reliability and security. This section delves into the key tools—Ansible, NAPALM, SaltStack, Chef, Entuity, Junos Space, NSO, and Jinja—that facilitate effective configuration management in network automation.

Configuration management in network automation involves maintaining the desired state of network devices such as routers, switches, firewalls, and servers. This process includes initial setup, configuration changes, and continuous monitoring to ensure compliance with organizational policies. Automating these tasks reduces manual errors, increases efficiency, and ensures consistent application of configurations across all devices.

5.3.2.1 Ansible

We have discussed about Ansible from the perspective of security in network automation. It is to be noted that Ansible is a widely used configuration management tool that employs an agentless architecture to push configurations to network devices via SSH or device APIs. This eliminates the need for additional software on managed devices, reducing the attack surface and simplifying deployment. Ansible playbooks, written in YAML, describe the desired network state, including system resources like services, packages, and files. These playbooks can be repeatedly executed to maintain device compliance with the desired configuration.

Ansible interacts with network devices through various methods tailored to the specific device and the chosen module. These modules are pre-written scripts designed to perform tasks such as configuring interfaces, routing protocols, VLANs, firewall rules, and more. Among the most frequently used network modules are network_cli, netconf, and httpapi. The following diagram depicts Ansible connecting with network device using Secure Shell (SSH) connection to run a set of instructions over the remote SSH (Fig. 5.7).

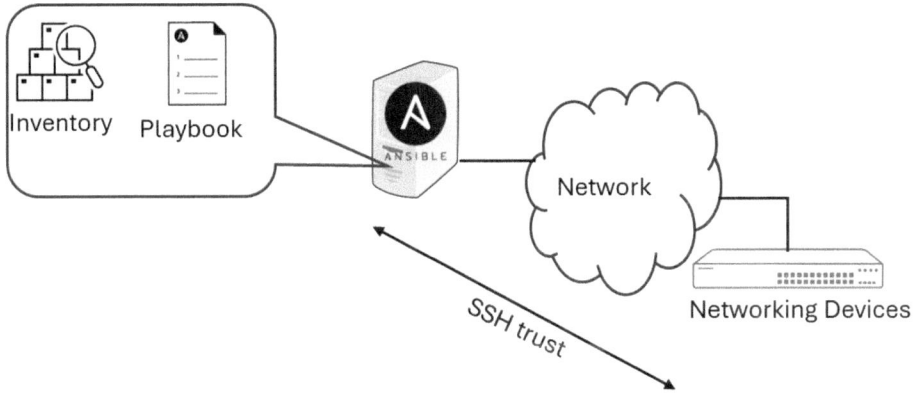

Fig. 5.7   Ansible connects to network device using SSH

The network_cli module connects to network devices using SSH and executes CLI commands as depicted in the diagram above. This module is suitable for devices that support SSH and have a command-line interface, such as HPE Aruba AOSCX, Cisco IOS, Juniper Junos, and Arista EOS. Through network_cli, administrators can automate tasks traditionally managed via the CLI, ensuring consistent and efficient operations. The netconf module leverages the NETCONF protocol to connect to devices, facilitating the exchange of XML data. It is compatible with devices that support NETCONF and use a YANG data model, such as Cisco IOS-XR and Nokia SR OS. This module allows for advanced configurations and state validations, utilizing the structured data models provided by YANG. The httpapi module uses HTTP or HTTPS to interface with devices, interacting with their REST APIs. This module is effective for devices that support REST APIs and use JSON or XML data formats, like Cisco NX-OS and F5 BIG-IP. By using httpapi, Ansible can manage modern network devices with comprehensive RESTful interfaces, providing flexible and powerful automation capabilities.

These modules enable Ansible to interface with a wide range of network devices, streamlining complex network management tasks and ensuring compatibility and efficiency across various network environments.

5.3.2.2 NAPALM

NAPALM (Network Automation and Programmability Abstraction Layer with Multi-vendor support) is an essential tool in the configuration management landscape. It is an open-source Python library that provides a unified API to interact with various network device configurations, allowing for easy and consistent management of devices from multiple vendors. NAPALM supports retrieving device configurations, managing operational states, and ensuring that network devices are configured according to defined policies. It can be integrated with popular automation frameworks, such as Ansible, SaltStack, and

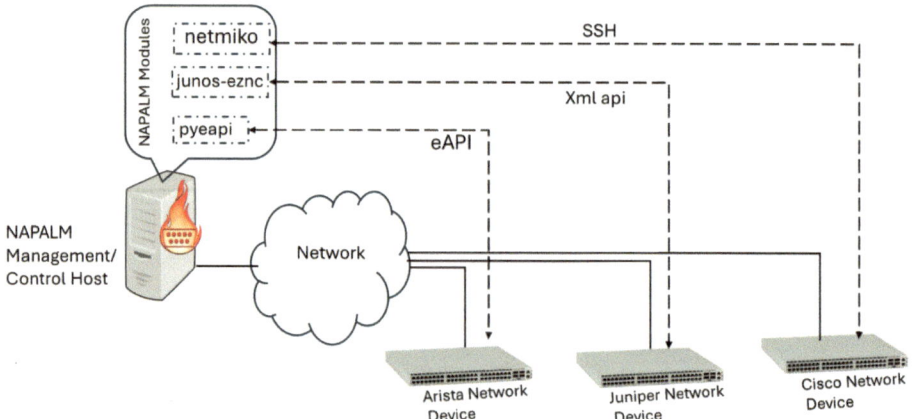

Fig. 5.8 NAPALM control host connecting to multi-vendor network devices using abstract drivers and associated methods of connectivity for each

StackStorm. For example, the napalm-ansible plugin for Ansible contains modules that use NAPALM to retrieve data or modify configurations on networking devices.

NAPALM provides an easy-to-use platform to programmatically configure and manage network devices, such as routers and switches, using native protocols like NETCONF, SSH and RESTAPI, etc. It supports various operating systems, including Cisco IOS, Arista EOS, and Juniper Junos. The following diagram depicts NAPALM connectivity to multi-vendor network devices (Fig. 5.8).

As shown in the diagram, the NAPALM control or management host is capable of connecting to multi-vendor network devices using abstracted drivers and communication methods associated with each. These drivers abstract the underlying complexity of connecting to and interacting with the devices. For Arista, NAPALM uses "pyeapi," which is a Python library for communicating with Arista's eAPI (a REST-like API for controlling and monitoring Arista switches). Similarly, NAPALM uses "Junos PyEZ" to connect to Juniper network devices using Juniper's "XML API." For Cisco, NAPALM uses "netmiko" to establish an SSH connection to Cisco devices.

In summary, NAPALM streamlines network automation by providing a consistent interface across different vendors, making it a powerful tool for network engineers looking to automate their network management tasks.

5.3.2.3   SaltStack

SaltStack, commonly known as Salt, is a robust automation and configuration management tool designed to handle complex IT environments, including network infrastructures. SaltStack excels in real-time configuration management and automation, leveraging a high-speed communication bus that makes it particularly effective for network automation due to its speed, scalability, and flexibility.

SaltStack operates on a master-minion architecture, where the master is the central server managing configuration tasks and issuing commands, and the minions are the nodes or devices executing these commands. This architecture supports both push and pull models, allowing for real-time command execution and state enforcement. Salt States, written in YAML, define the desired state of network devices, ensuring consistent configuration across the network. The Salt Pillar securely stores sensitive data, such as passwords and API keys, which minions can access as needed. The following diagram depicts SaltStack agent-based architecture and communications between salt master and minion (Fig. 5.9).

SaltStack offers both agent-based and agentless operation. The server-based component is known as the "Salt Master", while the agent component is called the "Salt Minion". While it primarily uses agents (minions) for managing devices, it can also operate agentlessly for devices that cannot run a minion. This flexibility makes SaltStack adaptable to diverse network environments. Its event-driven architecture enables real-time responses to network changes, which is essential for dynamic environments where quick reactions to events like device failures or configuration changes are necessary.

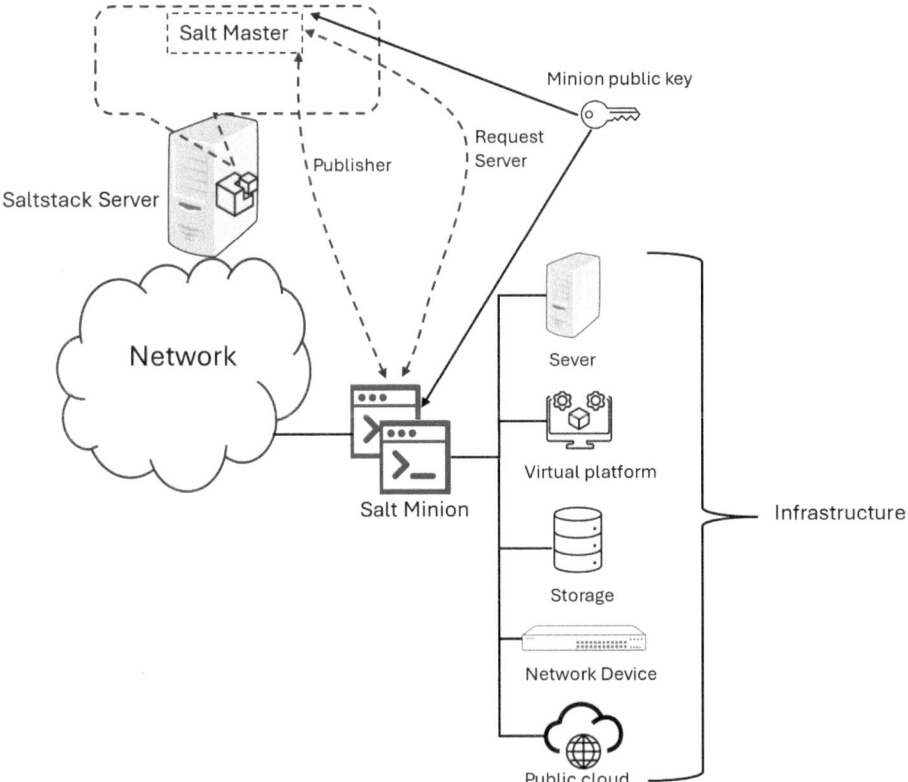

Fig. 5.9 SaltStack architecture and server-agent communication model

In the server-agent communication model, the Salt Master is responsible for sending commands to Salt Minions and then aggregating and displaying the results of those commands. A single Salt Master can manage thousands of systems. Communications use a publish-subscribe pattern, as illustrated in the diagram above. The Salt Minion initiates the connection with the Salt Master, which uses ports 4505 and 4506 to accept incoming connections. Port 4505 is considered the publisher port, and all Salt Minions establish a persistent connection to this port to listen for messages. Commands are sent asynchronously to all connections over this port, enabling commands to be executed over large numbers of systems simultaneously. Similarly, port 4506 is considered the "server request" port, and Salt Minions connect to the request server as needed to send results to the Salt Master and securely request files and minion-specific data values (called Salt Pillar). Connections to this port are 1:1 between the Salt Master and Salt Minion, rather than asynchronous.

Scalability is a key strength of SaltStack, facilitated by its high-speed communication bus, ZeroMQ. This allows it to manage thousands of devices simultaneously, making it suitable for large-scale network automation. SaltStack supports remote execution, enabling administrators to run commands across multiple devices in parallel, significantly reducing the time needed for network-wide updates and maintenance tasks.

Configuration management in SaltStack is achieved through Salt States, which define the desired configurations for network devices. These states ensure consistent application and automatic remediation of any deviations from the desired configuration. SaltStack's orchestration capabilities allow for the definition of complex workflows spanning multiple devices and systems, which is particularly useful for coordinated network updates and deployments. Integration with other network management and monitoring tools enhances SaltStack's utility, as it can use data from these systems to trigger automated responses, improving overall network reliability and performance. Common use cases include provisioning, where SaltStack automates the initial setup and configuration of network devices; compliance, where it regularly checks and enforces network policies to ensure devices adhere to security and operational guidelines; and monitoring and remediation, where it integrates with monitoring tools to automatically respond to network issues, such as triggering recovery procedures when a device goes offline.

SaltStack also excels in managing updates and patches across the network. Its orchestration capabilities ensure that updates are applied consistently and with minimal disruption to network operations. By automating repetitive tasks and maintaining consistent configurations, SaltStack enhances network reliability, security, and efficiency.

5.3.2.4 Chef

In the previous section, we discussed Chef within the context of security in network automation, touching on its architecture and security features. Here, we will delve deeper into Chef as a configuration management tool for network automation, focusing on its technical details and capabilities in managing and automating network configurations.

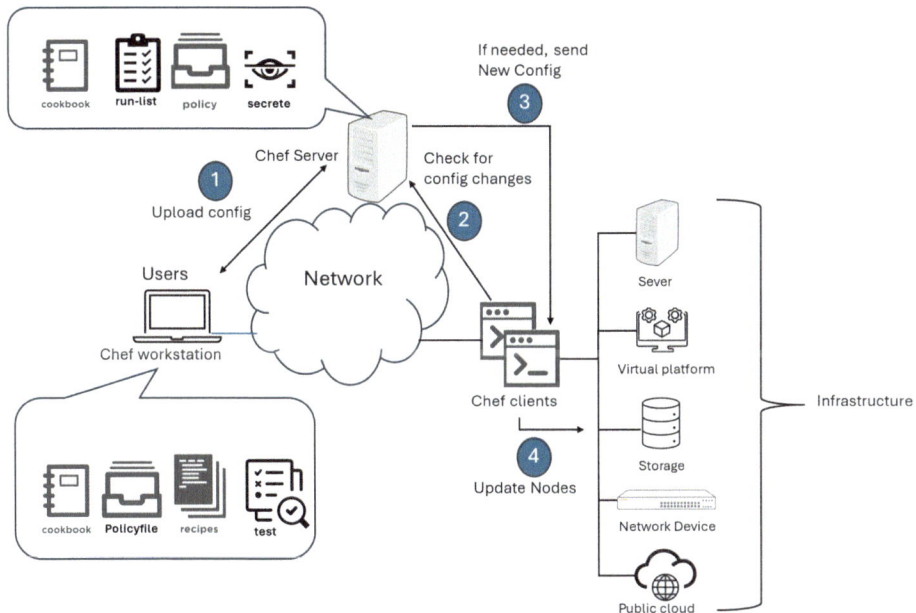

Fig. 5.10 Configuration management using Chef network automation tool

Chef operates on a client–server architecture, consisting of the Chef server, Chef clients (agents), and Chef Workstation. The Chef server acts as the central repository for all configuration data and policies. Chef clients, installed on network devices, periodically pull configuration updates from the Chef server, ensuring that all devices are consistently configured according to the defined policies. The diagram below illustrates the various components of Chef and their interactions in the network automation process for configuration management (Fig. 5.10).

At the heart of Chef's configuration management capabilities are Cookbooks and Recipes. Cookbooks are collections of Recipes, which are Ruby scripts defining the desired state and configuration of network devices. These scripts can specify various settings, such as interface configurations, routing protocols, VLANs, and firewall rules. By defining these configurations as code, Chef enables version control, testing, and repeatability, significantly reducing the risk of configuration drift and unauthorized changes. Chef uses a domain-specific language (DSL) based on Ruby to write these Cookbooks and Recipes. This allows for highly customizable and complex configurations. For instance, a Recipe can include conditional logic, loops, and data-driven configurations to handle various network scenarios and device types. The config for network devices is uploaded to chef server (please refer to the diagram above) which in turn checks for config changes with chef client and if needed, chef server pushes config to the network devices using chef client.

To ensure consistency and compliance, Chef employs an idempotent model. This means that Recipes can be applied multiple times without changing the system beyond the initial application, ensuring that network devices consistently achieve the desired state. This idempotency is crucial for maintaining stability and predictability in network operations.

Chef Automate is an integrated platform that includes Chef, InSpec, and Habitat, providing a comprehensive solution for continuous automation across infrastructure, applications, and compliance. It offers features for visibility, collaboration, and compliance, ensuring that all configurations meet organizational standards. With Chef Automate, network administrators can monitor the compliance of network devices, automate remediation of non-compliant configurations, and integrate with other network management tools for a holistic automation strategy.

In addition to Cookbooks and Recipes, Chef uses Encrypted Data Bags to securely store and manage sensitive information, such as passwords, API keys, and certificates. This ensures that sensitive data remains protected throughout the automation process and is only accessible to authorized entities. For managing updates and patches, Chef's orchestration capabilities ensure that these are applied consistently and with minimal disruption to network operations. Chef can automate the deployment of software updates and patches, monitor compliance of network devices with these updates, and automatically remediate any issues that arise during the update process. Chef's scalability is another significant advantage. It can manage thousands of devices simultaneously, making it suitable for large-scale network environments. Its integration with various monitoring and logging tools allows for automated responses based on network events, further enhancing network reliability and performance.

In summary, Chef is a robust configuration management tool for network automation. Its capabilities in defining configurations as code, ensuring compliance and security, and automating updates and patches make it an essential tool for maintaining a reliable, secure, and efficient network infrastructure. By leveraging Chef, organizations can achieve greater consistency, reduce the risk of errors, and streamline their network management processes.

5.3.2.5 Entuity

Entuity is a comprehensive network management solution that includes robust configuration management capabilities, making it a valuable tool for network automation. As an integrated platform, Entuity provides real-time monitoring, configuration management, and automated network operations, enabling organizations to maintain efficient, reliable, and secure network infrastructures. Entuity's configuration management features are designed to ensure that network devices are consistently and accurately configured according to predefined policies and standards. This is achieved through several key functionalities:

- **Automated Configuration Backup and Restore:** Entuity automatically backs up network device configurations, providing a reliable way to restore configurations in the event of device failure or misconfiguration. This ensures that network operations can quickly recover from disruptions, maintaining continuity and minimizing downtime.
- **Policy-Based Configuration Management:** With Entuity, administrators can define configuration policies that specify the desired state of network devices. These policies can include settings for interfaces, routing protocols, VLANs, security rules, and more. By enforcing these policies, Entuity helps prevent configuration drift and unauthorized changes, ensuring that all devices adhere to organizational standards.
- **Change Management and Auditing:** Entuity tracks all configuration changes across the network, providing detailed logs and audit trails. This capability is crucial for identifying and investigating unauthorized changes, understanding the impact of configuration modifications, and maintaining compliance with regulatory requirements.
- **Compliance Checking:** Entuity includes tools for automated compliance checking, comparing device configurations against predefined policies and industry standards. Non-compliant configurations are flagged, and remediation actions can be automated to ensure that all devices meet security and operational guidelines.
- **Integration with Other Tools:** Entuity seamlessly integrates with other network management and automation tools, enhancing its capabilities and providing a unified approach to network operations. This integration allows Entuity to use data from monitoring systems to trigger automated configuration changes, improving network performance and reliability.
- **Real-Time Configuration Monitoring:** Entuity provides real-time monitoring of network device configurations, allowing administrators to detect and respond to configuration issues promptly. This proactive approach helps prevent potential problems before they impact network performance.
- **User-Friendly Interface:** The Entuity platform includes a user-friendly interface that simplifies configuration management tasks. Administrators can easily view, modify, and apply configurations across multiple devices, streamlining the management process and reducing the potential for human error.
- **Scalability:** Entuity is designed to scale with large and complex network environments. It can manage configurations for thousands of devices, making it suitable for organizations with extensive and diverse network infrastructures.

In summary, Entuity is a powerful tool for configuration management within network automation. Its features for automated backups, policy enforcement, change management, compliance checking, and real-time monitoring provide a comprehensive solution for maintaining consistent and secure network configurations. By leveraging Entuity, organizations can enhance their network automation strategies, ensuring greater efficiency, reliability, and security in their network operations.

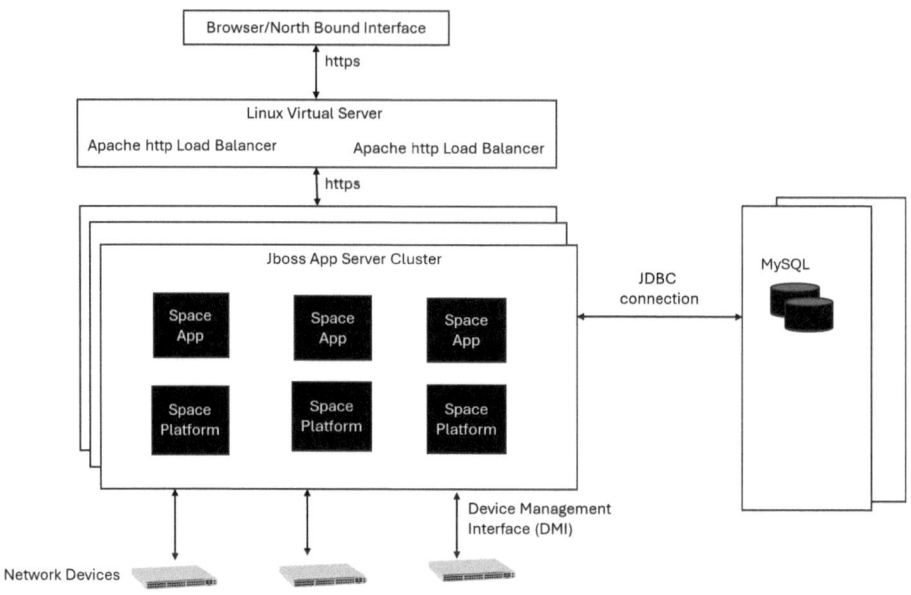

Fig. 5.11   Junos Space architecture (Courtesy: Juniper Networks, Inc[1])

5.3.2.6   Junos Space

Junos Space, developed by Juniper Networks, is a powerful network management platform designed to enhance configuration management within the realm of network automation. It simplifies network operations by offering centralized management, enabling administrators to handle configurations and monitor network devices from a single, unified interface. This centralization streamlines the complexity of managing large-scale networks and improves overall visibility across the infrastructure. The following diagram depicts Junos Space architecture (Fig. 5.11)

Junos Space uses DMI (Device Management Interface) an open API to connect with network devices. Network admin can use GUI and other NBI (North Bound Interface) clients to connect with Junos Space through virtual IP.

One of the standout features of Junos Space is its capability for automated configuration deployment. This allows administrators to create configuration templates that define the desired states for various device types. These templates can then be deployed across the network with minimal manual intervention, ensuring consistency and reducing the risk of configuration errors. This automation is crucial for maintaining a uniform network environment, particularly in large and dynamic networks.

---

[1] Juniper [1].

Junos Space also excels in change management by tracking all configuration changes and providing a detailed audit trail. This is essential for compliance with internal policies and external regulations, as it enables administrators to review historical changes and understand their impact on network performance and security. Such tracking is invaluable for troubleshooting and forensic analysis, as it helps identify the root cause of network issues. Compliance and policy enforcement are integral to Junos Space, which automatically checks device configurations against predefined standards and policies. It flags any deviations and can take corrective actions as needed, helping maintain the security and operational integrity of the network. This ensures that all devices adhere to organizational standards, reducing vulnerabilities and enhancing reliability.

Role-Based Access Control (RBAC) in Junos Space enhances security by defining specific roles and permissions for different users. This ensures that only authorized personnel can make configuration changes, preventing unauthorized modifications and protecting the network from potential security breaches. Integration with other Juniper tools is seamless in Junos Space. It works well with applications like Junos Space Network Director and Junos Space Security Director, offering a holistic approach to network management that combines configuration management with monitoring, security, and orchestration capabilities. This integrated approach improves the overall management and automation of the network.

Real-time configuration monitoring is another key feature of Junos Space. Administrators receive alerts and notifications about configuration changes, compliance violations, and other critical events. This proactive monitoring helps quickly identify and resolve issues, ensuring network stability and performance. Backup and restore functions in Junos Space automate the process of saving and recovering device configurations. Regular backups ensure that configurations can be quickly restored in case of device failure or misconfiguration, minimizing downtime, and maintaining business continuity. Scalability is a strong suit of Junos Space, as it is designed to manage thousands of devices, making it suitable for extensive network environments. Its architecture supports high availability and redundancy, ensuring reliable operation even in large-scale deployments.

The user-friendly, web-based interface of Junos Space simplifies configuration management tasks. Administrators can easily navigate through different modules, view network topologies, and manage device configurations, making the platform accessible even for less experienced users.

In conclusion, Junos Space is a robust configuration management tool that significantly enhances network automation. Its capabilities in centralized management, automated configuration deployment, change management, compliance enforcement, real-time monitoring, and seamless integration with other Juniper tools make it an essential solution for maintaining a consistent, secure, and efficient network infrastructure. By leveraging Junos Space, organizations can improve their network automation strategies, achieving greater operational efficiency, reliability, and security.

5.3.2.7 Puppet

Puppet is an open-source configuration management tool that plays a significant role in network automation, offering robust solutions for automating the management and configuration of network devices. Puppet's architecture is based on a client–server model, where the Puppet master server communicates with Puppet agents installed on network devices. This model ensures that configurations are consistently applied across the network, reducing the risk of configuration drift and maintaining uniformity. The diagram below illustrates Puppet client–server model for network-level configuration management (Fig. 5.12).

At the core of Puppet's configuration management capabilities are files known as 'manifests', which are written in Puppet's declarative language. These manifests define the desired state of network devices, specifying configurations such as interface settings, routing protocols, VLANs, and security policies. By defining configurations as code, Puppet enables version control, repeatability, and testing, which are crucial for maintaining stable and secure network operations. This approach also allows for automated remediation, ensuring that any deviations from the desired state are corrected automatically, thus maintaining compliance and consistency.

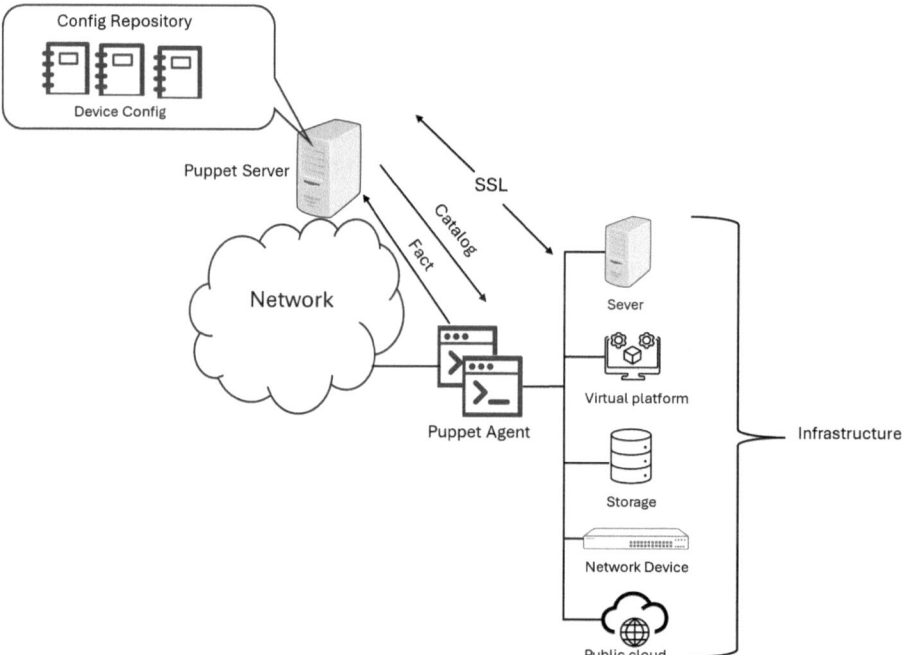

Fig. 5.12 Puppet client–server model for configuration management

Puppet employs a pull-based approach, where agents periodically pull configurations from the master server. This method ensures that even if a device goes offline, it will automatically update its configuration the next time it connects to the master server. This is particularly useful in large, distributed network environments where manual updates would be cumbersome and error-prone. Puppet's scalability is enhanced by its ability to manage thousands of devices simultaneously, making it suitable for complex network infrastructures. In addition to configuration management, Puppet supports detailed reporting and auditing, providing administrators with insights into the state of network devices and any changes made to them. This transparency is essential for compliance with internal policies and external regulations, as it allows for easy tracking of configuration changes and understanding their impact on the network.

Puppet integrates seamlessly with various network management and monitoring tools, enhancing its utility in network automation. This integration allows Puppet to use data from these tools to trigger automated responses, improving overall network reliability and performance. For example, Puppet can automatically adjust configurations based on real-time network conditions, ensuring optimal performance and minimizing downtime. Security is a critical aspect of Puppet's configuration management capabilities. Puppet uses SSL/TLS for secure communication between the master server and agents, ensuring that configuration data is transmitted securely. It also supports role-based access control, allowing administrators to define specific roles and permissions for different users. This ensures that only authorized personnel can make configuration changes, protecting the network from unauthorized access and potential security breaches.

Puppet's flexibility extends to its support for multiple platforms and device types, including network switches, routers, firewalls, and virtual devices. This broad compatibility makes it a versatile tool for managing diverse network environments. Furthermore, Puppet's orchestration capabilities enable the automation of complex workflows across multiple devices, simplifying tasks such as network-wide updates and configuration rollouts.

5.3.2.8   NSO (Network Services Orchestration)

Cisco Network Services Orchestrator (NSO) is a comprehensive configuration management and network automation tool designed to streamline the management of network devices and services. NSO's primary strength lies in its ability to provide a unified, model-driven approach to network automation, supporting both traditional and modern, software-defined networks. At the core of NSO is its model-driven architecture, which uses standardized YANG data models to represent network services and device configurations. This approach abstracts the complexities of underlying network devices, allowing network administrators to define services and configurations in a consistent and device-agnostic manner. The use of YANG models ensures that configurations are uniform across different device types and vendors, reducing the risk of configuration errors and inconsistencies. The diagram below depicts Cisco NSO architecture (Fig. 5.13).

Fig. 5.13 Cisco NSO
architecture (Courtesy: Cisco,
Inc[2])

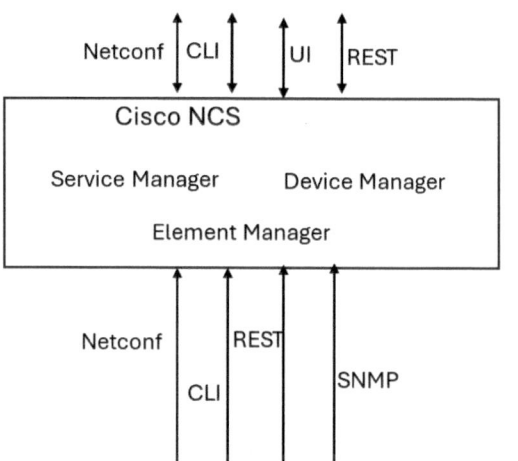

NSO operates using a transactional database that ensures atomicity, consistency, isolation, and durability (ACID) for all configuration changes. This means that any change
made to the network configuration is either fully completed or fully rolled back, ensuring
that partial configurations do not disrupt network operations. This transactional approach
enhances the reliability and stability of network management, as administrators can be
confident that changes will not leave the network in an inconsistent state. One of the
key features of NSO is its support for both CLI-based and API-based device configurations. For devices that support modern APIs, NSO can directly interact with these devices
using REST, NETCONF, or other API protocols. For traditional devices that rely on CLI
commands, NSO can automate configuration changes by translating high-level service
definitions into device-specific CLI commands. This dual support ensures that NSO can
manage a wide range of network devices, regardless of their interface capabilities.

NSO's service orchestration capabilities allow for the automation of complex network
services that span multiple devices and network segments. By defining these services
in YANG models, NSO can automatically provision, update, and decommission services

---

[2] Rayak [2].

across the network. This capability is particularly useful for service providers and large enterprises that need to manage dynamic and complex network environments.

In addition to configuration management, NSO provides powerful monitoring and compliance features. It can continuously monitor the state of network devices and services, ensuring that they remain compliant with predefined policies and configurations. If any deviations are detected, NSO can automatically remediate these issues, restoring the network to its desired state. This continuous compliance checking is essential for maintaining network security and operational integrity.

NSO's integration capabilities further enhance its utility in network automation. It can integrate with various network management and orchestration tools, such as Ansible, Puppet, and Chef, providing a cohesive automation ecosystem. This integration allows NSO to leverage the strengths of these tools while providing a unified interface for managing network configurations and services.

For network operators, NSO offers a high degree of scalability and flexibility. It can manage thousands of devices across multiple locations, making it suitable for large-scale deployments. Its flexible deployment options, including on-premises and cloud-based implementations, allow organizations to choose the best deployment model for their needs.

5.3.2.9 Jinja

Jinja is a templating engine for Python widely used in network automation for configuration management. Its flexibility and simplicity make it an essential tool for generating dynamic configurations and automating network device management.

In the context of network automation, Jinja is primarily used to create templates for network device configurations. These templates are written in a format that combines static text with dynamic placeholders, which can be populated with data at runtime. This approach allows network administrators to define a generic configuration template that can be customized for different devices and scenarios by substituting variables with actual values.

Jinja templates support a wide range of features, including variables, expressions, loops, and conditionals. This allows for highly customizable and complex configurations. For instance, a Jinja template can include logic to generate different configurations based on device types, interfaces, or other parameters. This flexibility is crucial for managing diverse network environments where each device may require a slightly different configuration.

To integrate Jinja with network automation workflows, it is typically used in conjunction with automation tools like Ansible, SaltStack, or custom Python scripts. For example, in Ansible, Jinja templates are used to generate configuration files dynamically based on the inventory and variables defined in playbooks. Ansible's built-in support for Jinja allows for seamless integration, making it easy to automate the deployment of configurations across multiple devices.

The process generally involves the following steps:

- **Template Creation:** A Jinja template is created to define the desired configuration, using placeholders for variable parts of the configuration. These placeholders can be populated with data at runtime.
- **Data Definition:** Variables and their values are defined in an external source, such as Ansible inventory files, SaltStack pillar data, or a custom Python script. These values provide the necessary data to fill in the placeholders in the template.
- **Template Rendering:** The Jinja template is rendered by the automation tool, which replaces the placeholders with the actual values from the defined variables. This generates the final configuration file specific to each device.
- **Deployment:** The rendered configuration files are then deployed to the network devices using the automation tool, ensuring that each device is configured according to the desired state.

One of the key advantages of using Jinja for configuration management is its ability to enforce consistency and reduce errors. By using templates, administrators can ensure that configurations adhere to a standardized format, minimizing the risk of configuration drift and inconsistencies. This is particularly important in large-scale network environments where manual configuration changes can lead to errors and operational issues. Jinja also enhances the scalability of network automation efforts. Templates can be reused across multiple devices and scenarios, reducing the time and effort required to create and manage configurations. This makes it easier to manage large networks with diverse device types and configurations.

In addition to generating configuration files, Jinja can be used for other automation tasks, such as generating reports, documentation, or any other text-based output that requires dynamic content. This versatility makes Jinja a valuable tool in the network administrator's toolkit.

5.3.3   Infrastructure as Code (IaC)

Infrastructure as Code (IaC) is a fundamental practice in modern network automation that involves managing and provisioning computing infrastructure through machine-readable configuration files rather than through physical hardware configuration or interactive configuration tools. IaC enables consistent, repeatable, and automated deployment of infrastructure resources. Here, we will focus on three prominent IaC tools: Terraform, Pulumi, and VMware vRealize Automation.

5.3.3.1 Terraform

Terraform, developed by HashiCorp, is an open-source tool that allows you to define infrastructure as code using a high-level configuration language called HashiCorp Configuration Language (HCL). It supports a wide range of cloud providers and services, making it highly versatile for network automation. Terraform's primary strength lies in its declarative approach, where users define the desired state of infrastructure, and Terraform takes care of creating and maintaining that state. This includes provisioning network devices, setting up virtual networks, configuring firewall rules, and more. Terraform's plan and apply workflow ensures that changes are predictable and easily reviewable before they are executed, reducing the risk of errors.

Terraform's extensive module ecosystem and provider support enable seamless integration with various network equipment and cloud services, allowing for the automation of complex network topologies and configurations. Its state management features keep track of the current infrastructure state, enabling efficient updates and rollbacks. The following diagram illustrates Terraform's architecture and usage in network automation. Typically, Terraform integrates with network management and orchestration platforms. However, in standalone deployments, users may install it on their workstations or on a CI/CD server.

As depicted in the diagram, Terraform architecture consists of several key components: Terraform Core, Providers, Provisioners, Terraform execution file, and the state file. Terraform Core is a statically compiled binary written in Go, which serves as the entry point for users and is also known as the Terraform CLI. It generates the command-line tool "terraform," which acts as the primary interface for users. Providers and Provisioners are plugin modules that are installed alongside Terraform Core. Providers enable Terraform to interact with a wide range of services and resources, including cloud providers, databases, and DNS services. For network resources, providers may connect with network devices using RESTCONF or NETCONF, depending on the vendor-specific implementation. The user execution file allows network administrators to create an execution plan and preview the changes Terraform intends to make to the infrastructure (Fig. 5.14).

This plan helps ensure that changes are deliberate and controlled. The state file is a critical component of Terraform's functionality. It is a JSON file that stores information about the resources managed by Terraform, their current state, and dependencies. Terraform uses the state file to determine the necessary changes to the infrastructure when a new configuration is applied, ensuring resources are not unnecessarily recreated across multiple runs.

Terraform Provisioners are built-in functionalities that allow users to execute commands, scripts, or other configuration actions on local or remote machines. They can also transfer files from a local environment to a remote one. Provisioners are utilized at various stages of the Terraform plan-apply-destroy cycle, such as during resource creation or deletion, adding flexibility and control over the automation process.

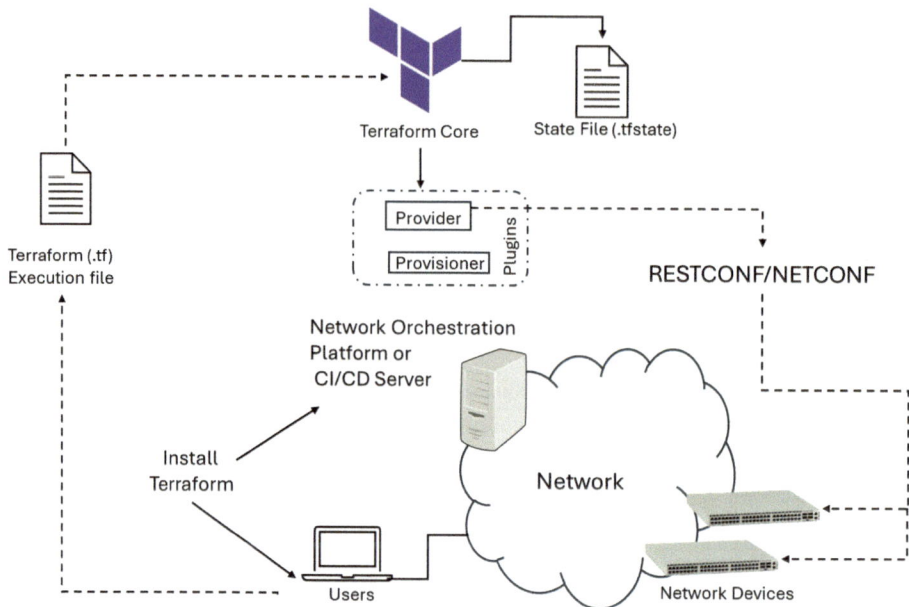

Fig. 5.14   Terraform architecture and deployment scenario

In summary, Terraform's architecture and capabilities make it a powerful tool for automating network configurations, enabling seamless integration, precise state management, and flexible provisioning, which together enhance the efficiency and reliability of network operations.

5.3.3.2   Pulumi

Pulumi is an infrastructure as code tool that enables infrastructure management using general-purpose programming languages such as Python, TypeScript, JavaScript, Go, and C#. Unlike Terraform's domain-specific language, Pulumi allows developers to use familiar programming constructs and libraries, making it an attractive option for those who prefer using traditional programming languages for infrastructure tasks.

Pulumi's approach to IaC allows for more complex and dynamic configurations, leveraging the full power of programming languages. For instance, users can define intricate logic, loops, and conditionals directly within their infrastructure code, providing greater flexibility for network automation tasks.

Pulumi supports a wide range of cloud providers and services, similar to Terraform, and facilitates the automation of network configurations across diverse environments. Pulumi's state management and preview features provide insight into changes before they are applied, ensuring safe and predictable updates.

5.3.3.3   VMware vRealize Automation

VMware vRealize Automation (vRA) is an enterprise-grade automation platform that integrates with VMware's ecosystem to provide comprehensive infrastructure management. vRA allows users to define infrastructure and application blueprints that automate the deployment and management of virtual machines, network configurations, and other IT services.

vRealize Automation's strength lies in its deep integration with VMware's suite of products, making it particularly suitable for environments heavily invested in VMware technologies. vRA enables the automation of complex multi-cloud environments, supporting hybrid cloud deployments and network automation across on-premises and cloud infrastructures. vRA's infrastructure as code capabilities are facilitated through its Blueprint Designer, where users can visually or programmatically define the desired state of their infrastructure. These blueprints can include network configurations, security policies, and application deployments. vRA also integrates with configuration management tools like Ansible and Puppet, enhancing its automation capabilities.

vRA's policy-driven governance and compliance features ensure that all automated deployments adhere to organizational standards and security policies. This is particularly important in large enterprises where compliance and auditability are critical.

5.3.4   Automation Hub and CI/CD in Network Automation

Automation Hub and Continuous Integration/Continuous Deployment (CI/CD) are essential components in modern network automation, providing centralized control, streamlined workflows, and enhanced deployment capabilities. This section explores how Ansible Tower, Puppet Enterprise, SaltStack, Rundeck, and Jenkins contribute to network automation through technical details.

5.3.4.1   Ansible Tower

In previous sections, we discussed the architecture of Ansible, so we will now focus on Ansible Tower, the enterprise version of Ansible. Ansible Tower builds on the capabilities of open-source Ansible by adding a range of features that provide a centralized platform for managing and scaling automation across the network. It offers a web-based interface and REST API, which allow administrators to manage configurations, monitor jobs, and orchestrate workflows more effectively.

Key features are as follows:

- **Job Scheduling:** Automate the execution of Ansible playbooks at scheduled times, ensuring timely updates and maintenance tasks.
- **Role-Based Access Control (RBAC):** Define user roles and permissions to control access to resources and playbooks, enhancing security.

- **Real-Time Reporting:** Monitor the status of automation jobs and view detailed logs to track changes and troubleshoot issues.
- **Inventory Management:** Manage dynamic inventories, automatically updating device lists and configurations.

5.3.4.2 Puppet Enterprise

In earlier sections, we discussed Puppet. The Puppet Enterprise is the commercial version of the Puppet tool. It is built on top of open-source puppet and a powerful configuration management solution designed to automate the management of network devices. It uses a declarative language to define the desired state of network configurations, ensuring consistency and compliance across the network infrastructure.

Key features:

- **Node Classification:** Group and manage devices based on their roles and characteristics, simplifying configuration management.
- **Event Inspector**: Track changes and audit events to ensure compliance and security.
- **Orchestration:** Automate complex workflows across multiple devices and services, improving efficiency and reducing the risk of human error.
- **Environment Isolation:** Use environments to manage different stages of deployment, such as development, testing, and production, ensuring safe and controlled rollouts.

5.3.4.3 SaltStack

As discussed in Sect. 5.3.2.3, SaltStack's architecture and use cases highlight its adaptability and robust capabilities in network automation. When applied within the context of an automation hub and CI/CD pipeline, SaltStack proves to be an invaluable tool for streamlining and enhancing network operations.

SaltStack operates with an event-driven architecture, enabling real-time responses to network changes, which is essential for dynamic environments where swift reactions to events like device failures or configuration changes are necessary. Its server-agent model, consisting of the Salt Master and Salt Minions, supports both agent-based and agentless operations, making it flexible and suitable for various network setups. In an automation hub, SaltStack can serve as the central orchestrator, managing and coordinating the various components of the network. It integrates seamlessly with other tools, allowing for comprehensive automation workflows that include configuration management, compliance checks, and real-time monitoring. SaltStack's high-speed communication bus, ZeroMQ, ensures efficient management of large-scale networks, facilitating the simultaneous handling of thousands of devices.

In a CI/CD pipeline, SaltStack plays a crucial role in automating the deployment and testing of network configurations. It supports remote execution, enabling the execution of commands across multiple devices in parallel, thus significantly reducing the time needed for network-wide updates and maintenance tasks. SaltStack's idempotent state management ensures that configurations are applied consistently, maintaining the desired state of the network even as updates are deployed.

By leveraging SaltStack within an automation hub and CI/CD framework, organizations can achieve higher efficiency, consistency, and reliability in their network operations. Its ability to integrate with other network management and monitoring tools enhances its utility, enabling automated responses to network events and ensuring continuous compliance with security and operational standards.

5.3.4.4 Rundeck

Rundeck is an open-source tool and a valuable addition to the network automation hub and CI/CD ecosystem, offering comprehensive job scheduling, orchestration, and automation capabilities. It is designed to facilitate the automation of routine tasks, enable seamless integration with various systems and tools, and improve operational efficiency across network environments. Rundeck's ability to integrate with CI/CD pipelines makes it a valuable component in modern network management strategies.

Rundeck provides a centralized platform where administrators can define, manage, and execute workflows that span different systems and network devices. It supports various automation tasks such as configuration changes, software deployments, and routine maintenance operations. Rundeck's intuitive web-based interface allows users to create jobs using a simple graphical interface or via scripting, making it accessible to both technical and non-technical users. The following diagram depicts Rundeck architecture, which consists of several key components that work together to provide a flexible and scalable automation platform:

- **Rundeck Server:** The core component that hosts the web-based interface, processes job execution requests, and manages workflows.
- **Job Definitions:** YAML or XML files that define the steps, scripts, and commands to be executed for each job.
- **Node Definitions:** Descriptions of the network devices and systems on which jobs will be executed, including details like IP addresses and access credentials.
- **Plugins:** Extensible modules that integrate Rundeck with various systems and tools, providing additional functionality and connectivity (Fig. 5.15).

In the context of network automation, Rundeck can be used to automate a wide range of tasks, from simple device reboots to complex configuration changes across multiple devices. For example, a network administrator can define a job in Rundeck to automatically back up configurations from all network devices every night, apply firmware updates in a staged manner, or monitor network performance and automatically adjust configurations in response to detected issues.

Rundeck's ability to execute jobs based on specific triggers makes it particularly valuable for network operations centers (NOCs) where quick and automated responses to

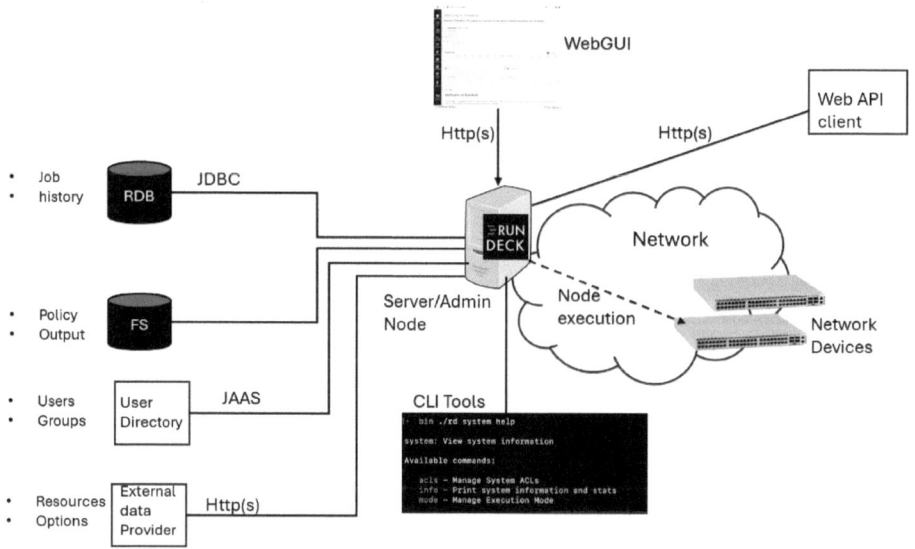

Fig. 5.15 Rundeck automation[3]

network events are critical. By integrating Rundeck into a CI/CD pipeline, network teams can ensure that all configuration changes are thoroughly tested and deployed in a controlled and repeatable manner, significantly reducing the risk of network outages and performance degradation.

5.3.4.5 Jenkins

Jenkins is a widely used open-source automation server that plays a crucial role in the network automation hub and CI/CD ecosystem. It enables continuous integration and continuous delivery (CI/CD) by automating the building, testing, and deployment of applications. Jenkins can be extended with a vast number of plugins to support a wide range of tasks, including those specific to network automation. In the context of network automation, Jenkins automates the deployment and management of network configurations. By integrating with various network automation tools like Ansible, Puppet, and SaltStack, Jenkins can orchestrate complex workflows that involve updating network devices, applying configuration changes, and monitoring the network for compliance and performance.

Jenkins supports pipeline as code, where complex workflows are defined using a simple, readable language. This feature allows network administrators to create repeatable, version-controlled processes for managing network infrastructure, ensuring consistency and reducing the risk of human error.

---

[3] Rundeck.com [3].

Jenkins follows a master-agent architecture, which enables scalable and distributed automation. Here is an overview of its components:

- **Jenkins Controller (Formerly Master):** This is the central control unit of Jenkins. It schedules jobs, dispatches builds to agents, monitors agents, and collects and presents the build results. The master is also responsible for serving the Jenkins UI, where users can configure jobs and view results.
- **Jenkins Agents:** Agents are responsible for executing the tasks delegated by the master. These tasks can include building code, running tests, and deploying applications. Agents can run on various platforms, allowing Jenkins to be flexible and scalable.

Users interact with the Jenkins controller through a web interface or API to create job configurations, specifying the source control repository (e.g., GitHub) and defining pipelines. By leveraging scripts stored in version control for consistency and scalability, this approach streamlines network management tasks and enhances operational efficiency. Plugins play a crucial role in Jenkins, extending its capabilities to integrate with various tools and platforms used in network automation. For instance, plugins can enable Jenkins to communicate with Ansible, Puppet, or other automation tools, allowing for seamless orchestration across different environments.

Once the job configuration and pipelines are set up, users can trigger tasks manually or automatically based on specified conditions. The pipeline, which is defined in a Jenkinsfile (a script that defines the CI/CD process), outlines the stages of the workflow, such as fetching the latest code from the repository, executing scripts to configure network devices, running tests to verify the configurations, and deploying the changes to the production environment. Jenkins' robust plugin ecosystem supports various steps within the pipeline. For instance, the Git plugin can pull the latest configuration scripts from a Git repository, while the Ansible plugin can run playbooks to automate network tasks. These plugins ensure that each step of the pipeline is executed efficiently and reliably.

As tasks and pipelines execute, Jenkins provides real-time feedback through its web interface, displaying logs and results for each stage. This visibility allows network administrators to monitor progress, identify issues, and take corrective actions promptly. By automating repetitive tasks and integrating with powerful network automation tools, Jenkins enhances the agility and reliability of network operations.

In summary, Jenkins' workflow for network automation involves creating job configurations and pipelines, utilizing plugins to extend functionality and ensure seamless integration with various tools, and executing tasks efficiently to streamline network management. This comprehensive approach not only reduces manual effort but also ensures consistency and scalability in managing network configurations.

5.3.5 Monitoring and Logging

Monitoring and logging are crucial components of network automation, ensuring that network operations remain efficient, reliable, and secure. These processes provide real-time visibility into network performance, facilitate proactive issue detection, and enable comprehensive analysis of network health. In the context of network automation, several tools stand out for their capabilities in monitoring and logging. The ELK Stack, consisting of Elasticsearch, Logstash, and Kibana, offers powerful search, analysis, and visualization functionalities. Telegraf is a versatile data collection agent that integrates seamlessly with various back-end systems. Grafana provides advanced visualization and dashboard creation capabilities, making it a popular choice for monitoring solutions. Dynatrace leverages artificial intelligence to deliver deep insights and proactive monitoring. InfluxDB, a time-series database, is well suited for handling metrics and events, while Prometheus offers reliable and scalable monitoring and alerting functionalities. SolarWinds provides a comprehensive suite of network management and monitoring tools, and Splunk excels in handling large volumes of machine-generated data for real-time analysis and alerting. Each of these tools brings unique strengths to the table, and together, they form a robust ecosystem for monitoring and logging in network automation. The following sections will delve into each of these tools, exploring their specific features and how they contribute to effective network management.

5.3.5.1 ELK Stack

The ELK stack, consisting of Elasticsearch, Logstash, and Kibana, is a powerful combination of tools designed for real-time search, analysis, and visualization of log data. It plays a crucial role in network automation by providing comprehensive logging and monitoring capabilities that help network administrators quickly identify and troubleshoot issues, ensuring efficient and reliable network operations. The following diagram depicts various components of the ELK stack and how they work together to provide effective logging and monitoring services for network automation. The ELK stack comprises three essential components: Elasticsearch, Logstash, and Kibana. Elasticsearch functions as a distributed, RESTful search and analytics engine, adept at storing and indexing large volumes of log data. Within network automation, Elasticsearch serves as the backbone of the ELK stack, efficiently managing and retrieving log data from various network devices and systems. Its support for complex queries enables network administrators to conduct detailed searches and analyses on log data (Fig. 5.16).

The scalability and performance of Elasticsearch are crucial for handling the extensive logs generated by automated network systems. Its distributed architecture ensures high availability and fault tolerance, vital for continuous monitoring and logging in large network environments. By indexing log data, Elasticsearch facilitates quick retrieval and analysis, aiding administrators in identifying issues and analyzing network performance trends.

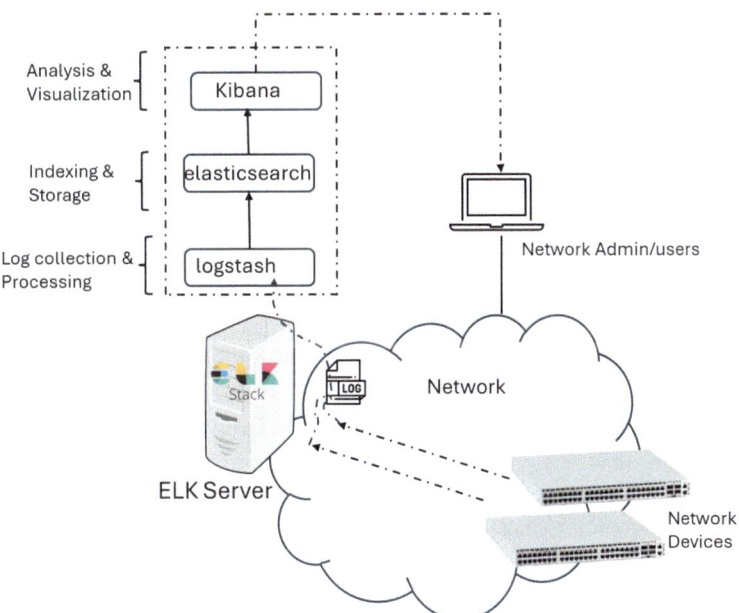

Fig. 5.16 Network monitoring and logging using ELK stack

Logstash acts as a dynamic data processing pipeline within the ELK Stack, ingesting data from multiple sources, transforming it, and sending it to a storage "stash" such as Elasticsearch. In network automation, Logstash serves as a versatile tool for collecting and processing log data from various network devices. Supporting a wide range of input plugins, Logstash can ingest data from diverse sources like syslogs, SNMP traps, and API endpoints. Its powerful filtering and transformation capabilities enable the normalization and enrichment of log data before transmission to Elasticsearch. This preprocessing step is vital for ensuring structured and standardized log data, simplifying analysis and visualization. By managing data ingestion and transformation complexities, Logstash ensures that only relevant and meaningful log data is stored in Elasticsearch, optimizing storage and search performance.

Kibana serves as the visualization layer of the ELK Stack, offering a web-based interface for exploring and visualizing data stored in Elasticsearch. For network automation, Kibana provides robust tools for creating real-time dashboards, visualizations, and reports based on log data collected by Logstash and indexed in Elasticsearch. Its intuitive interface enables network administrators to customize dashboards, providing insights into network performance, security events, and operational metrics. These dashboards support various visualizations such as line graphs, bar charts, heat maps, and tables, enabling administrators to monitor key performance indicators (KPIs) and detect anomalies quickly.

Moreover, Kibana supports advanced querying and filtering capabilities, facilitating root cause analysis by allowing administrators to drill down into specific log events. Real-time correlation and visualization of log data enhance network visibility, enabling proactive monitoring and quicker issue response.

Together, Elasticsearch, Logstash, and Kibana form a comprehensive logging and monitoring solution that empowers network administrators with the tools they need to maintain efficient and reliable network operations. By providing real-time insights into network performance and security, the ELK Stack helps administrators to.

- **Identify and Troubleshoot Issues:** The powerful search and analysis capabilities of Elasticsearch, combined with the visualization tools in Kibana, allow administrators to quickly identify the root cause of network problems and take corrective action.
- **Monitor Network Performance:** Real-time dashboards and visualizations in Kibana provide a clear view of network health, enabling administrators to track performance metrics and detect potential issues before they escalate.
- **Enhance Security:** The ELK Stack helps in monitoring security logs and detecting suspicious activities, contributing to improved network security posture.
- **Optimize Operations:** By automating the collection and analysis of log data, the ELK Stack reduces the manual effort required for monitoring and troubleshooting, allowing administrators to focus on strategic tasks.

In conclusion, the ELK stack is an essential tool for network automation, providing robust capabilities for logging, monitoring, and visualizing network data. Its ability to handle large volumes of log data, combined with real-time analysis and visualization, makes it an invaluable asset for network administrators seeking to ensure the optimal performance and security of their automated network environments.

5.3.5.2 Telegraf

Telegraf stands out as a versatile open-source data collection agent crucial for monitoring and logging in network automation contexts. Its robust support for a diverse range of input and output plugins renders it highly adaptable, capable of seamlessly gathering metrics from various sources across networks. Telegraf's lightweight design and extensibility make it an optimal choice for monitoring network performance, health, and beyond.

At its core, Telegraf excels in its ability to collect data from an array of network devices, applications, and other pertinent sources. Whether it's capturing metrics from routers, switches, servers, or applications, Telegraf ensures comprehensive data collection across the network infrastructure. Moreover, its support for numerous input plugins enables seamless integration with different types of data sources, including System Metrics, SNMP, Http APIs, JMX, database queries, and Log files. This flexibility allows

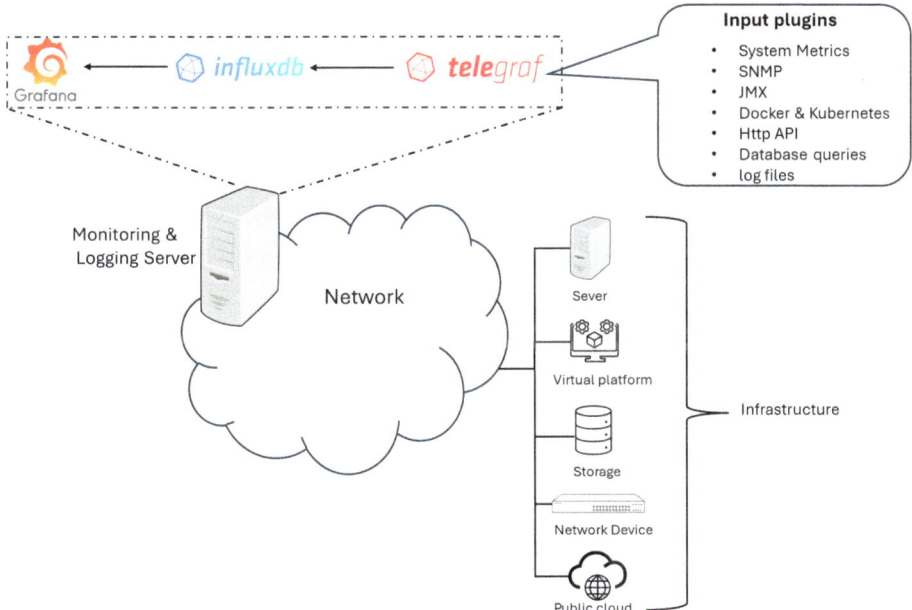

Fig. 5.17 Typical Telegraf deployment scenarios for network automation in terms of logging and monitoring

network administrators to monitor a wide range of metrics critical for ensuring network reliability, performance, and security. The following diagram depicts typical Telegraf deployment scenarios (Fig. 5.17).

Once collected, Telegraf efficiently sends this data to various back-end systems, such as InfluxDB or Prometheus, for storage, analysis, and visualization. By seamlessly integrating with these systems, Telegraf facilitates the creation of comprehensive monitoring solutions tailored to the specific needs of network automation environments. Whether it's storing time-series data in InfluxDB for real-time monitoring or leveraging Prometheus for alerting and trending analysis, Telegraf ensures that network administrators have access to the insights needed to maintain optimal network operations.

Telegraf's lightweight nature is another significant advantage, minimizing resource consumption on network devices and systems while maximizing efficiency in data collection and transmission. This lightweight footprint is particularly beneficial in distributed network environments where resources may be constrained, ensuring minimal impact on network performance.

Moreover, Telegraf's extensibility further enhances its utility in network automation scenarios. With the ability to develop custom plugins tailored to unique monitoring requirements, Telegraf offers unparalleled flexibility in capturing and analyzing network metrics. Whether it's integrating with proprietary monitoring solutions or extending func-tionality to capture specialized metrics, Telegraf empowers network administrators to tailor their monitoring and logging workflows to suit specific network automation needs.

In summary, Telegraf emerges as an indispensable tool in the realm of network automa-tion, offering unparalleled flexibility, efficiency, and extensibility in data collection for monitoring and logging purposes. Its support for diverse input and output plugins, seam-less integration with back-end systems, lightweight design, and extensibility makes it a preferred choice for network administrators seeking comprehensive visibility into network performance and health.

5.3.5.3 Grafana

Grafana stands as a beacon in the realm of network automation, offering a comprehensive suite of tools tailored to the intricate demands of monitoring and logging. At its core, Grafana epitomizes flexibility and adaptability, integrating seamlessly with an array of data sources such as Prometheus, InfluxDB, Elasticsearch, and beyond. This interoper-ability grants users the freedom to curate a monitoring environment that aligns precisely with their infrastructure's unique requirements.

In the context of network automation, Grafana emerges as a linchpin for observability, providing a centralized hub for aggregating and analyzing vital network metrics. Through its intuitive interface, users can effortlessly craft dynamic dashboards, transforming raw data streams into actionable insights with unparalleled clarity. Whether scrutinizing band-width utilization, latency patterns, or device health, Grafana empowers operators to dissect network performance at a granular level, facilitating swift troubleshooting and informed decision-making.

Crucially, Grafana's prowess extends beyond mere visualization, affording users the means to orchestrate proactive monitoring strategies. By configuring bespoke alerts, stake-holders can preemptively address burgeoning issues before they escalate, safeguarding network integrity and preserving operational continuity. This proactive stance is indispens-able in the fast-paced realm of network automation, where swift responses to anomalies can mitigate downtime and bolster resilience.

Moreover, Grafana's adaptability fosters a culture of innovation, enabling organizations to evolve their monitoring frameworks alongside technological advancements. Whether embracing emerging protocols or integrating novel telemetry solutions, Grafana serves as a flexible canvas upon which network architects can paint their vision of tomor-row's infrastructure. This forward-looking ethos positions Grafana as a cornerstone in the perpetual quest for network optimization and resilience.

In essence, Grafana transcends the confines of traditional monitoring platforms, tran-scending into a realm where data becomes actionable intelligence, and insights drive strategic imperatives. By harnessing its open-source ethos and robust feature set, network automation practitioners can navigate the complexities of modern infrastructure with confidence, secure in the knowledge that Grafana stands as a stalwart ally in their quest for operational excellence.

5.3.5.4   InfluxDB

InfluxDB, revered in the realm of time-series data management, serves as a cornerstone for organizations grappling with the deluge of metrics and events generated by modern network infrastructures. Engineered with a keen focus on scalability, performance, and flexibility, InfluxDB emerges as a beacon of reliability in the tumultuous seas of data ingestion and querying.

At its core, InfluxDB is purpose-built to excel in scenarios where time is of the essence. By eschewing traditional relational database paradigms in favor of a specialized time-series data model, InfluxDB streamlines the ingestion and retrieval of temporal data points with unparalleled efficiency. This architectural decision not only minimizes storage overhead but also facilitates lightning-fast query execution, enabling operators to extract insights from vast volumes of time-stamped data with remarkable expediency.

In the context of network performance monitoring, InfluxDB's prowess shines brightest. Its innate ability to effortlessly ingest and index metrics and events—ranging from bandwidth utilization and latency measurements to device statuses and error logs—renders it an ideal repository for capturing the pulse of network health. Whether tracking the ebb and flow of traffic volumes across myriad interfaces or scrutinizing the minutiae of packet loss occurrences, InfluxDB stands poised to capture the essence of network performance with unwavering fidelity.

However, InfluxDB's utility transcends mere data storage; it forms the linchpin of a holistic monitoring ecosystem when paired with Telegraf and Grafana. Telegraf, with its arsenal of input plugins tailored to collect system metrics, network telemetry, and application statistics, serves as the vanguard of data acquisition, ferrying real-time insights from the trenches of network infrastructure to the sanctum of InfluxDB's data reservoirs. Through this symbiotic relationship, Telegraf ensures that no metric goes unrecorded, no event slips through the cracks, thus fortifying the foundation of network observability.

Meanwhile, Grafana ascends as the maestro of visualization, wielding a palette of dynamic dashboards and interactive charts to transform raw data streams into actionable insights. By seamlessly integrating with InfluxDB, Grafana breathes life into the troves of temporal data, allowing operators to traverse time and space effortlessly in pursuit of performance trends, anomalies, and emergent patterns. Through its intuitive interface and rich customization capabilities, Grafana empowers stakeholders to craft bespoke monitoring dashboards tailored to their unique perspectives and priorities, fostering a culture of data-driven decision-making and operational excellence.

In summary, the triumvirate of InfluxDB, Telegraf, and Grafana represents more than a mere aggregation of tools; it embodies a paradigm shift in network monitoring, wherein the convergence of cutting-edge technology and thoughtful design fosters a newfound era of visibility, resilience, and efficiency. Armed with this formidable arsenal, organizations stand poised to navigate the turbulent seas of network operations with confidence, secure in the knowledge that their infrastructure's heartbeat is captured, analyzed, and visualized with unwavering precision.

5.3.5.5 Prometheus

Prometheus is a robust open-source monitoring and alerting toolkit designed specifically for reliability and scalability. It is particularly well suited for network automation due to its powerful features and integrations. Prometheus collects and stores metrics as time-series data, recording information with timestamps and optional key-value pairs called labels. This data model is particularly effective for monitoring network devices and infrastructure. Prometheus operates by periodically scraping metrics from instrumented applications and services, including network devices, through HTTP endpoints. This pull-based approach ensures that Prometheus can efficiently gather data from a diverse set of sources, making it ideal for complex network environments where devices and services are constantly changing.

At the core of Prometheus is its multi-dimensional data model. By using labels, Prometheus can tag metrics with metadata, allowing for granular analysis and filtering. This is particularly useful in network automation, where metrics from various devices and services need to be correlated and analyzed to understand the overall health and performance of the network. Prometheus also features a powerful query language called PromQL, which enables flexible and precise queries for real-time monitoring and alerting. Network administrators can use PromQL to create sophisticated dashboards and alerts that provide insights into network performance, detect anomalies, and trigger automated responses to potential issues.

For long-term storage and visualization, Prometheus integrates seamlessly with Grafana, an open-source analytics and monitoring platform. Grafana can ingest data from Prometheus and other sources, providing a unified view of the network's health and performance. With Grafana, network administrators can create detailed, interactive dashboards that display real-time and historical data, helping to identify trends and pinpoint issues quickly.

In addition to its built-in capabilities, Prometheus supports a wide range of exporters and integrations that extend its functionality. Exporters are lightweight programs that translate metrics from various services, including network devices, into a format that Prometheus can scrape. Commonly used exporters for network automation include node_exporter for hardware and OS metrics, and snmp_exporter for Simple Network Management Protocol (SNMP) data from network devices. Alertmanager is another critical component of the Prometheus ecosystem, responsible for handling alerts generated by Prometheus' monitoring rules. Alertmanager can route alerts to various destinations, such as email, Slack, PagerDuty, or custom webhooks, ensuring that network administrators are promptly notified of any critical issues.

In summary, Prometheus is a powerful tool for monitoring and logging in network automation. Its ability to efficiently collect, store, and query time-series data, combined with its integrations with Grafana and various exporters, makes it an essential component of a modern network automation monitoring and alerting solution. By leveraging Prometheus, organizations can achieve real-time visibility into their network infrastructure, proactively address potential issues, and maintain optimal network performance and reliability.

5.3.5.6 SolarWinds

SolarWinds is a comprehensive suite of network management tools designed to provide detailed monitoring, logging, and analysis of network infrastructure. It is widely used in network automation due to its robust capabilities in tracking network performance, detecting issues, and optimizing operations. SolarWinds offers various modules tailored to specific aspects of network management, making it a versatile solution for network automation environments.

At the core of SolarWinds' network monitoring capabilities is the Network Performance Monitor (NPM). NPM provides real-time visibility into network performance by continuously monitoring devices, interfaces, and network paths. It uses Simple Network Management Protocol (SNMP), ICMP, and other protocols to gather metrics such as bandwidth usage, packet loss, latency, and device status. This continuous monitoring allows network administrators to quickly identify and respond to performance issues and potential bottlenecks. SolarWinds also features the Network Configuration Manager (NCM), which automates the management of network device configurations. NCM allows administrators to schedule configuration backups, track changes, and enforce compliance with network policies. This capability is crucial for network automation, as it ensures that configurations are consistent and adherent to predefined standards, reducing the risk of misconfigurations that can lead to network downtime or security vulnerabilities.

For logging, SolarWinds offers the Log Analyzer module. This tool collects, analyzes, and correlates log data from various network devices and systems. It supports syslog, SNMP traps, and Windows Event Log data, providing a centralized repository for all log information. Log Analyzer enables administrators to search and filter logs, set up alerts for specific log events, and generate reports, helping to quickly diagnose issues and understand network events. Another important aspect of SolarWinds in network automation is its integration with other IT management tools. SolarWinds' suite of products, such as Server & Application Monitor (SAM) and Database Performance Analyzer (DPA), can work together to provide a holistic view of the entire IT infrastructure. This integration allows for seamless data sharing and correlation, enhancing the ability to automate and optimize network operations.

SolarWinds uses a web-based interface that is intuitive and easy to navigate, allowing users to customize dashboards, set up alerts, and generate detailed reports. This user-friendly design helps network administrators efficiently manage and monitor their networks without the need for extensive training. In addition to its robust feature set, SolarWinds provides comprehensive alerting capabilities. Administrators can configure custom alerts based on a wide range of criteria, such as performance thresholds, configuration changes, and specific log events. Alerts can be sent via email, SMS, or integrated with other communication platforms, ensuring that critical issues are promptly addressed.

SolarWinds' scalability is another key benefit. It can handle large, complex networks with thousands of devices, making it suitable for enterprises of all sizes. Its modular architecture allows organizations to start with essential monitoring and logging capabilities and expand as needed by adding more modules and functionalities.

5.3.5.7 Splunk

Splunk is a powerful platform widely used for network monitoring and logging, particularly in the context of network automation. It excels in collecting, analyzing, and visualizing vast amounts of machine-generated data, providing valuable insights that help in optimizing network performance and ensuring reliability. Splunk is particularly useful for the following functions:

- **Data Collection and Ingestion:** Splunk collects data from various sources, including network devices, servers, applications, and security systems. It supports multiple data input methods, such as syslog, SNMP, and REST APIs, enabling it to gather logs, performance metrics, and events from a diverse range of network equipment. This comprehensive data ingestion capability makes Splunk particularly effective in complex network environments where data comes from numerous and varied sources.
- **Indexing and Searching:** Once the data is ingested, Splunk indexes it in a way that makes it easy to search and analyze. This indexed data can be queried using Splunk's Search Processing Language (SPL), allowing network administrators to perform detailed searches, create complex queries, and correlate events across different data sources. The ability to rapidly search through extensive log files and performance data is crucial for troubleshooting and diagnosing network issues.
- **Real-Time Monitoring and Alerting:** Splunk provides real-time monitoring capabilities that are essential for network automation. It continuously analyzes incoming data and can trigger alerts based on predefined thresholds or specific conditions. For example, if Splunk detects unusual traffic patterns, excessive latency, or potential security threats, it can immediately notify network administrators. This proactive monitoring helps prevent downtime and quickly addresses performance issues before they escalate.
- **Visualization and Dashboards:** One of Splunk's strengths lies in its visualization capabilities. It allows users to create custom dashboards that provide a real-time overview of the network's health and performance. These dashboards can include a variety of visualizations, such as graphs, charts, and heat maps, which help in identifying trends, anomalies, and potential problem areas. By visualizing network

data, administrators can gain a clearer understanding of network behavior and make informed decisions.

- **Integration with Network Automation Tools:** Splunk integrates seamlessly with various network automation tools and platforms. For example, it can work alongside configuration management tools like Ansible, Puppet, and Chef to provide a comprehensive automation and monitoring solution. By integrating Splunk with these tools, organizations can automate network configurations while continuously monitoring their effectiveness and compliance. This integration ensures that any changes made through automation are immediately reflected in the monitoring data, allowing for rapid detection and remediation of issues.

- **Machine Learning and Advanced Analytics:** Splunk leverages machine learning and advanced analytics to enhance network monitoring and logging. Its Machine Learning Toolkit (MLTK) can analyze historical data to identify patterns and predict future network behavior. For instance, it can predict potential failures or performance degradation based on historical trends, enabling proactive maintenance and optimization. This predictive capability is particularly valuable in automated environments where early detection of potential issues can significantly reduce downtime and operational costs.

- **Security and Compliance:** In addition to performance monitoring, Splunk is also widely used for security monitoring and compliance. It can collect and analyze security-related logs, such as firewall logs, intrusion detection system (IDS) alerts, and authentication logs. Splunk's security capabilities help in detecting and responding to security incidents in real-time. Furthermore, it aids in compliance reporting by providing detailed audit trails and generating reports that demonstrate adherence to regulatory requirements.

- **Scalability and Flexibility:** Splunk is highly scalable, making it suitable for large enterprises with extensive network infrastructures. It can handle large volumes of data without compromising performance, ensuring that even as the network grows, monitoring and logging capabilities remain robust and effective. Its flexible architecture allows it to be deployed on-premises, in the cloud, or in hybrid environments, providing organizations with the flexibility to choose the deployment model that best suits their needs.

To summarize, Splunk is an indispensable tool for network monitoring and logging in network automation. Its robust data collection, real-time monitoring, powerful search capabilities, and advanced analytics make it an ideal solution for managing complex network environments. By providing comprehensive visibility into network operations and enabling proactive management, Splunk helps organizations achieve greater efficiency, reliability, and security in their network infrastructures.

5.3.6   Source of Truth

Network automation requires accurate, consistent, and up-to-date information about the network's topology, devices, and configurations. This is where the concept of a "Source of Truth" (SoT) becomes crucial. A SoT is a centralized repository that maintains the definitive data about the network, ensuring that automation processes have access to reliable information. By using a SoT, organizations can streamline network management, reduce errors, and improve overall efficiency. Several tools play significant roles as Sources of Truth in network automation, including NSoT, NetBox, SolarWinds, and Infoblox. In the following sections, we will discuss each of these tools in detail.

5.3.6.1   Network Source of Truth (NSoT)

Network Source of Truth (NSoT) is an open-source tool designed to serve as a centralized repository for network infrastructure data. It acts as a reliable source of truth for network automation processes, ensuring that all tasks are based on accurate and up-to-date information. NSoT is particularly effective at managing IP address allocation, device inventory, and metadata, which are crucial for maintaining consistent and error-free network configurations.

The architecture of NSoT is designed to facilitate integration, scalability, and reliability. At its core is a RESTful API, which provides a programmatic interface for interacting with the stored data, enabling seamless data retrieval and updates. This API allows for easy integration with various network automation tools and scripts. NSoT uses a backend database to store all network-related data, including IP addresses, device information, and metadata. The database is designed to handle large amounts of data and provide high availability, ensuring resilience and scalability.

NSoT also offers a web-based interface that allows network administrators to manage and visualize data intuitively. This interface provides a user-friendly way to interact with the data, perform searches, and generate reports. The data model in NSoT is flexible and can be customized to meet the specific needs of an organization, supporting various types of network devices, configurations, and relationships to provide a comprehensive view of the network. Security is a critical aspect of NSoT, which includes robust authentication and authorization mechanisms to ensure that only authorized users can access and modify data, protecting the integrity of the network's source of truth. Additionally, NSoT includes an event system that can trigger actions based on changes to the data, which is useful for automating responses to specific events such as IP address allocations or device status changes.

Using NSoT provides several benefits. It ensures consistency by providing a single source of truth, reducing the risk of errors and misconfigurations. It is designed to scale with the network, making it suitable for large and complex environments. The RESTful API and flexible data model make it easy to integrate NSoT with existing tools and workflows. The event system and API enable extensive automation capabilities, allowing for real-time updates and automated responses to network changes. The web interface and reporting features provide comprehensive visibility into the network's state, helping administrators make informed decisions.

NSoT is particularly useful in scenarios like IP address management, where it can automate the allocation and tracking of IP addresses to ensure efficient use of address space and avoid conflicts. In device inventory management, NSoT maintains an accurate inventory of network devices, streamlining device provisioning, configuration, and decommissioning processes. For configuration management, NSoT provides a reliable data source for automated configuration management tools, ensuring that all devices are configured according to the latest standards and policies. In compliance and auditing, NSoT's detailed logging and reporting capabilities aid in compliance audits and track changes to the network over time.

Network Source of Truth (NSoT) plays a crucial role in network automation by providing a centralized, accurate, and up-to-date repository of network information. Its robust architecture, including a RESTful API, scalable database, web interface, flexible data model, and security features, ensures that it meets the needs of modern network environments. By using NSoT, organizations can achieve greater consistency, scalability, and efficiency in their network management processes.

5.3.6.2   Netbox

NetBox is another open-source tool specifically designed to act as a Source of Truth (SoT) for network infrastructure. It provides comprehensive capabilities for documenting and managing various aspects of the network, such as IP addresses, devices, racks, circuits, and virtual machines. In the context of network automation, NetBox's ability to serve as a centralized, reliable repository of network data is invaluable for ensuring accuracy and consistency across automated processes. The following diagram depicts a typical deployment of Netbox with Ansible for network automation in which Netbox serves as SoT. In NetBox, one can store device information, IP addresses, VLANs, interfaces, custom attributes, and much more and access the data through APIs from anywhere. NetBox features a web-based user interface that allows network administrators to easily input, manage, and visualize network data. This interface provides detailed views and edit capabilities for all network components, making it straightforward to keep the documentation up to date (Fig. 5.18).

The backend is powered by a PostgreSQL database, which stores all the network data and ensures high availability and reliability. One of the key components of NetBox is its RESTful API, which enables seamless integration with various network automation tools and scripts. This API allows external systems to query and update the network data stored in NetBox, facilitating dynamic and automated workflows. For example, network configuration management tools can retrieve the latest device configurations and IP allocations from NetBox to ensure consistency in deployments.

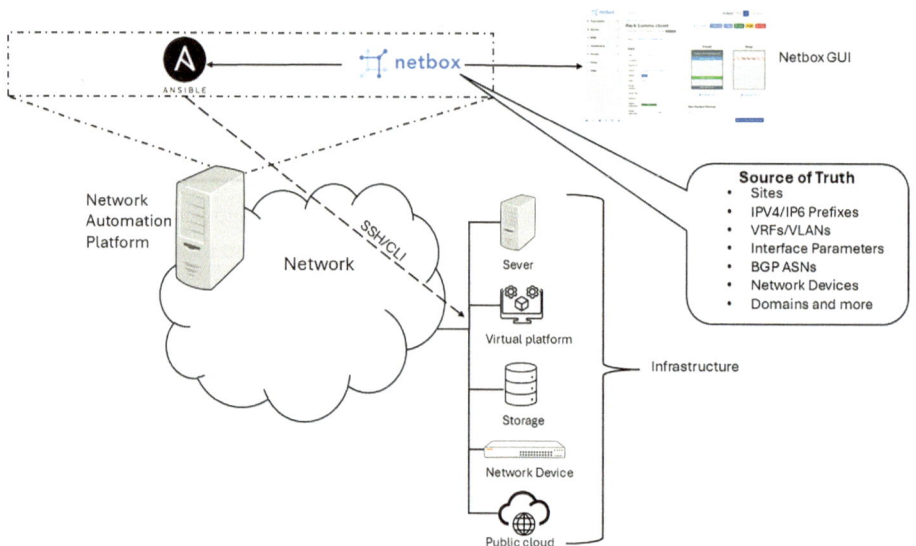

Fig. 5.18 Typical network automation deployments where netbox act as SoT

NetBox also includes a robust data model that is highly extensible, allowing users to define custom fields and relationships to suit specific organizational needs. This flexibility makes it possible to document a wide range of network assets and their interdependencies, providing a holistic view of the network. From a security standpoint, NetBox includes authentication and authorization features to control access to the network data. These mechanisms ensure that only authorized personnel can make changes, thereby protecting the integrity of the source of truth.

The use of NetBox in network automation offers several significant benefits. By maintaining a centralized repository of network information, NetBox ensures that all automation tasks are based on accurate and up-to-date data, reducing the risk of errors and inconsistencies. This centralization streamlines network management processes, as administrators and automation tools can rely on a single source for all network-related information. NetBox's event-driven capabilities further enhance its utility in network automation. Through webhooks and integrations with external systems, NetBox can trigger automated workflows in response to changes in the network data. For instance, when a new device is added to NetBox, it can automatically trigger a configuration management tool to provision the device with the correct settings.

In practical terms, NetBox is used for tasks such as IP address management, where it tracks allocations and prevents conflicts. It manages device inventories, providing detailed records of all network hardware and their configurations. For configuration management, NetBox serves as the authoritative source for network device configurations, ensuring that all changes are properly documented and compliant with organizational standards.

It also aids in compliance and auditing by maintaining detailed logs of all changes to the network infrastructure, making it easier to track modifications and ensure compliance with regulatory requirements.

NetBox plays a critical role in network automation by acting as a reliable, centralized source of truth. Its web-based interface, RESTful API, robust data model, and security features make it an ideal solution for managing and documenting network infrastructure. By integrating NetBox into their automation workflows, organizations can achieve greater accuracy, efficiency, and consistency in their network management practices.

5.3.6.3 SolarWinds

We have already discussed SolarWinds in Sect. 5.3.5; therefore, we will now focus on its applicability as a source of truth (SoT). Specifically, we will explore how SolarWinds can be utilized to maintain accurate and reliable data within an organization's IT infrastructure. This includes examining its capabilities for real-time monitoring, data integrity, and integration with other systems to ensure it serves as a dependable SoT.

SolarWinds plays a crucial role in serving as a source of truth (SoT) for IT infrastructure by providing comprehensive tools and functionalities that ensure accurate and reliable data management. SolarWinds offers real-time monitoring capabilities through its Network Performance Monitor (NPM) and Server & Application Monitor (SAM), which provide continuous visibility into the health and performance of network devices and servers. This real-time data collection ensures that the information reflects the current state of the infrastructure, making it a reliable SoT.

Additionally, SolarWinds ensures data integrity and accuracy through automated discovery and mapping of the IT environment. This process reduces the risk of human error and ensures that the network's structure and components are accurately represented. The Network Configuration Manager (NCM) maintains consistent and compliant device configurations, further enhancing the reliability of the SoT by ensuring that configuration data is accurate and up to date. SolarWinds also supports a dependable SoT through its robust reporting and analytics capabilities. Customizable dashboards and detailed reports aggregate data from various sources, providing clear and actionable insights. Historical data analysis enables the identification of trends and validation of changes, ensuring that the SoT is maintained over time. These reports and analytics tools ensure that the data presented is comprehensive and reflects the true state of the IT infrastructure.

Integration with other systems is another critical aspect of SolarWinds' contribution to a reliable SoT. Its API and third-party integrations allow for seamless data sharing and synchronization across different platforms. This ensures consistency in the data across the entire IT ecosystem. The unified IT management approach of SolarWinds, which covers network, server, database, and application performance, ensures that all aspects of the IT environment are monitored and managed from a single source, maintaining a consistent SoT. Proactive alerting and incident management within SolarWinds help maintain system integrity and accuracy. Advanced alerting mechanisms notify IT teams of potential issues, enabling quick resolution and documentation. This proactive approach ensures that any deviations from the expected state are promptly addressed, keeping the SoT accurate.

Security and compliance are also critical components of maintaining a reliable SoT. SolarWinds Security Event Manager (SEM) provides real-time threat detection and automated responses to security incidents, ensuring that the data remains accurate and secure. Compliance reporting tools help organizations adhere to regulatory requirements, and accurate compliance reports enhance the credibility and reliability of the SoT.

5.3.6.4 Infloblox

Infoblox is another commercial tool that serves as a source of truth (SoT) for IT infrastructure, though it positions itself as a Single Source of Truth (SSoT). Infoblox claims that an SSoT is a data management strategy that provides a centralized, comprehensive data environment shared by applications and the organization's user base. This concept ensures that there is only one primary reference source for data, which is stored only once, eliminating the potential for inconsistency and duplication.[4]

Infoblox plays a significant role in serving as an SSoT by offering advanced solutions for DNS, DHCP, and IP Address Management (DDI). These core network services are essential for the accurate management and reliable operation of an organization's network. Infoblox's tools ensure that the data associated with these services is precise and up to date. The platform provides comprehensive DNS, DHCP, and IP Address Management (DDI) solutions, centralizing and automating these critical network functions. The following diagram depicts Infoblox DDI integration with Ansible for network automation.

By centralizing control of IP address management, Infoblox ensures a single, authoritative source of truth for IP address allocation and management, reducing errors and inconsistencies that can arise from manual processes and disparate systems (Fig. 5.19).

Automation capabilities are crucial for maintaining data accuracy. Infoblox's automated IP address discovery and management ensure that IP address information is always current and accurate, reducing the likelihood of human errors, such as IP address conflicts or misconfigurations, which can disrupt network operations and lead to inaccurate data.

Infoblox's grid architecture enhances reliability and resilience, ensuring continuous availability of network services and the data associated with them. The grid architecture provides a distributed, synchronized database, ensuring all network services are consistently updated and available, even in the event of failures. This resilience is essential for maintaining a reliable SSoT, as it guarantees that the data is always accessible and up to date. Security features integrated into Infoblox's solutions contribute to a trustworthy SSoT. Infoblox offers DNS security extensions (DNSSEC) and advanced threat protection, ensuring the integrity and security of DNS data. By protecting against DNS-based attacks and ensuring the authenticity of DNS responses, Infoblox maintains the integrity of the data, which is crucial for a reliable SSoT.

---

[4] He [4].

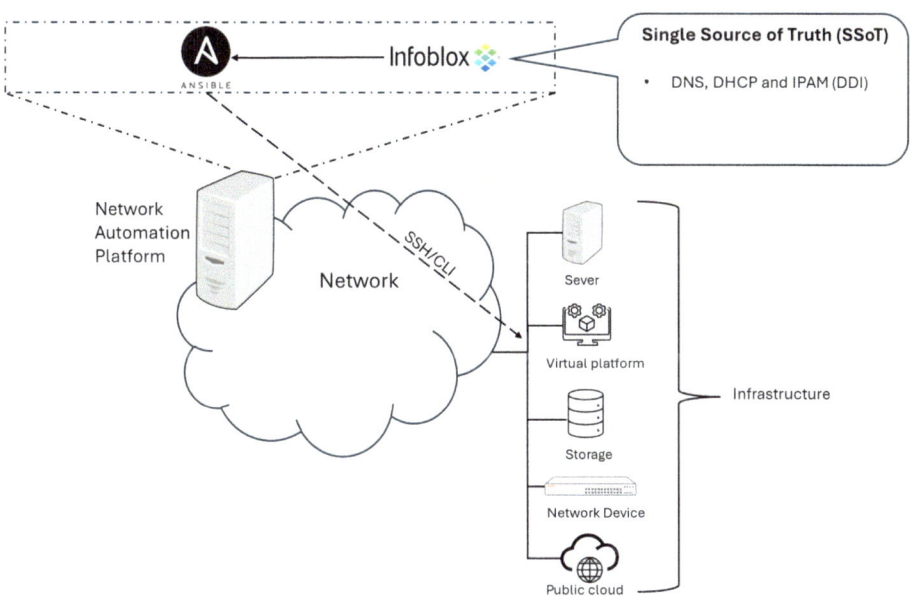

Fig. 5.19 Infoblox DDI integration with Ansible for network automation

The platform's reporting and analytics capabilities provide deep visibility into network operations and usage. Detailed reporting and real-time analytics allow IT teams to monitor and analyze network performance, detect anomalies, and make informed decisions. These insights help maintain an accurate SSoT by providing a clear and comprehensive view of the network's state and performance over time. Integration with other IT management and security tools is another vital aspect of Infoblox's contribution to a dependable SSoT. Infoblox integrates seamlessly with various network and security platforms, enabling data exchange and ensuring consistency across different systems. This integration supports a unified approach to network management and security, reinforcing the reliability of the SSoT.

Infoblox also supports compliance with regulatory requirements through robust auditing and reporting features. These features help organizations demonstrate compliance with industry standards and regulations, ensuring that the data is managed and documented according to best practices. Accurate compliance reporting enhances the credibility and reliability of the SSoT.

In summary, Infoblox supports a robust single source of truth for IT infrastructure by providing centralized and automated DNS, DHCP, and IP Address Management (DDI) solutions, ensuring data accuracy through automation, enhancing reliability with its grid architecture, securing DNS data, offering comprehensive reporting and analytics, integrating with other systems, and supporting regulatory compliance. These capabilities collectively ensure that the data used for network management and decision-making is accurate, reliable, and up to date, which is essential for effective IT operations and management.

5.3.7   Software Version Control

Software version control is a critical component in the realm of network automation, enabling organizations to manage and track changes to network configurations and automation scripts systematically. By utilizing version control systems, network administrators can ensure that changes are documented, reversible, and consistently applied across network devices. This section explores how various tools, including GitLab, Git, GitHub, and Puppet Enterprise, contribute to effective software version control within network automation. These tools not only enhance collaboration and workflow efficiency but also provide robust mechanisms for maintaining network reliability and security through meticulous version tracking and change management.

5.3.7.1   GitLab

GitLab is a comprehensive DevOps platform that integrates version control, continuous integration/continuous deployment (CI/CD), and project management. In the context of network automation, GitLab's version control capabilities allow network administrators to manage configuration files and automation scripts with precision. GitLab's repository management enables teams to collaborate on network automation scripts, ensuring that all changes are tracked and can be reviewed before implementation. The built-in CI/CD pipelines automate the testing and deployment of network changes, reducing the risk of errors. For instance, network configurations can be automatically tested in a simulated environment before being applied to production devices, ensuring that they do not introduce any issues. GitLab's merge request workflow facilitates peer review and approval processes, ensuring that only validated changes are integrated into the main branch. This diagram illustrates how GitLab integrates with Ansible to incorporate software version control and CI/CD in a network automation environment (Fig. 5.20).

In such scenarios, network administrators use GitLab to propose configuration changes. Ansible then deploys these changes into a virtual test environment and runs automated tests using open-source tools like pyATS to verify the results. If the tests are successful, the changes can be safely deployed to the production network environment.

5.3.7.2   Git

Git is a distributed version control system widely used for tracking changes in source code during software development. Its capabilities are highly applicable to network automation as well, providing a robust framework for managing network configuration files and automation scripts.

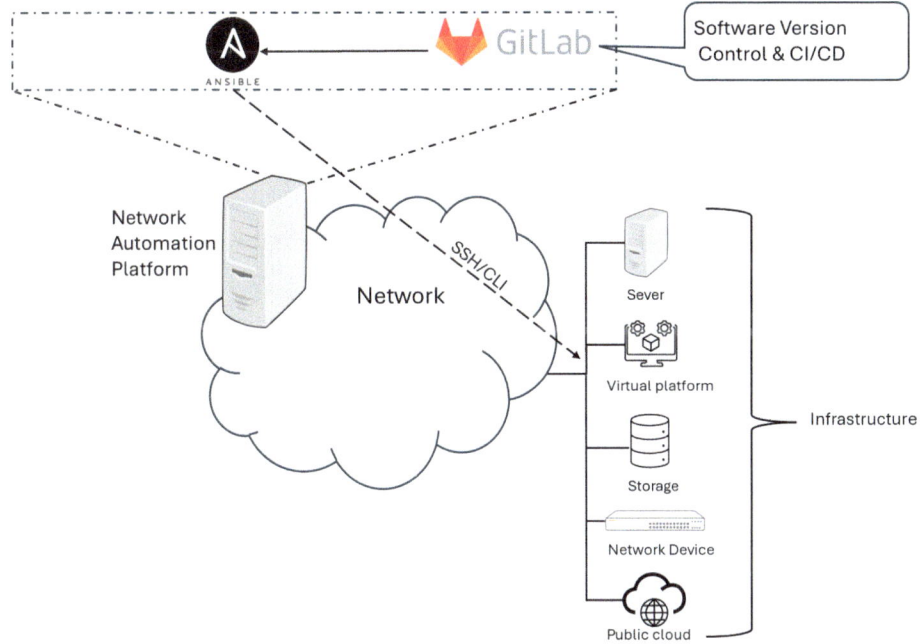

Fig. 5.20 Gitlab integrates well with Ansible to provide software version control and CI/CD for network automation environment

With Git, network engineers can create repositories to store and version control their network automation scripts. The distributed nature of Git allows multiple engineers to work on different parts of the network automation process concurrently, merging their changes seamlessly. Git's branching and merging features enable the development and testing of new automation scripts in isolated branches, ensuring that experimental changes do not affect the stable version of the scripts. Furthermore, Git's history tracking allows for easy rollback to previous versions in case a new configuration causes issues, enhancing the reliability and stability of network operations.

5.3.7.3   GitHub

GitHub, built on Git, adds a layer of collaboration and social coding features, making it an ideal platform for network automation projects. GitHub provides all the version control functionalities of Git, along with enhanced features for collaboration, code review, and project management.

In network automation, GitHub repositories can be used to store and version control network configuration files and automation scripts. GitHub's pull request mechanism allows for collaborative review and discussion of changes, ensuring that multiple eyes scrutinize each modification before it is merged. This collaborative approach helps in identifying potential issues early and fosters knowledge sharing among network engineers. GitHub Actions, the platform's CI/CD service, can automate the testing and deployment of network automation scripts, integrating seamlessly with the version control system to ensure that only tested and approved changes are deployed to the network.

5.3.7.4 Puppet Enterprise

We have discussed about Puppet Enterprise in previous sections and thus would be brief in this section. It is a configuration management tool that automates the provisioning, configuration, and management of infrastructure. While not a version control system per se, Puppet Enterprise integrates with version control tools like Git to manage the code that defines network configurations and automation tasks.

In the realm of network automation, Puppet Enterprise uses a declarative language to define the desired state of network devices. These configuration definitions are stored in version-controlled repositories, typically managed with Git or GitHub. Puppet Enterprise pulls the latest configurations from these repositories and applies them to the network devices, ensuring consistency and compliance with defined policies. The integration with version control systems allows for tracking changes to configuration definitions, facilitating audits, and enabling rollback to previous states if necessary. Puppet Enterprise also provides reporting and visibility into the state of network devices, helping administrators ensure that the network remains in the desired state as defined by the version-controlled configuration files.

5.3.8   Testing and Compliance

In the rapidly evolving landscape of network automation, ensuring that network configurations and changes are tested thoroughly and comply with established standards is crucial for maintaining network reliability and security. Testing and compliance frameworks play a significant role in verifying that automated processes and configurations function as intended and adhere to organizational policies and regulatory requirements. This section delves into several key tools used for testing and compliance in network automation: pyATS, Robot Framework, Batfish, and pytest.

5.3.8.1   pyATS

pyATS (Python Automation Test System) is a versatile, open-source test framework developed by Cisco for automating network testing and validation. It provides a comprehensive set of tools and libraries designed specifically for network engineers to create, manage, and execute test cases for network devices and services. With automated testing capabilities, pyATS ensures that network configurations and changes do not introduce errors or vulnerabilities. Its scalable architecture makes it suitable for large-scale network environments, and it integrates seamlessly with existing network management and automation tools, facilitating a cohesive workflow. Users can write custom test cases in Python, allowing for flexible and detailed testing of specific network scenarios. Applications of pyATS include regression testing, compliance verification, and performance testing, ensuring network configurations adhere to standards and perform optimally.

5.3.8.2 Robot Framework

Robot Framework is a generic, open-source automation framework widely used for acceptance testing, acceptance test-driven development (ATDD), and robotic process automation (RPA). Its keyword-driven testing approach allows users to write test cases in a high-level, human-readable language, making it accessible to non-programmers. The extensible nature of Robot Framework supports various libraries and tools, including those specific to network automation, and it easily integrates with other automation and CI/CD tools. The framework generates comprehensive reports and logs, providing insights into test results and compliance status. Robot Framework is ideal for end-to-end testing, compliance audits, and continuous integration, validating complete network workflows and processes while ensuring network configurations meet compliance requirements through automated audits.

5.3.8.3 Batfish

Batfish is an open-source network configuration analysis tool that helps network engineers and operators understand and verify network behavior. It focuses on static analysis of network configurations to identify potential issues before they impact live environments. Batfish analyzes network configurations to detect errors, inconsistencies, and potential security vulnerabilities. It simulates configuration changes to predict their impact on the network, ensuring that changes do not introduce problems. The tool's Network Query Language (NQL) allows users to query network configurations and states programmatically, and it can be integrated with other network management and automation tools for comprehensive network analysis. Batfish is particularly useful for pre-deployment validation, security audits, and policy compliance, ensuring network configurations are correct, safe, and adhere to organizational standards.

5.3.8.4 Pytest

Pytest is a mature, full-featured Python testing framework widely used for writing simple unit tests as well as complex functional testing for applications and libraries. Its simple syntax allows for easy creation of test cases, while its wide range of plugins and extensions enhance its capabilities, including those tailored for network testing. pytest's powerful fixture mechanism efficiently manages test setups and teardowns, and it easily integrates with other CI/CD tools, enabling automated testing workflows. It is suitable for unit testing, verifying individual components of network automation scripts; integration testing, ensuring different parts of the network automation system work together correctly; and regression testing, detecting issues introduced by new code changes to maintain stability and reliability.

5.3.8.5 Conclusion

Testing and compliance are critical components of network automation, ensuring that automated processes and configurations perform as expected and comply with standards. Tools like pyATS, Robot Framework, Batfish, and pytest provide robust solutions for automating testing and compliance tasks, enhancing the reliability, security, and efficiency of network operations. By leveraging these tools, organizations can maintain high standards of network performance and compliance, mitigating risks and ensuring smooth network operations.

5.4 Summary

In this chapter, we explored the transformative power of network automation in managing modern, complex network infrastructures. We began by tracing the evolution of network automation, from the early days of manual configuration to the sophisticated, software-driven approaches of today. Automation has emerged as a vital tool, offering increased efficiency, reliability, and scalability while reducing human error and operational costs.

We discussed the numerous advantages of network automation, such as improved agility, proactive maintenance, and enhanced security, while acknowledging its initial implementation challenges and the need for regular updates. The chapter also presented a comprehensive framework for deploying network automation, focusing on key tools within the different categories of the automation framework.

Furthermore, we highlighted the critical role of security in network automation, emphasizing how tools like Ansible and Chef enhance security through features like encrypted communications, role-based access control, and real-time compliance monitoring. Looking ahead, the future of network automation is poised to be shaped by advancements in artificial intelligence, machine learning, intent-based networking, and the expanding landscapes of cloud and edge computing. These technologies promise even greater automation capabilities, driving intelligent and adaptive network management.

In conclusion, network automation is an indispensable component of modern network management strategies, enabling organizations to stay ahead in a rapidly evolving digital landscape. By leveraging the principles, tools, and best practices discussed in this chapter, organizations can harness the full potential of automation to achieve more efficient, reliable, and secure network operations.

References

1. Juniper (2024) Junos space high availability software architecture overview. https://www.juniper.net/documentation/us/en/software/junos-space22.2/junos-space-high-availability/topics/concept/junos-space-ha-architecture-overview.html
2. Rayak (2021) What is Cisco NSO and NSO architecture. https://rayka-co.com/lesson/what-is-cisco-nso-and-nso-architecture/
3. Rundeck.com (2023) Process automation. https://docs.rundeck.com/docs/about/enterprise/
4. He J (2023) What is the source of truth, and why is it important? Kyligence Inc. https://kyligence.io/blog/what-is-the-source-of-truth-and-why-is-it-important/#:~:text=While%20the%20Source%20of%20Truth,and%20the%20organization's%20user%20base

In the rapidly evolving landscape of information technology, networks serve as the nervous system of modern enterprises, ensuring seamless communication and the uninterrupted flow of data. As organizations increasingly depend on complex, distributed networks to drive critical operations, the need for robust mechanisms to monitor and manage these networks has never been more pressing. This brings us to network monitoring, a foundational aspect of network management that focuses on the continuous observation of network operations to detect issues and measure performance metrics. Real-time tracking of network health indicators such as bandwidth usage, latency, packet loss, and uptime provides immediate insights and alerts, enabling quick responses to potential problems and helping maintain network stability.

Advances in network monitoring have paved the way for more sophisticated and comprehensive approaches to network management. Modern monitoring systems not only detect faults but also provide detailed analytics and predictive insights. This evolution in network monitoring has laid the groundwork for more advanced concepts like network telemetry and observability. Network telemetry involves the collection, transmission, and analysis of data from network devices. It encompasses a wide range of metrics, including traffic patterns, device health, and network performance indicators. This data serves as the foundation for understanding the state of the network at any given time. Telemetry systems collect and transmit vast amounts of data in real-time, offering a granular view of network operations. This detailed data collection is crucial for proactive troubleshooting and performance optimization.

While telemetry provides the raw data, observability builds on it by offering the tools and methodologies needed to interpret this data, identify anomalies, diagnose issues, and

D. D. Chowdhury, *Future of Networks*, Synthesis Lectures on Communications,
https://doi.org/10.1007/978-3-031-71440-5_6

predict future trends. Observability integrates data from various sources, including metrics, logs, and traces, to provide a holistic view of the network's health and performance. Advanced analytics and machine learning are often employed to gain deep insights and make informed decisions. Network visibility complements both monitoring and telemetry by providing a comprehensive view of all traffic and activities occurring within the network. It is essential for understanding how data flows through the network, identifying potential bottlenecks, and ensuring that security policies are enforced. Tools for network visibility analyze traffic patterns and provide detailed insights into network behavior, which is critical for both performance management and security.

The integration of network monitoring, telemetry, observability, and visibility represents a significant advancement in network insights. These combined approaches offer a multi-faceted view of network health and performance. Comprehensive data collection, from basic performance metrics to detailed logs and traces, ensures that information is gathered from every corner of the network. Real-time analysis and alerting, facilitated by continuous monitoring and telemetry, ensure that issues are detected and addressed promptly, minimizing downtime and service disruptions. Observability tools analyze this data to provide deep insights into network behavior, correlating events and identifying root causes of issues. Additionally, advanced analytics and machine learning models predict future trends, enabling proactive maintenance and capacity planning. Enhanced security postures are achieved through full visibility into network traffic and activities, allowing organizations to quickly identify and respond to security threats.

By leveraging these advanced network insights, organizations can ensure that their networks remain robust, secure, and capable of supporting the increasing demands of the digital age. This chapter explores each of these components in detail, providing a comprehensive guide to understanding and implementing effective network management strategies. As networks continue to grow in complexity and importance, the ability to monitor, analyze, and optimize their performance becomes ever more critical. The combined power of network visibility, monitoring, telemetry, and observability provides the comprehensive insights needed to maintain a resilient and high-performing network infrastructure. This chapter serves as a detailed exploration of these concepts, offering practical guidance for leveraging them to their fullest potential.

6.1   The Timeline

Once the Internet, originally known as ARPANET, was established and mainframes began connecting to form larger network structures, the need for effective network monitoring became apparent. Network devices continuously generate data about their operational status and activity, necessitating a standardized approach to manage and monitor them within increasingly complex and expansive networks. This quest led to the definition of SNMP

(Simple Network Management Protocol) in 1988, providing a basic framework for collecting and organizing information about network devices such as routers, switches, and servers. SNMP enabled network administrators to monitor network performance, detect faults, and manage configurations. The following diagram depicts the timeline of the advances in network insights.

SNMPv1 marked the beginning of a new era in network insights. In those early days of network monitoring, SNMP proved very useful by providing device-level information that aided in network performance monitoring and management, yet it lacked the capability to provide the traffic pattern information necessary for comprehensive network visibility.

In 1992, the IETF introduced RMON to provide deeper network insights at the packet level for network traffic monitoring. Despite focusing primarily on data link and physical layer parameters and being resource-intensive without real-time capabilities, both SNMP and RMON remain integral elements of network monitoring solutions (Fig. 6.1).

During the same period, syslog was introduced in Unix systems as a method to collect event logs, thereby offering additional system information. This development marked the inception of network visibility through both protocol-based network monitoring and log-based monitoring. As networks grew increasingly complex, there arose a greater demand for insights. Packet-based technologies such as DPI (Deep Packet Inspection), port mirroring, network TAP, and Network Packet Broker (NPB) emerged as logical next steps in network monitoring.

Sniffer and Wireshark tools are key elements of DPI, providing detailed packet information and the ability to identify network security issues and potential cybercriminal activity. These tools have played a crucial role in meeting the growing need for network visibility by enabling thorough analysis of network traffic and behavior. The evolution from basic protocol and log monitoring to advanced packet inspection technologies has

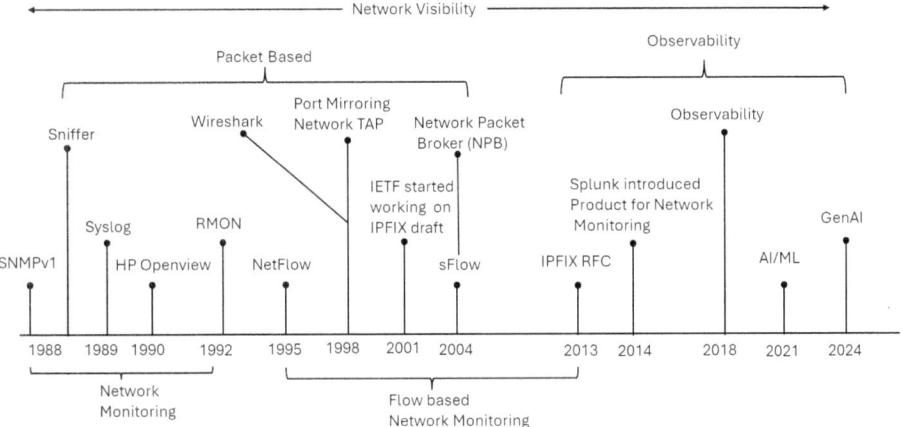

Fig. 6.1   A timeline depicting the evolution of network insights

significantly enhanced network administrators' ability to monitor, analyze, and secure their networks effectively. These advancements continue to be critical as networks become more intricate and cyber threats more sophisticated.

The next trend to emerge was the adoption of flow-based monitoring, spearheaded by Cisco's introduction of NetFlow in 1996. NetFlow enabled the collection of network traffic data for analysis and monitoring purposes. Subsequently, the IETF (Internet Engineering Task Force) developed a draft in 2001 aimed at enhancing real-time data streaming capabilities, which would provide more comprehensive network insights compared to the periodic sampling inherent in traditional NetFlow.

In addition to NetFlow, sFlow emerged as another significant protocol for flow-based monitoring. Developed by InMon Corporation, sFlow provides a scalable solution for monitoring high-speed networks by sampling packets at wire speed and sending summarized traffic data to collectors. This approach allows for real-time visibility into network traffic patterns and anomalies, complementing the capabilities of NetFlow and extending the scope of flow-based monitoring.

Flow-based monitoring technologies like NetFlow, sFlow, and their successors (such as IPFIX—Internet Protocol Flow Information Export) revolutionized network visibility by offering continuous, detailed analysis of network traffic flows. Unlike traditional SNMP-based monitoring, which focuses on device-level statistics, flow-based monitoring provides granular insights into application usage, traffic patterns, and potential security threats.

These advancements have been crucial for network administrators and security professionals, enabling them to proactively manage network performance, optimize resource allocation, and detect and mitigate security incidents in real-time. As networks continue to evolve with increasing complexity and data volumes, flow-based monitoring remains a cornerstone of modern network management practices.

Splunk's introduction of a network monitoring product in 2014 marked a significant advancement in network observability. Leveraging its expertise in data analytics and visualization, Splunk brought a new approach to monitoring and managing network performance and security. By ingesting and analyzing vast amounts of network data in real-time, Splunk enabled organizations to gain deep insights into their network operations, application performance, and user behaviors.

In parallel with flow-based monitoring technologies like NetFlow and sFlow, Splunk's network monitoring product enhanced the ability to achieve comprehensive observability across complex networks. This integration of network telemetry data with advanced analytics empowered IT teams to detect anomalies, troubleshoot issues faster, and optimize network resources more effectively. The combination of flow-based monitoring and Splunk's observability solutions has transformed network management practices, providing a holistic view of network health and performance. This approach not only enhances operational efficiency but also strengthens cybersecurity posture by identifying and responding to threats in real-time.

Splunk's entry into network monitoring also underscores the evolution toward more sophisticated and integrated approaches to network visibility and management. Flow-based monitoring technologies laid the groundwork by offering detailed traffic analysis, while Splunk's analytics capabilities added a layer of contextual understanding and proactive management. Together, these advancements continue to drive innovation in network monitoring, ensuring networks remain resilient, efficient, and secure amidst growing complexity and digital transformation.

As the demand for real-time insights and proactive network management continues to grow, observability tools like Splunk play a crucial role in ensuring the reliability, security, and scalability of modern networks. However, Splunk is not the only provider in this space; many network vendors have also introduced similar capabilities in their network orchestration platforms. Open-source tools like the ELK stack provide vendor-neutral capabilities for deeper network insights. The ELK stack can ingest IPFIX data to provide real-time network insights and analysis. Some vendors have even integrated AI/ML-based analytics into their orchestration platforms. As of 2024, there is a growing momentum to integrate Generative AI (GenAI) to provide even greater insights into network operations, marking a trend toward mainstream adoption.

6.2 Network Monitoring

From the timeline presented in earlier sections, we can clearly see the evolution of network monitoring, shaped by the needs and complexities of networks over time. Given that network monitoring is a fundamental building block for network insights, it warrants deeper exploration.

Network monitoring involves the continuous observation of a network's performance, security, and operational status to ensure smooth and efficient functioning. The primary goals of network monitoring are to detect faults, optimize performance, ensure compliance with security policies, and provide valuable insights for network management and planning. It is a systematic approach to identifying slow or failing components within a network before they cause significant problems. For example, crashed, frozen, or overloaded servers, failed switches, failing routers, and other troublesome components can potentially cause outages or network failures. If an issue arises and triggers an outage, the network monitoring system's role is to alert the network administrator promptly (Fig. 6.2).

This figure above depicts a simplistic purview of network monitoring using SNMP. Whether it is protocol based or flow based, setup is typical and remains the same.

The following table depicts various network monitoring tools available that use different methods for network monitoring the primary goal remains unchanged that is monitor

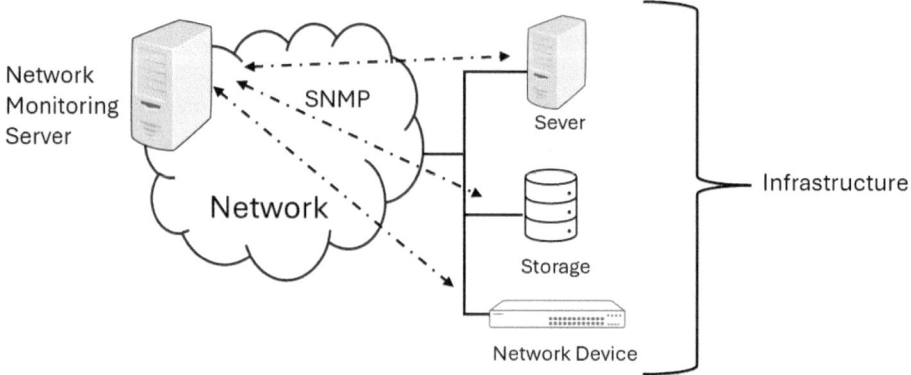

Fig. 6.2 Network monitoring using SNMP

network, health performance and compliance with security policies and help with network management and planning (Table 6.1).

Network monitoring is an indispensable practice for maintaining the health, performance, and security of IT environments. By leveraging various monitoring techniques and tools, organizations can gain deep insights into their networks, detect and resolve issues promptly, and ensure optimal performance and security. From SNMP and flow-based monitoring to advanced packet analysis and observability tools, the landscape of network monitoring is rich with technologies designed to meet the diverse needs of modern networks. As networks continue to evolve, the importance of robust and comprehensive network monitoring solutions will only grow, making them a critical component of IT infrastructure management.

6.2.1   SNMP

Simple Network Management Protocol (SNMP) is a widely used protocol for network management and monitoring. It enables network administrators to collect information, configure devices, and manage network performance remotely. SNMP was developed in the late 1980s and has since become the standard protocol for managing and monitoring network devices such as routers, switches, servers, and workstations. It operates based on a client–server model where the network management system (NMS) acts as the client, and the managed devices (network devices like routers, switches, etc.) act as servers. The protocol uses a simple set of operations to retrieve and manipulate data on the managed devices.

The following diagram illustrates various components of SNMP and how they work in tandem to provide network insights. Her are some of the key components of SNMP for Network Monitoring:

Table 6.1 Network monitoring technologies and tools

Type	Technology	Overview	Tools
Protocol based	SNMP	SNMP is a widely used protocol that provides a standardized framework for monitoring and managing network devices	Nagios, Zabbix, PRTG network monitor, and SolarWinds network performance monitor (NPM)
	RMON	RMON (Remote Network Monitoring) provides detailed insights into network performance, traffic patterns, and device health by leveraging the capabilities defined in the RMON standard	Solar NPM, ManageEngine OpManager, PRTG network monitor, WhatsUp gold, Cisco prime infrastructure, Nagios XI, Zabbix, and observer analyzer
Packet based	DPI technologies	Packet sniffing involves capturing and analyzing network packets, while DPI goes deeper by examining the contents of these packets	Wireshark, Tcpdump, SolarWinds Network Performance monitor (with DPI capabilities), and suricata
	Port Mirroring (SPAN) and Network TAPs	These methods replicate network traffic to a monitoring device, providing visibility without impacting network performance	Hardware TAPs from vendors like garland technology and profitap, and SPAN configurations on network switches
Log and event monitoring	Syslog	Collects and analyzes logs from network devices, servers, and applications to monitor events and detect issues	Splunk, ELK stack (Elasticsearch, Logstash, Kibana), Graylog, and ManageEngine eventLog analyzer
Endpoint monitoring	PKI certificates, SMB, http/htps, TCP/UDP and ICMP	Focuses on monitoring devices connected to the network, such as computers, mobile devices, and IoT devices	Microsoft system center configuration manager (SCCM), symantec endpoint Protection, and Cisco AnyConnect
Flow-based monitoring	NetFlow, sFlow and IPFIX	Flow-based monitoring involves analyzing data flows within the network to understand traffic patterns and application usage	SolarWinds NetFlow traffic analyzer, plixer scrutinizer, ntopng, and ManageEngine NetFlow analyzer

(continued)

Table 6.1 (continued)

Type	Technology	Overview	Tools
Application performance monitoring	software tools and telemetry data	Monitors the performance of applications to ensure optimal user experience and operational efficiency	Dynatrace, new relic, AppDynamics, and SolarWinds server and application monitor
Synthetic monitoring	ICMP	Involves simulating user interactions with applications or websites to measure performance and availability	Pingdom, uptrends, and catchpoint
Real user monitoring (RUM)	Java scripts, HTTP requests, browser extensions, and log files	Tracks actual user interactions with applications to measure performance from the end-user perspective	Google analytics, new relic browser, and Dynatrace RUM

- **Managed Devices**: These are network devices equipped with an SNMP agent, such as routers, switches, servers, printers, etc.
- **SNMP Agents**: Software modules residing on managed devices that collect and store management information. They respond to requests from the NMS and generate alerts called traps.
- **Network Management Station (NMS)**: A system responsible for communicating with the SNMP agents to gather data and control the managed devices (Fig. 6.3).

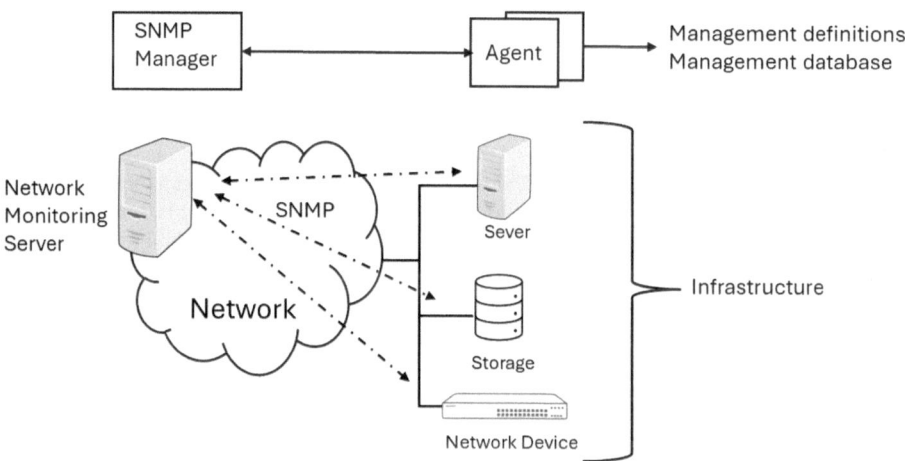

Fig. 6.3 SNMP based network monitoring

The SNMP protocol uses a simple set of operations to retrieve and manipulate data on managed devices. Each device in the network contains a Management Information Base (MIB), which is a collection of information organized hierarchically. These MIBs are used by SNMP to manage devices within a network, containing definitions of network objects and their relationships. Each object in the MIB is identified by an Object Identifier (OID), a globally unique identifier.

In SNMP operations, a management station gets and sets objects in the MIB, and an agent notifies the management station of significant but unsolicited events called traps. SNMP performs several basic operations to manage network devices effectively. The GET operation retrieves one or more values from the managed device, while the GET-NEXT operation retrieves the next variable in the data sequence. The GET-BULK operation is used to efficiently retrieve large blocks of data. The SET operation allows the NMS to set a value on the managed device. SNMP agents can send asynchronous notifications called TRAPs to the NMS about certain events. The INFORM operation is an acknowledged version of TRAP, ensuring the NMS receives the notification. Lastly, the RESPONSE operation includes replies from agents to the NMS in response to GET, SET, and other requests.

6.2.1.1   SNMP Version

SNMP has evolved through several versions, each enhancing the protocol's security and capabilities. The original version, SNMPv1, provided basic features but lacked security. SNMPv2c improved performance and introduced the GET-BULK operation but still lacked robust security. SNMPv3 added security features, including message integrity, authentication, and encryption, making it suitable for use in more sensitive environments.

6.2.1.2   SNMP in Network Monitoring

SNMP is integral to network monitoring, enabling administrators to track and manage the performance and status of network devices. It is essential for fault management, where SNMP traps notify administrators of critical events, such as device failures or threshold breaches, and log events for post-incident analysis. For performance management, SNMP monitors CPU, memory, and bandwidth usage on devices, tracks data transmission rates and delays, and monitors packet error rates and interface statuses. In configuration management, SNMP can change configurations on network devices and automate the backup and restoration of device configurations. For security management, SNMPv3 ensures secure access to network devices through authentication and encryption and monitors security events and tracks unauthorized access attempts.

6.2.1.3   SNMP Tools

Several network monitoring tools leverage SNMP for comprehensive network management. SolarWinds Network Performance Monitor (NPM) provides real-time monitoring, automated network discovery, and customizable alerts, making it ideal for comprehensive performance monitoring and fault management. ManageEngine OpManager offers network and server monitoring, automated workflows, and detailed reporting, suitable for both small and large enterprises. PRTG Network Monitor features a user-friendly interface, customizable sensors, and real-time alerting, making it suitable for diverse network environments. Nagios XI is a scalable and highly configurable solution that supports SNMP traps and polling, ideal for enterprises needing detailed, customizable monitoring. Zabbix is an open-source and scalable monitoring solution that supports SNMP polling and traps, suitable for cost-effective, large-scale network monitoring. WhatsUp Gold offers interactive maps, robust alerting, and comprehensive reporting, great for small to medium-sized businesses. Cisco Prime Infrastructure integrates with Cisco devices to provide deep insights and advanced analytics, ideal for enterprises with extensive Cisco network infrastructure.

6.2.2   Remote Network Monitoring (RMON)

Remote Network Monitoring (RMON) is a standard specification defined by the Internet Engineering Task Force (IETF) that allows network administrators to monitor and analyze network traffic remotely. RMON extends the capabilities of the Simple Network Management Protocol (SNMP) by providing more detailed and granular information about network traffic and device performance, particularly at the data link and physical layers. The protocol was introduced in 1992 as part of the effort to provide deeper insights into network performance and troubleshooting. It was designed to address some of the limitations of SNMP, particularly in terms of real-time traffic analysis and detailed packet-level monitoring. The primary objective of RMON is to enable network administrators to gain better visibility into network traffic and performance, allowing for proactive management and troubleshooting.

6.2.2.1   RMON Groups
RMON defines a set of monitoring and control functions, organized into different groups, each providing specific types of network data. The following table depicts the most commonly used RMON groups (Table 6.2).

6.2.2.2   Benefits of RMON
RMON (Remote Network Monitoring) provides several significant benefits for network management. It enhances visibility by offering detailed insights into network traffic and performance at the packet level. This allows network administrators to have a comprehensive view of network activity and detect issues more effectively. Another key advantage is the ability to collect and analyze historical performance data, which is crucial for trend

Table 6.2 The RMON groups

RMON group	Purpose	Metrics
Statistics	Provides basic statistics on network traffic, such as packet counts, error counts, and utilization rates	Packet and byte counts, error counts, collision counts, and other basic network statistics
History	Records historical data on network performance over time	Periodic sampling of statistics like utilization and error rates, stored for trend analysis and troubleshooting
Alarm	Sets thresholds for various metrics and generates alarms when thresholds are crossed	Any RMON metric can trigger alarms, such as high error rates or excessive utilization
Host	Provides detailed statistics on individual hosts (devices) on the network	Packet and byte counts for each host, as well as error counts and other host-specific metrics
HostTopN	Identifies the top N hosts based on various metrics	Ranking hosts by traffic volume, error rates, or other statistics
Matrix	Monitors traffic between pairs of hosts	Detailed statistics on traffic flows between specific host pairs, including packet and byte counts
Filter	Allows for the creation of filters to capture specific types of traffic	User-defined filters can be set to capture packets based on various criteria, such as protocol type or source/destination addresses
Capture	Captures a subset of packets for detailed analysis	Packet capture based on defined filters, used for in-depth protocol analysis and troubleshooting
Event	Manages events and log entries generated by other RMON groups	Tracks events such as threshold crossings, filter matches, and other significant occurrences

analysis and capacity planning. This historical data enables administrators to understand long-term network behavior and make informed decisions about future network needs.

RMON also supports proactive management by enabling the setting of alarms and thresholds. These alerts notify administrators of potential issues before they escalate into critical problems, allowing for timely intervention and resolution. Additionally, RMON offers detailed host information, providing granular statistics on individual network hosts. This detailed information helps in identifying problematic devices or applications that may be affecting network performance. Finally, RMON facilitates traffic analysis by allowing the examination of traffic patterns between host pairs. This analysis helps identify network bottlenecks and optimize traffic flow, ensuring a more efficient and effective network operation.

6.2.2.3   RMON Tools

Several tools and platforms support RMON for network monitoring, providing detailed traffic analysis and performance insights. SolarWinds Network Performance Monitor integrates RMON capabilities alongside its comprehensive network performance monitoring features, allowing administrators to effectively track and manage network health. ManageEngine OpManager also supports RMON, enabling detailed traffic analysis and efficient network management. PRTG Network Monitor includes RMON sensors to monitor network traffic and performance metrics, offering a robust solution for maintaining network efficiency. WhatsUp Gold leverages RMON-based monitoring to deliver in-depth network insights and traffic analysis, helping administrators to identify and resolve network issues promptly.

6.2.3   Syslog

Syslog, short for "System Logging Protocol", is a standard protocol used for message logging. It allows network devices such as routers, switches, firewalls, and servers to send event messages to a centralized log server (Syslog server). Developed in the 1980s by Eric Allman as part of the Sendmail project, Syslog has become a crucial tool in network monitoring and management due to its ability to log a wide range of events from different types of devices.

Syslog operates using a client–server architecture. The devices configured to generate logs act as Syslog clients, while the system designated to collect and store these logs acts as the Syslog server. The communication typically occurs over User Datagram Protocol (UDP) on port 514, though Transmission Control Protocol (TCP) can also be used for more reliable delivery (please refer to the diagram below) (Fig. 6.4).

Syslog messages consist of a standardized format that includes a timestamp, hostname or IP address of the device, message severity level, and the actual log message. The severity levels range from 0 (emergency) to 7 (debug), indicating the urgency of the event. In recent years, the demand for TCP and TLS-based Syslog communications is growing driven by the need for reliable and secure log data transport. TCP ensures the reliability and integrity of log messages through its connection-oriented and error-correcting mechanisms. TLS adds a layer of security by encrypting log data and enabling mutual authentication, which is crucial for protecting sensitive information and meeting compliance requirements. As networks continue to evolve and security threats grow, the demand for these enhanced Syslog communication protocols will only increase, making them essential components of modern network monitoring and management strategies.

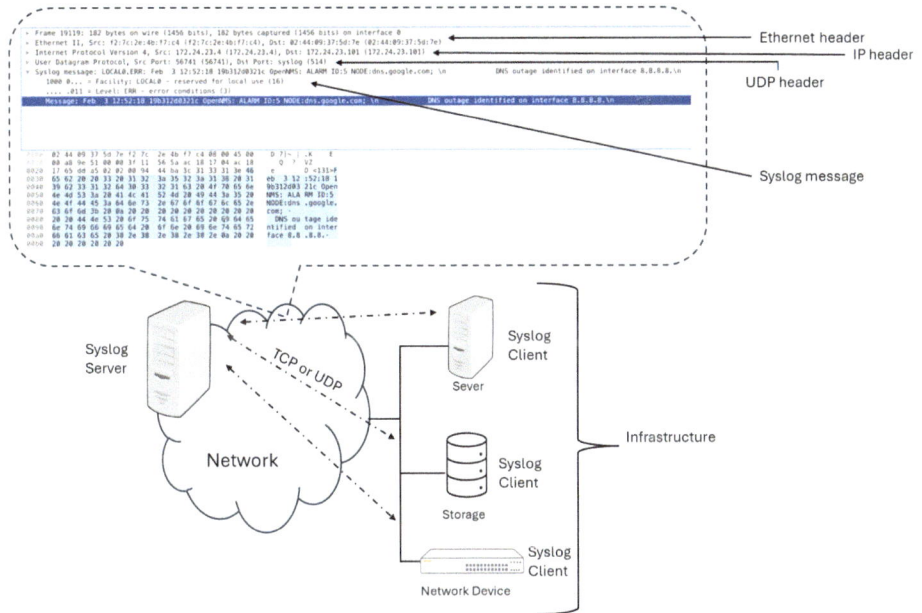

Fig. 6.4 Syslog communications and packet structure

6.2.3.1 Use of Syslog in Network Monitoring

Syslog plays a pivotal role in network monitoring by providing a centralized method for logging and analyzing events across the network. Here are several ways Syslog is used in network monitoring:

- **Centralized Logging:** Syslog collects logs from various network devices and consolidates them in a single location. This centralization simplifies the monitoring process by allowing administrators to review all logs from a unified interface.
- **Real-time Alerts:** Syslog can be configured to send real-time alerts for specific events or conditions, such as failed login attempts, device reboots, or hardware failures. These alerts enable administrators to respond promptly to potential issues.
- **Troubleshooting:** By analyzing Syslog messages, network administrators can identify and diagnose network problems. Logs provide detailed information about the events leading up to an issue, making it easier to pinpoint the root cause.
- **Performance Monitoring:** Syslog can log performance-related data, such as bandwidth usage, CPU load, and memory utilization. Monitoring these metrics helps administrators ensure that network devices are functioning optimally.
- **Security Auditing:** Syslog is crucial for security auditing as it records events related to access control, configuration changes, and other security-related activities.

This information is vital for detecting and investigating security breaches or policy violations.

- **Compliance:** Many industries have regulatory requirements for logging and monitoring. Syslog helps organizations meet these compliance standards by maintaining a comprehensive record of network activities.
- **Historical Analysis:** Storing Syslog data over time allows for historical analysis, which is essential for identifying long-term trends, planning capacity upgrades, and improving network reliability.

6.2.3.2 Syslog Tools

Various tools and platforms support Syslog and enhance its functionality for network monitoring:

- **SolarWinds Log Analyzer:** Provides real-time log collection, analysis, and alerting. It integrates with other SolarWinds products for comprehensive network monitoring.
- **Splunk:** Offers powerful log management capabilities, including real-time search, analysis, and visualization of Syslog data.
- **ELK:** The Elastic Stack, also known as ELK Stack (Elasticsearch, Logstash, and Kibana), provides a powerful and flexible solution for managing Syslog data.
- **Graylog:** An open-source log management tool that supports Syslog, providing flexible data collection, analysis, and alerting.
- **Loggly:** A cloud-based log management service that simplifies the collection and analysis of Syslog data from various sources.
- **Papertrail:** Another cloud-based solution that provides real-time Syslog monitoring and analysis with an easy-to-use interface.

6.2.4 Port Mirroring

Port mirroring is a network monitoring technique used to capture and analyze traffic passing through a network switch. By duplicating network traffic from one or more switch ports (or an entire VLAN) to another port where a monitoring device is connected, administrators can inspect and analyze the data for performance, troubleshooting, and security purposes. Port mirroring is particularly useful in network diagnostics and forensic investigations.

Three primary implementations of port mirroring are Switch Port Analyzer (SPAN), Remote Switch Port Analyzer (RSPAN), and Encapsulated Remote Switch Port Analyzer (ERSPAN). These methods differ mainly in how they handle the mirrored traffic and where it is sent. Please refer to the diagram below that shows how those port mirroring techniques are used in typical enterprise networks (Fig. 6.5).

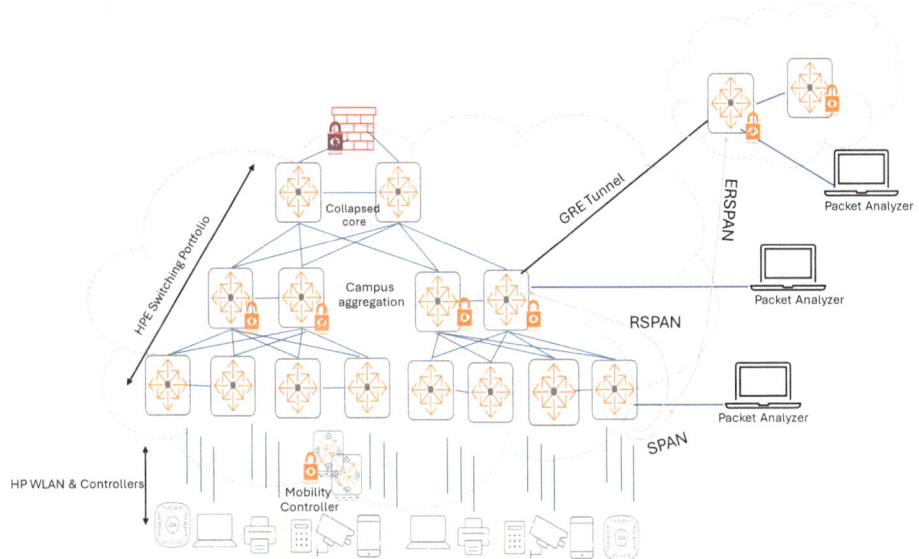

Fig. 6.5 Different types of port mirroring: SPAN, RSPAN, and ERSPAN

As depicted in the diagram above, the SPAN technique is used to mirror traffic from specified ports or VLANs within the same switch, sending the mirrored traffic to a designated monitoring port on that switch. Operating within a single switch, SPAN is straightforward to configure and deploy, making it ideal for scenarios where the monitoring device is in the same physical location as the network switch. Network administrators configure SPAN by specifying the source ports or VLANs to be monitored and the destination port for the mirrored traffic. This configuration is done through the switch's management interface or command-line interface (CLI). SPAN can mirror various types of traffic, including ingress (incoming), egress (outgoing), or both, allowing administrators to capture the specific traffic needed for analysis. SPAN is commonly used for network performance monitoring, troubleshooting connectivity issues, and detecting anomalies in traffic patterns, providing real-time visibility into the network without requiring additional hardware beyond the existing switch infrastructure.

In contrast, RSPAN extends the capabilities of SPAN by allowing traffic mirroring across multiple switches within the same network. It can mirror traffic from one switch to another within the same network, which is useful for centralized monitoring in a distributed environment. RSPAN uses a dedicated VLAN (RSPAN VLAN) to carry the mirrored traffic between switches, segregating it from regular network traffic. Configuring RSPAN involves setting up source ports or VLANs on the source switch, specifying the RSPAN VLAN, and configuring the destination port on the remote switch where the monitoring device is connected. This approach is beneficial in larger network environments where it is necessary to monitor traffic from multiple locations using a centralized monitoring device.

There are circumstances where the network administrator needs to monitor traffic from a distance, typically from different locations. In this case, the ERSPAN technique is used. It extends the capabilities of both SPAN and RSPAN by enabling traffic mirroring across different network locations using IP tunnels to transport mirrored traffic to a remote monitoring device. ERSPAN can send mirrored traffic to a monitoring device on a different switch or even across different networks, making it suitable for distributed network environments where central monitoring is required. The mirrored traffic is encapsulated in Generic Routing Encapsulation (GRE) packets, which are routed over the IP network to the remote destination, preserving the original packet structure and providing additional flexibility in traffic routing. Configuring ERSPAN involves specifying the source and destination IP addresses for the GRE tunnel, along with the source ports or VLANs to be mirrored, through the switch's management interface or CLI. Similar to SPAN, ERSPAN can mirror ingress, egress, or both types of traffic, with the encapsulated packets then decapsulated by the monitoring device for analysis. ERSPAN is particularly useful for centralized monitoring in large or geographically dispersed networks, allowing network administrators to consolidate traffic analysis and monitoring in a single location, thereby improving efficiency and oversight.

6.2.4.1 Port Mirroring Tools

Port mirroring tools are essential for effective network monitoring and analysis. These tools enable administrators to capture and analyze network traffic for performance optimization, security monitoring, and troubleshooting. Various software solutions and hardware devices support port mirroring, including SPAN, RSPAN, and ERSPAN, providing diverse functionalities to cater to different network environments and requirements.

One widely used tool is SolarWinds Network Performance Monitor (NPM), which offers robust support for SPAN and RSPAN. SolarWinds NPM provides comprehensive network performance monitoring, real-time traffic analysis, and detailed reporting, making it a preferred choice for many organizations. Wireshark, another popular tool, excels in packet analysis by working seamlessly with SPAN, RSPAN, and ERSPAN. Wireshark captures and inspects packets at a granular level, offering deep insights into network behavior and aiding in the identification of issues such as latency, packet loss, and security breaches.

ManageEngine OpManager is also a notable tool that supports port mirroring through SPAN and RSPAN. OpManager provides detailed traffic analysis, performance monitoring, and automated workflows to enhance network management. It is suitable for both small and large enterprise environments, ensuring efficient network operations. Cisco's suite of network management tools, including Cisco Prime Infrastructure, integrates effectively with Cisco's own switches to leverage SPAN, RSPAN, and ERSPAN functionalities.

These tools offer advanced analytics, detailed network visibility, and real-time monitoring, tailored for networks with extensive Cisco infrastructure.

Another valuable tool is PRTG Network Monitor, which includes built-in sensors for monitoring SPAN and RSPAN traffic. PRTG provides real-time alerting, customizable dashboards, and comprehensive traffic analysis, making it a versatile solution for various network monitoring needs.

WhatsUp Gold is known for its interactive maps and robust alerting capabilities, supporting SPAN and RSPAN to provide detailed network insights and traffic analysis. It is particularly useful for small to medium-sized businesses seeking an efficient and user-friendly network monitoring solution. Lastly, open-source tools like ntopng and Zeek (formerly known as Bro) offer support for port mirroring, providing cost-effective options for detailed network analysis and security monitoring. These tools leverage SPAN, RSPAN, and ERSPAN to capture and analyze network traffic, aiding in performance optimization and threat detection.

In summary, port mirroring tools are integral to network monitoring, offering a range of functionalities from performance analysis to security monitoring. Tools like SolarWinds NPM, Wireshark, ManageEngine OpManager, Cisco Prime Infrastructure, PRTG Network Monitor, WhatsUp Gold, and open-source solutions like ntopng and Zeek, all play crucial roles in leveraging port mirroring techniques to maintain efficient and secure network operations.

6.2.4.2 Benefits and Challenges of Port Mirroring

Port mirroring, through techniques such as SPAN, RSPAN, and ERSPAN, offers significant benefits for network monitoring. It provides real-time visibility into network traffic, allowing immediate detection and response to network issues. By analyzing mirrored traffic, administrators can identify performance bottlenecks, optimize resource allocation, and improve overall network efficiency. Additionally, port mirroring is invaluable for security monitoring, helping to detect suspicious activities, unauthorized access, and potential security breaches by providing detailed insights into network traffic. Another advantage of port mirroring is its non-intrusive nature; it captures traffic without affecting the performance of the source devices, making it an effective monitoring solution.

However, there are challenges and considerations associated with port mirroring. Resource utilization is a key concern, as port mirroring can consume significant switch resources and potentially impact performance if not managed properly. It is essential to balance the load to avoid overburdening the network infrastructure. Bandwidth limitations also pose a challenge, particularly in high-volume networks, where the mirrored traffic can be substantial. Ensuring that the monitoring port and network links can handle the additional load is crucial. The security of mirrored traffic, especially with ERSPAN, is another important consideration. The encapsulated traffic traverses the IP network, which could expose it to interception or tampering. Using encryption and secure transport methods can mitigate this risk. Lastly, the accuracy of network analysis depends on the correct configuration of port mirroring. Misconfigurations can lead to incomplete or misleading data, affecting the quality of monitoring.

6.2.5 Network Tap

A network tap, also known simply as a tap, is a hardware device essential for real-time monitoring and capturing of network traffic in computer networks. Unlike traditional network switches that selectively forward traffic based on MAC or IP addresses, a tap duplicates all traffic passing through it. This allows monitoring devices to receive a complete and unaltered stream of network data for analysis. Network taps operate passively, meaning they do not interfere with network traffic or introduce delays. This characteristic is crucial for monitoring applications requiring uninterrupted real-time analysis without impacting network performance.

The diagram below illustrates a typical data center network (DCN) configuration with a network tap installed on each link. While this setup may incur higher costs, it offers customers complete network visibility. Network taps provide several key advantages, notably their ability to offer comprehensive visibility into network activity. They achieve this by capturing all packets, even those not specifically addressed to the monitoring device. This ensures thorough monitoring of traffic between devices, thereby enabling the detection of anomalies and potential security threats. In network monitoring, taps are deployed at critical points within the network architecture to capture relevant traffic effectively (Fig. 6.6).

They are used extensively for tasks such as network performance monitoring, where they help identify bottlenecks and optimize network configurations. Taps also play a vital role in security monitoring by detecting intrusions, malware, and unauthorized access attempts. This data is invaluable for incident investigation and strengthening network defenses.

Furthermore, taps are integral in meeting compliance requirements and facilitating forensic investigations. Their continuous operation and reliability ensure consistent capture and delivery of network traffic data to monitoring tools, supporting auditing and reconstructing network events. Overall, network taps are indispensable tools in modern network management and security operations. Their passive nature, reliability, and ability to provide comprehensive network visibility make them essential for maintaining efficient and secure network environments.

6.2.5.1 Network Tap Solutions

Several vendors offer network taps, each with unique product features catering to various network monitoring and security needs. Garland Technology provides aggregation taps that consolidate multiple network links into a single monitoring port, reducing the number of required monitoring tools. Their filtering taps allow for selective traffic filtering based

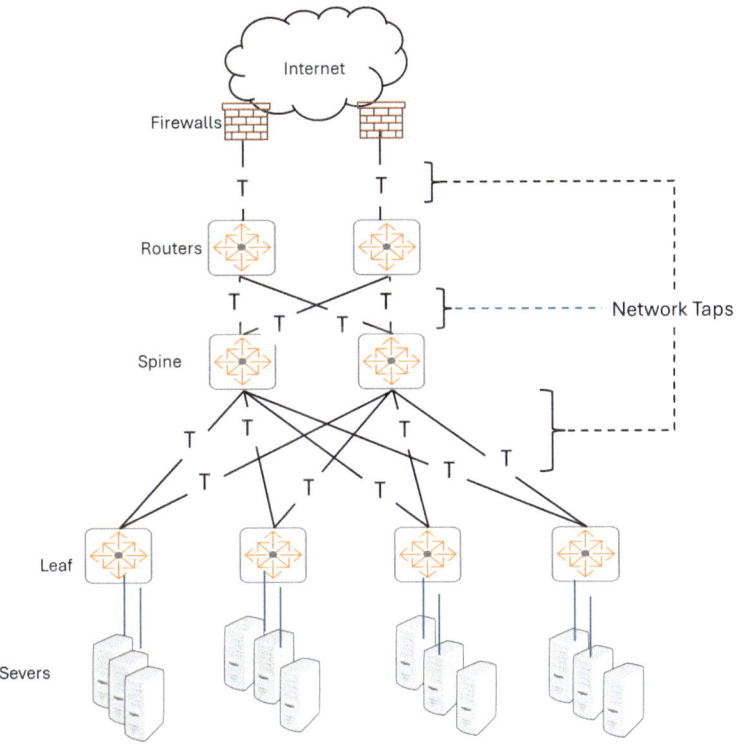

Fig. 6.6 Typical data center network (DCN) with network taps

on criteria like VLAN tags or MAC addresses, optimizing data sent to monitoring tools. Additionally, their bypass taps ensure continuous network availability by automatically redirecting traffic in case of power loss or device failure.

Gigamon offers a scalable platform known as the Visibility Fabric, which aggregates, filters, and optimizes network traffic for monitoring tools. Their inline bypass solutions provide seamless failover options, ensuring network resilience by bypassing security tools during maintenance or failures. Gigamon also enhances visibility into application performance and user experience with its application intelligence features.

Ixia (Keysight) delivers matrix taps with high-density configurations suitable for monitoring large-scale networks, supporting multiple ports, and advanced filtering capabilities. Their cloud visibility solutions are tailored for monitoring cloud environments, ensuring visibility into virtualized and distributed network architectures. Additionally, Ixia provides SSL decryption capabilities to inspect encrypted traffic, enhancing threat detection.

NetScout's smart data sources offer enriched metadata and contextual information about network traffic, facilitating deeper analysis and troubleshooting. Their packet broker solutions provide centralized management and distribution of network traffic to multiple monitoring tools, optimizing efficiency and performance. NetScout also integrates threat intelligence feeds to enhance security monitoring, enabling proactive threat detection and response.

VSS Monitoring, part of Cubro, provides advanced filtering options to capture specific types of traffic based on protocol, port, or application, optimizing tool performance. Their packet slicing capabilities enable the extraction of specific packet data for detailed analysis, reducing storage and processing requirements. VSS Monitoring also offers precise time synchronization across captured packets, which is crucial for correlation and analysis in distributed monitoring environments.

These vendors offer a range of network tap solutions designed to meet diverse requirements in scalability, performance, and functionality. Whether optimizing network performance, enhancing security monitoring, or ensuring compliance, selecting the right network tap depends on the specific network architecture and monitoring objectives.

6.2.6   Network Packet Broker (NPB)

A Network Packet Broker (NPB) is a crucial device in modern network monitoring and management. It acts as an intermediary that manages and directs network traffic from multiple sources to the appropriate monitoring, security, and analysis tools. NPBs aggregate, filter, and distribute data, ensuring that the right information reaches the right tools in an efficient and organized manner. The following diagram depicts how NPB aggregate traffic and distributes data for network operations, application operations, security administration, and network forensics (Fig. 6.7).

It provides centralized visibility and control over network traffic, which is vital for comprehensive monitoring and security operations. They can aggregate traffic from various network links and consolidate it into fewer streams, which simplifies monitoring and reduces the load on individual tools. This aggregation is particularly useful in environments with high-traffic volumes, as it ensures that no data is missed due to overload or bandwidth limitations. One of the primary functions of an NPB is to filter network traffic. By using specific criteria such as IP addresses, protocols, or application types, NPBs can selectively direct only relevant traffic to monitoring tools. This targeted approach helps in reducing the volume of data that needs to be processed, allowing monitoring tools to operate more efficiently and effectively.

NPBs also support traffic optimization through load balancing and deduplication. Load balancing distributes network traffic evenly across multiple tools, preventing any single tool from becoming overwhelmed. Deduplication removes redundant data packets, further reducing the processing burden on monitoring and security tools. These capabilities enhance the performance and reliability of network monitoring solutions.

In addition to these functions, NPBs provide advanced features such as packet slicing and SSL decryption. Packet slicing involves capturing only the necessary parts of a packet,

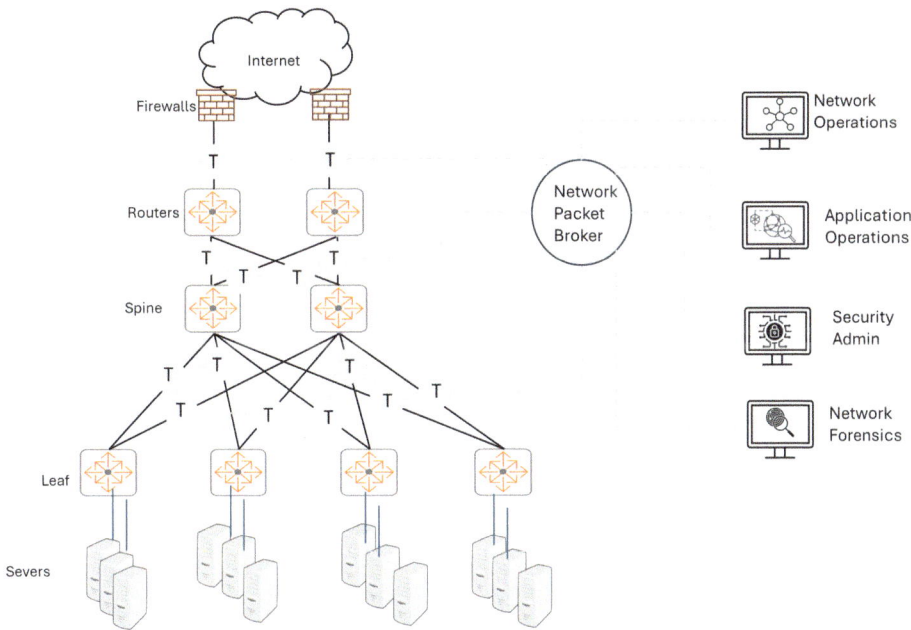

Fig. 6.7 Network packet broker (NPB) aggregating and distributing traffic for network and application operations, security, and forensics

which minimizes data processing and storage requirements. SSL decryption allows for the inspection of encrypted traffic, ensuring that potential threats hidden within encrypted data can be detected and addressed. The deployment of NPBs enhances the overall security posture of an organization. By enabling efficient and thorough monitoring of network traffic, NPBs help in the early detection of anomalies and potential threats. They also facilitate compliance with regulatory requirements by ensuring that all necessary data is captured and available for audit purposes.

6.2.6.1 NPB Solutions

Network Packet Broker (NPB) solutions are critical for modern network infrastructure, enabling efficient and effective traffic management, monitoring, and security. Several vendors offer a variety of NPB solutions, each with distinct features and capabilities to address diverse network requirements.

Gigamon is a leading provider of NPB solutions known for their GigaVUE H Series. Their products excel in providing intelligent traffic visibility and control, offering features like traffic aggregation, replication, filtering, and load balancing. Gigamon's solutions also integrate advanced capabilities such as application intelligence and SSL decryption, which enhance the visibility into encrypted traffic and application-level data. This makes Gigamon a preferred choice for organizations seeking comprehensive monitoring and security across both physical and virtual environments.

Ixia, now part of Keysight Technologies, offers a robust range of NPB solutions through their Vision Series. Ixia's solutions are designed to handle high-density environments with their high port counts and support for multiple protocols. The Vision Series NPBs provide advanced packet processing features like deduplication, packet slicing, and time stamping, ensuring optimized and accurate data delivery to monitoring tools. Ixia's focus on high-performance and scalability makes their NPBs suitable for large-scale enterprise networks and data centers.

NetScout is another prominent player in the NPB market, with their InfiniStreamNG and nGenius series. NetScout's solutions are known for their comprehensive visibility into network traffic and deep packet inspection capabilities. Their NPBs offer enriched metadata generation and contextual analysis, which are essential for detailed network performance monitoring and security analysis. NetScout's emphasis on smart data and integration with their broader nGeniusONE platform provides a holistic approach to network and application performance management.

Arista Networks offers the DANZ Monitoring Fabric, which provides high-performance NPB solutions tailored for data centers and cloud environments. Arista's NPBs leverage their advanced EOS (Extensible Operating System) for enhanced programmability and automation. Their solutions are known for low-latency traffic processing and extensive support for high-speed interfaces, making them ideal for modern, high-bandwidth network environments. Arista's approach to network visibility integrates seamlessly with their broader network infrastructure solutions, providing a cohesive and scalable monitoring strategy.

Big Switch Networks, acquired by Arista Networks, offers Big Monitoring Fabric, a versatile NPB solution that leverages SDN (Software-Defined Networking) principles. Their approach focuses on simplifying network visibility and monitoring through centralized management and automation. Big Monitoring Fabric provides features like dynamic traffic steering, load balancing, and extensive API support for integration with third-party tools. This solution is particularly beneficial for organizations looking to implement agile and programmable network monitoring strategies.

In summary, NPB solutions from vendors like Gigamon, Ixia (Keysight), NetScout, Arista Networks, and Big Switch Networks provide a range of capabilities to meet the varying needs of modern networks. These solutions enhance network visibility, optimize monitoring tool performance, and strengthen security measures by efficiently managing and directing network traffic. Each vendor offers unique features and integration capabilities, allowing organizations to select an NPB solution that aligns with their specific network architecture and operational objectives.

6.3 Telemetry

In the realm of modern networking, telemetry has emerged as a critical component for maintaining optimal network performance, ensuring security, and facilitating effective management. Telemetry involves the automated collection, transmission, and analysis of data from various network devices to provide real-time or near-real-time insights into network operations. By continuously monitoring and analyzing network metrics, telemetry allows network administrators to detect anomalies, troubleshoot issues, and optimize network performance proactively.

Telemetry in network monitoring can be broadly categorized into four main types: Hardware or Flow Telemetry, Synthetic Telemetry, Software Telemetry and INT (In Band Network Telemetry). Each type has its own methods, use cases, and benefits, which together provide a comprehensive view of network performance and health.

6.3.1 Hardware or Flow Telemetry

In Sect. 6.1, we presented a timeline for flow-based protocols for network monitoring. These protocols, NetFlow, sFlow and IPFIX (IP Flow Information Export) are integral to the concept of telemetry. Since Flow Telemetry involves collecting data directly from network devices such as routers, switches, and firewalls, the concept is also known as hardware streaming telemetry. This type of telemetry typically focuses on capturing flow data, which includes information about the traffic passing through the network devices.

6.3.1.1 NetFlow

Developed by Cisco, NetFlow captures flow data from routers and switches, providing detailed insights into IP traffic patterns. By exporting this data to a central collector, NetFlow enables real-time analysis of network usage, identification of traffic trends, and detection of anomalies. As a telemetry tool, NetFlow allows network administrators to gain a comprehensive understanding of network behavior, facilitating proactive management and troubleshooting. The following diagram illustrates Netflow deployment in typical data center networks (Fig. 6.8).

A Flow in Netflowconsists of packets that share common attributes such as source and destination IP addresses, ports, and protocols. The operational workflow of NetFlow can be broken down into several essential stages.

- **Data Collection:** NetFlow-enabled routers and switches generate flow records for each unique traffic flow passing through them. These records typically include details like IP addresses, port numbers, protocol types, quality of service (QoS) indicators, input interfaces, and packet and byte counts. Flow records are stored temporarily in a flow cache within the device.

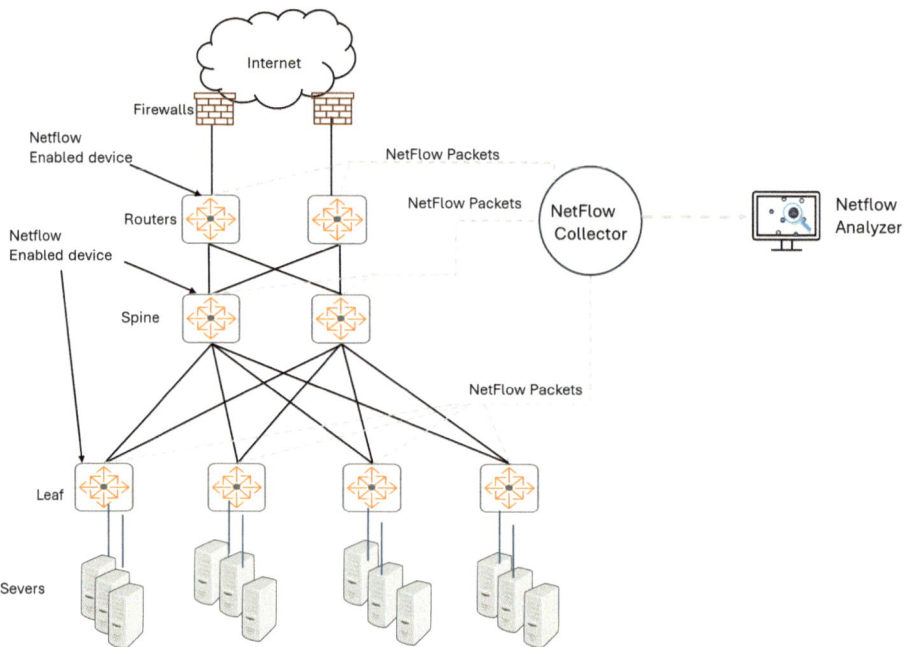

Fig. 6.8 Netflow deployment in a typical data center network

- **Flow Export:** At intervals or when a flow concludes, the flow records are transmitted from the flow cache to a designated NetFlow collector. This export process utilizes the NetFlow Export Protocol to convey flow data to the collector, where it is accumulated for centralized storage and analysis.
- **Data Analysis:** The NetFlow collector aggregates and stores flow records received from multiple network devices. It serves as a centralized repository for comprehensive analysis of network traffic patterns. Network administrators utilize specialized tools to interpret and visualize the flow data, enabling them to identify trends, anomalies, and potential issues affecting network performance.

There are several benefits to using Netflow for network monitoring. These include:

- **Granular Traffic Analysis:** Provides detailed insights into individual IP traffic flows, including key attributes such as IP addresses, ports, protocols, and traffic volumes. This granularity facilitates precise traffic analysis and deepens understanding of network utilization patterns.
- **Real-Time Monitoring:** Enables continuous monitoring of network traffic by capturing and exporting flow data in near real–time. This capability allows administrators to

promptly detect and respond to network incidents, ensuring optimal performance and security.

- **Historical Data Analysis:** Supports the storage and analysis of historical flow data over extended periods. This capability aids in capacity planning, trend analysis, and retrospective troubleshooting of network issues.
- **Scalability:** Designed to handle large-scale networks, NetFlow can efficiently monitor traffic across multiple routers and switches. It consolidates flow data from diverse network segments into a unified monitoring framework.
- **Application Identification:** Advanced implementations of NetFlow can classify and characterize network traffic based on application signatures. This functionality assists in managing bandwidth allocation, optimizing application performance, and enforcing network policies.

In summary, NetFlow remains a crucial element of network telemetry, offering administrators comprehensive visibility and insights into IP traffic flows. Its capability to capture detailed flow data, along with robust analytics and reporting features, facilitates proactive network management, security monitoring, and resource optimization. As networks evolve, NetFlow continues to serve as a foundational tool for ensuring the reliability, security, and efficiency of modern network infrastructures.

However, it's important to note that unlike IPFIX, NetFlow lacks the flexibility to customize data fields and templates, which limits its adaptability to specific monitoring needs and newer network technologies. Additionally, NetFlow typically uses sampled packet data rather than capturing every packet in real-time. This approach can restrict its ability to provide precise insights into network traffic compared to IPFIX, which supports full packet capture and streaming. This distinction impacts the granularity and accuracy of traffic analysis each protocol can offer.

6.3.1.2 sFlow

sFlow was originally developed by InMon Corporation and later became an open-source standard. It is a network monitoring protocol designed to provide real-time visibility into network traffic. Unlike traditional flow-based protocols like NetFlow and IPFIX, which focus on sampling individual flow records, sFlow samples packets at the interface level. This approach allows sFlow to capture detailed information about network traffic across the entire infrastructure without relying solely on flow creation. The following diagram illustrates typical sFlow deployment for telemetry (Fig. 6.9).

As depicted in the figure above, sFlow agents are embedded within network devices and hosts, sending sFlow telemetry data to the sFlow-RT analytical engine, developed by InMon Corporation. sFlow-RT converts this raw data into measurable metrics, revealing network insights and making this data available to other open-source tools through a REST API. Open-source tools like Logstash, InfluxDB, and Grafana can be used to further

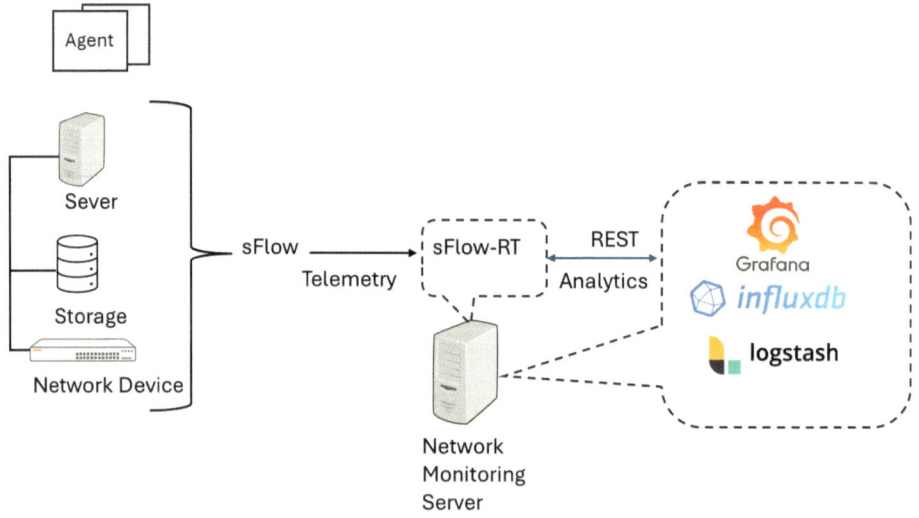

Fig. 6.9   sFlow telemetry deployment with sFlow-RT

process this data and present it through Grafana dashboards. These tools are discussed in detail in Chap. 5.

Additionally, sFlow supports a wide range of network devices and vendors, making it a versatile choice for heterogeneous network environments. It can be implemented on switches, routers, and other network infrastructure devices from different manufacturers, providing consistent monitoring capabilities across the entire network. In terms of use cases, sFlow is commonly used for network performance monitoring, troubleshooting network congestion, and identifying bandwidth utilization patterns. It can also assist in security monitoring by detecting anomalies in traffic behavior that may indicate malicious activity or unauthorized access attempts.

Overall, sFlow offers a flexible and efficient approach to network monitoring by leveraging packet sampling to provide real-time visibility and scalability across diverse network environments. Its ability to monitor traffic at the interface level makes it a valuable tool for both performance optimization and security management in modern networks.

6.3.1.3   IPFIX (IP Flow Information Export)

IP Flow Information Export (IPFIX) is a network protocol standardized by the Internet Engineering Task Force (IETF) and defined in RFC 7011. Additionally, RFC 5470 outlines the architecture of IPFIX, and RFC 6526 specifies its use with Stream Control Transmission Protocol (SCTP). IPFIX is designed to export flow information from routers, switches, and other network devices, providing a robust and flexible framework

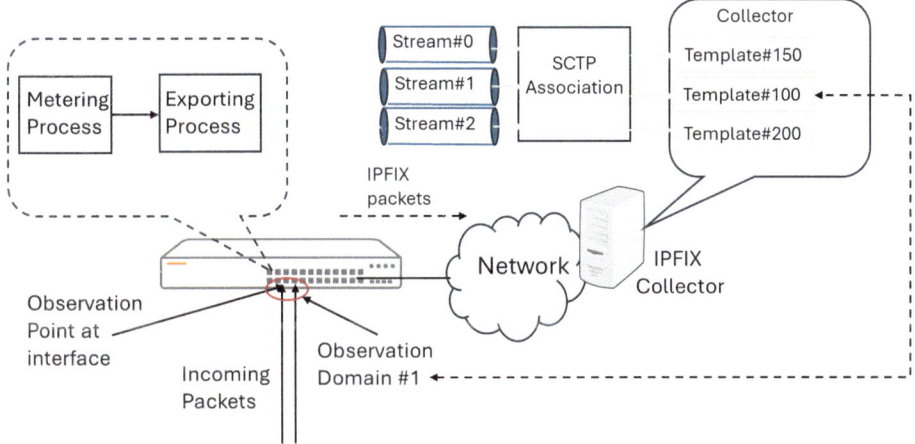

Fig. 6.10 IPFIX telemetry components

for network monitoring and management. This protocol offers detailed insights into traffic patterns, network usage, and performance metrics, addressing several limitations of its predecessor, NetFlow, while introducing new capabilities that enhance network visibility and control. The diagram below illustrates different components of IPFIX-based telemetry solutions (Fig. 6.10).

In an IPFIX setup, network devices or routers serve as IPFIX devices where administrators can define observation points tailored to monitoring needs. These points can encompass single interfaces or multiple interfaces grouped into observation domains (refer to figure above). Each observation domain in IPFIX is identified by a locally unique ID within the exporting process's templates. This ID allows the export process to precisely indicate where flows were metered. Administrators are advised to configure unique IDs for each IPFIX flow; a value of 0 denotes no specific observation domain, which may limit the interpretation of related information elements and hinder device identification by collectors.

IPFIX solutions support flow streams over Reliable Stream Control Transmission Protocol (SCTP) and Transmission Control Protocol (TCP) transport session protocols, ensuring secure and reliable data transport. SCTP enables flow records to be transmitted with varying degrees of reliability across multiple logical channels or streams. Typically, stream 0 is used for transmitting fully reliable templates and data, while other streams handle flow records in a fully reliable, partially reliable, or unreliable manner.[1]

An IPFIX packet comprises several components: a header, set headers, template sets, data sets, and options template sets. The header contains essential metadata such as version number, packet length, export timestamp, and sequence number. Each set within an

---

[1] Kobayashi [1].

IPFIX message includes a set header specifying its type and length. Template sets define the structure of subsequent data records, specifying fields like IPv4/IPv6 addresses, ports, protocols, and timestamps. Data sets contain the actual flow records formatted according to these templates, encapsulating detailed information about individual flows. Options template sets extend IPFIX's capabilities by defining additional metadata such as sampling rates or system-level information.

IPFIX operates through a process of data collection, template definition, data exportation, and subsequent analysis. Network devices gather flow information by examining packets traversing the network, generating flow records based on attributes like source/destination IP addresses and transport protocols. These records are temporarily stored and then sent to a central IPFIX collector after template sets ensure data interpretation accuracy. Flow records are encapsulated into IPFIX packets and transmitted via UDP, TCP, or SCTP to the collector, where they are aggregated and stored. Network administrators utilize various tools to analyze this data, gaining insights into traffic patterns, performance metrics, and network issues for tasks like capacity planning, security monitoring, and troubleshooting.

Comparing IPFIX to NetFlow and sFlow highlights their respective strengths and weaknesses. IPFIX, an evolution of NetFlow standardized by the IETF, offers customizable templates that enhance flexibility compared to NetFlow's fixed templates. IPFIX ensures interoperability across vendors' devices and supports reliable transport protocols, whereas NetFlow, initially proprietary to Cisco, traditionally uses less reliable UDP. In contrast, sFlow, developed by InMon Corporation, samples packets at the interface level to provide a broader view of traffic with less granularity than IPFIX.

In summary, IPFIX distinguishes itself with flexibility and detailed insights, making it a potent tool for network monitoring. It excels in detailed traffic analysis and security monitoring, whereas sFlow offers scalable network visibility with minimal performance impact, making it preferable for high-speed networks. When choosing a telemetry protocol, network managers should consider their specific monitoring needs, network scale, and performance requirements. IPFIX is ideal for environments requiring precise traffic analysis and security monitoring, while sFlow suits environments prioritizing high-level performance monitoring and scalability. For mixed requirements, combining IPFIX for detailed analysis with sFlow for broader visibility provides a balanced approach to network telemetry.

6.3.2 Synthetic Telemetry

The "synthetic telemetry" as a concept doesn't have a widely recognized definition or established use in the field of network monitoring and telemetry. Typically, telemetry refers to the automated process of collecting and transmitting data from remote sources. However, "synthetic telemetry" may refer to synthetic monitoring or testing within the context of network performance and application monitoring. It involves simulating user interactions with network services or applications to measure and analyze their performance. This approach allows network administrators to proactively monitor the health and performance of critical services without waiting for actual user traffic or events. Synthetic telemetry, if understood in this context, focuses on generating synthetic data or traffic to assess network conditions and application responsiveness. Examples of synthetic telemetry solutions are Cisco IPSLA (Internet Protocol Service Level Agreement) and HPE NAE (Hewlett Packard Enterprise's Network Analytics Engine).

Synthetic telemetry tools create artificial transactions or traffic patterns that mimic real user behavior. These tools send requests to network services or applications and measure their response times, latency, throughput, and other performance metrics. By comparing these synthetic measurements against predefined thresholds or benchmarks, administrators can identify potential issues before they affect real users.

Synthetic telemetry offers a number of benefits that includes:

- **Proactive Monitoring:** Synthetic telemetry allows administrators to detect performance degradation or anomalies in network services before they impact end-users.
- **Continuous Monitoring:** Unlike passive monitoring methods like flow-based telemetry, synthetic telemetry actively generates traffic at regular intervals to ensure ongoing assessment of network and application performance.
- **Isolation of Issues:** By generating controlled synthetic traffic, administrators can isolate specific network segments, devices, or applications for targeted performance analysis.

6.3.2.1 Tools and Setups

Several tools and platforms provide synthetic telemetry capabilities for network monitoring and performance testing:

- **CISCO IPSLA:** Cisco IPSLA is a feature in Cisco IOS software that allows network administrators to collect real-time metrics on the performance of IP services and applications. IPSLA generates synthetic traffic between Cisco devices and measures various performance metrics, such as latency, jitter, packet loss, and more.
- **HPE NAE:** HPE NAE is a feature of the HPE Aruba network infrastructure that provides advanced network analytics and insights. It uses synthetic telemetry to continuously monitor network health and performance, identifying issues and providing recommendations for optimization.
- **Pingdom:** Offers synthetic monitoring to check uptime, page speed, and transactional performance of websites and web applications.
- **New Relic Synthetics:** Provides synthetic monitoring to simulate user interactions with web applications and APIs, measuring performance and availability from global locations.

- **Dynatrace Synthetic Monitoring:** Offers synthetic traffic generation to monitor application availability, performance, and functionality across different network environments.
- **GTMetrix:** Provides synthetic monitoring for websites, offering performance metrics based on synthetic tests from various global locations.

Considerations for implementing synthetic telemetry include defining baseline performance metrics and benchmarks to evaluate synthetic telemetry results effectively. It's crucial to ensure that synthetic telemetry tools can scale with the network's size and complexity to maintain accurate performance measurements. Integrating synthetic telemetry with other monitoring tools and platforms is important for achieving comprehensive network and application visibility. Security measures must be taken to safeguard synthetic telemetry tests and results to prevent them from impacting real users or exposing sensitive data during testing.

In conclusion, while the term "synthetic telemetry" might not be commonly used, synthetic monitoring plays a vital role in network performance management. By simulating user interactions and generating controlled traffic, synthetic telemetry enables proactive monitoring and optimization of network services and applications, ensuring optimal performance and user experience.

6.3.3   Software Telemetry

Software telemetry involves the automatic collection, transmission, and analysis of data from software applications and systems. This data can include metrics, logs, traces, and other forms of information that provide insights into the performance, health, and behavior of the software. Software telemetry is essential for modern DevOps practices, enabling proactive monitoring, troubleshooting, and optimization of applications and services.

Software telemetry encompasses several key components. Metrics are quantitative measures that provide insights into the performance of an application, such as CPU usage, memory consumption, request latency, and error rates. These metrics are typically collected at regular intervals to monitor the overall health and performance trends of a system. Logs are detailed records of events occurring within a software system, including error messages, transaction records, and user activities. They are crucial for diagnosing issues and understanding the sequence of events leading up to a problem. Traces capture the execution path of a request as it travels through different components of a distributed system, providing a detailed view of how a request is processed and highlighting latency and bottlenecks at each step. Tracing is particularly useful for debugging complex, microservices-based architectures. Events represent specific occurrences within a system that are noteworthy, such as the completion of a significant transaction, a security breach, or the deployment of a new software version. Events can trigger alerts and automated responses.

6.3.3.1 Tools for Software Telemetry

Several tools are commonly used for implementing software telemetry. Prometheus is an open-source monitoring system that collects and stores metrics as time-series data, offering powerful querying capabilities and integrating well with Grafana for visualization. The Elasticsearch, Logstash, and Kibana (ELK) stack is widely used for log management, where Elasticsearch indexes and stores the logs, Logstash processes and forwards logs, and Kibana provides a web interface for visualizing and querying logs. Jaeger is an open-source, end-to-end distributed tracing tool that helps monitor and troubleshoot microservices-based architectures, providing insights into latency issues and assisting in root cause analysis. Datadog is a cloud-based monitoring and analytics platform that offers metrics, traces, and log collection, providing a unified view of the infrastructure and applications. New Relic is a cloud-based observability platform that provides detailed telemetry data for applications and infrastructure, offering APM (Application Performance Monitoring), infrastructure monitoring, and log management (Fig. 6.11).

The diagram above depicts an example of software telemetry deployment using Prometheus, the New Relic Data Telemetry Platform, New Relic One™, and Grafana. Prometheus's write capability forwards metrics from existing Prometheus servers to the New Relic Telemetry Data Platform. Time series data can be plotted through the New Relic One™ UI, while the Grafana dashboard can provide log insights. Grafana can be deployed on the same telemetry or network monitoring server as Prometheus.

Additionally, Splunk offers comprehensive log management and real-time monitoring for machine-generated data, providing robust searching, monitoring, and analyzing functions.

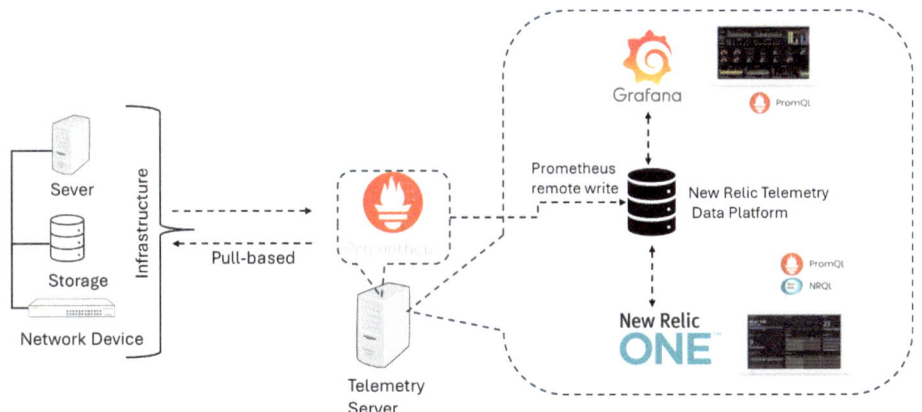

Fig. 6.11 Example of software telemetry using Prometheus, Grafana and New Relic One™

6.3.3.2 Summary

Software telemetry is a cornerstone of modern application monitoring and management, providing the insights needed to ensure applications run smoothly, perform optimally, and remain secure. With a wide range of tools available, organizations can implement robust telemetry solutions tailored to their specific needs. By leveraging software telemetry, businesses can achieve greater visibility into their systems, improve operational efficiency, and deliver a better user experience.

6.3.4   In Band Network Telemetry (INT)

In Band Network Telemetry (INT) represents a significant advancement in network monitoring and management. The P4.org forum, which defined programmability for ASICs through the P4 language, also defined INT. The following timeline depicts the work on INT (Fig. 6.12)

Unlike traditional out-of-band telemetry, which collects data through separate monitoring channels, INT embeds telemetry information directly within the data packets as they traverse the network. This approach allows for real-time monitoring and analysis of network performance and behavior without introducing additional overhead or delays. INT provides granular visibility into network conditions, making it a powerful tool for network administrators and engineers.

INT can be defined as a method of collecting detailed network data by embedding telemetry information directly within the data packets. This information can include metrics such as latency, packet loss, jitter, and other performance indicators. By embedding this data within the packets, INT enables real-time analysis and immediate feedback on network conditions, helping network administrators quickly identify and address issues. The following diagram depicts the concept of INT. In this diagram, the INT header is inserted at the first switch where the watchlist instruction is programmed. Each traversing switch adds its path-related attributes, such as latency, packet loss, jitter, and other performance parameters. This information is decapsulated at the switch connected to the telemetry monitoring station. The event is detected and forwarded to the telemetry monitoring station (Fig. 6.13).

The initial concept of INT architecture consists of the INT control plane and INT data plane. When P4.org defined the specification, the idea was that an SDN controller would implement the INT control plane and program the devices for data plane functionalities.

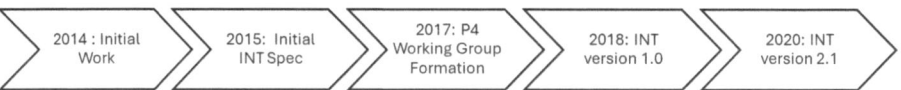

Fig. 6.12 INT specification timeline

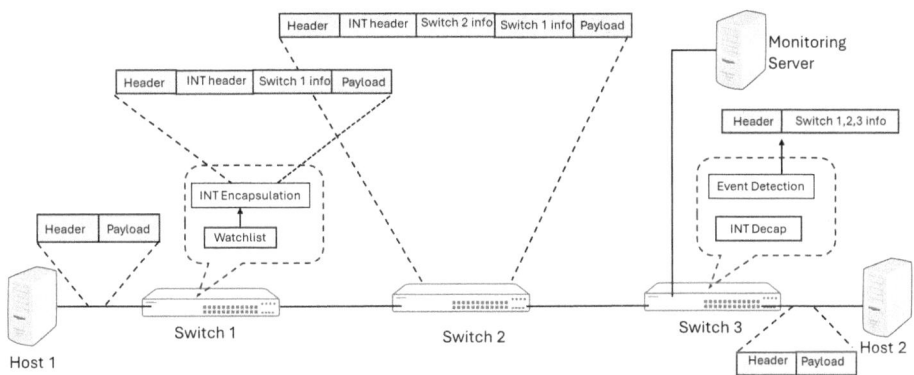

Fig. 6.13 The in band network telemetry (INT) concept

Today, CLI and orchestration platforms can implement INT control plane information regarding configuration and other data plane functionalities directly at the switch. For example, HPE's CX10K switch implements a Data Processing Unit (DPU) and can receive P4 instructions directly to the switch through CLI or Policy Services Manager (PSM). The INT data plane specification version 2.1, developed by P4.org, defines the INT header and various packet headers depending upon the mode of INT operations. INT spec 2.1 supports three modes of operations: INT-XD (eXport Data), INT-MX (eMbed instructions), and INT-MD (eMbed Data).

6.3.4.1   INT-XD (eXport Data)
INT-XD focuses on exporting telemetry data from the network devices to external collectors for analysis. This mode emphasizes the transmission of telemetry data out of the network devices to a centralized or distributed telemetry analysis system.

Key Features are listed as follows:

- **Data Export:** Telemetry data is exported from network devices to external collectors.
- **Centralized Analysis:** Enables centralized analysis of telemetry data, facilitating comprehensive performance and security monitoring.
- **Interoperability:** Ensures compatibility between network devices and external telemetry collectors.

The diagram below illustrates the operation of INT-XD: in this mode, each network switch incorporates watchlist and event detection functionalities. Metadata is directly exported from the switch to the monitoring system (Fig. 6.14).

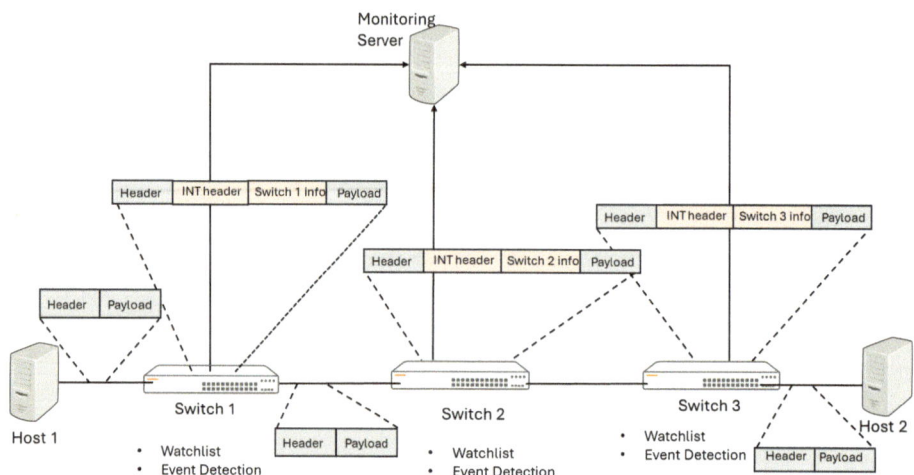

Fig. 6.14 INT-XD operation in which information is directly sent from each switch to the monitoring server

The specific data exported is determined by the INT instructions and the Flow Watchlist. Notably, this mode does not require packet modification, and it is also known as Postcard-based Telemetry.[2]

6.3.4.2 INT-MX (eMbed Instruct(X)ions)

INT-MX involves embedding specific instructions within the data packets as they travel through the network. These instructions can dictate how telemetry data should be collected, processed, and reported, enabling more dynamic and flexible telemetry operations.

Key Features are listed as follows:

- **Instruction Embedding**: Embeds instructions within data packets to guide telemetry operations.
- **Dynamic Telemetry:** Allows for dynamic adjustments to telemetry data collection based on real-time network conditions.
- **Flexibility:** Provides flexibility in how telemetry data is collected and processed.

The diagram below illustrates the operation of INT-MX: in this mode, INT instructions are embedded into packets by the INT source node (Fig. 6.15).

The INT source, each INT transit hop, and the INT sink send this metadata directly to the monitor by following the embedded instructions in the packets. Finally, before forwarding the packets to the final host, the INT sink removes the INT instructions[2].

---

[2] Joshi [2].

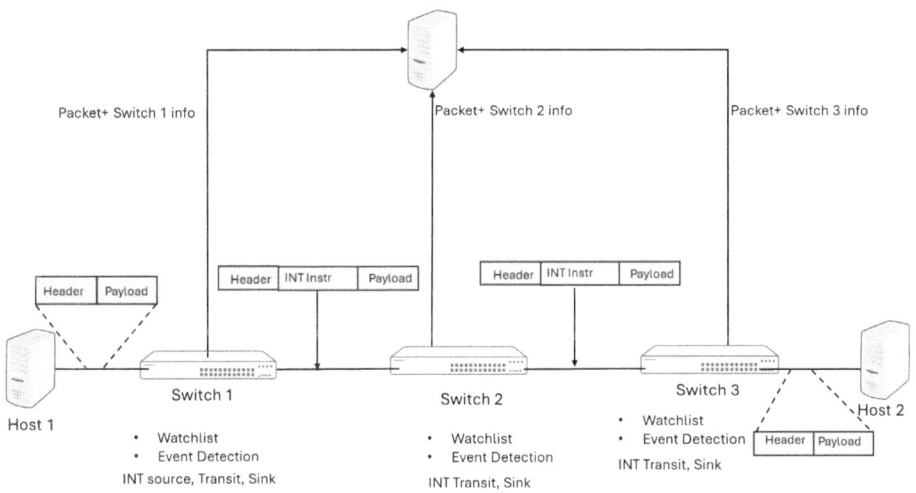

Fig. 6.15 INT-MX operations: switches sent information directly to the monitoring server based on instructions

6.3.4.3 INT-MD (eMbed Data)

INT-MD refers to embedding telemetry data directly into the packets themselves as they traverse the network. This mode ensures that telemetry data is collected and reported in real-time, providing immediate insights into network performance and conditions.

Key Features are listed as follows:

- **Data Embedding:** Embeds telemetry data within the packets as they travel through the network.
- **Real-Time Insights:** Provides real-time visibility into network performance and issues.
- **Detailed Monitoring:** Enables detailed monitoring of network conditions at each hop along the packet's path.

The following diagram depicts how INT-MD operations work. You will notice that the diagram has a similarity with the INT concept diagram presented in Fig. 6.13. In this mode, both INT Instructions and INT Metadata are written into packets as shown in the figure (Fig. 6.16).

This process can be visualized in three steps: First, the INT source embeds the INT instructions into the packets. Next, the INT source and INT transit hops embed the INT metadata as the packets traverse the network. Finally, the INT sink removes the INT instructions and the collected metadata, sending this information to the monitoring system.

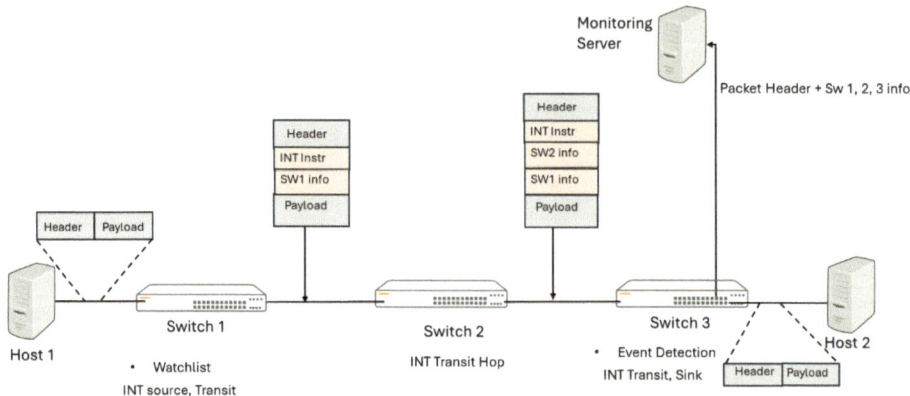

Fig. 6.16 INT-MD operation: INT Sink sends INT meta data to the monitoring server

6.3.4.4 Tools and Technologies Supporting INT Modes

Several tools and technologies support these INT modes, leveraging the capabilities of modern network devices and programmable networks:

- **P4 Language:** A programming language designed to enable the customization of packet processing to implement INT functionalities, including embedding instructions and data.
- **Intel Tofino:** A programmable switch ASIC that supports P4 and INT, facilitating advanced telemetry implementations.
- **Cisco Nexus 9000 Series:** Switches that support INT through Cisco's P4-based implementations, providing comprehensive telemetry capabilities.
- **Broadcom Trident and Tomahawk:** Chipsets that offer support for INT features, enabling detailed telemetry across various network segments.
- **Telemetry Streaming Platforms:** Platforms like Google Cloud's Packet Mirroring and Amazon Web Services' VPC Traffic Mirroring, which integrate INT data into cloud-based monitoring and analytics solutions.

6.3.4.5 Summary

INT-XD (eXport Data), INT-MX (eMbed Instructions), and INT-MD (eMbed Data) are distinct modes of In-band Network Telemetry that enhance its capabilities to meet the diverse needs of modern, complex network environments. INT-XD focuses on exporting telemetry data for centralized analysis, INT-MX embeds instructions within packets to guide telemetry operations dynamically, and INT-MD embeds telemetry data directly within packets for real-time insights. Together, these modes extend the applicability of INT, making it a powerful tool for ensuring network performance, reliability, and security in a variety of contexts. The adoption of these INT modes, supported by advanced tools and technologies, empowers network administrators to achieve comprehensive, real-time insights into network operations, facilitating proactive management and optimization.

6.4 Network Observability

Network observability is the ability to deeply understand a network's operations by collecting, analyzing, and visualizing comprehensive telemetry data from the entire network infrastructure, including routers, switches, firewalls, servers, and applications. The goal is to deliver real-time visibility and a detailed contextual understanding of the network's performance, behavior, and health. Unlike traditional network monitoring, which often focuses on predefined metrics and alerts, observability offers a more nuanced view of the network. It enables IT professionals to quickly identify, diagnose, and address issues before they impact service quality. A key part of this process is the ability to correlate vast amounts of data from diverse network elements, creating a holistic view that can uncover underlying problems or inefficiencies.

The principles of network observability involve continuously collecting telemetry data from network devices and applications to monitor performance, behavior, and security. Real-time analysis of this data provides immediate insights into network health and performance. By integrating data from multiple sources, observability creates a comprehensive view of the network, which aids in better diagnosis and troubleshooting. This capability is essential for monitoring complex and distributed networks, ensuring they can accommodate growth and increasing complexity.

Network observability involves sophisticated data analysis and visualization, presenting telemetry data in an easily understandable format. It also focuses on proactive incident response, detecting and addressing issues before they escalate into significant problems. Automation and machine learning are leveraged to enhance issue detection, prediction, and resolution, making network management more efficient and effective. Additionally, it facilitates collaboration among different IT teams and systems, ensuring a unified approach to network management. The following diagram depicts a typical Network Observability Framework (Fig. 6.17).

The main components of network observability include data collection, a data lake, and a network observability platform. As discussed earlier, network telemetry plays a crucial role in providing real-time data insights by involving the collection, analysis, and transmission of data. Depending on the type of network telemetry in use, network administrators gain a wealth of network insights that can be further processed for network observability.

However, network telemetry provides one set of information. The network observability framework, as presented in the figure above, allows the ingestion of data through various methods such as log data via syslog and packet capture. All these data are then

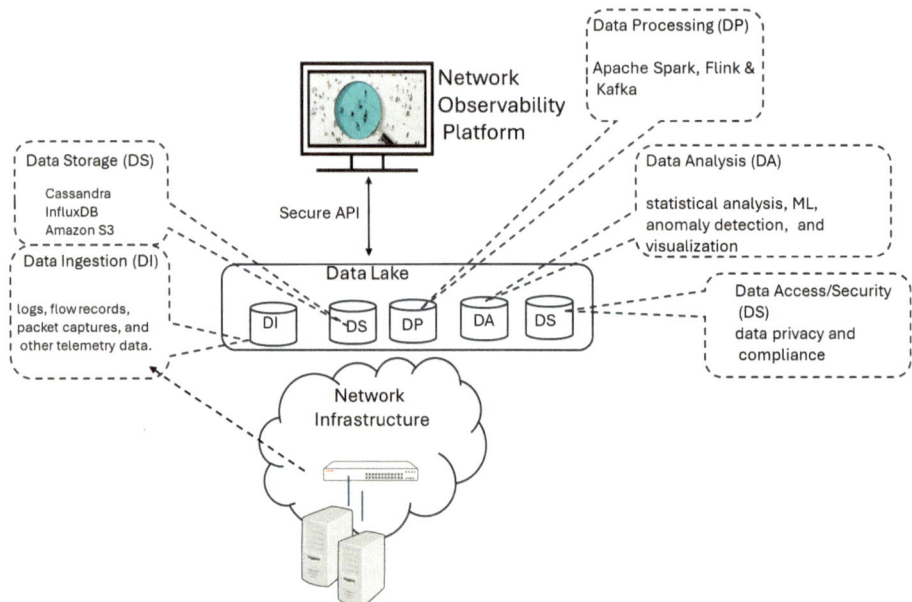

Fig. 6.17 Network observability framework

fed into the data lake system, which in turn enables further manipulation, visualization, and intervention through the network observability platform, e.g., Arista CloudVision.

6.4.1   Data Collection

We have discussed network telemetry in detail in Sect. 6.3, and we will now focus on the data collection aspects of network observability for which network telemetry is a subset. Data collection involves gathering metrics, logs, traces, and events from various network devices, applications, and services. This process is typically achieved through:

- **Metrics:** Quantitative data points such as CPU usage, memory consumption, network latency, and bandwidth utilization. These metrics provide a high-level overview of network performance.
- **Logs:** Detailed records of events and transactions within the network, capturing information about device operations, traffic flows, and security incidents. Logs are essential for troubleshooting and forensic analysis.
- **Traces:** Detailed records of the execution path of network requests or transactions, helping identify performance bottlenecks and understanding data flow within complex systems.

- **Events:** Notifications about significant occurrences or state changes within the network, such as security alerts, hardware failures, or traffic anomalies.

6.4.2 Data Lake

A Network Data Lake is a centralized repository that allows you to store, manage, and analyze large volumes of structured and unstructured network data at any scale. It is designed to handle the diverse and high-volume data generated by various network devices and services, providing a foundation for advanced analytics, machine learning, and real-time monitoring.

Key Components of a Network Data Lake are:

- **Data Ingestion:** The process of collecting data from multiple sources such as routers, switches, firewalls, servers, and IoT devices. This data can include logs, flow records, packet captures, and other telemetry data.
- **Data Storage:** Utilizing scalable storage solutions to accommodate the growing volume of network data. Common storage technologies include distributed file systems (e.g., HDFS, Amazon S3) and databases (e.g., NoSQL databases like Cassandra or time-series databases like InfluxDB).
- **Data Processing:** Transforming and preparing data for analysis. This can involve cleaning, normalizing, enriching, and aggregating data using batch or stream processing frameworks (e.g., Apache Spark, Flink, Kafka).
- **Data Analysis:** Applying various analytical techniques to extract insights from the data. This includes statistical analysis, machine learning, anomaly detection, and visualization.
- **Data Access and Security:** Ensuring secure and efficient access to data for different stakeholders while maintaining data privacy and compliance with regulations.

6.4.2.1 How Telemetry Fits into a Network Data Lake

Network telemetry is the continuous collection of data from network devices to provide visibility into the network's performance, behavior, and security. Telemetry data is crucial for the functioning of a network data lake. Here's how telemetry fits into the network data lake architecture:

- Data Collection: Telemetry protocols like IPFIX, NetFlow, sFlow, and INT collect data from network devices in real-time or near-real-time. This data is then ingested into the network data lake.
- Data Variety: Telemetry data includes various types of information such as traffic flows, packet statistics, performance metrics, and event logs, all of which are valuable for different analyses.

- Real-Time Processing: Stream processing frameworks can process telemetry data in real-time to detect anomalies, generate alerts, and provide actionable insights immediately.
- Historical Analysis: The stored telemetry data can be used for historical analysis, trend analysis, capacity planning, and forensic investigations.
- Machine Learning: Machine learning models can be trained on historical telemetry data to predict future network behavior, identify potential issues, and optimize network performance.
- Visualization and Reporting: Telemetry data can be visualized in dashboards and reports to provide network administrators with clear insights into the network's health and performance.

Benefits of Integrating Telemetry into a Network Data Lake

Integrating Network Telemetry into a Network Data Lake system help in the followings:

- Comprehensive Visibility: By integrating telemetry data, a network data lake provides a holistic view of network performance, enabling better decision-making.
- Scalability: Network data lakes can scale to handle the massive volumes of telemetry data generated by large networks.
- Advanced Analytics: Combining telemetry data with other data sources allows for more sophisticated analytics, leading to deeper insights.
- Real-Time Insights: Real-time processing of telemetry data helps in proactive network management and quicker response to issues.
- Cost-Efficiency: Storing and analyzing data in a centralized repository can be more cost-effective than traditional network management solutions.

In summary, a network data lake serves as a powerful platform for managing and analyzing network telemetry data, providing network administrators with the tools needed to ensure optimal network performance and security.

6.4.3 Network Observability Platforms

Network Observability Platform is the crucial element of Network Observability framework. These platforms provide comprehensive visibility into network operations by collecting, analyzing, and visualizing data from a multitude of network devices and systems. They offer real-time insights, predictive analytics, and automated actions that help network administrators proactively manage and optimize their networks. Network observability platforms integrate with data lakes to leverage vast amounts of network data for advanced analysis and decision-making, ultimately ensuring a robust and resilient network infrastructure. There are a number of these platforms available; we present a few examples here. This is not an exclusive list, but rather a sample of what is available in the market.

6.4.3.1   Arista CloudVision

Arista CloudVision is a comprehensive network observability platform designed to provide end-to-end visibility and control of network infrastructure. It integrates various data sources to deliver real-time insights and management capabilities. CloudVision collects telemetry data from Arista switches and routers, including flow data, device health metrics, and configuration changes. This data can be ingested into a centralized data lake for long-term storage, aggregation, and analysis. The platform provides dashboards and analytics tools to visualize network performance and detect anomalies in real-time. Additionally, it includes automation capabilities for network configuration, monitoring, and troubleshooting, leveraging the insights derived from the data lake.

6.4.3.2   Cisco DNA Center

Cisco DNA Center is another powerful network observability platform that provides centralized management and analytics for Cisco-based networks. It collects telemetry data, logs, and network statistics from Cisco devices, sending this data to a data lake for historical analysis, machine learning, and big data processing. With AI and machine learning, Cisco DNA Center offers predictive insights and automation for network operations. It features advanced visualization tools for monitoring and troubleshooting network issues, providing a comprehensive view of network performance and health.

6.4.3.3   Juniper Networks' Paragon Insights

Juniper Networks' Paragon Insights offers a robust solution for network observability and assurance, focusing on delivering proactive insights and automated actions. It collects data from Juniper devices and third-party sources, including telemetry, logs, and SNMP data. This data is ingested into a data lake for enhanced analytics and correlation with other data sources. Paragon Insights uses machine learning to identify potential issues and recommend proactive measures. It provides intuitive dashboards for monitoring network health and performance, enabling early detection of network anomalies and performance issues.

6.4.3.4   Elastic Observability

Elastic Observability is an open-source observability solution built on the Elastic Stack (Elasticsearch, Logstash, Kibana, and Beats). It provides unified visibility across your infrastructure, applications, and network. Elastic Observability can ingest data from various sources, including network telemetry, logs, and metrics, using Beats and Logstash. The Elastic Stack can function as a data lake itself or integrate with other data lakes for deeper analysis and storage. Elasticsearch enables powerful search and real-time analysis of ingested data, while Kibana provides comprehensive visualization capabilities, including customizable dashboards and alerting.

6.4.3.5 Connecting to a Data Lake

All these network observability platforms share a common approach to connecting with a data lake. They ingest data from network devices and other sources in real-time or near-real-time, storing this data in a data lake built using technologies like Hadoop, Amazon S3, or Elasticsearch. The data lake enables batch and stream processing for analytics, machine learning, and other advanced analyses. These platforms provide interfaces and tools to access, visualize, and interact with the data stored in the data lake, delivering actionable insights and improving network operations.

6.5   Summary

In this chapter, we explored the critical components of network insights, including network monitoring, telemetry, observability, and visibility, which collectively provide a comprehensive understanding of network performance, health, and security. The chapter begins by emphasizing the importance of network monitoring as the foundational aspect of network management, focusing on real-time tracking of performance metrics such as bandwidth usage, latency, packet loss, and uptime. This real-time data provides immediate insights and enables quick responses to potential problems, helping maintain network stability. We then examined the evolution of network monitoring technologies, starting with SNMP (Simple Network Management Protocol) in the 1980s, which provided a basic framework for collecting information from network devices. The introduction of RMON (Remote Network Monitoring) in the 1990s allowed for deeper insights at the packet level, followed by the development of flow-based monitoring technologies like NetFlow and sFlow. These innovations provided granular visibility into traffic patterns, enabling detailed analysis and real-time detection of anomalies.

The chapter also delves into network telemetry, which involves the collection, transmission, and analysis of data from network devices, offering a granular view of network operations. Telemetry provides the raw data that powers network observability, integrating data from metrics, logs, and traces to offer a holistic view of network health. Observability uses advanced analytics and machine learning to interpret this data, identify anomalies, diagnose issues, and predict future trends, enhancing proactive management. Network visibility, another key aspect discussed, complements monitoring and telemetry by providing a comprehensive view of all traffic within the network. Tools that provide network visibility analyze traffic patterns, detect bottlenecks, and ensure security policies are enforced, which is crucial for both performance management and security.

The integration of these approaches—network monitoring, telemetry, observability, and visibility—represents a significant advancement in network management. Real-time data collection, continuous monitoring, and advanced analytics provide deep insights into network behavior, facilitating proactive maintenance, capacity planning, and enhanced security postures. By leveraging these advanced network insights, organizations can ensure their networks remain robust, secure, and capable of supporting the growing demands of the digital age.

Overall, this chapter provides a detailed exploration of these network insights, offering practical guidance for implementing effective network management strategies. As networks continue to grow in complexity, the ability to monitor, analyze, and optimize their performance becomes increasingly critical, making the combined power of these technologies essential for maintaining resilient and high-performing network infrastructures.

References

1. Kobayashi A (2012) A study of flow-based IP traffic measurements for large-scale networks. The University of Electro Communications
2. Joshi M (2021) Implementation and evaluation of inband network telemetry in P4. Degree Project in Information and Communication Technology. KTH Royal Institute of Technology, Sweden

Innovations in Network Security: Embracing a Security First Approach

In today's interconnected world, network security has become a paramount concern for organizations of all sizes. As businesses increasingly rely on digital infrastructure to operate, the importance of safeguarding sensitive data and ensuring the integrity, availability, and confidentiality of network communications cannot be overstated. Network security encompasses a broad range of practices and technologies designed to protect networks from unauthorized access, misuse, malfunction, modification, destruction, or improper disclosure. The evolution of network security has been driven by the ever-changing landscape of threats. Early security measures, such as basic firewalls and antivirus programs, were once sufficient to counteract the relatively simple attacks of the past. However, as technology has advanced, so too have the methods employed by malicious actors. Today, network security faces sophisticated threats that can bypass traditional defenses, necessitating the development of more advanced and integrated security solutions.

This chapter delves into both traditional and cutting-edge network security technologies, providing a comprehensive understanding of the tools and strategies available to protect modern networks. We will explore established technologies such as firewalls, intrusion detection and prevention systems, and virtual private networks, which form the backbone of traditional network security. Additionally, we will examine the latest advancements in the field, including zero trust architecture, artificial intelligence, machine learning and behavioral analytics, which offer new ways to detect and mitigate threats.

To discuss how the security-first paradigm works in modern networking infrastructure is to share some specific examples of how vendors are offering this. HPE seems to be leading in this space with the AI-powered Security First Networking concept. I thought it would be appropriate to show how HPE pulling together its discrete technologies to offer a comprehensive solution in the space. The discussion may refer to some HPE specific solutions that may construe proprietary. However, the solutions offer a new look at how network security improves. It is a paradigm shift I the network infrastructure design and worth highlighting such advances to the readership. From the network architect, such solutions offer a new perspective that may simply network deployments and improve network performance and TCO.

Ultimately, the goal of this chapter is to provide readers with a holistic view of network security, combining the strengths of traditional methods with the innovations of modern technology. By understanding and integrating these diverse approaches, organizations can build resilient, secure networks that are capable of withstanding the challenges of today and tomorrow.

7.1   Evolution of Network Security

The evolution of network security is a journey marked by significant milestones and technological advancements, driven by the escalating complexity and frequency of cyber threats. Understanding this evolution provides critical insights into the development of both traditional and modern security measures, highlighting the ongoing battle between security professionals and malicious actors. The following diagram depicts a representative timeline for Network Security. The timeline presents representative milestones and is not intended to be an exhaustive list of events that occurred (Fig. 7.1).

In the early days of network computing during the 1970s and 1980s, security was relatively straightforward. Networks were closed systems with limited external connectivity, and the primary concerns revolved around physical security and simple password protections. The first cyberattack is believed to have occurred in 1971 when Bob Thomas, a computer programmer at BBN, created and deployed a virus as a security test. Although it was not malicious, it highlighted areas of vulnerability and security flaws in what would

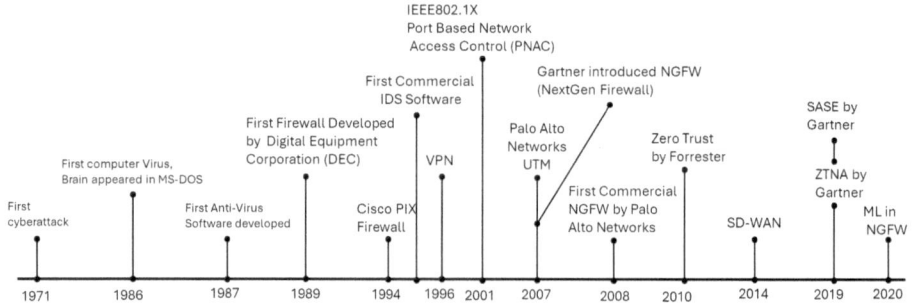

Fig. 7.1 Network security timeline

become "the Internet."[1] In response, Ray Tomlinson developed the "Reaper" program to find and eliminate this virus, named "Creeper." This was the first non-commercial antivirus software.

In 1986, the first computer virus, "Brain," appeared on personal computers (PCs) and affected Microsoft DOS (MS-DOS). In 1987, John McAfee created the first commercial antivirus software to detect and remove the "Brain" virus. As computer networks expanded and the Internet emerged in the late 1980s and early 1990s, the need for more robust security measures became apparent. The introduction of firewalls by Digital Equipment Corporation (DEC) in 1989, followed by Cisco in 1994, marked a significant milestone. Firewalls provided a first line of defense by filtering incoming and outgoing traffic based on predefined security rules. In 2001, IEEE 802.1X introduced Port-Based Network Access Control (PNAC) to network switches, providing authentication for network devices to protect against hacking.

As the Internet continued to grow, the mid to late 1990s saw the development of more sophisticated threats, such as viruses and worms, leading to the proliferation of antivirus software. During this period, Intrusion Detection Systems (IDS) were also introduced, capable of monitoring network traffic for suspicious activities and alerting administrators to potential breaches. The first commercial IDS software, "NetRanger," was released by WheelGroup in 1995. The late 1990s also brought the advent of Virtual Private Networks (VPNs), which allowed secure communication over public networks by encrypting data transmissions. An engineer at Microsoft developed the first VPN technology based on the Point-to-Point Tunneling Protocol (PPTP) in 1996. This era also witnessed the rise of Intrusion Prevention Systems (IPS), which not only detected but also actively prevented security breaches. Around the same time, Security Information and Event Management (SIEM) systems emerged, providing comprehensive visibility into network activities by aggregating and analyzing log data from various sources.

In 2007, Palo Alto Networks introduced advanced firewalls known as Unified Threat Management (UTM), integrating IPS into its firewall product. This created momentum toward advanced firewall technology that Gartner termed as Next-Generation Firewall (NGFW). In 2008, Palo Alto Networks introduced the first NGFW featuring IPS, full-stack visibility, and the ability to enforce security policies based on applications, users, and content.

The 2010s marked a significant shift toward more dynamic and proactive security measures. The concept of Zero Trust Architecture gained prominence, advocating that no entity, whether inside or outside the network, should be trusted by default. Forrester Research introduced the Zero Trust concept in 2010, creating a paradigm shift that led to the development of technologies focused on continuous verification of user identities and access permissions. In 2014, SD-WAN technology was introduced to the market, providing connectivity between remote sites while introducing advanced security for the traffic path between the sites. However, there remained a need for securing remote user

---

[1] Monroe [1].

access to IT resources, which led to the momentum toward Zero Trust Network Access (ZTNA) technologies. The concept of ZTNA was first introduced by Gartner in 2019, followed by Secure Access Service Edge (SASE) as a cloud-native security and network architecture.

Simultaneously, the integration of Artificial Intelligence (AI) and Machine Learning (ML) into security solutions began to revolutionize threat detection and response. These technologies enabled more accurate identification of anomalies and proactive threat mitigation. Behavioral Analytics, specifically User and Entity Behavior Analytics (UEBA), emerged as a powerful tool for detecting insider threats and sophisticated attacks by analyzing patterns of normal behavior and identifying deviations.

The latter part of the 2010s and early 2020s saw the rise of Software-Defined Security, which leverages the principles of Software-Defined Networking (SDN) to create more flexible and responsive security policies. Deception technologies, such as honeypots and deception grids, also gained traction as a means to mislead and trap attackers, thereby enhancing threat intelligence.

In recent years, the focus has shifted toward comprehensive, integrated security frameworks. HPE's Security First Networking exemplifies this approach by embedding security into every layer of the network infrastructure. By combining advanced technologies such as HPE Aruba ClearPass for network access control and HPE Aruba EdgeConnect for secure SD-WAN, organizations can create a robust and adaptable security posture. The evolution of network security is a testament to the ever-changing nature of cyber threats and the continuous innovation required to counter them. From the early days of basic firewalls and antivirus programs to the sophisticated, AI-driven, and zero trust-based solutions of today, network security has transformed into a complex, multi-faceted discipline that is critical for safeguarding the digital world.

7.2 Traditional Network Security Technologies

In the ever-evolving landscape of cyber threats, traditional network security technologies remain the cornerstone of any robust security strategy. These time-tested technologies have been developed and refined over decades, each addressing a unique set of security challenges and laying the groundwork for more advanced solutions. As we navigate through this chapter, we will explore these foundational technologies in detail, understanding their critical role in protecting network infrastructure and data.

Traditional network security technologies encompass several key areas, each playing a vital role in maintaining the security and integrity of networks.

Antivirus and Antimalware solutions were among the first tools developed to combat malicious software. Initially focused on identifying and removing known viruses through signature-based detection, these solutions have evolved to include heuristic and behavioral analysis to detect a broader range of threats, including spyware, ransomware, and other forms of malware. These tools provide essential protection for endpoints, ensuring that known threats are swiftly identified and neutralized.

Access Control mechanisms are crucial for managing who can access network resources and what actions they are permitted to perform. This includes traditional methods like Role-Based Access Control (RBAC), where permissions are assigned based on user roles, and more dynamic approaches such as Network Access Control (NAC), which evaluates the security posture of devices before granting network access. Proper access control helps prevent unauthorized access and reduces the risk of insider threats.

Virtual Private Networks (VPNs) play a pivotal role in securing remote communications. By creating encrypted tunnels over public networks, VPNs ensure that data transmitted between remote users and corporate networks remains confidential and secure. This technology is especially important in today's work-from-anywhere environment, providing secure access to corporate resources for employees, contractors, and partners.

Intrusion Detection and Prevention Systems (IDS/IPS) are designed to identify and respond to potential security breaches. IDS passively monitor network traffic, generating alerts when suspicious activity is detected, while IPS actively intercept and block malicious traffic. These systems use a combination of signature-based detection, which relies on known attack patterns, and anomaly-based detection, which identifies deviations from normal network behavior. IDS/IPS are vital for detecting and mitigating attacks before they can cause significant harm.

Firewalls are one of the oldest and most fundamental components of network security. These devices control the flow of traffic between different parts of a network, enforcing security policies through packet filtering, stateful inspection, and application-level gateways. Next-Generation Firewalls (NGFWs) extend these capabilities by incorporating advanced features such as deep packet inspection, intrusion prevention, and application awareness. Firewalls serve as the first line of defense, preventing unauthorized access and blocking malicious traffic.

Security Information and Event Management (SIEM) systems provide a centralized platform for collecting, correlating, and analyzing security-related data from various sources across the network. By aggregating logs and events, SIEMs enable security teams to gain comprehensive visibility into network activities, detect potential threats, and coordinate incident response efforts. SIEM solutions are essential for maintaining situational awareness and ensuring a rapid and effective response to security incidents.

In this section, we will explore each of these traditional network security technologies in greater detail. We will examine their functionalities, benefits, and the ways they can be effectively implemented to create a secure network environment. Understanding these foundational tools is essential for building a comprehensive security strategy that can effectively mitigate risks and respond to incidents in an increasingly complex digital landscape.

7.2.1 Antivirus and Antimalware Solutions

Antivirus and antimalware software have long been the frontline defense against malicious software threats. Over the years, these tools have evolved from simple signature-based scanners to sophisticated, multi-layered security solutions capable of defending against a wide range of cyber threats. Antivirus programs protect against more established threats, like traditional worms, viruses, and Trojans. Antimalware specializes in newer exploits, like polymorphic malware and zero-day malware. Antivirus programs are good at protecting against more predictable, dangerous malware. Both antivirus and antimalware software are host-based meaning those are installed directly in the computer system. Antivirus detection can be classified into three main approaches:

- **Static Detection**: It is one of the simplest antivirus detection techniques, relying on predefined signatures of known malicious files. This method uses pattern-matching techniques, such as identifying unique strings, checksums (CRC), sequences of byte-code or hexadecimal values, and cryptographic hashes (like MD5 and SHA1). The antivirus software compares files in the operating system against a database of these signatures. If a match is found, the file is flagged as malicious, please refer to the diagram (Fig. 7.2).
  This approach is particularly effective against static malware, which does not change its signature frequently.
- **Dynamic Detection**: The dynamic detection approach is advanced and more complicated than static detection. Dynamic detection is focused more on checking files at runtime using different methods (Fig. 7.3).

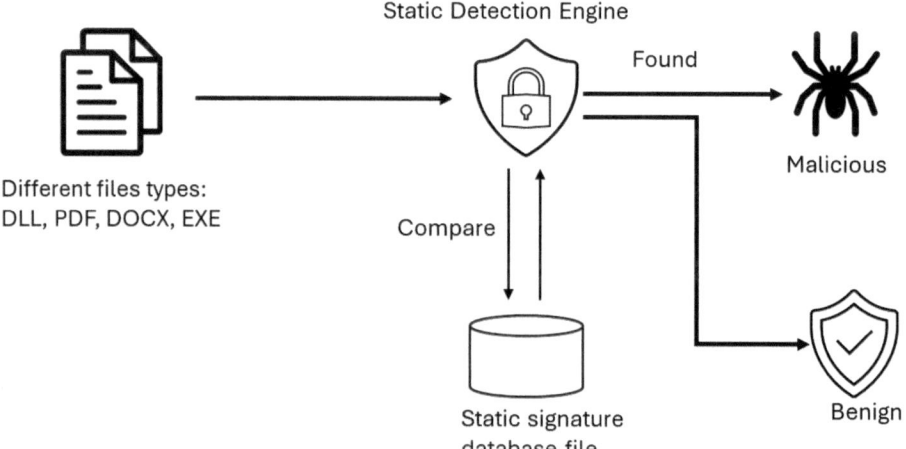

Fig. 7.2 The static detection process in antivirus software

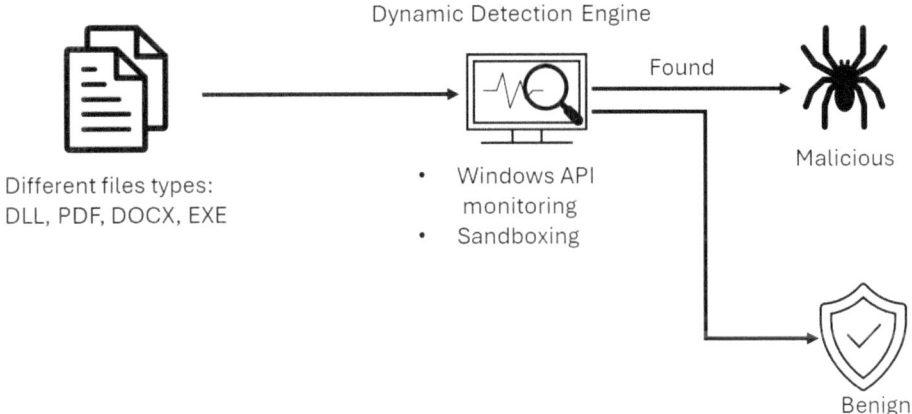

Fig. 7.3 Dynamic detection process in antivirus software

There are different types of dynamic detection methods such as windows API monitoring and Sandboxing. For the former, the detection engine inspects Windows application calls and monitors Windows API calls using Windows Hooks. Sandboxing involves running potentially malicious files in a controlled, isolated environment to observe their behavior without risking the main system. If the file exhibits malicious behavior, it is prevented from executing on the actual system.

- **Heuristic and Behavioral Detection**: The heuristic analysis examines the behavior of files and programs. By analyzing the code for suspicious characteristics or behaviors, heuristic techniques can detect new, previously unknown malware variants (Fig. 7.4).

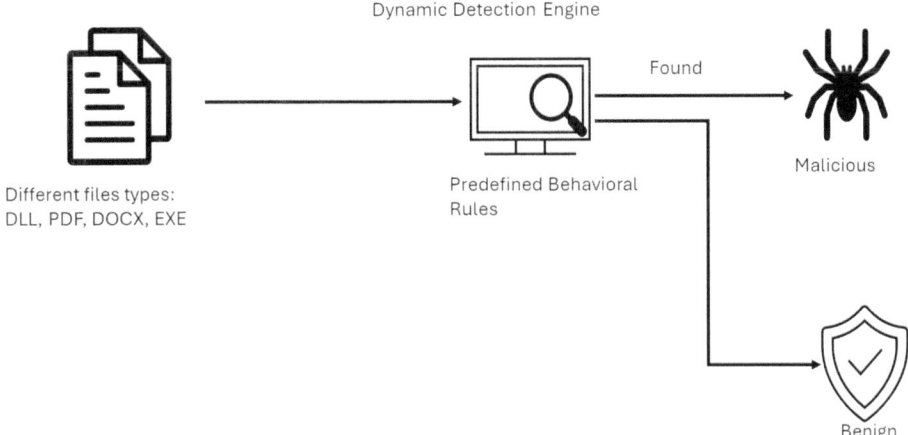

Fig. 7.4 Heuristic analysis process in antivirus software

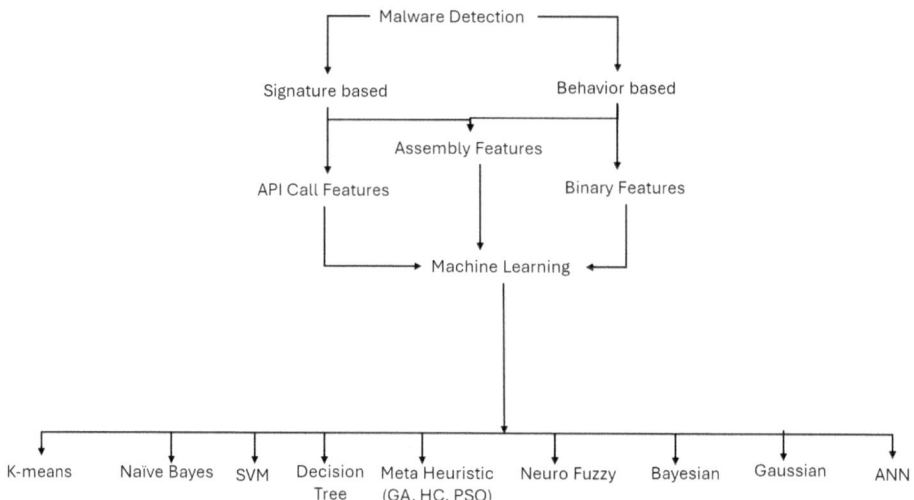

Fig. 7.5 Antimalware detection process including approaches in using machine learning[2]

Researchers operate in isolated environments to develop behavioral rules that identify malicious activities performed by software on a target machine. For example, one such rule could be if a process opens a listening port and waits to receive commands from a Command and Control server.

Antimalware software also uses similar methods as described above, in addition, antimalware software may use machine learning in the detection process. The diagram below depicts a taxonomy of the antimalware detection process. The detection process includes signature based, behavioral based and machine learning approaches. Signature-based detection remains a crucial component of antimalware strategies, particularly for identifying and mitigating known threats quickly and efficiently. While it has limitations, especially against emerging threats, it serves as a foundational layer of protection when combined with other detection methods such as heuristic and behavioral analysis. By integrating API call features in antimalware solutions enhance the traditional signature-based detection process and provide deeper insights into application behaviour (Fig. 7.5).

Behavior-based detection in antimalware solutions focuses on identifying malicious activities by observing the runtime behavior of programs rather than relying solely on known malware signatures. This approach involves monitoring programs for suspicious actions, such as unauthorized file modifications, unusual network activity, or attempts to exploit system vulnerabilities. By analyzing these behaviors, antimalware software can detect new, unknown, or polymorphic malware that may evade signature-based detection.

---

[2] Souri, and Hosseini [2].

The binary feature enhances the behavior-based process by allowing the antimalware solution to monitor and analyze the specific API calls made by a program during execution. This detailed insight into the program's interactions with the operating system and other software components helps identify malicious behaviors more accurately. For instance, if a binary file makes API calls associated with known malware actions, such as opening unauthorized network connections or modifying system settings, the antimalware solution can flag it as potentially harmful.

By combining behavior-based detection with binary feature analysis, antimalware solutions can provide a more comprehensive defense. This dual approach improves the identification of sophisticated threats, reduces false positives, and ensures real-time protection against emerging malware.

7.2.1.1   Machine Learning Approaches

Machine learning approaches have significantly enhanced the capabilities of antimalware software by enabling more dynamic and adaptive detection methods. Here's a discussion on how various machine learning algorithms are utilized in antimalware software:

K-means Clustering

- **Application**: K-means clustering is used in antimalware for grouping similar types of malware based on behavioral patterns or characteristics.
- **Functionality**: It identifies clusters of malware samples based on features extracted from their behavior or code, helping to classify and categorize new malware.
- **Benefit**: Enables quick classification of new samples into existing malware families, aiding in rapid response and mitigation.

Naive Bayes

- **Application**: Naive Bayes classifiers are used in antimalware for probabilistic classification of files as either benign or malicious based on extracted features.
- **Functionality**: It calculates the probability of a file being malicious given its features, assuming independence between features.
- **Benefit**: Provides fast and efficient classification, particularly useful in email filtering and initial triage of potentially malicious files.

Support Vector Machines (SVM)

- **Application**: SVMs are employed in antimalware for binary classification tasks, distinguishing between benign and malicious files.
- **Functionality**: SVMs find the optimal hyperplane that separates malicious from benign samples in a high-dimensional feature space.

- **Benefit**: Effective in handling complex, non-linear relationships between features, leading to accurate classification of malware.

Decision Trees

- **Application**: Decision trees are used in antimalware for hierarchical classification of files based on extracted features.
- **Functionality**: They partition the feature space into segments based on thresholds, making decisions at each node to classify files.
- **Benefit**: Provides interpretable models and insights into decision-making processes, useful for understanding malware behaviors.

Meta-heuristic Algorithms (GA, HC, PSO)

- **Application**: Genetic Algorithms (GA), Hill Climbing (HC), and Particle Swarm Optimization (PSO) are used in antimalware for feature selection, optimizing parameters, or improving classification accuracy.
- **Functionality**: They iteratively search for optimal solutions in large search spaces, adjusting parameters or selecting features to enhance detection capabilities.
- **Benefit**: Adaptive nature allows for effective optimization and customization of antimalware models based on evolving threats.

Neuro Fuzzy Systems

- **Application**: Neuro Fuzzy systems are used in antimalware for modeling complex relationships and patterns in malware behavior.
- **Functionality**: They combine fuzzy logic with neural networks to handle uncertainty and imprecision in data, improving detection accuracy.
- **Benefit**: Provides robust modeling capabilities for capturing subtle variations and behaviors in malware samples.

Bayesian Networks

- **Application**: Bayesian Networks are used in antimalware for probabilistic reasoning and inference based on observed behaviors or features.
- **Functionality**: They model relationships between variables, allowing for reasoning about the likelihood of a file being malicious given its characteristics.
- **Benefit**: Enables effective handling of uncertainty and varying degrees of evidence, aiding in decision-making processes.

Gaussian Processes

- **Application**: Gaussian Processes are used in antimalware for modeling complex, non-linear relationships in malware data.
- **Functionality**: They provide a probabilistic framework for regression and classification tasks, accommodating uncertainty in predictions.
- **Benefit**: Effective in scenarios where data distributions are not well-defined, allowing for flexible modeling of malware behaviors.

Artificial Neural Networks (ANN)

- **Application**: ANNs are used in antimalware for learning complex patterns and features from malware samples.
- **Functionality**: They consist of interconnected neurons organized in layers, capable of learning hierarchical representations of malware behaviors.
- **Benefit**: Provides high flexibility and adaptability in detecting new and evolving malware threats, based on learned patterns from historical data.

Antimalware software integrates these machine learning approaches to create multi-layered defense mechanisms. For example, combining SVMs for initial classification, decision trees for detailed analysis, and neural networks for learning complex behaviors. These methods of combining various approaches of machine learning (ML) enhance detection accuracy, reduce false positives, and improve response times to new threats by leveraging adaptive learning and reasoning capabilities.

7.2.2   Access Control

Access control is a cornerstone of network security, employing cybersecurity principles like authentication and authorization to protect sensitive data and resources. It ensures that only authorized users, based on predefined identity and access policies, can access specific information. This involves defining user roles, assigning appropriate permissions, implementing authentication mechanisms, and continuously monitoring access to detect unauthorized attempts or anomalies. Effective access control systems are essential for maintaining the confidentiality, integrity, and availability of organizational data, mitigating risks associated with unauthorized access, and ensuring compliance with security protocols. In the following sections, we will discuss different types of access control used in IT infrastructures.

7.2.2.1 Mandatory Access Control (MAC)

Mandatory Access Control (MAC) is a stringent access control model employed in environments where data confidentiality and integrity are paramount. Unlike discretionary access control (DAC), where resource owners determine access permissions, MAC imposes centralized control over access decisions based on security classifications and labels. In MAC systems, each user and resource is assigned a security label that specifies their sensitivity level or clearance. These labels determine the access rights users have to resources. The access control decisions are made by the system administrator or security policy administrator, who defines rules and policies governing access.

MAC systems enforce a hierarchical model where higher-level security policies dictate access rights. This approach ensures that users can only access information classified at or below their clearance level, preventing unauthorized data exposure or modification. Although such systems cannot be considered true network security per se, they utilize available security mechanisms inclusive of network access control to provide access to IT resources. From that perspective, such mechanisms are worth discussing in the context of network security.

The application of MAC is mostly in government sectors; however, it is also used in other mission-critical infrastructures. For example, in banks, MAC is used to manage highly sensitive customer and financial records, ensuring that only authorized personnel with the appropriate clearance levels can access or modify those records (Fig. 7.6).

Implementing of MAC involves "security labels" in which labels are assigned to subjects (users, processes) and objects (files, directories) based on sensitivity and security classifications. The decision mechanism for access control is done using rules and policies. A centralized authority (e.g., security policy administrator) is responsible for defining and managing security policies across the system.

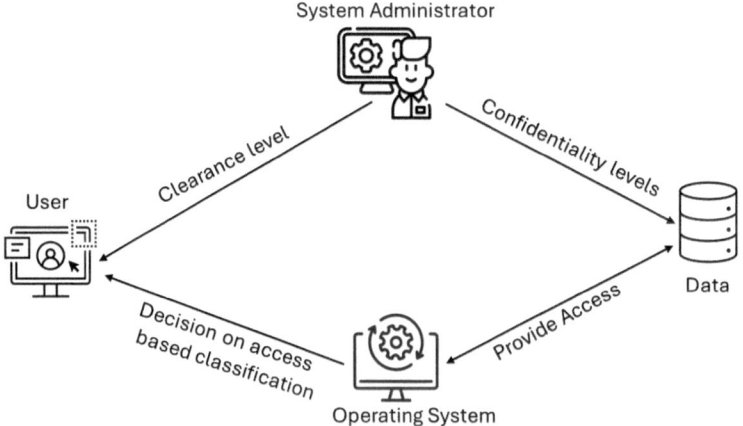

Fig. 7.6 Example of MAC deployment in government IT systems

The MAC provides a robust defense against unauthorized access attempts and ensures strict adherence to security policies. However, MAC systems can be complex to implement and manage due to the stringent control requirements and hierarchical nature of access decisions.

7.2.2.2 Discretionary Access Control (DAC)

Discretionary Access Control (DAC) is a flexible access control model widely used in various organizational settings to manage access to resources based on the discretion of resource owners or administrators. Unlike Mandatory Access Control (MAC), which imposes centralized control over access decisions, DAC allows resource owners to determine who can access their resources and what level of access they have.

In a DAC system, each resource (such as files, directories, or devices) has an associated Access Control List (ACL) that specifies which users or groups are granted permissions and what actions they can perform (e.g., read, write, execute). These permissions are typically based on the identity of the requesting user and their relationship to the resource owner (Fig. 7.7).

DAC systems are characterized by their flexibility and decentralized administration. Key aspects include:

- **Access Control Lists (ACLs)**: These lists define the permissions granted to users or groups for specific resources. ACLs can be modified by resource owners or administrators to reflect changing access requirements.
- **Owner-Based Control**: Resource owners have the autonomy to grant or revoke access permissions as needed. This allows for tailored access management based on organizational policies and operational needs.
- **Ease of Implementation**: DAC systems are relatively straightforward to implement and manage compared to more rigid access control models like MAC. They support a user-centric approach where access decisions can be based on relationships and responsibilities within the organization.

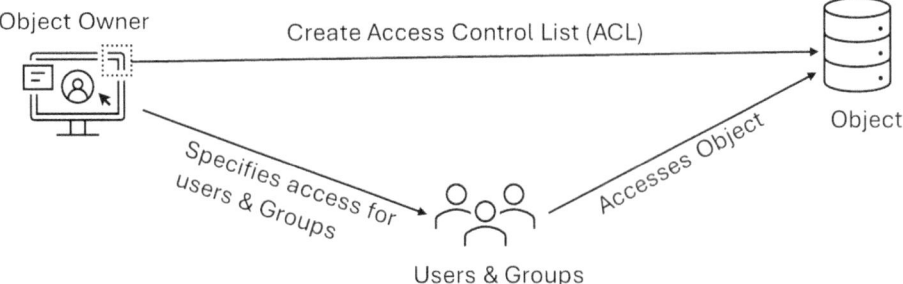

Fig. 7.7 Discretionary access control

While DAC offers flexibility and simplicity, it also presents security challenges. Since access decisions rely heavily on individual resource owners, there is a risk of inconsistent or overly permissive access permissions if not managed carefully. Additionally, DAC may not be suitable for environments where strict regulatory compliance or high levels of data confidentiality are required, as it lacks the centralized control and granularity provided by other access control models.

In conclusion, Discretionary Access Control remains a widely adopted approach in organizational settings due to its flexibility and ease of administration. It empowers resource owners to manage access permissions effectively while accommodating varying operational requirements and user needs within the organization.

7.2.2.3 Role-Based Access Control (RBAC)

Role-Based Access Control (RBAC) is a robust and scalable access control model that assigns permissions to users based on their roles within an organization. Unlike discretionary or mandatory access control models, RBAC simplifies the management of permissions by grouping users into roles and associating access rights with those roles, rather than with individual users.

In RBAC, roles represent job functions or responsibilities within an organization, each associated with a set of permissions defining the actions that can be performed and the resources that can be accessed. Users, in turn, are assigned to one or more roles based on their job functions. This setup allows for efficient management of access rights, as permissions are controlled at the role level rather than individually for each user. Implementing RBAC in network devices involves defining roles and their corresponding permissions, assigning users to these roles, and ensuring that network devices enforce access control policies based on the assigned roles. For instance, a Network Administrator might have full access to configure and manage network devices, while a Guest User would only have access to basic network services. Role hierarchies can be established, allowing roles to inherit permissions from other roles, further streamlining the management process. When users attempt to perform actions, the network devices check the roles and associated permissions to determine whether the actions are allowed. This enforcement mechanism ensures that users only access resources necessary for their roles, thereby enhancing security. The diagram below illustrates a typical RBAC implementation using HPE's Aruba ClearPass Policy Manager as an example of how RBAC policies are applied to users and devices to gain access to network resources (Fig. 7.8).

In this implementation, it is assumed that devices will identify and authenticate themselves in order to gain access to the network. Once device identification and authentication are completed, ClearPass will apply specific RBAC policies to the devices, thereby restricting or allowing their access to resources within the network. In addition to RBAC policy administration, HPE Aruba Clearpass also provides IAM (Identity and Access Management) functions. The ClearPass Policy Manager supports various AAA (Authentication, Access, and Accounting) technologies such as RADIUS, TACACS, 802.1X, MAC

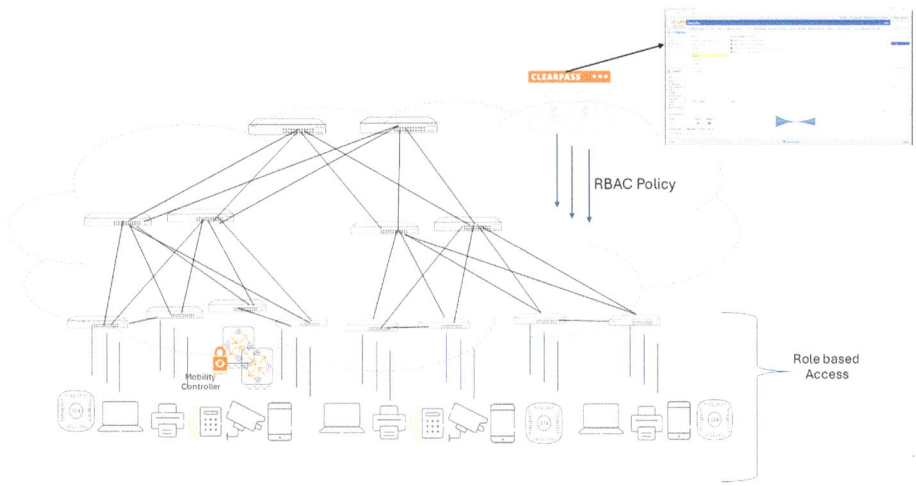

Fig. 7.8 Implementation of RBAC policy in a network using HPE aruba clearpass manager

address authentication, and Web authentication. RBAC-equipped network devices can log access attempts and user activities, providing essential data for auditing and monitoring. This feature helps administrators detect unauthorized access attempts and ensure compliance with security policies. By managing permissions at the role level, RBAC reduces the complexity of access control, making it easier to implement and update policies. It also scales efficiently as organizations grow, allowing for the addition of new roles and adjustments to permissions without modifying individual user settings.

Role-Based Access Control (RBAC) can be implemented in various ways, each with different features and levels of complexity. Here's a discussion on the different types of RBAC:

- **Core RBAC**: Core RBAC is the simplest form of RBAC and serves as the foundation for more complex models. It involves:
  - User Assignment: Users are assigned to roles.
  - Permission Assignment: Permissions are assigned to roles.
  - Role Authorization: Users gain permissions through their roles.

  In Core RBAC, the relationship between users and roles, and between roles and permissions, is many-to-many. This allows a user to be assigned multiple roles and a role to be assigned to multiple users.

- **Hierarchical RBAC**: Hierarchical RBAC introduces role hierarchies to Core RBAC. In this model, roles can inherit permissions from other roles. There are two types of hierarchies:

- General Hierarchies: Roles can inherit permissions from multiple parent roles and can also have multiple child roles.
- Limited Hierarchies: Roles inherit permissions from a single parent role only, forming a simpler tree structure.

This hierarchical structure simplifies permission management by allowing higher-level roles to inherit permissions from lower-level roles, reflecting the organizational structure more accurately.

- **Constrained RBAC**: Constrained RBAC introduces additional constraints to Core and Hierarchical RBAC models to enhance security and ensure segregation of duties. There are two main types of constraints:
  - Mutually Exclusive Roles: Ensures that users cannot be assigned to certain combinations of roles simultaneously. This prevents conflicts of interest and reduces the risk of fraud.
  - Cardinality Constraints: Limits the number of users that can be assigned to a specific role, ensuring that critical roles are not overloaded with too many users.

Constrained RBAC enhances security by enforcing rules that prevent inappropriate access combinations and ensuring that critical roles are managed properly.

- **Attribute-Based RBAC (ABAC-RBAC):** Attribute-Based RBAC combines the principles of RBAC with Attribute-Based Access Control (ABAC). In this model, access decisions are based on the roles assigned to users and additional attributes such as user location, time of access, or device type. This hybrid approach allows for more dynamic and context-aware access control.

ABAC-RBAC enhances the flexibility of access control policies by incorporating contextual information, making it suitable for environments where access needs can change based on varying conditions.

Implementation of RBAC in Network Devices
Implementing these RBAC models in network devices involves configuring roles, permissions, and constraints to reflect organizational policies and security requirements. For instance:

- **Core RBAC Implementation**: Network administrators define basic roles such as Administrator, User, and Guest, and assign permissions accordingly.
- **Hierarchical RBAC Implementation**: More complex organizations may implement role hierarchies where Senior Administrators inherit permissions from Junior Administrators, reflecting their broader responsibilities.

- **Constrained RBAC Implementation**: Critical roles like Security Officer might be constrained to prevent users from holding other potentially conflicting roles, ensuring segregation of duties.
- **ABAC-RBAC Implementation**: Organizations with dynamic access needs can implement ABAC-RBAC to ensure access decisions consider additional attributes like time of day or location, providing a more flexible and secure access control mechanism.

In summary, the different types of RBAC—Core, Hierarchical, Constrained, and Attribute-Based RBAC—offer varying levels of complexity and security features. By choosing the appropriate RBAC model and implementing it effectively in network devices, organizations can streamline access management, enhance security, and ensure compliance with organizational policies and regulatory requirements.

7.2.2.4 Network Access Control (NAC)

Network Access Control (NAC) is a comprehensive security solution that manages and enforces policies on devices and users trying to access a network. It plays a crucial role in maintaining network security by ensuring that only compliant and authorized devices can connect and interact with network resources. NAC encompasses various technologies and protocols to achieve these objectives. The following diagram illustrates a framework for NAC implementation. This framework will provide readership an understanding of how NAC objectives are achieved for IT infrastructure (Fig. 7.9).

This framework will help readers understand how NAC objectives are achieved in IT infrastructure. The tools listed in the figure above represent a sample and not an exhaustive list. In this section, we will explore selected protocols and tools for NAC

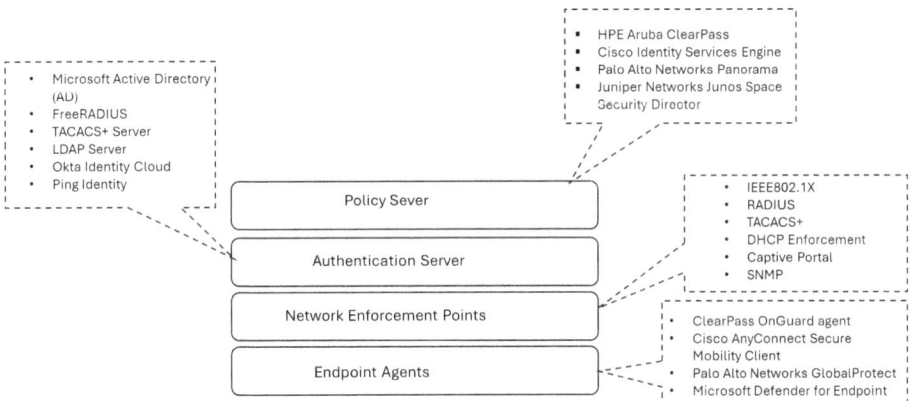

Fig. 7.9 Network Access Control (NAC) framework

discussion. Through this discussion, we aim for readers to gain a comprehensive understanding of how NAC operates, enabling them to develop implementation strategies for their respective deployments.

Endpoint Agents

The Endpoint Agents are software agents that are installed on endpoint devices such as laptops, desktops, and mobile devices. Endpoint agents collect information about the device's security posture, such as antivirus status, operating system patches, and configuration settings. This information is then sent to the policy server, which uses it to determine whether the device meets the organization's security requirements. Examples of Endpoint Agents are as follows:

- **HPE Aruba ClearPass OnGuard**: OnGuard offers both agent-based and agentless deployment options. The agent-based approach involves installing a lightweight software agent on endpoints, which provides continuous health monitoring and compliance checks. The agentless approach uses standard network protocols and does not require software installation, making it suitable for environments where deploying agents is challenging. ClearPass OnGuard performs detailed health checks on endpoints (laptops, desktops, and mobile devices) before allowing them to connect to the network. It assesses various security parameters such as antivirus status, operating system patches, firewall status, and device configuration settings. This ensures that only compliant devices are granted network access.
- **Cisco AnyConnect Secure Mobility Client**: This is a widely used endpoint agent that provides secure VPN access, posture assessment, and network visibility. It integrates with Cisco's Identity Services Engine (ISE) to enforce security policies and ensure that devices meet the required compliance standards before being granted network access.
- **Microsoft Defender for Endpoint**: This agent offers comprehensive endpoint protection, including antivirus, anti-malware, and threat detection and response capabilities. It integrates with Microsoft's security ecosystem to enforce security policies and ensure that endpoints comply with network access requirements.
- **Symantec Endpoint Protection**: This agent offers comprehensive security features including antivirus, anti-malware, firewall, and intrusion prevention. It also includes device and application control features to enforce security policies and ensure endpoint compliance.
- **McAfee Endpoint Security:** McAfee provides an endpoint security agent that offers protection against malware, ransomware, and other threats. It also includes firewall and web control features to enforce security policies and ensure that endpoints meet compliance requirements.

- **Palo Alto Networks GlobalProtect**: This agent provides secure VPN access and enforces security policies on endpoints. It integrates with Palo Alto Networks' security platform to ensure that endpoints meet compliance standards before being granted network access.
- **Fortinet FortiClient**: This agent offers endpoint protection, secure VPN access, and compliance enforcement. It integrates with Fortinet's Security Fabric to provide comprehensive security and ensure that endpoints meet the required security policies.

These endpoint agents play a crucial role in ensuring that devices meet security requirements before being granted access to network resources, thereby enhancing overall network security and compliance.

Network Enforcement Points

Network enforcement points are essential components in a network access control (NAC) system, responsible for implementing and enforcing the security policies defined by the policy server. These enforcement points include switches, routers, firewalls, and other network devices that inspect traffic and make real-time decisions to allow, block, or quarantine devices based on their compliance with security policies. Here is a detailed explanation of various protocols and mechanisms used in network enforcement:

- **IEEE 802.1X**: It is a standard for port-based network access control (PNAC), providing a robust authentication mechanism for devices attempting to connect to a Local Area Network (LAN) or Wireless Local Area Network (WLAN). The following diagram depicts the components of IEEE802.1X framework and identity servers (Fig. 7.10).

In this framework, three key components interact to ensure secure network access: the supplicant, the authenticator, and the authentication server. The supplicant is the client device, such as a laptop or smartphone, seeking network access. It runs software capable of communicating with the network authenticator. The authenticator, typically a network switch or wireless access point, acts as an intermediary between the supplicant and the authentication server. Until the device is authenticated, the authenticator blocks all traffic from the supplicant except for authentication traffic. The supplicant devices use EAP (Extensible Authentication Protocol) for authentication, with the type of EAP method being negotiated between the supplicant and the authentication server. Options include LEAP, EAP-TLS, EAP-MD5, EAP-FAST, EAP-GTC, and PEAP. Non-supplicant devices, such as printers, can utilize MAC Authentication Bypass (MAB), an extension of IEEE 802.1X, to gain network access. In this mechanism, the IEEE 802.1X framework uses a device's MAC address to grant network access.

The authentication server, often a Remote Authentication Dial-In User Service (RADIUS) server, validates the credentials provided by the supplicant. The diagram above depicts HPE Aruba ClearPass providing authentication services as the authentication

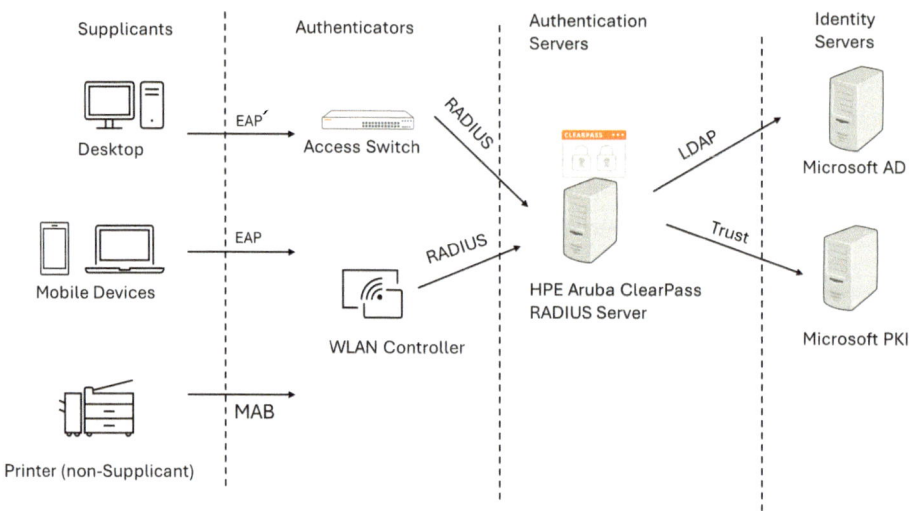

Fig. 7.10   A typical network setup using IEEE 802.1X components

server. When a device attempts to connect, the authenticator initiates the 802.1X authentication process. The supplicant sends its credentials, such as a username and password, to the authenticator, which forwards them to the authentication server, in this case, the Aruba ClearPass RADIUS Authentication Server. The ClearPass authentication server then verifies these credentials. If the credentials are valid, the server instructs the authenticator to grant network access to the device. HPE Aruba ClearPass can further extend this process by connecting to identity servers like Microsoft Active Directory (AD) and Microsoft Public Key Infrastructure (PKI) to validate device trust and apply policies accordingly.

- **RADIUS (Remote Authentication Dial-In User Service)**: RADIUS is a networking protocol that provides centralized authentication, authorization, and accounting (AAA) management for users connecting to and using a network service. When a user requests access to a network resource, the network device, acting as an authenticator, sends the request to the RADIUS server. The RADIUS server checks the user's credentials against a database and responds with either an accept or reject message. If the credentials are accepted, the server can also return configuration information specifying access privileges. This process ensures that the RADIUS server handles authentication by verifying user credentials, authorization by determining accessible resources, and accounting by keeping track of network activity for billing, auditing, and reporting purposes.
- **TACACS+ (Terminal Access Controller Access Control System Plus)**: TACACS+ is a protocol developed by Cisco to provide detailed control over access to network devices. Similar to RADIUS, TACACS+ handles authentication by verifying user credentials

but separates this process from authorization, allowing more granular control over the commands a user can execute on a network device. Accounting in TACACS+ involves logging the commands executed by the user, which is useful for auditing. When a user connects to a network device, the device sends the user's credentials to the TACACS+ server, which authenticates the user and returns information about the user's privileges while logging each command executed.

- **DHCP Enforcement**: Dynamic Host Configuration Protocol (DHCP) Enforcement is used in Network Access Control (NAC) systems to assign IP addresses and enforce access control policies based on a device's compliance status. When a device connects to the network and requests an IP address via DHCP, the DHCP server checks the device's compliance status, often integrating with a NAC solution. If the device is compliant, it receives an IP address granting access to the network. If the device is non-compliant, it may be assigned an IP address that only allows access to a remediation network where the device can be brought into compliance.
- **Captive Portal**: A Captive Portal is a web page that users must view and interact with before accessing the network, commonly used in guest networks or public Wi-Fi. When a device connects to the network, all HTTP/HTTPS requests are redirected to a captive portal where the user is required to log in or agree to terms and conditions. Once authenticated or after terms are accepted, the user is granted access to the network. This mechanism is often used to provide Internet access to guests in hotels or cafes and to ensure users accept terms of service or network usage policies.
- **SNMP (Simple Network Management Protocol)**: SNMP is used for network management, including monitoring and configuring network devices. In the context of NAC, SNMP can enforce policies by configuring devices to control access. SNMP agents installed on network devices allow the NAC system to query and configure these devices. Based on the compliance status of devices, the NAC system can use SNMP to change the configuration of network devices, such as restricting access ports or applying access control lists (ACLs).

Authentication Server

An authentication server plays a crucial role in Network Access Control (NAC) by managing user authentication, authorization, and sometimes accounting (AAA) functions. It verifies the identities of users and devices attempting to connect to a network, ensuring only authorized access based on predefined security policies. The following are authentication servers commonly utilized within NAC frameworks.

- Microsoft Active Directory (AD) is a widely used authentication and directory service in enterprise environments. It centralizes network management and allows administrators to authenticate and authorize users and computers in a Windows domain. Active Directory integrates with NAC systems to validate user credentials and enforce access policies based on user roles, group memberships, and organizational units.

- FreeRADIUS is an open-source RADIUS server that provides AAA services, including authentication, authorization, and accounting. It supports various authentication protocols and can integrate with NAC solutions to authenticate users and devices accessing the network. FreeRADIUS offers flexibility and customization options, making it popular in environments requiring scalable and customizable AAA solutions.
- TACACS+ Server (Terminal Access Controller Access Control System Plus) is another authentication protocol developed by Cisco. Unlike RADIUS, TACACS+ separates authentication, authorization, and accounting functions, offering finer control over access permissions and command-level authorization on network devices. TACACS+ servers are commonly used in environments where detailed auditing and access control are critical.
- LDAP (Lightweight Directory Access Protocol) Server works as a directory service for managing user identities and access permissions. LDAP servers store and retrieve directory data, including user credentials and attributes. NAC solutions can leverage LDAP directories for user authentication and authorization, providing a centralized repository for user information across the network.
- Okta Identity Cloud and Ping Identity are cloud-based identity and access management (IAM) platforms that offer comprehensive authentication, single sign-on (SSO), and identity governance capabilities. These platforms integrate with NAC systems to extend identity-based access control across cloud and on-premises applications. They support various authentication protocols and provide robust identity verification mechanisms to enforce security policies based on user context and device posture.

In summary, authentication servers such as Microsoft Active Directory, FreeRADIUS, TACACS+ servers, LDAP servers, Okta Identity Cloud, and Ping Identity are integral to NAC implementations. They facilitate secure access to network resources by verifying user identities, enforcing access policies, and ensuring compliance with security standards across diverse IT environments.

Policy Servers

Policy servers are integral components of Network Access Control (NAC) architectures, playing a pivotal role in defining, enforcing, and managing access policies across enterprise networks. These servers act as centralized hubs for policy management, ensuring that only authorized users and devices gain access to network resources while maintaining security and compliance.

HPE Aruba ClearPass, Cisco Identity Services Engine (ISE), Palo Alto Networks Panorama, and Juniper Networks Junos Space Security Director are among the leading policy servers used in NAC deployments. These platforms provide sophisticated capabilities for policy creation, enforcement, and monitoring, tailored to meet the complex security requirements of modern networks.

HPE Aruba ClearPass offers comprehensive policy management and enforcement capabilities, integrating seamlessly with existing network infrastructure to enforce policies based on user roles, device types, and contextual information. It facilitates dynamic policy decisions that adapt to changing network conditions and user behaviors, ensuring secure access across wired, wireless, and remote environments.

Cisco Identity Services Engine (ISE) centralizes policy management and enforcement across heterogeneous network environments, enabling organizations to define policies based on user identity, device posture, and application usage. It integrates tightly with Cisco network infrastructure to enforce policies dynamically and provide real-time visibility into network access and security events.

Palo Alto Networks Panorama serves as a centralized management platform for security policies across Palo Alto Networks' firewall and NAC solutions. It enables administrators to define and enforce consistent policies based on user identity, application behavior, and threat intelligence. Panorama supports policy orchestration across distributed network environments, enhancing security posture and compliance.

Juniper Networks Junos Space Security Director provides policy management and enforcement capabilities for Juniper Networks' security products. It allows organizations to define policies based on user roles, application usage, and threat intelligence, and integrates seamlessly with Juniper Networks' ecosystem to streamline policy deployment and management.

These policy servers empower organizations to implement granular, adaptive access control policies that align with business needs and security best practices. By centralizing policy management and enforcement, they enhance network visibility, mitigate security risks, and optimize compliance efforts, ensuring secure and efficient access to critical resources across the enterprise network.

7.2.3   Virtual Private Network (VPN)

A Virtual Private Network (VPN) establishes an encrypted connection, typically over the Internet, securely linking a device to a network. Additionally, VPN technologies are employed in site-to-site interconnects and Data Center Interconnects (DCI). This encryption ensures the safe transmission of sensitive data and prevents unauthorized interception of traffic. VPN technology is extensively used in corporate settings. VPNs use encryption protocols such as IPSec, SSL/TLS, and WireGuard to ensure that transmitted data remains confidential and protected from unauthorized access or interception.

There are several types of VPN connectivity: Remote Access VPN, Site-to-Site VPN, Data Center Interconnect (DCI), and Multi-cloud VPNs. Remote Access VPNs allow individual users to securely connect to a corporate network from remote locations, making them ideal for telecommuting or remote work. Site-to-Site VPNs connect entire networks, such as branch offices to a central corporate network, enabling seamless communication between geographically dispersed locations. The DCI VPN primarily uses IPSec tunneling to transport encrypted data between two or more data centers. Similarly, IPSec tunneling can connect multiple cloud providers' VPN gateways.

The main components of a VPN include the VPN client, VPN server or gateway, and tunneling protocols. The VPN client is software installed on the user's device that initiates and maintains the VPN connection. The VPN server or gateway hosts the VPN endpoint on the corporate network, managing authentication, encryption, and user access. Tunneling protocols, such as PPTP, L2TP/IPSec, and OpenVPN, define how data is encapsulated and encrypted for transmission over the VPN.

The diagram below illustrates a typical VPN setup in an enterprise network environment. Site-to-site connectivity, whether between branch offices and the main office or the main office and colocation facilities, uses secure IPSec VPN tunnels. Remote workers can use either IPSec or SSL VPN tunnels to connect to the main office, depending on the setup. IPSec tunnels are typically terminated at VPN Gateway, whereas SSL VPN tunnels are terminated at the host and are generally used for secure connectivity to specific applications. For SSL VPN tunnels, there is no need for a network device or software agent for tunnel termination; instead, a standard web browser can be used to establish endpoint connectivity to SSL VPN gateways or servers (Fig. 7.11).

VPNs offer several benefits, including enhanced security through data encryption, increased privacy by masking the user's IP address, and improved access control with

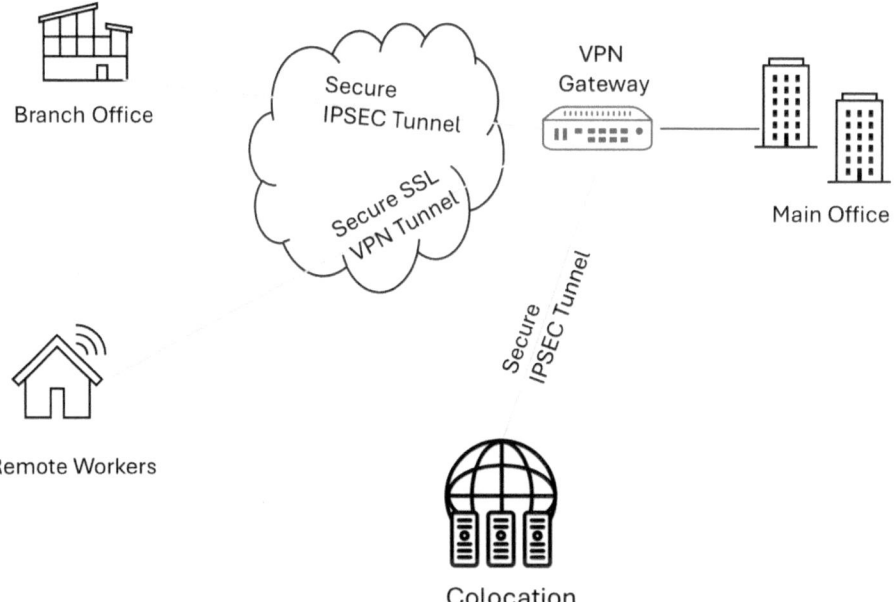

Fig. 7.11 Typical VPN setup in enterprise networks

granular management of resource access within the private network. They also provide flexibility, supporting remote work and allowing employees to securely access corporate resources from any location.

However, VPNs come with challenges and considerations. Performance can be affected by latency introduced by encryption overhead, which may impact real-time applications. Setting up and managing VPN infrastructure can be complex, requiring expertise in network security and configuration. Additionally, scaling VPNs to accommodate a growing number of users or sites may require additional resources and careful planning. VPNs can be deployed in various models, such as client-based VPNs and network-based VPNs. Client-based VPNs are installed on individual devices and are commonly used for remote access. Network-based VPNs, established between network devices like routers and firewalls, are suitable for site-to-site connections. These deployment models cater to different use cases and requirements, making VPNs versatile tools for securing network communications.

7.2.3.1 VPN Technologies

VPN (Virtual Private Network) technologies use various protocols and methods to establish secure and encrypted connections over public networks like the Internet. One of the most widely used protocols is IPSec (Internet Protocol Security), which secures IP communications by authenticating and encrypting each IP packet of a communication session. IPSec consists of two main protocols: Authentication Header (AH), which provides data integrity, authentication, and anti-replay protection, and Encapsulating Security Payload (ESP), which offers confidentiality, integrity, and optional authentication for the packet payload. IPSec is commonly used in both site-to-site VPNs and remote access scenarios with client software support.

Another key technology is SSL/TLS (Secure Sockets Layer/Transport Layer Security), originally designed for secure web browsing but also used to create VPN tunnels known as SSL VPNs. These use SSL/TLS protocols to encrypt data transmitted between the client and the server, typically via a web browser or a dedicated client application, making SSL VPNs versatile for providing remote access to applications and resources without requiring client software on the user's device.

L2TP/IPSec combines the best features of L2TP (Layer 2 Tunneling Protocol) and IPSec for creating secure tunnels. L2TP provides the tunneling mechanism, while IPSec handles encryption and authentication, making it useful for site-to-site VPNs and remote access scenarios where platform compatibility is important. PPTP (Point-to-Point Tunneling Protocol), one of the earliest VPN protocols, was developed for creating VPNs over dial-up networks and encapsulates PPP frames in IP packets for Internet transmission. However, it is now considered less secure compared to newer protocols like IPSec and SSL/TLS.

OpenVPN is an open-source VPN solution that uses SSL/TLS for key exchange and authentication. It supports flexible configurations, including client-to-server, site-to-site, and hybrid setups, and is known for its security, scalability, and cross-platform compatibility, making it favored for various VPN deployments.

WireGuard is a relatively new VPN protocol designed to be simpler, faster, and more secure than traditional protocols. It aims for better performance using modern cryptography techniques while maintaining a minimal codebase and is growing in popularity for both client-to-server and site-to-site VPN deployments due to its efficiency and simplicity.

IKEv2 (Internet Key Exchange version 2) handles the negotiation of security associations (SA) and key exchange in IPSec VPNs, simplifying the process of establishing IPSec tunnels. It supports mobility features, making it suitable for mobile and remote access scenarios. IKEv2 is often used in mobile device VPN clients because of its ability to maintain VPN connections when switching between networks, such as from Wi-Fi to cellular.

Each VPN technology has its strengths and ideal use cases, depending on factors such as security requirements, compatibility, ease of deployment, and performance considerations. Choosing the right VPN technology involves assessing these factors against specific use case requirements.

7.2.4   Intrusion Detection System (IDS)

An Intrusion Detection System (IDS) is a security technology designed to detect unauthorized access or deviations from standard behavior within a network or system. IDS works by monitoring network traffic, system activities, and analyzing data for signs of potential security breaches. It does not take action to block or prevent these activities but instead alerts administrators to take appropriate measures. The following diagram depicts IDS methods of operation (Fig. 7.12).

IDS operates through two primary detection methods: signature-based detection and anomaly-based detection. Signature-based detection relies on a database of known threat signatures and patterns. When network traffic or system behavior matches one of these signatures, the IDS triggers an alert. This method is effective against known threats but can struggle with new or unknown threats. Anomaly-based detection, on the other hand, establishes a baseline of normal behavior for the network or system. It then monitors for deviations from this baseline, which could indicate a potential security issue. Anomaly-based detection is more adept at identifying new and unknown threats but can generate false positives if normal behavior changes unexpectedly. Additionally, IDS may use hybrid methods of detection by combining both signature-based and anomaly-based detection techniques to detect both known and unknown attacks.

IDS can be deployed as either a Network Intrusion Detection System (NIDS) or a Host Intrusion Detection System (HIDS). NIDS monitors network traffic at strategic points

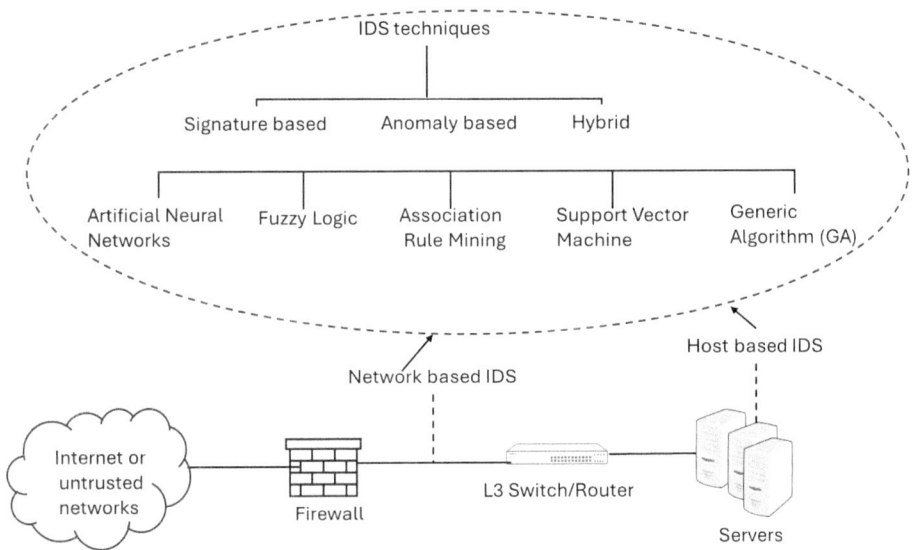

Fig. 7.12 IDS methods of operation

within the network, analyzing the data flowing between devices. It is placed in critical network paths to detect malicious activities such as port scanning, DoS attacks, and unusual traffic patterns. HIDS operates on individual hosts or devices, monitoring system calls, application logs, file system modifications, and other host-specific activities. It is particularly effective at identifying insider threats and unauthorized changes to system files.

Several tools are available for implementing IDS. Snort is one of the most widely used open-source NIDS, known for its robust signature-based detection capabilities and extensive rule set. Suricata, another popular NIDS, offers high performance and scalability with advanced features like multi-threading. OSSEC is a well-known HIDS that provides comprehensive monitoring and alerting capabilities for host-based activities.

In practical usage, IDS is employed in various scenarios to enhance network security. For example, it can be used to monitor critical infrastructure networks for signs of cyber-attacks, provide early warnings of potential security incidents in enterprise environments, and ensure compliance with regulatory requirements by logging and analyzing security events. IDS is often integrated into a broader security strategy, complementing other tools and technologies to provide a layered defense against cyber threats.

7.2.5   Intrusion Prevention System (IPS)

An Intrusion Prevention System (IPS) is an advanced network security technology that not only detects potential threats but also takes proactive measures to prevent them. Unlike IDS, which is primarily passive, IPS is an active defense mechanism that can block or mitigate malicious activities in real-time. It is designed to intercept and respond to threats before they can cause harm to the network or systems. IPSs can be implemented as either hardware or software solutions and are frequently integrated into next-generation firewall (NGFW) or unified threat management (UTM) systems.

IPS operates similarly to IDS in that it employs both signature-based and anomaly-based detection methods. Signature-based IPS uses a database of known threat signatures to identify malicious activities, while anomaly-based IPS establishes a baseline of normal behavior and monitors for deviations. When a potential threat is detected, IPS can take various actions such as dropping malicious packets, blocking traffic from suspicious IP addresses, resetting connections, and alerting administrators.

IPS can be deployed inline within the network, meaning it is positioned directly in the path of network traffic. This allows it to actively monitor and control traffic flow, making real-time decisions to block or allow packets based on the detection of malicious activities. The inline deployment ensures that IPS can immediately respond to threats, providing a robust layer of defense against network-based attacks such as DoS attacks, worms, and other forms of malware.

Several tools and technologies are used to implement IPS. Cisco Firepower is a comprehensive network security solution that includes IPS functionality, offering deep packet inspection, advanced threat detection, and automated response capabilities. Palo Alto Networks provides Next-Generation Firewalls (NGFW) with integrated IPS features, leveraging machine learning and advanced analytics to detect and prevent threats. Suricata, while primarily an IDS, also offers IPS functionality when configured to operate inline, providing a versatile open-source option for intrusion prevention.

In terms of practical usage, IPS is crucial for protecting sensitive data and critical infrastructure. It is widely used in enterprise networks to safeguard against sophisticated cyber threats, prevent data breaches, and ensure business continuity. IPS can also be deployed in data centers to protect virtualized environments and cloud-based services, providing a critical layer of security in modern IT infrastructures. Additionally, IPS is essential for regulatory compliance, helping organizations meet security standards and protect against data breaches.

In summary, IDS and IPS are essential components of a comprehensive security strategy. While IDS provides valuable insights and alerts on potential threats, IPS goes a step further by actively preventing malicious activities. Both technologies are critical for detecting and responding to cyber threats, ensuring the security and integrity of networked systems.

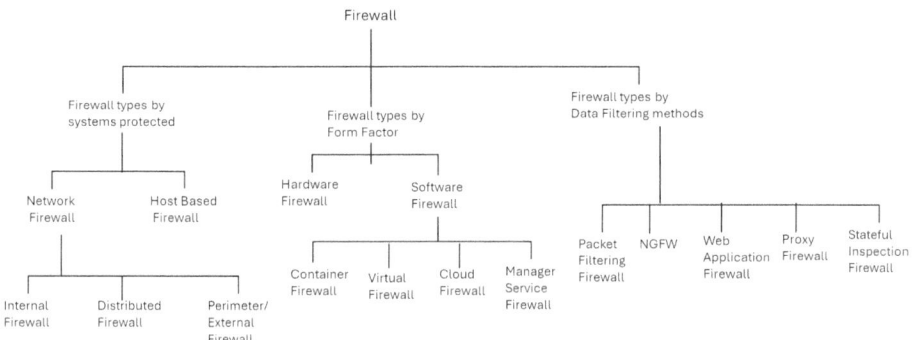

Fig. 7.13 Different types of firewalls

7.2.6 Firewall

Firewalls are essential components of network security, serving as a barrier between trusted internal networks and untrusted external networks. They monitor and control incoming and outgoing network traffic based on pre-determined security rules. The primary goal of a firewall is to block unauthorized access while permitting legitimate communication. Technological advancements have resulted in the development of various types of firewalls, each with its own unique functions and characteristics. The diversity of terminology and available options can often be overwhelming. To differentiate between the various types of firewalls, one can look at the specific functions they perform. Firewalls are commonly categorized based on the systems they protect, their form factor, their placement within the network infrastructure, and their methods of data filtering.[3] The following diagram depicts different types of firewalls available. To achieve effective network security, organizations may need to utilize multiple types of firewalls. Additionally, a single firewall product can often encompass multiple types of firewall functionality (Fig. 7.13 and 7.14).

In the following sections, we will explore different types of firewalls based on their respective categories.

7.2.6.1 Firewall by Systems Protected

Firewalls can be classified based on the systems they are designed to protect. The two primary categories are network-based firewalls and host-based firewalls. Each type serves a distinct purpose within a security infrastructure, providing protection at different levels of the network architecture.

---

[3] Paloalto [3].

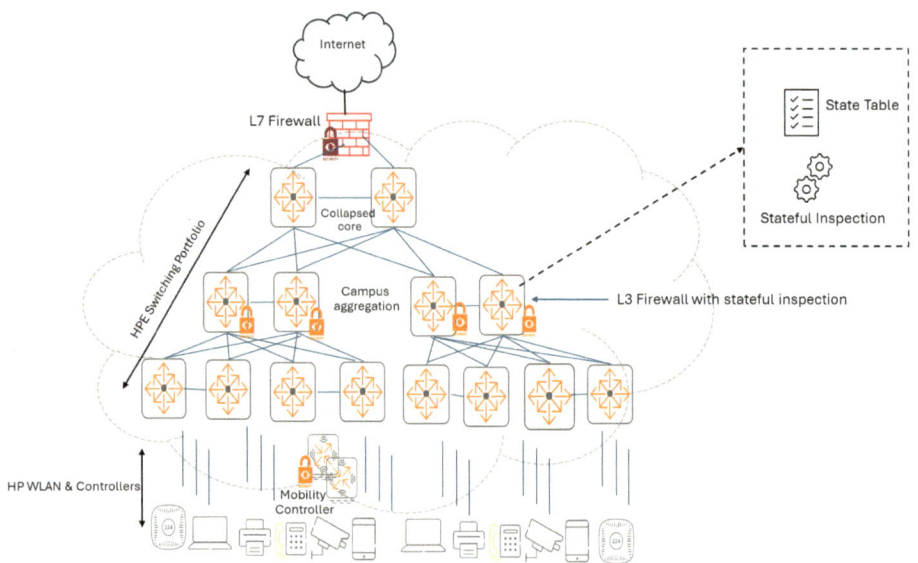

Fig. 7.14 Example of enterprise campus deployment with stateful inspection firewall

Network-Based Firewalls

Network-based firewalls are typically deployed at strategic points within a network, such as at the boundary between an internal network and the Internet. They are designed to protect entire networks by monitoring and controlling traffic between different segments based on predefined rules. These rules prevent unauthorized access and ensure network integrity. Network firewalls operate by scrutinizing each data packet that traverses the network. They compare packet attributes, such as source and destination IP addresses, protocols, and port numbers, against established rules to block potential threats or unwanted data flows. Network firewalls can be implemented as hardware appliances, software solutions, or a combination of both, ensuring comprehensive traffic screening at critical network junctions.

Beyond regulating traffic, network firewalls offer extensive logging capabilities. These logs are invaluable for administrators, providing insights that help track and investigate suspicious activities.

Cisco ASA (Adaptive Security Appliance) is a popular network-based firewall that provides robust security features, including stateful packet inspection, VPN support, and intrusion prevention. It is deployed at the network perimeter to protect internal network resources from external threats. Other firewalls in this category include products offered by Palo Alto Networks, Fortinet, and Barracuda Networks.

Network-based firewalls can be further categorized into three subcategories: internal firewalls, distributed firewalls, and perimeter or external firewalls. Internal firewalls are deployed within internal networks to segregate and secure different segments or departments. They help control and monitor traffic between various parts of the internal network, enhancing security and preventing lateral movement of threats.

Distributed firewalls are deployed across multiple points within a network, often within virtualized or cloud environments. They provide localized security and traffic control at each point of deployment, offering flexibility and scalability in protecting distributed infrastructures.

Perimeter firewalls, also known as external firewalls, are positioned at the network boundary between internal networks and external networks, such as the Internet. They serve as the first line of defense, filtering and monitoring incoming and outgoing traffic to protect internal resources from external threats.

These categories reflect different deployment scenarios and strategic placements within a network architecture, each serving specific roles in ensuring comprehensive network security.

Host-Based Firewalls

Host-based firewalls, also known as personal firewalls, are installed directly on individual devices or hosts. They provide security for the specific device on which they are installed, controlling incoming and outgoing traffic based on security policies. These firewalls are crucial for protecting devices that may not always be behind a network-based firewall, such as laptops used by remote employees. An example of a host-based firewall is Windows Defender Firewall, which is included with the Windows operating system. Windows Defender Firewall offers protection for individual devices by monitoring and controlling network traffic according to predefined rules. Its functionality includes application control, allowing or blocking applications from accessing the network based on security policies. It also monitors incoming and outgoing traffic to detect and block suspicious activity, provides users with the ability to configure rules and exceptions for specific applications and ports, and integrates with antivirus software to offer comprehensive protection against malware and network threats.

7.2.6.2 Firewall by Form Factor

Firewalls can be categorized based on their form factors, which describe their physical or virtual structure and deployment characteristics. These form factors play a crucial role in determining where and how firewalls are deployed within a network architecture. There are mainly two types of firewalls by form factor: hardware and software firewalls.

Hardware firewalls are physical devices dedicated to firewall functionality. They often come as standalone appliances that are purpose-built for high-performance network security. Hardware firewalls typically offer robust processing power and specialized hardware components optimized for deep packet inspection and security operations. They are commonly deployed at network entry points, such as the perimeter between internal networks and external networks like the Internet. Examples include Cisco ASA (Adaptive Security Appliance) and Palo Alto Networks' hardware firewall appliances.

Software firewalls, also known as host-based firewalls, are implemented as software applications that run on general-purpose hardware, such as servers or endpoint devices. They provide firewall capabilities directly on individual devices or hosts, allowing for granular control over traffic entering and leaving the device. Software firewalls are versatile and can be installed on various operating systems, including Windows, macOS, and Linux. They are particularly useful for protecting devices that frequently connect to different networks or operate remotely. Examples include Windows Defender Firewall and iptables for Linux. Software firewalls encompass various types, including container firewalls, virtual firewalls (often referred to as cloud firewalls), and managed service firewalls.

- **Container Firewall**: Container firewalls are designed to secure containerized applications and microservices within container orchestration platforms like Kubernetes. They provide security controls at the container level, ensuring that communication between containers and with external networks adheres to defined security policies. Container firewalls offer visibility and control over network traffic within dynamic and distributed containerized environments, enhancing security without compromising agility.
- **Virtual Firewall**: Virtual firewalls, also known as cloud firewalls, are software-based firewall solutions deployed within virtualized or cloud environments. They operate as virtual appliances and provide network security functionalities similar to traditional hardware firewalls. Virtual firewalls protect virtual machines (VMs), cloud instances, and network segments within cloud platforms by filtering and monitoring incoming and outgoing traffic based on predefined rules. They offer scalability, flexibility, and integration with cloud-native security services, making them suitable for dynamic and elastic cloud environments.
- **Managed Services Firewall**: Managed service firewalls are firewall solutions managed and maintained by a third-party provider. Organizations subscribe to managed firewall services to delegate firewall management, monitoring, and updates to experienced security professionals. Managed service firewalls can be implemented as hardware or software solutions and are typically offered as part of a broader managed security service portfolio. They provide organizations with expert security management, proactive threat monitoring, and rapid response capabilities, helping to enhance overall security posture while reducing the burden on internal IT teams.

Each type of software firewall—container firewalls, virtual firewalls, and managed service firewalls—addresses specific security challenges and operational requirements, offering organizations flexibility in choosing the most suitable firewall solution based on their infrastructure, scalability needs, and security strategy.

7.2.6.3 Firewall Types by Data Filtering Method

Firewalls can be classified based on the method they use to filter data. These methods determine how the firewall inspects and controls traffic entering or leaving a network. The primary types of firewalls by data filtering method include packet-filtering firewalls, stateful inspection firewalls, proxy firewalls, next-generation firewalls (NGFWs), and web application firewalls (WAFs).

Packet Filtering Firewalls

Packet filtering firewalls operate at the network layer (Layer 3) of the OSI model and make decisions based on individual packets. They inspect packet headers, focusing on source and destination IP addresses, port numbers, and protocols. Based on predefined rules, the firewall either allows or blocks the packets.

- Advantages: Simple and efficient with low resource consumption. They can quickly make decisions without deep inspection, which results in minimal impact on network performance.
- Disadvantages: Limited in scope as they do not inspect the payload of packets. Vulnerable to IP spoofing and cannot detect more sophisticated threats.

Example: Access Control Lists (ACLs) used in routers and switches for basic traffic filtering.

Stateful Inspection Firewalls

Stateful inspection firewalls, also known as dynamic packet filtering firewalls, monitor the state of active connections. They operate at the network and transport layers (Layers 3 and 4), keeping track of the state and context of network connections in a state table. The diagram below shows stateful firewall operation (Fig. 7.13).

Stateful inspection firewalls can be deployed as standalone devices or integrated with Layer 3 (L3) switches. The example presented in the diagram above is a DPU technology-based L3 switch known as the HPE Aruba CX10K. This switch provides L3 switching capabilities along with stateful inspection firewall at near line rate.

Advantages of stateful inspection firewalls include enhanced security compared to packet-filtering firewalls, as they understand the state and context of connections. They can track sessions and recognize legitimate packets within the context of a valid session. However, these firewalls are best suited for east-west traffic within an internal network and are not intended for use cases where perimeter firewalls are more appropriate, such as managing north-south traffic that traverses the Internet.

NextGen Firewall (NGFW)

Next-Generation Firewalls (NGFWs) combine traditional firewall capabilities with advanced features such as application awareness and control, integrated intrusion prevention systems (IPS), and cloud-delivered threat intelligence. NGFWs operate across multiple layers of the OSI model.

- Advantages: Comprehensive security features that go beyond traditional firewalls, including the ability to identify and control applications regardless of port, protocol, or evasive tactics. They can detect and mitigate advanced threats through integrated IPS and threat intelligence.
- Disadvantages: Higher cost and complexity. They require more powerful hardware and sophisticated management to handle the advanced features effectively.

Example: Palo Alto Networks NGFW, which offers granular control over applications, user-based policies, and integrated threat prevention.

Web Application Firewalls (WAFs)

Web Application Firewalls (WAFs) operate at the application layer and are specifically designed to protect web applications by filtering and monitoring HTTP/HTTPS traffic between a web application and the Internet. WAFs can detect and block various attacks that target web applications, such as SQL injection, cross-site scripting (XSS), and other OWASP (Open Web Application Security Project) Top 10 threats.

- Advantages: Provides specialized protection for web applications by focusing on application-specific threats. Can be deployed quickly and provide immediate protection against common web attacks. WAFs can also help comply with industry regulations and standards like PCI-DSS.
- Disadvantages: Limited scope as they only protect web applications. They can also add latency to web traffic and may require regular updates and tuning to handle evolving threats effectively.

Example: AWS WAF, which offers customizable web security rules to protect web applications deployed on Amazon Web Services.

7.3   Advanced Network Security Technologies

In today's hyper-connected world, network security has become more crucial and complex than ever. The rise of sophisticated cyber threats, coupled with the increasing digitization of critical infrastructure, necessitates the deployment of advanced security technologies. Traditional perimeter-based defenses are no longer sufficient in a landscape where threats can emerge from both external and internal sources, and where the attack surface continues to expand due to the proliferation of Internet of Things (IoT) devices, cloud computing, and remote work. Advanced network security technologies are essential for safeguarding sensitive information, ensuring the integrity and availability of critical services, and maintaining trust in digital interactions. These technologies are designed to address the limitations of traditional security measures by providing enhanced detection, protection, and response capabilities. They leverage cutting-edge innovations in fields such as artificial intelligence, machine learning, and behavioral analytics to stay ahead of evolving threats.

The paradigm shift toward more dynamic and proactive security frameworks is embodied in concepts like Zero Trust Architecture and Software-Defined Security. These approaches recognize that security must be woven into the very fabric of the network, with continuous monitoring and adaptive controls that respond to threats in real-time. Meanwhile, technologies like Endpoint Detection and Response (EDR) and Deception Technologies focus on protecting the most vulnerable entry points and tricking adversaries into revealing their tactics.

Artificial Intelligence and Machine Learning play a pivotal role in modern network security by enabling automated, intelligent threat detection and response. These technologies can process vast amounts of data at speeds and accuracies unattainable by human analysts, identifying subtle indicators of compromise that might otherwise go unnoticed. Behavioral Analytics further enhances this capability by establishing baselines of normal activity and highlighting deviations that may signify malicious behavior.

Encryption and Data Protection remain fundamental pillars of network security, ensuring that sensitive information remains confidential and untampered. With data breaches becoming increasingly common, robust encryption protocols and effective key management practices are critical for protecting data both in transit and at rest. As cyber threats become more advanced and persistent, organizations must adopt a multi-layered security strategy that incorporates these advanced technologies. This holistic approach not only strengthens defenses but also provides the agility needed to adapt to new threats and vulnerabilities. The following sections will delve into each of these advanced network security technologies in detail, exploring their principles, functionalities, and benefits, as well as their roles in a comprehensive security strategy.

By understanding and implementing these advanced security measures, organizations can better protect their assets, ensure regulatory compliance, and maintain a resilient security posture in an ever-evolving threat landscape.

7.3.1 Zero Trust Architecture

Zero Trust Architecture (ZTA) is a security framework that assumes that threats could exist both inside and outside the network. It operates on the principle of "never

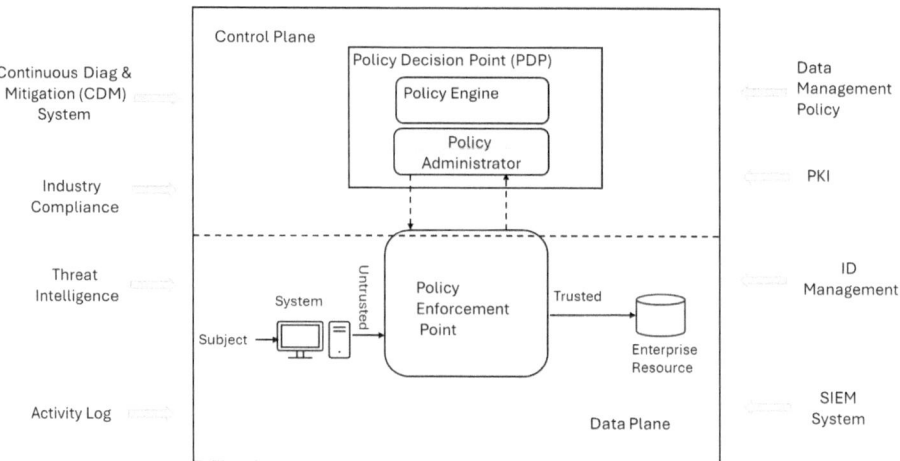

Fig. 7.15 NIST zero trust reference architecture[4]

trust, always verify." Every access request is thoroughly authenticated, authorized, and encrypted before being granted, regardless of its origin. The diagram below illustrates NIST (National Institute of Standard and Technology) reference framework for ZTA (Fig. 7.15).

Though there are numerous reference architecture for zero trust available from different vendors, NIST is considered ideal for organization seeking to achieve zero trust security capabilities for their network. The NIST has outlined comprehensive guidelines for implementing ZTA in the document NIST Special Publication (SP) 800-207. These guidelines provide a framework to help organizations adopt Zero Trust principles effectively. These guidelines include:

- **All data sources and computing services are considered resources**: Recognizes that resources can exist both on-premises and in the cloud.
- **All communication is secured regardless of network location**: Treats all network traffic as untrusted and ensures that data is encrypted both in transit and at rest.
- **Access to individual enterprise resources is granted on a per-session basis**: Adopts the principle of least privilege and requires re-authentication and re-authorization for each access request.
- **Access to resources is determined by dynamic policy**: Uses contextual data such as user identity, device health, and location to make access decisions.
- **The enterprise monitors and measures the integrity and security posture of all owned and associated assets**: Continuously monitors the security posture of devices and infrastructure to detect and respond to threats.

---

[4] NIST [4].

- **All resource authentication and authorization are dynamic and strictly enforced before access is allowed**: Implements real-time authentication and authorization processes to adapt to changing threat landscapes.
- **The enterprise collects as much information as possible about the current state of network infrastructure and communications**: Uses telemetry and analytics to gain insights into network activities and improve security policies.

The core components of NIST ZTA include:

- **Policy Engine (PE)**: Makes access decisions based on enterprise policy and risk analysis.
- **Policy Administrator (PA)**: Establishes and terminates the communication path between the subject and resource.
- **Policy Enforcement Point (PEP)**: Enforces policy decisions, often at the point where access requests occur.

For the NIST ZTA implementation, several logical components interact to enforce strict access controls and ensure security. The PEP ensures that the subject here in users, end devices, or applications cannot gain unauthorized access to network or enterprise resources. The elements within Policy Decision Point (PDP) enforce roles and policies through PEP to the endpoint devices, users, or applications.

Implementing Zero Trust Architecture involves deploying several models aimed at enhancing security and control throughout the network:

- Enhanced Identity Governance strengthens identity management and authentication processes, ensuring that only authorized users can access resources. It encompasses Public Key Infrastructure (PKI), which provides the cryptographic foundation for secure communications and authentication. PKI issues and manages digital certificates to verify identities and secure data exchanges, enhancing overall identity management.
- Microsegmentation and Network Isolation divide the network into smaller segments, isolating resources and allowing for more granular control over access.
- Device Health and Posture Assessment continuously evaluates the security status of devices accessing the network, proactively preventing potential vulnerabilities. It integrates with Security Information and Event Management (SIEM) systems to aggregate and analyze security event data, providing real-time monitoring, detecting policy violations, and enabling rapid response to security incidents.

By aligning data management policies, PKI, SIEM, and ID Management with NIST ZTA principles, organizations can effectively implement a Zero Trust Architecture that enhances overall cybersecurity resilience and mitigates modern threats.

7.3.2   Artificial Intelligence (AI) and Machine Learning (ML)

In an era defined by interconnected systems and digital dependency, the landscape of cybersecurity has evolved dramatically. As organizations and individuals alike navigate increasingly complex networks and face sophisticated cyber threats, the integration of Artificial Intelligence (AI) and Machine Learning (ML) has emerged as a pivotal paradigm shift. These technologies not only augment traditional cybersecurity measures but also introduce transformative capabilities that redefine how we detect, analyze, and respond to cyber threats.

AI and ML empower cybersecurity professionals with the ability to process vast amounts of data in real-time, identifying patterns and anomalies that may signal potential risks. Through advanced algorithms, these technologies enable predictive analytics, enhancing the proactive identification of vulnerabilities before they are exploited. Moreover, AI-driven systems automate response mechanisms, swiftly mitigating threats and minimizing the impact of cyber incidents. Beyond reactive defense strategies, AI and ML facilitate a paradigm of adaptive security. By continuously learning from data and adapting to evolving threats, these technologies bolster resilience against dynamic cyber landscapes. From behavioral analysis to intelligent authentication and anomaly detection in network traffic, AI and ML provide robust tools that fortify defenses and optimize cybersecurity operations.

However, like any technological advancement, the integration of AI and ML in cybersecurity presents its own set of challenges. These include adversarial attacks targeting AI models and ethical concerns regarding data privacy, which underscore the importance of thoughtful implementation and continuous refinement. Despite these challenges, the adoption of AI and ML signifies a transformative approach in cybersecurity. It promises a future where proactive protection and adaptive resilience redefine the cybersecurity landscape. In the following sections, we will delve deeper into specific applications of AI and ML particularly in network security.

7.3.2.1 Threat Detection and Mitigation

AI and machine learning systems are adept at processing extensive datasets in real-time, enabling them to discern patterns that signal potential cyber threats effectively. They autonomously identify and categorize malicious activities, including malware infections, phishing attacks, and anomalous network behavior. This capability reduces dependency on human operators, who may find it challenging to handle the sheer volume and complexity of emerging threats.

Anomaly Detection
Anomaly detection involves identifying unusual patterns or behaviors within network data that deviate from the norm. This is particularly useful in spotting potential security threats that do not match known attack signatures.

- **AI Models Used**: Unsupervised learning algorithms, such as K-means clustering, DBSCAN (Density-Based Spatial Clustering of Applications with Noise), and Principal Component Analysis (PCA), are commonly employed. These models analyze data without predefined labels to detect outliers that might indicate malicious activities.
- **Application in Network Security:** AI-driven anomaly detection systems continuously monitor network traffic and user activities to establish a baseline of normal behavior. When significant deviations from this baseline are detected, they are flagged for further investigation. For example, an unexpected spike in data transfer from a specific user account could indicate a data exfiltration attempt.

Example: A PCA model can be used to reduce the dimensionality of network traffic data, highlighting unusual patterns that might be missed by traditional methods. If the model identifies traffic patterns significantly different from established norms, it can alert security teams to investigate potential threats.

Behavioral Analysis

Behavioral analysis focuses on understanding and modeling the normal behavior of users, devices, and applications within a network to detect anomalies that may indicate security breaches.

- **AI Models Used:** Supervised learning models like Decision Trees, Random Forests, and Neural Networks, as well as unsupervised models such as Autoencoders, are utilized to analyze behavioral patterns and detect deviations.
- **Application in Network Security:** By learning the typical behavior of users and devices, AI systems can detect subtle changes that might indicate malicious activity. For instance, if an employee's account suddenly starts accessing large volumes of sensitive data outside of regular working hours, the system can flag this as suspicious.

Example: An Autoencoder can be trained to reconstruct normal user behavior patterns. When new behavior significantly deviates from what the Autoencoder has learned, it indicates a potential threat. This approach is effective for detecting insider threats, compromised accounts, and unauthorized access.

Natural Language Processing (NLP)

NLP is used to analyze and interpret human language, enabling AI systems to understand and respond to text-based data. In cybersecurity, NLP can be applied to various tasks, including threat intelligence and detecting phishing attacks.

- **AI Models Used**: Techniques such as text classification, sentiment analysis, and entity recognition using models like BERT (Bidirectional Encoder Representations from Transformers), LSTM (Long Short-Term Memory) networks, and CNNs (Convolutional Neural Networks) are employed for processing and understanding text data.

- **Application in Network Security**: NLP models can analyze email content to detect phishing attempts by identifying malicious links, suspicious sender addresses, and unusual language patterns. Additionally, NLP can process threat intelligence reports, extracting relevant information and correlating it with internal data to identify potential threats.

Example: A BERT-based model can be trained to classify emails as legitimate or phishing based on their content. By analyzing the language used in emails, the model can detect phishing attempts that traditional signature-based systems might miss, thus enhancing email security.

7.3.2.2 Benefits of AI in Threat Detection

AI systems can handle vast amounts of data and scale effortlessly, a critical feature given the exponential growth in data traffic and the complexity of modern networks. Additionally, AI models process and analyze data at high speeds, enabling real-time threat detection and response, which is crucial for minimizing the impact of cyberattacks. By learning from historical data and continuously refining their algorithms, AI models improve their accuracy in identifying genuine threats while reducing false positives. These systems can adapt to new and evolving threats by updating their learning models and staying ahead of cybercriminals who constantly develop new attack methods.

However, there are implementation challenges, including data quality and adversarial attacks. The effectiveness of AI models heavily depends on the quality and quantity of training data. Incomplete or biased data can lead to inaccurate models. Implementing AI systems requires significant expertise and resources; organizations need skilled personnel to develop, deploy, and maintain these systems. Cybercriminals can attempt to deceive AI models by feeding them manipulated data, necessitating robust defenses against such adversarial attacks.

AI algorithms, especially deep learning models, are often seen as "black boxes" with decisions that are difficult to interpret. Ensuring transparency and explainability in AI-driven threat detection is crucial for building trust and accountability.

In summary, AI and machine learning are revolutionizing network security by enabling real-time, scalable, and accurate threat detection and mitigation. Through anomaly detection, behavioral analysis, and NLP, these technologies provide robust tools to identify and respond to complex cyber threats. By leveraging advanced AI models, organizations can proactively defend against increasingly sophisticated cyberattacks, enhancing their overall security posture.

7.3.3   Deception Technologies

Deception Technologies represent a sophisticated and proactive approach to network security, designed to mislead and trap cyber attackers. By creating a landscape of decoys and false information, these technologies aim to detect and analyze malicious activities without alerting the attacker to their presence. This approach provides valuable intelligence on attack methods and motives while diverting threats away from critical assets.

The concept of deception in security is not new; it has been used in various forms, from military tactics to digital defenses. However, the evolution of Deception Technologies in cybersecurity marks a significant advancement in this field. Initially, simple honeypots were used—decoy systems designed to attract attackers and study their methods. Over time, these have evolved into complex, dynamic systems capable of mimicking entire networks and responding adaptively to threat activities. The diagram below shows typical lures and decoys deployment using a deception platform (Fig. 7.16).

Today's Deception Technologies encompass a range of tools and strategies, including honeypots, honey tokens, decoy networks, and more. These tools are designed to integrate seamlessly into existing security infrastructures, providing a robust layer of defense that works in tandem with traditional measures like firewalls and intrusion detection systems (IDS). A common way of deploying these deception technologies to a given network environment is to use a deception platform of choice.

The followings are a list of deception platforms:

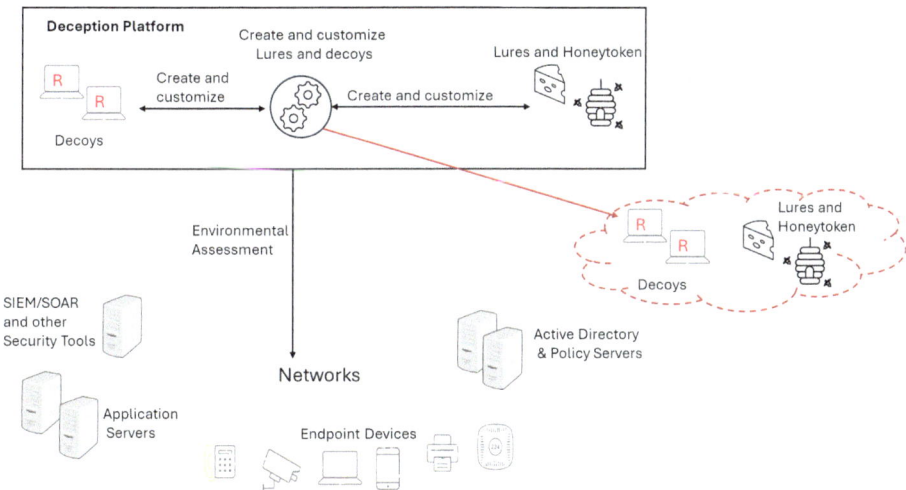

Fig. 7.16 Lures and decoy deployment using deception platform

- **Attivo Networks**: Their ThreatDefend® Deception and Response Platform is known for its comprehensive deception technology, covering endpoints, Active Directory, cloud environments, and networks. Attivo provides both on-premises and cloud-based solutions, focusing on detecting, analyzing, and responding to attacks in real-time.
- **TrapX Security (now part of CommVault)**: The TrapX DeceptionGrid is designed to deploy a network of traps and decoys that mislead attackers, capturing detailed attack data to help organizations understand and mitigate threats.
- **Illusive Networks**: Known for their Illusive Platform, which focuses on detecting and stopping lateral movement within networks. Their solutions include endpoint-based deception techniques to mislead attackers and provide actionable intelligence to security teams.
- **Acalvio Technologies**: ShadowPlex is their leading deception platform, utilizing advanced AI-driven deception techniques to create believable decoys and lures that detect, engage, and respond to attackers across various environments.
- **Rapid7**: The Rapid7 **InsightIDR** incorporates deception technology within its broader threat detection and response capabilities, offering decoys and honey credentials to detect intruders early in the attack lifecycle.
- **SentinelOne**: Their Singularity Hologram utilizes dynamic deception techniques and distributed decoy systems to transform the network into a trap for attackers, aiding in the detection and analysis of threats as they interact with decoy assets.
- **Morphisec**: The Morphisec Breach Prevention Platform focuses on preventing advanced attacks across endpoints, servers, and cloud environments. It uses lightweight deception techniques to detect and stop threats without prior knowledge of attack methods.
- **Zscaler**: Zscaler Deception is part of the Zscaler Zero Trust Exchange, utilizing decoys and honeypots to detect advanced threats that bypass other security measures, focusing on zero trust principles and extending protection across the network.
- **Smokescreen Technologies (now part of Zscaler)**: The IllusionBLACK solution is known for its simplicity and effectiveness, allowing for quick deployment of deception campaigns and integration with other security tools for comprehensive threat detection and response.

These platforms offer various features, such as automated deployment, integration with existing security infrastructures, and advanced threat intelligence capabilities, making them vital components of modern cybersecurity strategies.

7.3.4   SIEM and SOAR

In the ever-evolving landscape of cybersecurity, advanced technologies like Security Information and Event Management (SIEM) and Security Orchestration, Automation, and

Response (SOAR) have become essential components of robust network security strategies. These technologies address the increasing complexity and sophistication of cyber threats that modern organizations face.

SIEM technology has progressed far beyond its original function of log management and compliance reporting. Modern SIEM systems integrate advanced analytics, real-time monitoring, and automated responses to provide comprehensive threat detection and incident response capabilities. They gather and analyze data from various sources across the network, enabling security teams to identify, prioritize, and mitigate threats more effectively.

SIEM systems are considered advanced due to their ability to:

- Real-Time Threat Detection: Providing immediate insights into potential security incidents as they occur.
- Advanced Analytics: Utilizing machine learning and behavioral analysis to detect anomalies and sophisticated threats.
- Comprehensive Visibility: Aggregating data from diverse sources, including network devices, applications, and cloud environments, to offer a holistic view of the security landscape.
- Automated Response: Enabling quicker and more effective responses to incidents through predefined actions and workflows.

Prominent SIEM tools include Splunk Enterprise Security, IBM QRadar, ArcSight ESM (Enterprise Security Manager), LogRhythm NextGen SIEM Platform, and Microsoft Sentinel.

Unlike SIEM, which primarily focuses on gathering and analyzing security data, SOAR platforms are designed to enhance the efficiency and effectiveness of security operations by orchestrating and automating a wide range of security tasks. They integrate various security tools and processes into a unified platform, enabling seamless coordination and response to security incidents. SOAR systems help security teams manage the overwhelming volume of alerts and streamline the incident response process.

Key reasons SOAR is considered advanced include its ability to coordinate actions across multiple security tools, ensuring cohesive and efficient responses to threats. By automating repetitive tasks such as threat intelligence gathering and alert triage, SOAR reduces the manual workload on security teams. SOAR platforms provide comprehensive playbooks and automated workflows to respond to incidents swiftly and effectively, and offer detailed incident tracking and reporting features to ensure thorough analysis and resolution of security events.

Leading SOAR tools include Palo Alto Networks Cortex XSOAR, IBM Resilient, Splunk Phantom, ServiceNow Security Operations, and Swimlane.

The need for SIEM and SOAR technologies in today's network infrastructure is driven by several factors. Cyber threats are becoming more complex and frequent, requiring advanced tools to detect and respond to them effectively. The exponential growth of data generated by network devices, applications, and users necessitates systems that can process and analyze vast amounts of information in real-time. Organizations must also comply with stringent regulatory requirements that mandate comprehensive security monitoring and incident response capabilities. Additionally, security teams often face resource limitations, making automation and orchestration essential for managing workloads and responding to incidents efficiently.

7.3.4.1 Choosing Between SIEM and SOAR: Should You Use Both?

When deciding whether to implement SIEM, SOAR, or both, it's essential to understand the unique capabilities of each technology and how they complement each other. While both SIEM and SOAR play critical roles in a comprehensive cybersecurity strategy, their functions differ significantly, and many organizations find value in deploying both systems together.

SIEM solutions are primarily focused on log management and analysis, providing real-time threat detection by analyzing log data from various sources. They offer comprehensive visibility into the security landscape and help organizations meet regulatory compliance requirements through detailed reports and records of security events. Historical data analysis capabilities of SIEMs are particularly useful for forensic investigations and identifying patterns in past incidents.

SOAR platforms, on the other hand, are designed to enhance the efficiency and effectiveness of security operations by orchestrating and automating a wide range of security tasks. They integrate various security tools and processes into a unified platform, enabling seamless coordination and response to security incidents. SOAR systems help security teams manage the overwhelming volume of alerts and streamline the incident response process through automated workflows and playbooks, reducing the manual workload on security teams.

When choosing between SIEM and SOAR, several factors should be considered. If your primary need is to collect and analyze logs from various sources and meet compliance requirements, an SIEM solution may be sufficient. However, if you need to streamline and automate your incident response process, or if your security team is overwhelmed by the volume of alerts, a SOAR solution can be invaluable. Resource availability also plays a critical role; implementing and managing an SIEM can be resource-intensive, requiring skilled personnel to interpret data and manage the system. SOAR platforms can help alleviate some of this burden by automating tasks and orchestrating responses, making them suitable for organizations with limited security resources.

For many organizations, using both SIEM and SOAR can provide a more comprehensive and efficient security posture. A SIEM solution offers the foundational capabilities of log management, threat detection, and compliance reporting, while a SOAR platform enhances these capabilities by automating and orchestrating the incident response process. By integrating SIEM and SOAR, organizations can achieve enhanced visibility and context, improved efficiency, and faster response times, allowing security teams to focus on more strategic tasks and providing robust threat detection and streamlined incident response.

In summary, choosing between SIEM and SOAR—or deciding to implement both—depends on your organization's specific needs, resources, and security objectives. For comprehensive security coverage and enhanced operational efficiency, integrating both technologies can be highly effective, providing robust threat detection, streamlined incident response, and improved overall security posture.

7.3.5 Endpoint Detection and Response (EDR)

Endpoint Detection and Response (EDR) is another important advanced technology solution in cybersecurity. EDR solutions are designed to monitor, detect, and respond to security threats on endpoints, such as laptops, desktops, and servers that connect to a network. These systems provide enhanced visibility and control over endpoint activities, enabling organizations to quickly identify and mitigate threats before they cause significant damage. EDR solutions continuously monitor endpoint activities, capturing data in real-time to provide detailed visibility into what is happening on each device. This data includes process activity, network connections, and file changes. Using advanced analytics and machine learning, EDR systems analyze endpoint behavior to identify anomalies and detect potential threats. This capability helps in identifying sophisticated and previously unknown threats that traditional signature-based antivirus solutions might miss. The diagram below shows a typical EDR implementation with EDR agents installed on endpoints, providing data about the respective devices to the EDR server (Fig. 7.17).

The EDR server uses a combination of real-time monitoring, data analytics, and automated responses to detect and respond to threats like malware and ransomware. The EDR server can also help identify unknown threats and reduce the risk of future attacks. Enterprises may opt for third-party Security Operations Center (SOC) services, which can use cloud-based SIEM to collect threat intelligence and provide mitigation actions accordingly. They also alert the IT team about any incident observations for further action. Through these services, the IT team can effectively manage network security threats as they arise.

EDR tools provide robust incident response capabilities, enabling security teams to quickly investigate and respond to detected threats. Automated responses can be configured to contain and remediate threats, minimizing the impact on the organization. EDR platforms often include threat hunting capabilities, allowing security analysts to proactively search for indicators of compromise (IoCs) and other signs of malicious activity within the network. They typically integrate with other security tools, such as

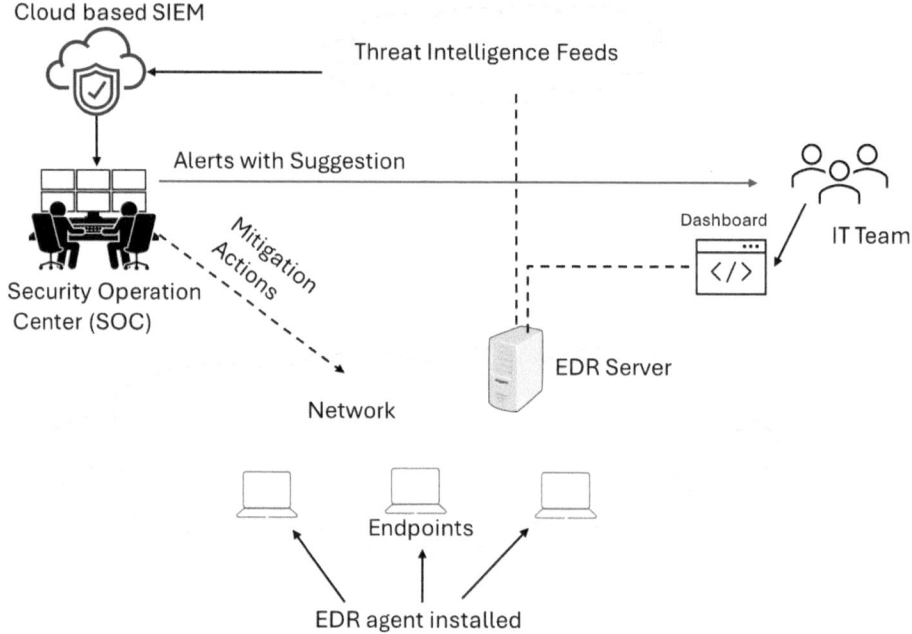

Fig. 7.17 Typical EDR setup in an enterprise network

SIEM (Security Information and Event Management) and SOAR (Security Orchestration, Automation, and Response) platforms, to provide a comprehensive security posture.

Additionally, EDR delivers deep visibility into endpoint activities, utilize sophisticated analytics for threat detection, and offer automated and effective incident response mechanisms. Traditional antivirus solutions focus mainly on preventing known threats, while EDR systems are designed to detect and respond to a wide range of attacks, including zero-day exploits, fileless malware, and advanced persistent threats (APTs).

In today's cybersecurity landscape, where threats are becoming increasingly sophisticated and frequent, EDR solutions are essential for several reasons. With the proliferation of remote work and the use of personal devices for business purposes, endpoints have become prime targets for cyberattacks. EDR provides the necessary visibility and control to secure these devices. EDR's ability to detect and respond to advanced threats that bypass traditional defenses is crucial for protecting sensitive data and maintaining business continuity. Many regulatory frameworks require organizations to have robust incident detection and response capabilities, which EDR solutions can help fulfill. Additionally, by automating many aspects of threat detection and response, EDR systems help organizations manage security more efficiently, even with limited resources.

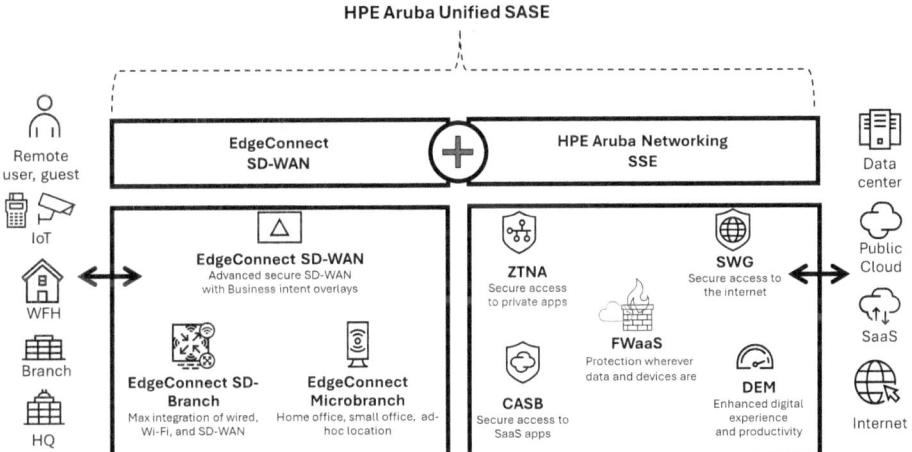

Fig. 7.18   HPE Aruba Unified SASE framework[5]

7.3.6   SASE and ZTNA

Secure Access Service Edge (SASE) is a network architecture concept that merges wide-area networking (WAN) with network security services into a single cloud-based service model. This integration includes secure web gateways (SWG), cloud access security brokers (CASB), firewalls as a service (FWaaS), and zero trust network access (ZTNA). SASE is designed to provide comprehensive security while optimizing access to applications and services. The diagram below depicts how HPE Aruba offers its SASE services to customers. The information presented in this diagram and explained herein is for educational purposes only. I believe this unique solution presented in the diagram would be useful for readers to understand how SASE and ZTNA interact, while also providing perspectives on the advanced technologies in use. It is by no means an endorsement of any product or service (Fig. 7.18).

The HPE Aruba Unified SASE framework includes Edge Connect SD-WAN solutions that cover secure Edge Connect SD-WAN for Business Internet, SD-WAN for Branch, and Micro Branch, along with HPE Aruba Secured Service Edge (SSE). SSE is a connectivity-as-a-service platform that provides secure access to users, devices, and applications. It is a security component of SASE (Secure Access Service Edge), which combines SSE with an advanced SD-WAN edge. SSE integrates several security capabilities into a single interface, including ZTNA (Zero Trust Network Access), Secure Web Gateway (SWG), Cloud Access Security Broker (CASB), Firewall as a Service (FWaaS), and device, app, and network performance monitoring (DEM).

---

[5] HPE Aruba [5].

At its core, SASE enhances WAN performance through Software-Defined WAN (SD-WAN), which directs traffic over the most efficient paths, improving connectivity to cloud services and remote locations. It protects against web-based threats and enforces Internet security policies via secure web gateways. Cloud access security brokers secure cloud service usage, offering visibility and control over cloud applications. Firewalls as a service delivers firewall capabilities from the cloud, simplifying management and scalability. The integration of zero trust network access ensures secure remote access to applications based on the principle of "never trust, always verify."

SASE provides several key benefits, including improved security through unified policies and centralized management, scalability via a cloud-based delivery model, cost efficiency by reducing the need for multiple on-premises solutions, and optimized performance by routing traffic through the most efficient paths.

7.3.6.1 ZTNA (Zero Trust Network Access)

Zero Trust Network Access (ZTNA) is a security approach grounded in the principle that no user or device should be trusted by default, whether inside or outside the network. Every access request is verified based on context, such as the user's identity, the device being used, and the requested resource. The diagram below shows ZTNA connectivity to enterprise networks (Fig. 7.19).

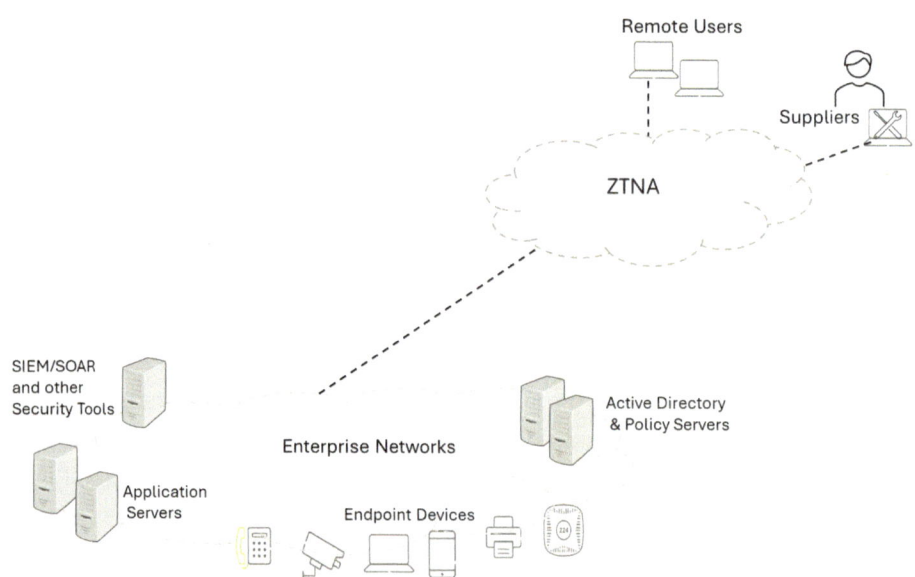

Fig. 7.19 Typical ZTNA connectivity to enterprise networks

ZTNA operates on several fundamental principles. It grants users the minimum level of access necessary to perform their tasks, thus limiting potential damage from compromised accounts. Microsegmentation divides the network into smaller, isolated segments to restrict lateral movement within the network. Continuous monitoring and validation of user and device identities, along with network activity, help detect anomalies and potential threats. Context-aware access considers various factors like location, device type, and time of day to enforce adaptive security policies. User and device authentication, often involving multi-factor authentication (MFA), ensures that only secure and verified devices gain access.

The benefits of ZTNA include enhanced security through continuous verification, a reduced attack surface by limiting access to specific applications and resources, improved compliance with data protection and access control regulations, and greater visibility into user activities and network traffic.

7.3.7 Security First Networking

When I referred to the security-first networking paradigm in the introduction of this chapter, I emphasized the need for comprehensive security measures that deviate from traditional approaches by delivering a solution that integrates all available tools. Today's networking technologies and hardware advancements enable offloading some security measures to networking devices, while policy and orchestration tools utilize AIOps capabilities to leverage AI in addressing network security concerns at the edge. This approach neither discounts traditional methods of network security nor omits technological advances in the field. Instead, it aims to combine the best of both worlds while integrating these capabilities into networking equipment.

Moreover, addressing network concerns at the edge using the existing network platform adds a powerful security envelope that adheres to the Zero Trust Architecture (ZTA) mentioned earlier. HPE's security-first, AI-powered networking stands out by incorporating sophisticated AI-driven techniques and advanced security measures, transforming network defense with end-to-end security measures. Let's delve into the technical specifics that make these solutions superior in addressing modern cyber threats.

7.3.7.1 Secured Edge to Cloud

HPE Aruba offers seamless connectivity from edge to cloud, providing effective security mechanisms to protect scalable network infrastructure while orchestrating it through Aruba Central. The solution integrates the entire HPE Aruba networking portfolio and tools under a comprehensive framework known as the Edge Services Platform (ESP). The foundational block of this framework is a security envelope that adheres to the Zero Trust construct and utilizes AIOps for enhanced visibility and optimization from day one.

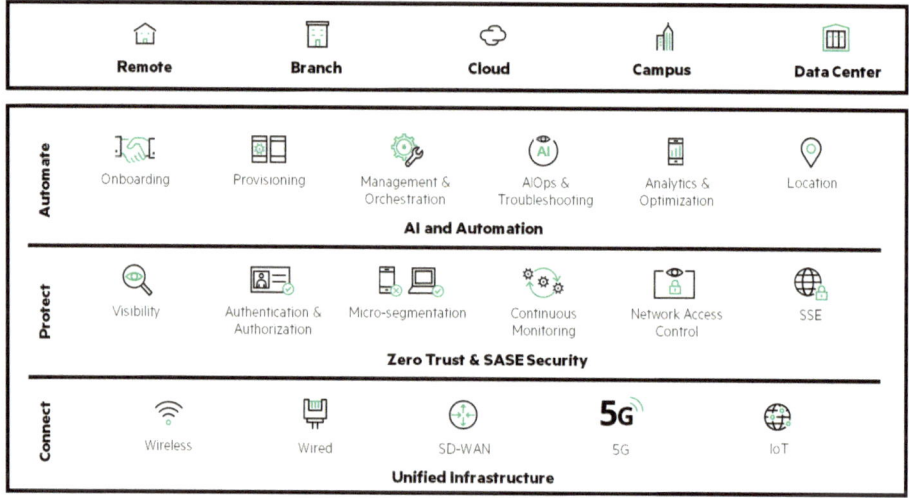

Fig. 7.20 HPE aruba Edge Services Platform (ESP) (Courtesy: HPE Aruba)[6]

From device onboarding to network operations, Aruba Central delivers unparalleled AIOps for network optimization and management. For example, the AI search capability of Aruba Central extends analytics and AI-based suggestions for improved design decisions at new and existing sites during device onboarding. Similarly, its AI engine automatically records support cases once an anomaly is detected that may result in adverse network conditions if not intervened (Fig. 7.20).

This AIOps capability extends over various types of connectivity—wired or wireless—from campus networks to private 5G and IoT. In the ESP architecture, this varied network connectivity is categorized under "Unified Infrastructure," with network security and assurance capabilities integrated directly into the network devices that HPE Aruba sells. For example, HPE Aruba mobility controllers and switches like the CX10000 integrate network services such as firewalls that provide stateful inspection at line rate. Similarly, HPE Aruba's AI-powered unified SASE, which incorporates EdgeConnect solutions, can extend the security envelope from remote users' endpoints to their respective enterprise networks. A flow that starts at a remote user's desktop can be tracked, with the appropriate security measures applied through an end-to-end policy that extends from the remote user's desktop to the applications they access.

---

[6] HPE Aruba [6].

7.3.7.2   AI-Driven Threat Detection and Response

AI-powered threat detection and response mechanisms are at the heart of HPE's security-first approach. These systems continuously analyze vast amounts of network data in real-time, identifying patterns and anomalies that may indicate potential security threats. By leveraging machine learning algorithms, these networks can:

- **Identify Zero-Day Threats**: AI models trained on diverse data sets can detect new, previously unknown threats based on their behavior, even before signature-based systems recognize them.
- **Automate Responses**: When a threat is detected, the system can automatically initiate predefined response actions, such as isolating affected network segments, blocking malicious traffic, or alerting security personnel, thus minimizing response times and reducing potential damage.

HPE Aruba has been a leader in behavioral and signature-based analysis through network insights since long. With the addition of AI and machine learning, now these insights are more dynamic allowing pattern modeling that can be shared across multiple operators (Fig. 7.21).

Additional AI-powered features designed to simplify time-to-resolution and improve administrator confidence include a Natural Language Processing (NLP)-based search, event-driven AI Assist, and Application Insights.

7.3.7.3   Dynamic Segmentation and Microsegmentation

HPE's dynamic segmentation and microsegmentation technologies ensure that security policies are enforced at the most granular level. This approach limits the lateral movement of attackers within the network by compartmentalizing different parts of the network into isolated segments. For example, dynamic segmentation utilizes HPE Aruba's award-winning ClearPass Policy Manager and Aruba Central NetConductor to enforce robust Network Access Control (NAC) while traffic is further examined through its firewall implemented in the mobility controller or other Aruba Gateways. Traffic is then tunneled to its destination, whether an application or other endpoints, through dynamic overlays (Fig. 7.22).

In addition, switches like the CX10000 bring microsegmentation and other network services capabilities to the edge of networks through the implementation of DPU technology within the switch. Microsegmentation can augment the security posture of campus and data center networks, limiting the blast radius to a specific microsegment within the network in case of a successful attack (Fig. 7.23).

The CX10000 (CX10K) also offers service chaining capabilities, adding features such as workload group policy for tagging applications and associated policies, DDoS protection, IPSEC, NAT, and firewall, to name a few.

Fig. 7.21 AI/ML based automated root cause analysis using Aruba Central (Courtesy: HPE Aruba)[6]

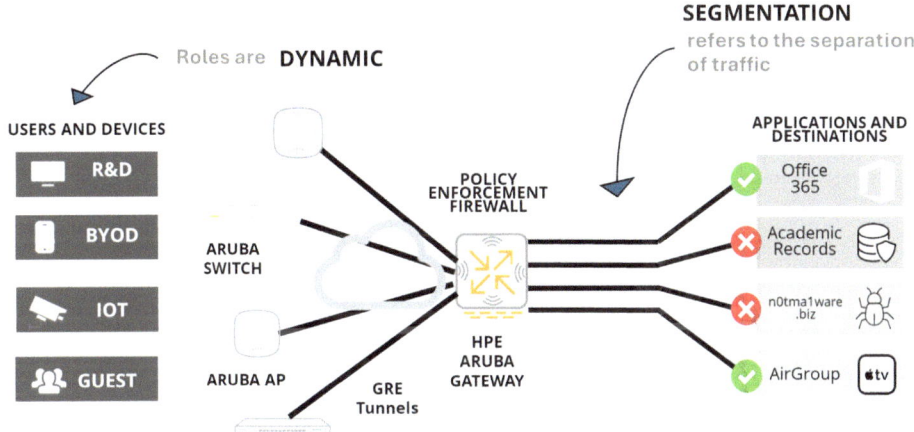

Fig. 7.22 Dynamic segmentation and overlay connectivity based on NAC policy profile (Courtesy: HPE Aruba)[7]

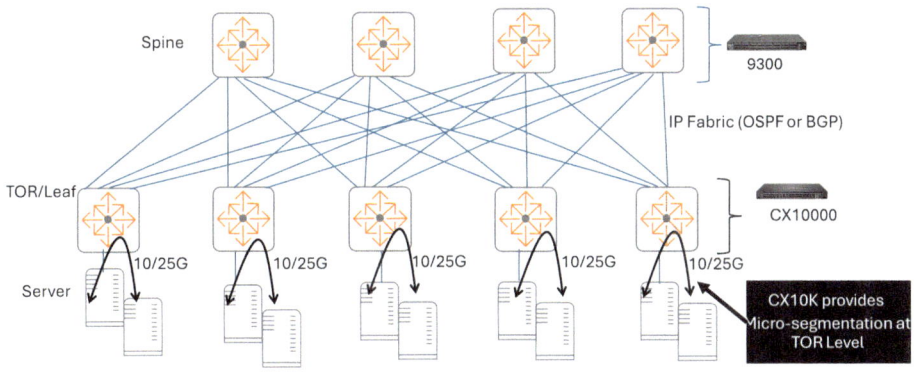

Fig. 7.23 CX10000 deployed as TOR offering microsegmentation at VM level

While security models like Zero Trust and centralized firewalls are foundational, they are often insufficient on their own against the sophisticated threats faced by modern networks. HPE's security-first, AI-powered networking solutions offer an integrated, adaptive, and scalable approach that provides comprehensive protection. This solution does not exclude centralized firewalls and other security tools; rather, these appliances and tools should be viewed as complementary to the solution offered by HPE.

---

[7] HPE Aruba [7].

7.3.8 Summary

In this chapter, I endeavored to provide a comprehensive overview of network security technologies, a topic so vast that it could warrant its own dedicated book. Nevertheless, the information presented here aims to equip readers with the foundational knowledge necessary to make informed decisions when formulating a network security strategy for their respective organizations.

The discussion encompassed both traditional and advanced network security technologies, offering valuable insights into each. Traditional technologies, such as firewalls and antivirus programs, were examined alongside more advanced solutions like Zero Trust Architecture, Artificial Intelligence and Machine Learning in security, Behavioral Analytics, Deception Technologies, Encryption and Data Protection, and Endpoint Detection and Response (EDR). By comparing these technologies, readers can better understand the strengths and weaknesses of various approaches and select the most appropriate tools for their needs.

Additionally, the chapter highlighted how specific vendors, such as HPE, integrate these technologies to create comprehensive and powerful security solutions. By leveraging a combination of traditional and advanced methods, HPE's Security First Networking exemplifies how a layered approach can enhance an organization's overall security posture.

In summary, this chapter aims to serve as a practical guide for understanding the diverse landscape of network security technologies. By exploring the capabilities and applications of both traditional and advanced solutions, readers are better prepared to develop robust security strategies that address the complex and evolving threats facing modern organizations.

References

1. Monroe (2024) Cybersecurity history: hacking & data breaches. Monroe College. https://www.monroecollege.edu/news/cybersecurity-history-hacking-data-breaches
2. Souri A, Hosseini R (2018) A state-of-the-art survey of malware detection approaches using data mining techniques. Human-centric computing & information Sciences. Spinger Nature
3. Paloalto (2024) Types of firewalls defined and explained. Palo Alto Networks. https://www.paloaltonetworks.com/cyberpedia/types-of-firewalls
4. NIST (2024) Zero trust networks. National Institute for Standard and Technology. https://www.nist.gov/programs-projects/zero-trust-networks
5. HPE Aruba (2024) AI-powered unified SASE. Award-winning SSE. Industry-leading SD-WAN. One single solution. https://www.arubanetworks.com/solutions/sase/
6. HPE Aruba (2024) Connect and secure your edges with a single unified platform. https://www.arubanetworks.com/solutions/aruba-esp/
7. HPE Aruba (2024) HPE aruba networking dynamic segmentation. https://www.arubanetworks.com/resource/dynamic-segmentation-solution-overview/

AI Infrastructure

In recent years, the intersection of artificial intelligence (AI) and networking has transformed the landscape of network management and security. The integration of AI and machine learning (ML) technologies into networking, often referred to as AI-Networking, encompasses a broad spectrum of innovations. These innovations can be categorized into AI-native networking, which leverages AI and ML to enhance the capabilities of network management, security, and operations. This includes advancements in AI-driven security measures, automated network management (AIOps), and predictive maintenance. We have explored how AI/ML are integrated into modern network infrastructure to provide proactive responses to network conditions in earlier chapters. The objective of such AI-networking is to create self-optimizing and self-healing networks that can adapt to changing conditions and threats in real-time. As discussed in previous chapters, AI/ML and AIOps play a crucial role in self-healing networks, a topic further elaborated in Chap. 9. AI-native networking has been progressing over the past few years and is now becoming mainstream due to increased security challenges and the complexities of network orchestration and management. Accurate data insights are essential for flawless network operations and proactive threat mitigation.

While AI-native networking is foundational for any data center (DC) and campus networks today, particularly those supporting AI/ML workloads, the networks designed specifically for AI/ML workloads have distinct definitions and objectives and thus should be considered a separate category. Networks for AI/ML workloads have emerged more recently, driven by the increasing adoption of generative AI (GenAI) and other AI-related workloads supporting ML, deep learning, and computer vision. These networks focus on optimizing infrastructure to handle the specific demands of AI and ML processes. Designing networks that support the high-performance computing (HPC) needs of AI/ML

D. D. Chowdhury, *Future of Networks*, Synthesis Lectures on Communications,
https://doi.org/10.1007/978-3-031-71440-5_8

workloads involves creating high-bandwidth, lossless, low-latency, and highly available networks. Additionally, GPU clustering and scale-out configurations using NVLink place additional demands on the network, bringing networking and compute infrastructure into close proximity. The design of such networks is heavily dependent on compute clustering configurations.

Given the complexity and interplay of AI-networking, compute clustering solutions, cluster management software, storage infrastructure, and the AI stack involved in modern AI/ML workload environments, this chapter will be approached from the broader perspective of AI infrastructure. This comprehensive approach will provide a holistic view of AI and HPC infrastructure, encompassing both software and hardware components. We will explore the various elements that constitute AI infrastructure, including underlying technologies of Compute clustering, architectural considerations thereof, and the technologies needed for network infrastructure to support Ai/ML workloads.

8.1   High-Performance Computing (HPC)

My first foray into HPC was in the early 1990s when I was tasked with phasing out the mainframe at Schlage Lock Company and transitioning to the HP T500 series of mini-computers. These minicomputers, each equipped with 26 processors, provided a parallel computing infrastructure. At the time, we used FDDI (Fiber Distributed Data Interface) ring technology, which offered a 100Mb/s transfer rate among the computing systems. Parallel processing was essential to support 65,000 transactions during working hours and manage enormous data catalogs based on Oracle RDBMS.

Decentralization, scalability, and the shift from batch processing on mainframes to real-time processing in distributed computing were some of the key reasons that led us to the parallel computing infrastructure deployment at Schlage. This fundamental concept is what drives the era of HPC today. Since then, HPC has become a norm for highly scalable compute clusters needed for advanced research, mind-blowing special effects and media rendering, oil and natural gas exploration, and financial services, to name a few. Today, HPC is a de facto requirement for AI infrastructure, making it imperative to explore HPC before delving into AI infrastructure.

Understanding High-Performance Computing (HPC) is crucial in AI infrastructure for several reasons. AI workloads, especially those involving deep learning, demand immense computational power to train and infer from complex models. HPC provides the necessary performance through parallel processing, leveraging multiple processors and accelerators such as GPUs (graphical processing units) and TPUs (Tensor Processing Units). This significantly reduces the time required for training large AI models, making it feasible to iterate quickly and improve model accuracy. AI applications often involve processing vast amounts of data. HPC systems are designed to handle large datasets efficiently, with high-speed storage solutions and advanced data management techniques. This ensures that data can be accessed and processed at the speeds necessary to meet the demands of AI workloads, minimizing bottlenecks and maximizing throughput.

As AI models grow in complexity and size, the infrastructure must scale accordingly. HPC systems are inherently scalable, allowing organizations to add more compute nodes, storage capacity, and networking bandwidth as needed. This scalability is essential for maintaining performance as AI workloads expand, ensuring that infrastructure can grow with the demands of the applications.

HPC environments are optimized for maximum efficiency, utilizing advanced scheduling, resource allocation, and parallel computing techniques. This optimization is critical for AI workloads, which can be resource-intensive and require precise coordination of computational tasks. By leveraging HPC, organizations can ensure that their AI infrastructure operates at peak efficiency, reducing costs and energy consumption. Many AI/ML frameworks, such as TensorFlow and PyTorch, are designed to take advantage of HPC capabilities. Understanding HPC enables organizations to fully leverage these frameworks, optimizing them for the available hardware and maximizing their performance. This integration is essential for achieving the best possible results from AI models.

High-Performance Computing (HPC) is at the forefront of scientific and technological research, driving innovations that directly benefit Artificial Intelligence (AI). Developments in HPC, from new algorithms to advanced hardware architectures, often translate into improvements in AI infrastructure. Keeping up with HPC trends and advancements ensures that organizations can incorporate the latest technologies and methodologies into their AI strategies. The following diagram illustrates a typical data center with HPC clusters (Fig. 8.1).

The first cluster in the figure is specifically designed for AI/ML workloads. The network topology supporting this large AI/ML workload cluster uses HPE Slingshot network

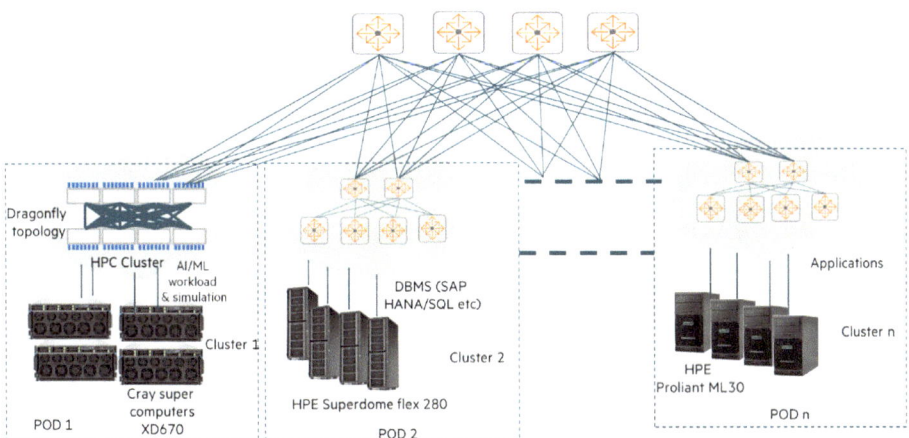

Fig. 8.1 Typical data center with HPC clusters

switches, which create a Dragonfly architecture that supports a highly ultra-low latency RDMA network. It is important to note that HPE is a leader in HPC, and its Slingshot product-based Dragonfly architecture is ideal for highly scalable AI/ML workloads. The Dragonfly architecture shown in the figure is an Ethernet architecture. Other HPC clusters for non-AI/ML workloads are using a spine-leaf architecture. However, HPC can also utilize InfiniBand network architecture. InfiniBand offers high throughput and low latency, making it well-suited for demanding HPC applications. It is widely used in supercomputing environments where performance and scalability are critical. InfiniBand provides a robust alternative to Ethernet, especially for applications that require efficient and fast data transfers across large clusters.

8.1.1   InfiniBand in HPC

InfiniBand is a high-performance network architecture designed specifically for HPC (High-Performance Computing) environments. The diagram below illustrates a typical InfiniBand (IB) network for HPC clusters. There are six main components of an IB network architecture: the Host Channel Adapter (HCA), the Target Channel Adapter (TCA), the IB switch, the IB subnet, the IB subnet manager, and the IB router (Fig. 8.2).

The HCA is installed in compute clusters, typically in each processor node, similar to an Ethernet Network Interface Card (NIC). IB HCAs are designed for high-throughput, low-latency communication using the InfiniBand protocol and connect to CPU memory via PCIe interfaces. HCAs act as end-nodes that connect to an IB network, such as a server, and can perform transport layer functions. On the other hand, the TCA (Target Channel Adapter) is the network interface for individual I/O devices (e.g., RAID and

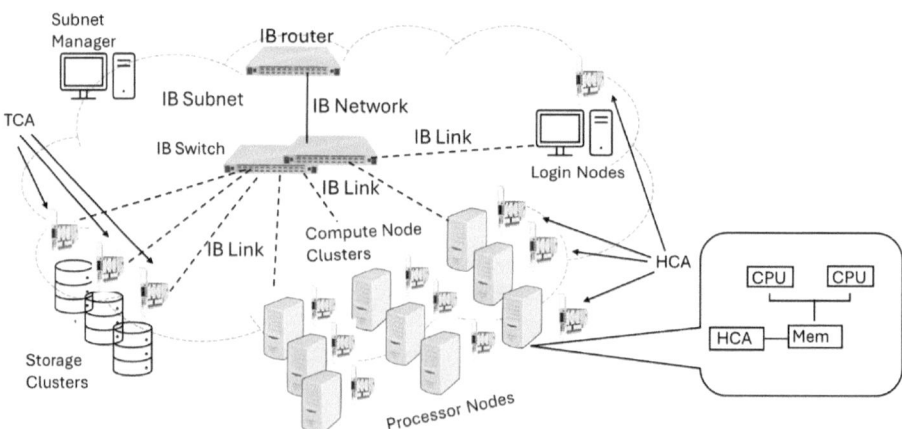

Fig. 8.2 Infiniband networks for HPC clusters

storage subsystems). The TCA is similar to the HCA but can be simplified according to the requirements of the attached device(s).

The function of an IB switch is to forward messages between InfiniBand networks, enabling efficient data transmission across connected devices. The link between the IB switch and devices in an IB subnet is referred to as IB links, whereas the link between an IB switch and an IB router is called the IB network. The IB router facilitates the transmission of messages between different InfiniBand subnets.

The IB subnet is the smallest complete unit of an InfiniBand network and is composed of switches, end-nodes, links, and subnet managers. To understand an IB subnet, one can apply the notion of an Ethernet VLAN. InfiniBand networks are made up of multiple subnets connected by routers to form a larger network. Each subnet is the smallest complete unit in the InfiniBand architecture and includes end-nodes, switches, links, and a subnet manager (SM) for managing the InfiniBand subnet. The SM can run on hosts, switches, or be deployed alongside the Unified Fabric Manager (UFM) for comprehensive management.

8.1.1.1 IB Architecture

The InfiniBand architecture has much similarity with the OSI layered concept and is divided into multiple layers where each layer operates independently. Specifically, the architecture is broken up into: physical, link, network, transport, and additional upper layers. The following diagram depicts IB layered concept and communication stack at HCA. IB Nodes implements here in servers implements physical layers to upper layers and the switch implements the physical layer and packet relay block of the network layer (Fig. 8.3).

At the base of IB layered concept is the physical layer, which manages the physical transmission of data across the InfiniBand network. This includes specifications for the physical media, connectors, and electrical signaling. The physical media can be copper or fiber optics, while the signaling involves encoding schemes and signaling rates such as SDR (Single Data Rate, 8Gbps), DDR (Double Data Rate, 10Gbps/16Gbps), QDR (Quad Data Rate, 40Gbps/32Gbps), FDR (Fourteen Data Rate, 56Gbps), EDR (Enhanced Data Rate, 100Gbps), HDR (High Dynamic Range, 200Gbps), NDR (Next Data Rate, 400Gbps) and XDR (eXtreme Data Rate, 800Gbps). Connectors and pin configurations are also detailed in this layer.

Above the physical layer is the link layer, which manages point-to-point communication between InfiniBand devices. It handles data framing, error detection and correction, and flow control. This layer ensures data is structured into frames for transmission, detects and corrects errors, and manages data flow to prevent congestion and ensure smooth data transfer.

The network layer is responsible for routing data between different nodes in the InfiniBand fabric. It supports both unicast and multicast communication, assigns unique

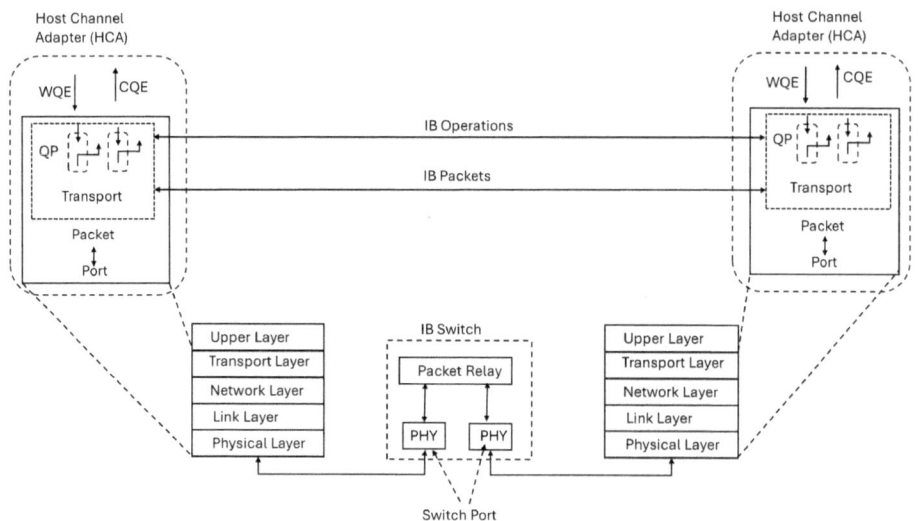

Fig. 8.3 Infiniband (IB) layered architecture with communication stack in HCA

addresses to each InfiniBand device, determines the optimal path for data transmission, and forwards data packets based on routing decisions.

The transport layer provides both reliable and unreliable data transfer services, handling packet sequencing, error detection, and retransmission in case of errors. It includes reliable connection (RC) for reliable, in-order delivery of data with error detection and retransmission, unreliable connection (UC) for connection-oriented communication without guarantees for reliable delivery, reliable datagram (RD) for reliable delivery of datagrams without requiring a connection, and unreliable datagram (UD) for connectionless communication with no guarantees for delivery.

Finally, the upper layers include protocols and services built on top of the transport layer, providing high-level services and APIs for applications to interact with the InfiniBand network. This includes the Verbs API, which offers a standardized interface for applications, MPI (Message Passing Interface) for parallel computing in HPC environments, and SMP (Subnet Management Protocol) for managing and configuring the InfiniBand subnet, including topology discovery and management.

The Host Channel Adapter (HCA) in an InfiniBand (IB) node is responsible for implementing functions that span from the physical layer up to the transport layer, as shown in the figure above. Each communication channel between devices is assigned a Work Queue Pair (WQP), which includes a send queue and a receive queue at both ends. During Send Queue operations, the HCA processes the Work Queue Element (WQE) by generating a request message, segmenting the message into multiple packets if needed, appending the necessary routing headers, and transmitting the packets through the appropriate port. The port logic sends the packet over the link to switches and routers that form the IB network fabric, relaying it until it reaches its destination. Upon receiving the packet, the port logic at the destination checks its integrity. The channel adapter then associates the packet with a specific Queue Pair (QP) and processes it according to the context of that QP. If required, the channel adapter creates a response or acknowledgment message and transmits it back to the sender.

8.1.1.2   IB Virtual Lanes (VL)

In InfiniBand, a Virtual Lane (VL) is a mechanism used to manage data traffic within the network by creating multiple independent data paths within a single physical link. This allows the network to handle various types of traffic with different service requirements simultaneously. The use of VLs enhances network performance by isolating different traffic types, managing congestion, and ensuring quality of service (QoS). The following diagram illustrates how the port maintains separate flow control over each data VL as such that excessive traffic on one VL does not block the other (Fig. 8.4).

Virtual Lanes enable traffic isolation by allowing different traffic types to coexist on the same physical link without interference. This is particularly useful for implementing QoS policies, where different priorities can be assigned to different lanes. For example, high-priority traffic can be assigned to a specific VL, ensuring it receives the necessary bandwidth and reduced latency, even in congested conditions. This ability to prioritize traffic is crucial in maintaining the performance of critical applications. InfiniBand supports up to 15 VLs (VL0 to VL14) for data traffic, with an additional VL (VL15) reserved for management traffic, providing a total of 16 VLs per link. This flexibility allows for efficient congestion management, as traffic can be distributed across multiple VLs. If one VL

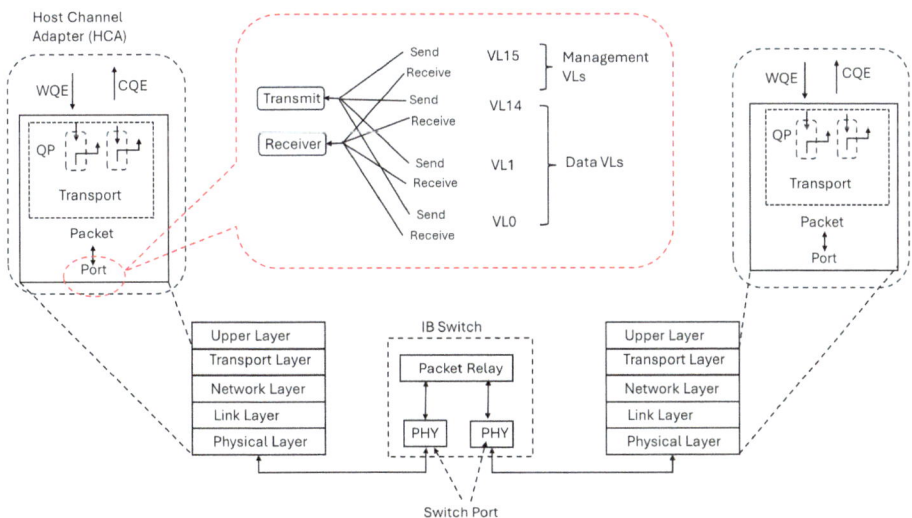

Fig. 8.4 Virtual lanes at port level in an HCA

becomes congested, others can still carry traffic, preventing bottlenecks and maintaining overall network performance.

Switches and routers in an InfiniBand network are designed to handle multiple VLs, making routing decisions based on the VL to ensure traffic is directed through the appropriate path with specified QoS parameters. Each VL has its own set of flow control credits, which helps manage data flow and prevents packet loss due to buffer overflow. Flow control ensures that the sender does not overwhelm the receiver by sending more data than it can process. In a practical HPC environment, different applications may have varying network performance requirements. For instance, a real-time simulation may require low-latency communication, while a bulk data transfer job may need high bandwidth but can tolerate higher latency. By assigning these two types of traffic to different VLs, Infini-Band ensures that the real-time simulation traffic is prioritized and experiences a minimal delay, while the bulk data transfer can use the remaining bandwidth without impacting the simulation's performance.

In summary, Virtual Lanes in InfiniBand are essential for optimizing network traffic management. They provide traffic isolation, enable QoS policies, and help manage congestion effectively. By leveraging multiple VLs, InfiniBand ensures efficient and reliable data transfer, which is crucial for high-performance computing and other data-intensive applications.

8.1.1.3 InfiniBand Verbs

In the context of InfiniBand, "verbs" refer to the set of operations or API calls that applications use to interact with the InfiniBand hardware, such as Host Channel Adapters (HCAs). These verbs are part of the programming interface provided by the InfiniBand software stack, enabling applications to perform network communication and remote memory operations efficiently. Verbs allow fine-grained control over the communication process, including data transfer, synchronization, and resource management.

InfiniBand verbs cover various aspects of network communication. Initialization verbs are used to configure the InfiniBand hardware and software environment, including creating protection domains, allocating memory regions, and setting up queue pairs (QPs). Queue Pair Management verbs handle the creation, modification, and destruction of QPs, which are essential for sending and receiving messages. Send/Receive verbs manage the posting of work requests to the send and receive queues, with operations like ibv_post_send and ibv_post_recv enqueuing send and receive tasks, respectively.

Completion Queue Management verbs deal with completion queues, which notify applications when work requests are completed. These verbs include creating, modifying, destroying completion queues, and polling for completions. Memory Management verbs involve memory registration and deregistration, allowing applications to register memory regions that the InfiniBand hardware can access for RDMA (Remote Direct Memory Access) operations. RDMA verbs execute RDMA read, write, and atomic operations, enabling direct memory access to remote nodes, with examples like ibv_post_rdma_write and ibv_post_rdma_read. We will discuss RDMA in detail in the subsequent section.

A typical workflow using InfiniBand verbs starts with initialization, such as creating a protection domain with ibv_alloc_pd and registering a memory region using ibv_reg_mr. Next, a queue pair is set up with ibv_create_qp and a completion queue is created with ibv_create_cq. Send and receive operations are then posted using ibv_post_send and ibv_post_recv. The completion queue is polled for finished operations using ibv_poll_cq. Finally, resources are torn down by deregistering the memory region with ibv_dereg_mr, destroying the queue pair with ibv_destroy_qp, and deallocating the protection domain with ibv_dealloc_pd.

InfiniBand verbs are fundamental for achieving high-performance networking in applications. They provide the necessary control for fine-tuning communication patterns and optimizing data transfer, which is critical for high throughput and low latency in high-performance computing (HPC) and data center environments.

8.1.1.4 Benefits of Infiniband in HPC

The IB network offers several key benefits that make it an attractive choice for supercomputing and other performance-sensitive applications:

- **High Throughput**: InfiniBand supports extremely high data transfer rates 10Gbps to 800Gbps, making it ideal for applications that require rapid movement of large volumes of data.
- **Low Latency**: InfiniBand provides minimal latency, ensuring fast communication between nodes. This is crucial for performance-sensitive applications where delays can significantly impact overall performance.
- **Scalability**: InfiniBand can efficiently scale across large clusters, maintaining high performance as the system grows. A single subnet up to 48000 nodes. This scalability is essential for expanding HPC environments.
- **Reliability**: InfiniBand includes enhanced error detection and correction mechanisms, ensuring data integrity and reliable communication even in demanding conditions.
- **RDMA Support**: InfiniBand natively supports Remote Direct Memory Access (RDMA), which allows direct memory access from one computer to another without involving the operating system. This leads to lower latency and reduced CPU overhead, further enhancing performance.

Due to these attributes, InfiniBand is widely used in supercomputing environments where the highest levels of performance and scalability are required. It is particularly well-suited for applications such as scientific simulations, complex data analysis, and large-scale machine learning tasks.

8.1.1.5   RDMA

Remote Direct Memory Access (RDMA) is an extension of Direct Memory Access (DMA) technology that allows computers to exchange data in memory without involving the CPU, OS, or cache. This technology enables high-throughput, low-latency networking, making it highly suitable for high-performance computing (HPC) and data center environments. It was first introduced by the Virtual Interface Architecture (VIA) Consortium, which included industry giants like Compaq, Intel, and Microsoft in the late 1990s. The initial goal was to develop a standard interface for high-speed communication in cluster computing environments. The introduction of RDMA marked a significant advancement in network technology by enabling direct memory-to-memory data transfer between computers, bypassing the traditional TCP/IP stack. The following diagram shows how RDMA technology facilitates direct memory-to-memory communication between the source host ant target host over the DCN (Data Center Network) bypassing kernel at respective hosts (Fig. 8.5).

RDMA is extensively used in HPC for several reasons. It provides low latency by eliminating the need for CPU involvement in data transfer, which is crucial in HPC applications where even microsecond-level delays can impact performance. It offers high throughput, essential for the massive data exchanges in HPC workloads. By offloading data transfer tasks from the CPU, RDMA frees up computational resources for application processing, which is particularly beneficial in HPC, where computational resources

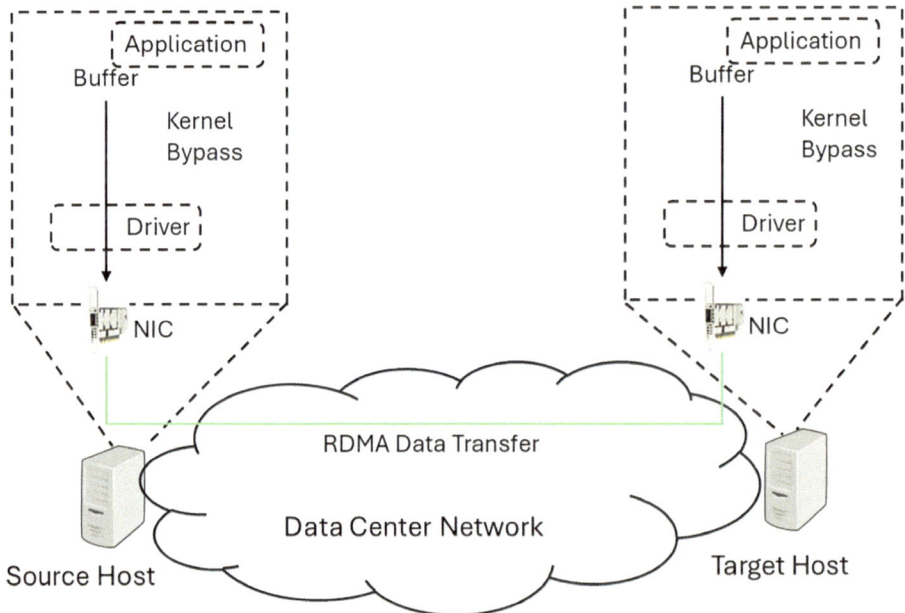

Fig. 8.5   RDMA communications between memory at source host and memory at target host

are at a premium. Additionally, RDMA reduces the need for data copying between user space and kernel space, leading to more efficient use of memory and processing power. Its efficient communication protocols also enable better scalability in large HPC clusters, allowing more nodes to be added without a linear increase in communication overhead.

As discussed earlier, InfiniBand (IB) utilizes RDMA for host-to-host data transfer in an IB network. It is the first major network technology that incorporated RDMA (Remote Direct Memory Access) as a fundamental feature, designed from the outset to leverage RDMA for efficient data transfer. The success of RDMA in InfiniBand also influenced the development of other network technologies and protocols, such as RoCE (RDMA over Converged Ethernet) and iWARP (Internet Wide Area RDMA Protocol).

8.1.2   Ethernet in HPC

Ethernet was first introduced in the context of High-Performance Computing (HPC) in the early 1990s. During this time, Ethernet began to gain popularity as a networking technology due to its cost-effectiveness and widespread adoption in commercial and academic settings. Ethernet's first significant use in High-Performance Computing (HPC) can be traced back to the early 1990s with the development of the Beowulf cluster at NASA. Donald Becker and Tom Sterling created this cluster using off-the-shelf PCs connected with 10 Mb/s Ethernet. This innovation demonstrated that cost-effective, commodity hardware and standard networking technology could achieve significant computational power, marking a pivotal moment in the use of Ethernet for HPC applications.[1] Initially, Ethernet was not considered ideal for HPC due to its higher latency and lower bandwidth compared to specialized HPC interconnects like InfiniBand and proprietary solutions such as Cray's interconnects. However, the introduction of Gigabit Ethernet in the late 1990s and 10 Gigabit Ethernet in the early 2000s significantly closed these gaps, making Ethernet a preferred choice in HPC for leveraging commodity hardware. Notably, vendors like Arista now offer switches with relatively low latency. Furthermore, Ethernet's advancement in HPC is significantly bolstered by Cray's Slingshot interconnect technology, contributing to Ethernet's success in this sector. Cray, a leader in supercomputing since 1976, has developed several proprietary interconnect technologies for its systems. One of the most notable is the Aries interconnect, used in the Cray XC series, designed for high bandwidth, low latency, and efficient scalability. Similarly, the Gemini interconnect in Cray XE and XK systems focused on high performance and efficient communication for large-scale HPC applications, featuring advanced routing and network management. In 2019, Cray introduced the Slingshot interconnect with the Shasta series supercomputers. Slingshot represents a significant evolution over Aries, featuring an Ethernet-based design that enhances interoperability with existing data center networks. It offers up to 200 Gb/s bandwidth per port and includes advanced features such as congestion control, adaptive

---

[1] Gordon [1].

routing, and enhanced QoS. Slingshot's Ethernet compatibility enables seamless integration with a broad range of networking hardware and software, making it versatile and future-proof.

8.1.2.1 Slingshot

Cray recognized the limitations and potential of Ethernet in HPC environments. To address this, Cray significantly modified the Ethernet protocol for HPC and AI applications. This included integrating congestion control techniques, which were enhanced by its 2013 acquisition of Gnodal, and improving upon its pioneering adaptive routing technology first introduced with the Cray T3D supercomputer in 1996. These innovations were incorporated into the Rosetta ASIC. Furthermore, Cray has a history with high-radix switches, evidenced by the "BlackWidow" YARC router in the 2007 X2 supercomputer, which featured a 64-port, folded Clos network topology. Additionally, Cray developed the dragonfly topology for its Aries XC systems, aiming to reduce the use of costly optical links in clusters while maintaining uniform scalability.[2]

Cray (now part of HPE) has utilized the technological advancements of the Rosetta ASIC to develop the Slingshot interconnect. Slingshot marks a significant evolution in networking, as Cray enhances standard Ethernet into what it terms "HPC Ethernet." This development not only aligns Ethernet more closely with traditional high-performance computing interconnects but also introduces substantial improvements in adaptive routing and congestion control. These enhancements surpass existing technologies like Aries, InfiniBand, and Omni-Path. Slingshot incorporates Cray's best innovations, including high-radix switching, dragonfly topology, and advanced congestion management, to deliver a robust and efficient networking solution. The following diagram shows Cray's slingshot ethernet switch with dragonfly topology for HPC interconnects (Fig. 8.6).

The Slingshot interconnect, developed with Cray's Rosetta ASIC, features 64 ports and a 12.8 Tbps backplane. It uses SerDes with 25 Gb/sec signaling that, through PAM4 encoding, doubles to 50 Gb/sec per lane. This setup allows each SerDes to support four lanes, creating a 200 Gb/sec port. The true signal rate is 28 Gb/sec, allowing for 56 Gb/sec per lane after accounting for encoding. Slingshot also demonstrates superior performance in managing small packets, handling over 1.2 billion packets per second per port. Within a dragonfly topology, slingshot solution can connect over 250,000 endpoints with a maximum of three hops between any two endpoints [2].

8.1.2.2 Dragonfly Topology

Given our discussion on Slingshot, it's crucial to understand the dragonfly topology, a sophisticated network architecture designed for high-performance computing (HPC). This topology is favored in HPC for its effective bandwidth management, reduced latency, and cost efficiency. By employing high-radix routers with numerous ports, dragonfly topology allows for a condensed network configuration that reduces the number of hops between

---

[2] Morgan [2].

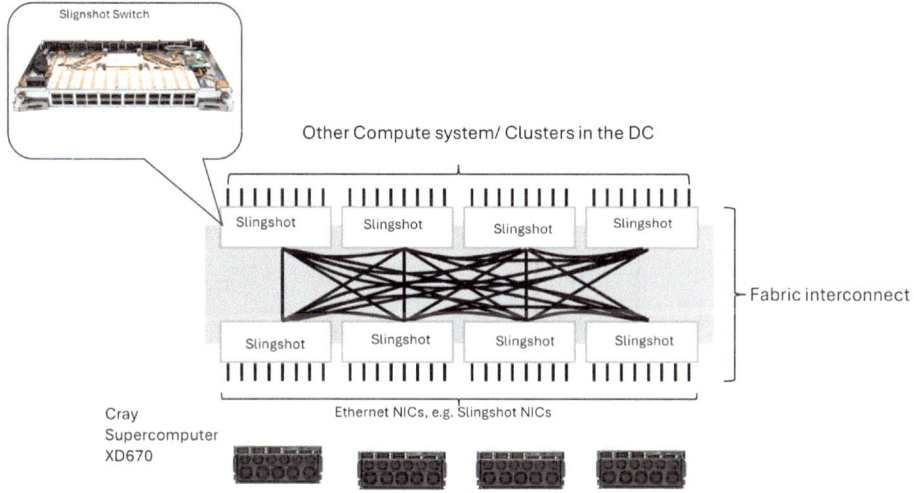

Fig. 8.6 Cray (now part of HPE) slingshot switch with dragonfly topology

nodes, significantly decreasing communication delays and enhancing overall system performance. The diagram provided illustrates the dragonfly topology featuring groups of virtual routers and their connections. Each router is interconnected with others within and across groups, and terminals are connected directly to endpoints, facilitating efficient communication pathways throughout the network (Fig. 8.7).

Dragonfly topology utilizes virtual router groups (refer to G1–G8 in the diagram), consisting of multiple high-radix routers that act collectively as a single, higher-radix virtual router (refer to R1–R3 in the diagram). This design increases the network's radix and allows the router to connect directly to many other routers or endpoints, promoting efficient and scalable networking configurations. This setup reduces the need for lengthy global channels, which are costly and typically increase latency. Instead, by minimizing these global connections, dragonfly topology achieves a more efficient and cost-effective network structure. The network is hierarchical, featuring three levels: routers, groups, and the entire system. Routers are linked to local terminals and other routers within the same group through local channels, while groups of routers are interconnected via global channels, forming the overarching system.

The dragonfly topology offers several advantages including a reduced network diameter, ensuring that packets traverse at most one global hop to reach their destination, thereby lowering latency and increasing network speed. It employs adaptive routing algorithms like minimal routing and Valiant's load-balancing algorithm, as well as the Universal Globally Adaptive Load-balanced (UGAL) algorithm, which optimizes performance by dynamically selecting the best routing path based on current network load. Additionally, the topology's scalability is particularly beneficial for HPC environments that handle large

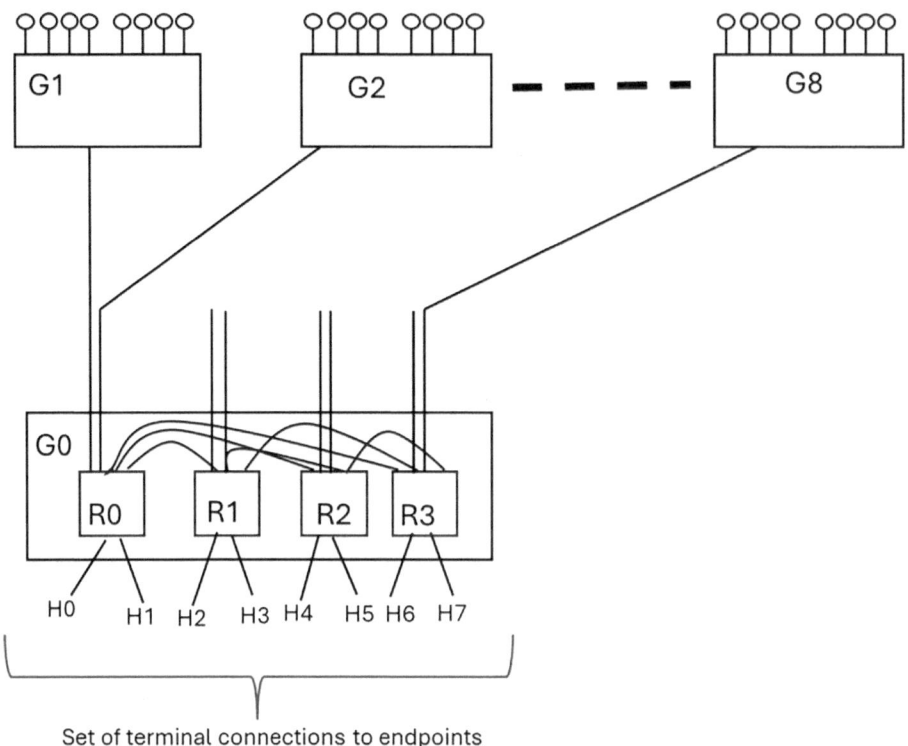

Set of terminal connections to endpoints

Fig. 8.7   Dragonfly topology with virtual router groups and router connections[3]

datasets and complex computations. Its ability to maintain a low network diameter while supporting a large number of nodes makes it a robust solution for HPC systems. The use of optical signaling technology for long-distance connections further enhances cost efficiency, particularly in configurations with many nodes, providing significant savings over other high-radix topologies like the flattened butterfly and folded Clos networks.

8.1.2.3   Adaptive Routing

Adaptive routing dynamically adjusts to network topology and traffic changes. It improves network performance by detecting link congestion early and preferentially selecting short and non-congested paths, enhancing throughput, resilience, and reducing latency. This approach is crucial in large supercomputing centers that use direct topology, allowing for expansive network access with minimal diameter. Under normal conditions, adaptive routing selects the shortest path; when congestion occurs, it shifts to less congested alternative routes. This strategy optimizes bandwidth use and supports high throughput and

---

[3] Kim et al. [3].

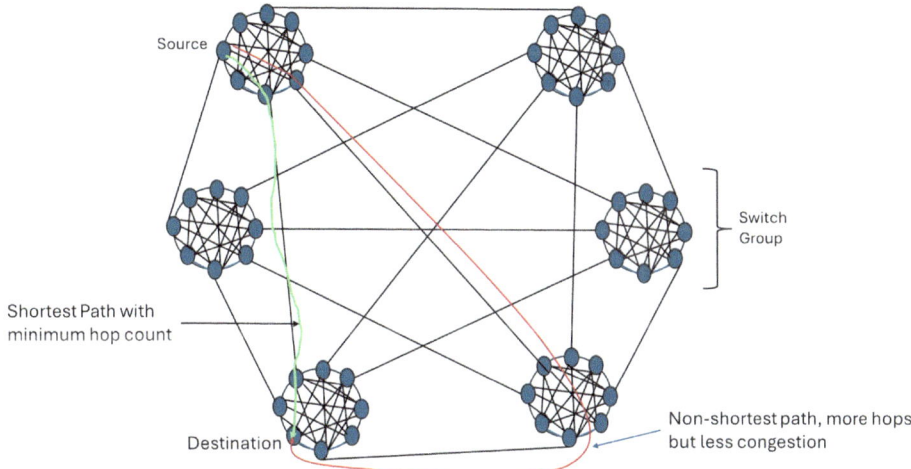

Fig. 8.8 Adaptive routing using shortest and non-shortest path for facilitating communication between source and destination hosts

low latency requirements efficiently. The included diagram illustrates how adaptive routing employs both shortest-path and non-shortest-path routing. Under moderate loads, the shortest path is used, while congestion triggers a switch to alternative paths, redistributing traffic to enhance performance and avoid bottlenecks. This dynamic process is vital for maintaining efficiency and reliability in complex networks (Fig. 8.8).

In adaptive routing, as illustrated in the diagram, each node has multiple potential forwarding paths. The source node uses an adaptive routing algorithm to select the most efficient path for packet entry into the network. Subsequent nodes forward packets based on pre-determined routing tables, minimizing the need for ongoing path reselection. These tables are categorized into three types: one for shortest paths, one for non-shortest paths, and a mixed table. The source node determines the optimal path and the local outbound interface for the packets, while non-source nodes use this preset path for routing. This system ensures dynamic, efficient routing tailored to current network conditions, thereby enhancing performance and reliability.

8.1.2.4   Valiant Load Balancing (VLB) Routing

Valiant Load Balancing (VLB) is a routing algorithm that is aimed at creating fault-tolerant and congestion-free networks with minimal over-provisioning. Unlike traditional routing protocols like OSPF, IS-IS, and MPLS, which struggle to ensure congestion-free operations during failures, VLB uses path diversity to maintain performance even with multiple failures. It was first proposed by L. G. Valiant for processor interconnection networks,[4] and has received recent interest for scalable routers with performance

---

[4] Valiant [4].

guarantees.[5,6] VLB operates on the principle of path diversity, where multiple paths are utilized between any two nodes, distributing traffic across all available paths. This approach significantly reduces the risk of congestion and enhances fault tolerance. One of the key advantages of VLB is its requirement for minimal over-provisioning. It can achieve fault tolerance with a significantly lower over-provisioning ratio compared to traditional networks. For instance, a 50-node network can continue to operate congestion-free with up to 5 link or router failures with just 10% over-provisioning. Another major benefit is instantaneous rerouting; since all paths are used continuously, VLB can reroute traffic immediately upon detecting a failure, eliminating the need for complex and time-consuming rerouting computations. The following is a diagram illustrating VLB operations in a fully meshed network (Fig. 8.9).

The mechanism of VLB in a typical network involves a full mesh topology where each node is connected with logical links of capacity 2r/N, where r is the capacity of each access node and N is the total number of nodes. Traffic is load-balanced across all nodes in two hops: in the first hop, incoming traffic is uniformly distributed across all

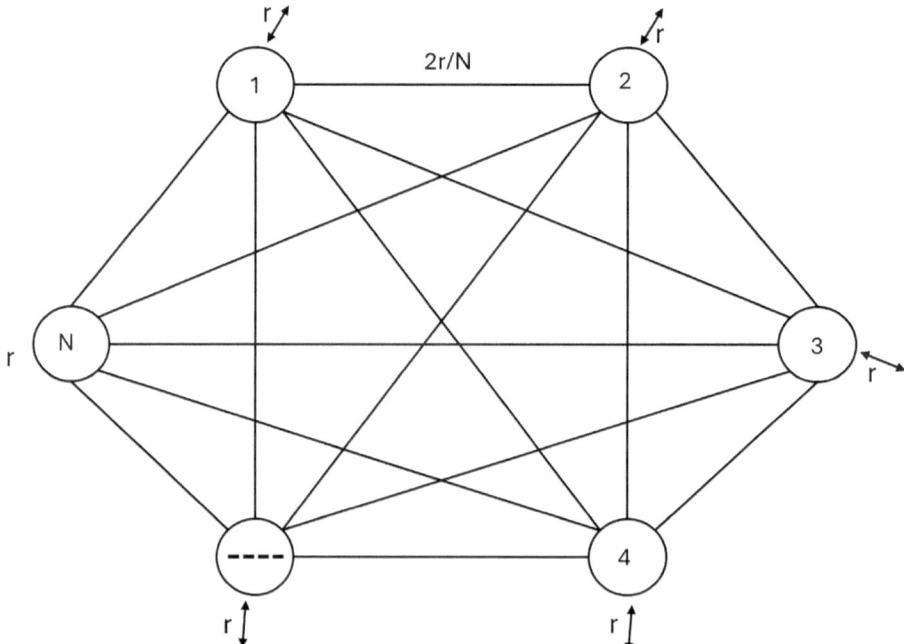

Fig. 8.9 Valiant Load Balancing (VLB) in a network of N identical nodes with capacity "r"[7]

---

[5] Chang et al. [5].
[6] Keslassy et al. [6].
[7] Zhang-Shen and McKeown [7].

nodes, and in the second hop, traffic from each node is directed to its final destination. This ensures that no single link is overwhelmed, and the network can support any traffic matrix that does not oversubscribe a node.

VLB's fault tolerance is built on its ability to maintain multiple working paths between nodes. To tolerate k failures, each link must have a capacity of $2r/(N - k)$ with the additional capacity required being relatively small, and the over-provisioning ratio being approximately k/N. When a node fails, it stops all traffic, and the network effectively becomes an $(N - K)$ node full mesh, with the required link capacity adjusting to $2r/(N - k)$. In the worst-case scenario of adversarial link failures, the capacity requirement is slightly higher, while for random link failures, the additional capacity needed is even smaller. VLB can also tolerate a combination of node and link failures by appropriately adjusting the link capacities.

In practical application, VLB allows network operators to design their networks with minimal extra capacity, ensuring efficient use of resources. For example, in a 50-node network with nodes having 1 Tb/s access capacity, each link should be designed with a capacity of 50 Gb/s to tolerate up to five arbitrary failures while maintaining acceptable utilization levels. To understand this, let's discuss the VLB formula for determining the required link capacity in a VLB network to tolerate k failure in an N-node network:

$$C(N, k) = \frac{2r}{(N - k)}$$

Where

- $C(N, k)$ is the required capacity of each link to tolerate k failures.
- r is the capacity of each access node.
- N is the number of nodes in the network.
- k is the number of failed nodes.

Given:

- $N = 50$(total number of nodes).
- r = 1 Tb/s (access capacity per node).
- k = 5 (number of failures to be tolerated).

we can write the calculation as

$$C(50, 5) = 2x1Tbs/(50 - 5)$$

or

$$C(50,5) = 2Tbs/45$$

By converting Tb/s to Gb/s, we can write

$$C(50,5) = \frac{2000Gbs}{45} = 44.44Gb/s$$

The calculation shows that each link should have a capacity of approximately 44.44 Gb/s to tolerate up to five arbitrary node or link failures in a 50-node network. Designing each link with a capacity of 50 Gb/s provides a margin of safety and accounts for any additional overhead or unexpected traffic variations, ensuring even better fault tolerance and network performance.

As indicated in this discussion, Valiant Load Balancing (VLB) presents a highly efficient and fault-tolerant network design strategy that addresses the limitations of traditional routing protocols. By leveraging path diversity and minimal over-provisioning, VLB ensures continuous, congestion-free operation even in the face of multiple failures. This makes it a compelling choice for modern backbone networks that require high reliability and efficiency.

8.1.2.5   UGAL Routing Scheme in Dragonfly Networks

Universal Globally Adaptive Load-balanced routing (UGAL) is a sophisticated routing strategy used in modern HPC networks, such as the Dragonfly topology. UGAL dynamically selects between minimal (MIN) and non-minimal (Valiant Load-Balanced, VLB) paths based on current network conditions, aiming to optimize overall network performance by balancing load and minimizing congestion.

Minimal Routing (MIN) selects the shortest possible path from source to destination, typically involving at most one global link between groups in the Dragonfly topology. MIN routing is efficient for uniform traffic but can suffer from congestion under adversarial traffic patterns where multiple nodes in one group communicate heavily with nodes in another group. Valiant Load-Balanced Routing (VLB), on the other hand, mitigates congestion by routing packets through a randomly selected intermediate node that is neither in the source nor the destination group. This method spreads the traffic load more evenly across the network but increases the path length and thus the overall latency.

UGAL has two main variants. UGAL-G (Global Information) is a theoretical variant that assumes global network state information is available, allowing for optimal path selection. UGAL-G has been shown to achieve high performance across different network topologies but is impractical for implementation due to the overhead of maintaining global state information. UGAL-L (Local Information) is a practical variant that uses local queue occupancy and hop count information to approximate path latency, making routing decisions based on this localized view. UGAL-L is more feasible for real-world networks but can make suboptimal routing decisions under certain conditions due to inaccurate latency approximations.

Several enhancements to UGAL-L have been proposed to improve its performance. UGAL-LE,[8] UGAL with enhanced local information, introduces a contention factor to better estimate the latency of minimal paths under imbalanced traffic conditions. This improvement helps in making more accurate routing decisions and reducing network congestion. Enhanced DGB (EDGB) [8] combines UGAL-LE with the Decoupled Gradient Descent-based Bias (DGB) routing algorithm, dynamically adjusting the bias value based on both local and global information. This hybrid approach further refines latency estimation and improves overall routing performance.

UGAL-LE enhances the basic latency estimation by introducing a contention term that reflects the effect of link congestion. This term is derived using information local to each router, allowing for a more accurate approximation of the minimal path latency. The algorithm makes routing decisions based on these parameters, choosing between minimal and non-minimal paths by comparing their estimated latencies, adjusted for the contention factor. If the source router has a direct global link to the destination group, it uses the actual queue occupancy for latency estimation. Otherwise, it approximates the global link queue occupancy using the contention factor.

Simulation experiments have shown that UGAL-LE and EDGB significantly outperform traditional UGAL-L, especially under traffic conditions with heavy inter-group communication. For instance, UGAL-LE improves communication times for various HPC applications compared to UGAL-L, demonstrating its robustness and effectiveness in maintaining network performance under diverse traffic patterns.

In summary, UGAL and its enhanced variants provide a dynamic and adaptive routing framework that efficiently manages load and reduces congestion in high-performance networks. These schemes are crucial for maintaining optimal performance in modern supercomputing environments.

8.1.2.6 RoCE

This section could ideally be placed under the RDMA section since RoCE is an RDMA technology developed by the InfiniBand Trade Association (IBTA) to extend RDMA support to Ethernet networks. However, given that RoCE (RDMA over Converged Ethernet) has significantly facilitated Ethernet's penetration into high-performance computing (HPC), it makes sense to include it as a section under Ethernet in HPC. RoCE is a network protocol that allows RDMA (Remote Direct Memory Access) to function over Ethernet networks. It aims to combine the efficiency of RDMA with the ubiquity of Ethernet, facilitating high-throughput, low-latency data transfer without the need for specialized InfiniBand hardware. The InfiniBand Trade Association (IBTA) first specified RoCE in 2010, seeking to extend the benefits of RDMA to Ethernet environments commonly found in data centers, and subsequently released version 2 of the protocol in 2014. The following diagram depicts RDMA network architecture for RoCE v1 and RoCE v2 protocol stack (Fig. 8.10).

---

[8] Chaulagain [8].

Fig. 8.10 RDMA network
architecture for RoCE v1 and
RoCE v2

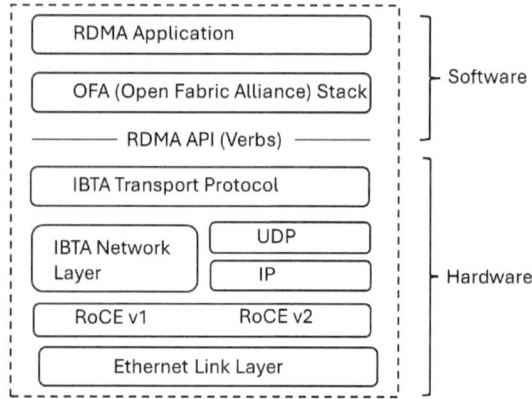

RoCE has undergone several iterations, each enhancing its capabilities and expanding its applicability. The initial version, RoCEv1, operates directly over Ethernet, similar to how InfiniBand operates within its domain. This version requires the network to be lossless, typically achieved through Data Center Bridging (DCB) enhancements to Ethernet, which provide priority flow control and other mechanisms to ensure no packet loss. RoCEv2, introduced later, encapsulates RDMA traffic within UDP/IP packets, allowing it to traverse standard Ethernet and IP networks without requiring a lossless fabric. This version brings the flexibility of RDMA to wider Ethernet deployments, overcoming some of the limitations of RoCEv1 by enabling routing across subnets and broader network configurations (Fig. 8.11).

The packet structure in RoCE as depicted in the figure above, is designed to support efficient and reliable data transfer. In RoCEv1, packets are encapsulated directly

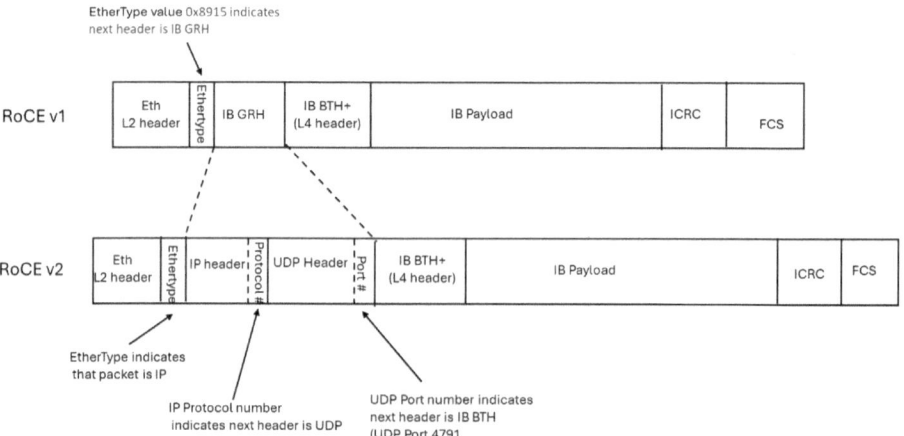

Fig. 8.11 Packet structure of RoCE v1 and RoCE v2

over Ethernet with an ethertype field indicating RDMA. The packet consists of an Ethernet header with ethertype field, RDMA header that includes InfiniBand Global Routing Header (IB GRH) and InfiniBand Base Transport Header (IB BTH) and the IB payload. The RDMA header (IB GRH and IB BTH) contains information crucial for RDMA operations, such as queue pair numbers, memory addresses, and operation codes. In RoCEv2, the RDMA packets are encapsulated within UDP/IP packets. The structure includes an Ethernet header, an IP header, a UDP header, and then the RDMA-specific header and payload. This encapsulation allows RoCEv2 to leverage the IP layer for routing and interoperability with existing IP-based network infrastructures. The IP and UDP headers are standard, enabling compatibility with existing network devices and protocols.

The architecture of RoCE incorporates both hardware and software elements to facilitate RDMA over Ethernet (Please refer Figs. 8.8, 8.9 and 8.10). On the hardware side, network interface cards (NICs) that support RoCE provide RDMA capabilities, including zero-copy data transfer, kernel bypass, and low-latency communication. These NICs handle RDMA operations offloaded from the CPU, enabling high performance. On the software side, RoCE implementations typically include drivers and libraries that enable applications to utilize RDMA over Ethernet, interfacing with the NICs to perform memory operations directly.

RoCE's architecture also relies on enhanced Ethernet features to ensure efficient operation. In RoCEv1, these enhancements include priority flow control and enhanced transmission selection, which help create a lossless Ethernet fabric. In RoCEv2, while these enhancements are beneficial, the protocol's ability to operate over standard IP networks provides greater flexibility and deployment options.

In summary, RoCE extends the benefits of RDMA to Ethernet environments, offering high-performance data transfer without specialized hardware. Specified by the InfiniBand Trade Association in 2010, RoCE has evolved through different versions to improve its capabilities and expand its applicability. The protocol's packet structure and architecture are designed to support efficient, low-latency communication, making it a valuable technology for modern data centers and high-performance computing environments.

8.1.2.7 iWARP

The success of RDMA over InfiniBand (IB) significantly influenced the Internet Engineering Task Force (IETF) to specify iWARP (Internet Wide Area RDMA Protocol). RDMA over InfiniBand demonstrated substantial benefits, such as low latency, high throughput, and reduced CPU overhead, which were highly desirable in broader networking contexts, including Ethernet. These performance improvements highlighted RDMA's potential to revolutionize network performance by providing direct memory access between hosts, bypassing the traditional TCP/IP stack, and thereby reducing latency and CPU utilization. These advantages made it clear that extending RDMA capabilities to more ubiquitous and widely deployed networking infrastructures, like Ethernet, could bring significant benefits to many industries.

To this end, the IETF developed iWARP to enable RDMA over standard TCP/IP networks. iWARP leverages TCP/IP protocols to provide RDMA capabilities, enabling remote direct memory access over a wide-area network (WAN) while maintaining compatibility with existing network infrastructure. This made RDMA technology more accessible and easier to deploy in diverse networking environments. By specifying iWARP in 2007, the IETF created an industry-standard protocol to ensure interoperability between different vendors and devices, which was crucial for the widespread adoption and integration of RDMA capabilities in various networked systems.

However, iWARP did not achieve the same level of industry adoption as RoCE (RDMA over Converged Ethernet). This was due to iWARP's higher latency and overhead, increased complexity, and misalignment with the performance requirements of modern data center environments. In contrast, RoCE's direct Ethernet-based approach, stronger vendor support, and more robust ecosystem made it a more attractive option for RDMA in these settings.

The following diagram depicts the iWARP protocol stack. iWARP introduces RDMAP (Remote Direct Memory Access Protocol), DDP (Direct Data Placement Protocol), and MPA (Marker PDU Aligned Framing for TCP) on top of TCP/IP to provide connectivity for HPC hosts. RDMAP supports operations to transfer Upper Layer Protocol (ULP) data between a local peer and a remote peer (Fig. 8.12).

The DDP (Direct Data Placement) allows data to be placed directly into the appropriate memory buffers at the target without intermediate data copying, while MPA (Marker

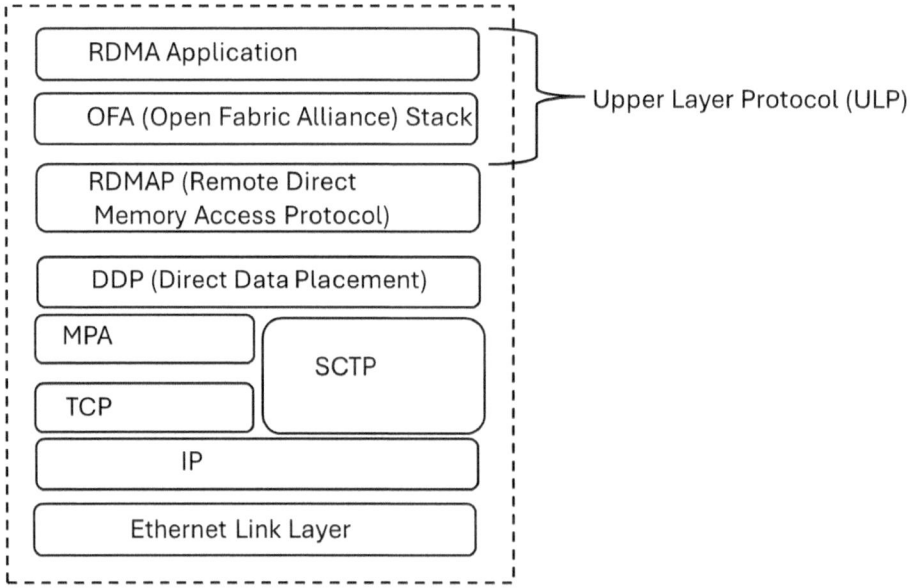

Fig. 8.12 iWARP Protocol stack as defined in RFC 5040

PDU Aligned Framing for TCP) ensures data alignment and framing on top of a transport protocol like TCP. The SCTP (Stream Control Transmission Protocol) in iWARP (Internet Wide Area RDMA Protocol) provides reliable data transport essential for RDMA operations over IP networks. It enhances iWARP by offering features like multi-streaming, which allows concurrent data flows without blocking, and multi-homing, which uses multiple network paths for fault tolerance and load balancing. SCTP ensures reliable, ordered delivery of messages and supports unordered delivery where needed.

In the iWARP stack, RDMA operations formatted by RDMAP (RDMA Protocol) are encapsulated within DDP packets, which SCTP then transports. The SCTP packet structure includes common headers and DATA chunks that carry DDP and RDMAP headers, along with the payload data. SCTP's selective acknowledgments and path management capabilities, including heartbeats to monitor network health, enhance the robustness and efficiency of iWARP deployments. This integration allows iWARP to leverage RDMA benefits over standard IP networks, ensuring low-latency, high-throughput data transfer with enhanced reliability.

In conclusion, iWARP effectively combines the benefits of RDMA with the ubiquity of IP networks, offering a robust solution for high-performance data transfer. The integration of DDP allows direct placement of data into memory buffers, bypassing intermediate copying, while MPA ensures proper data alignment and framing over TCP. The use of SCTP enhances iWARP with reliable data transport, multi-streaming, and multi-homing capabilities, ensuring efficient, low-latency, and high-throughput data transfer. Although iWARP did not achieve the same level of industry adoption as RoCE due to its higher complexity and overhead, it remains a valuable technology for enabling RDMA over standard IP networks, providing enhanced reliability and efficiency in diverse networking environments.

8.2   AI Infrastructure Framework

Now that we have an understanding of HPC interconnects and components, let's discuss the recent phenomenon of increasing AI/ML workload penetration in HPC environments. More and more enterprises are starting to deploy various AI workloads, including generative AI, vision AI, speech AI, computer vision, predictive AI, and more. These AI workloads can be deployed either on-premises or in the cloud, each offering unique benefits and challenges.

On-premises AI infrastructure involves deploying and managing AI workloads within an organization's data center. This approach provides several advantages:

- **Control and Customization**: On-premises solutions allow organizations to have full control over their hardware and software configurations. This enables tailored optimization for specific AI workloads, ensuring maximum performance and efficiency.

- **Security and Compliance**: For industries with stringent regulatory requirements, such as healthcare or finance, on-premises solutions offer enhanced security and compliance capabilities. Sensitive data remains within the organization's control, reducing risks associated with data breaches and ensuring adherence to regulatory standards.
- **Performance and Latency**: Deploying AI workloads on-premises can significantly reduce latency, which is critical for real-time applications. The proximity of compute, storage, and networking resources ensures faster data processing and decision-making.

On the other hand, cloud-based AI infrastructure offers a flexible and scalable alternative to on-premises solutions. By leveraging the resources of cloud service providers, organizations can quickly adapt to changing workload demands without significant capital investment:

- **Scalability**: Cloud solutions provide virtually unlimited scalability. Organizations can scale their resources up or down based on workload requirements, paying only for what they use.
- **Cost-Efficiency**: Cloud infrastructure eliminates the need for upfront capital expenditure on hardware. Operational expenses are predictable, with costs tied to actual usage, making it easier to manage budgets.
- **Accessibility and Collaboration**: Cloud platforms enable global accessibility, allowing teams to collaborate from anywhere. This is particularly beneficial for organizations with distributed teams or those involved in multi-site projects.

Understanding the network interconnect requirements for AI/ML workloads and the corresponding HPC requirements would be incomplete without exploring the AI infrastructure framework as a whole. For our discussion, we will focus on on-premises deployment and present a typical AI infrastructure framework. The diagram below depicts an on-premises AI infrastructure model (Fig. 8.13).

In this model, we see various layers and components essential for AI infrastructure, including:

1. **AI Hardware Infrastructure**: This includes compute, storage, and networking components that form the foundation of the infrastructure.
2. **Virtualization**: This layer involves the use of containerization and virtual machines (VMs) to provide a flexible and efficient environment for AI workloads.
3. **Data Management and Processing Layer**: This includes frameworks and libraries for managing and processing data, as well as model training platforms.
4. **AI and ML Framework**: This layer consists of model development, training validation, model evaluation, deployment and monitoring.
5. **AI Workload**: This encompasses various AI tasks such as NLP, generative AI, computer vision, data processing, and ML/deep learning.

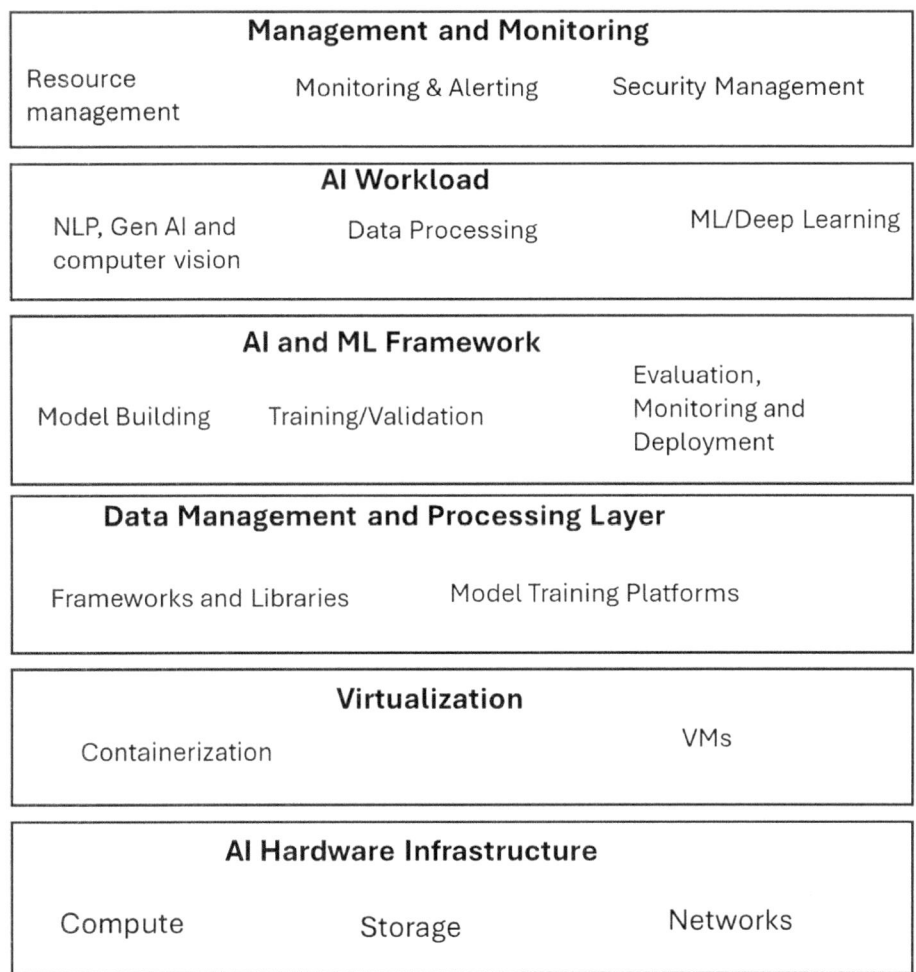

Fig. 8.13 AI infrastructure framework

6. **Management and Monitoring**: This top layer includes resource management, monitoring and alerting, and security management to ensure efficient and secure operation of the AI infrastructure.

This framework provides a comprehensive view of the elements involved in building and maintaining an on-premises AI infrastructure, highlighting the critical aspects of control, customization, security, performance, and latency. In the following sections, we will delve further into each of these layers.

8.2.1 AI Hardware Infrastructure

The AI Hardware Infrastructure forms the foundation of an on-premises AI deployment. It encompasses the critical compute, storage, and networking components required to support AI workloads. The robustness and efficiency of this layer determine the overall performance, scalability, and reliability of the AI infrastructure. Properly architecting this foundational layer is essential for ensuring that AI models can be trained and deployed effectively, meeting the high demands of modern AI and ML applications.

In an AI infrastructure, compute resources handle the intense processing tasks involved in training and inferencing AI models. Storage solutions manage vast amounts of data, ensuring quick access and efficient handling. Networking infrastructure connects these components, facilitating rapid data transfer and communication, which is vital for maintaining performance and minimizing latency. Investing in high-quality AI hardware infrastructure allows organizations to optimize their AI workflows, reduce operational costs, and achieve faster time-to-insight. This infrastructure must be scalable to accommodate growing data volumes and evolving computational needs, flexible to support diverse AI workloads, and resilient to ensure continuous operation without performance degradation.

Let's delve deeper into the technical details of each component within this layer, focusing on their roles, interconnects, and key elements.

8.2.2 Compute

The compute component of AI Hardware Infrastructure is pivotal in determining the performance and efficiency of AI workloads. In AI and ML applications, the need for massive computational power is evident, as tasks such as training deep neural networks and performing real-time inference are highly resource-intensive. This section delves into the various types of compute resources—GPUs, CPUs, TPUs, and FPGAs—each serving a unique role in the AI infrastructure.

Compute resources are designed to handle specific types of operations that are common in AI tasks. Selecting the right type of computing resource depends on the nature of the workload, performance requirements, and cost considerations. Understanding the strengths and applications of each compute resource helps in building a balanced and efficient AI infrastructure.

8.2.2.1 GPU (Graphics Processing Units)

GPUs are specialized processors designed to handle highly parallel tasks, making them ideal for the computational demands of AI and ML workloads. Their architecture consists of thousands of smaller cores that can perform many operations simultaneously, which

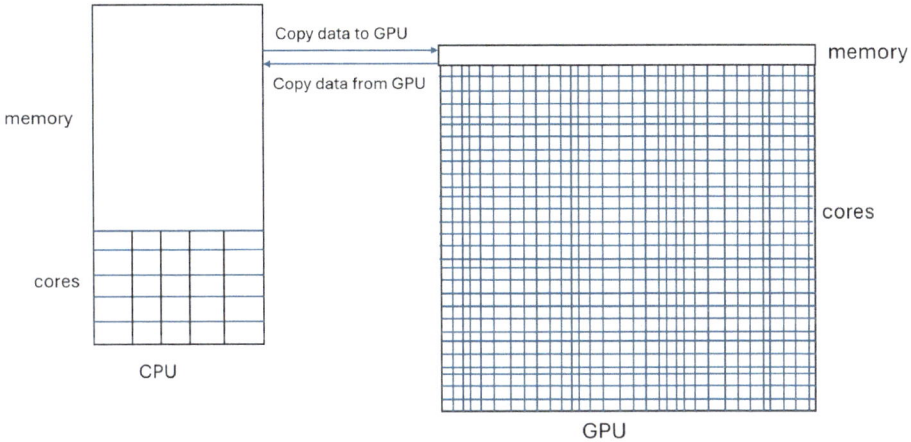

Fig. 8.14 A GPU accelerated task functions as auxiliary hardware, working in tandem with a CPU to quickly carry out numerically intensive operations[9]

is highly efficient for matrix operations and large-scale data processing tasks common in deep learning.

GPUs excel in parallel processing due to their massive number of cores. For example, while a CPU might have tens of cores, a GPU can have thousands. This allows GPUs to perform many operations simultaneously, vastly improving performance for tasks like training artificial neural networks and image processing. In scientific computing, GPUs are used alongside CPUs to handle numerically intensive operations. The CPU executes the main program, while the GPU is used to carry out specific functions, such as matrix computations in parallel, please refer to the diagram (Fig. 8.14).

As auxiliary hardware, GPUs work in tandem with CPUs to perform GPU-accelerated tasks through three main steps: copying input data from CPU memory to GPU memory, executing the GPU kernel on the GPU, and copying the results back to CPU memory. This process enables GPUs to serve as accelerators, handling specific numerically intensive tasks much faster than CPUs alone.

The integration of NVLink and NVSwitch technologies significantly enhances GPU performance by providing high-bandwidth, low-latency interconnects between GPUs. NVLink allows multiple GPUs to efficiently share data, which is essential for large-scale computations and deep learning tasks. The following diagram depicts GPU interconnect in a server using NVLink (Fig. 8.15).

NVLink has evolved through several generations, each offering higher data rates and improved features. The fifth generation of NVLink allows a single NVIDIA Blackwell Tensor Core GPU to support up to 18 connections, each at 100 gigabytes per second,

---

[9] Princeton [9].

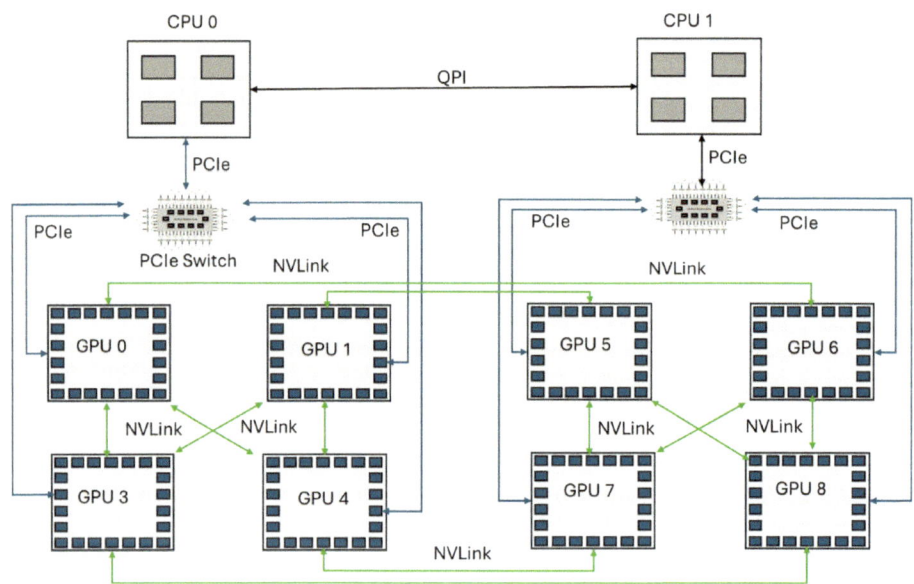

Fig. 8.15 NVLink interconnect with multiple GPUs allowing them to efficiently share data and perform large-scale computation

resulting in a total bandwidth of 1.8 terabytes per second. NVSwitch enables dense interconnects of GPUs within a server and across servers. It is a high-performance switch fabric designed to interconnect multiple GPUs using NVLink, enabling a massively scalable architecture. Each NVSwitch can connect up to 16 GPUs, each with robust 300 GB/s connectivity. This interconnect framework removes bottlenecks and intermediary steps, allowing the 16 GPUs to operate in unison (please refer to the diagram) (Fig. 8.16).

As a result of this interconnect framework and scale-out design capabilities, an impressive 2 petaFLOPS of deep learning computing power is unleashed, paving the way for training advanced AI networks. The NVSwitch stands out from traditional interconnect solutions due to its advanced hardware and software integration. It utilizes high-speed signaling, sophisticated error detection, and correction mechanisms, and intelligent routing algorithms to maximize data throughput and minimize latency. This combination ensures efficient data transfer and seamless operation of interconnected GPUs, making it a critical component for scalable AI infrastructure. Additionally, AI systems often require multiple GPUs (eight to sixteen) sharing their memory to simplify programming and make datasets accessible at memory speeds rather than network speeds. NVSwitch addresses this need by providing scalable GPU connectivity, as depicted in the diagram above.

It is noteworthy that as of writing this book, industry momentum is gathering around Ethernet to compete with Infiniband and NVIDIA's NVLink and NVSwitch technologies. The Ultra Ethernet Consortium and Ultra Accelerator Link (UALink) aim to provide

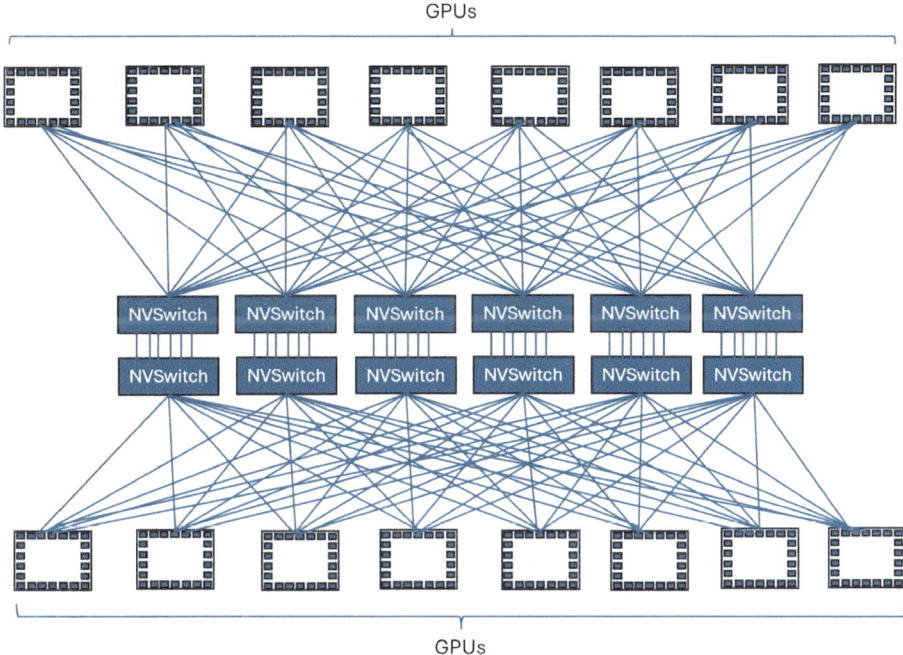

Fig. 8.16 NVSwitch providing GPU interconnects for scale-out design

Ethernet-based technologies for AI infrastructure and HPC, and an open shared memory accelerator interconnect, respectively.

UALink

The Ultra Accelerator Link (UALink) is part of the Ultra Ethernet Consortium's initiative to develop Ethernet-based technologies for AI infrastructure and high-performance computing (HPC). UALink aims to facilitate efficient communication between multiple GPUs and other accelerators over an Ethernet network, leveraging an open shared memory architecture that simplifies programming and ensures high-speed data access. The diagram below depicts proposed UALink solutions (Fig. 8.17).

By utilizing Ethernet, UALink offers a scalable and flexible interconnect solution that integrates seamlessly with existing network infrastructures. This contrasts with traditional, more proprietary interconnect technologies like Infiniband and NVLink/NVSwitch. UALink is designed to provide high bandwidth and low latency, essential for the performance of AI and ML tasks, ensuring efficient data transfer between accelerators and minimizing bottlenecks.

One of the significant advantages of UALink is its scalability. Ethernet infrastructure is typically more cost-effective than specialized interconnect solutions, making UALink an affordable option for expanding AI and HPC infrastructures. Additionally, being part

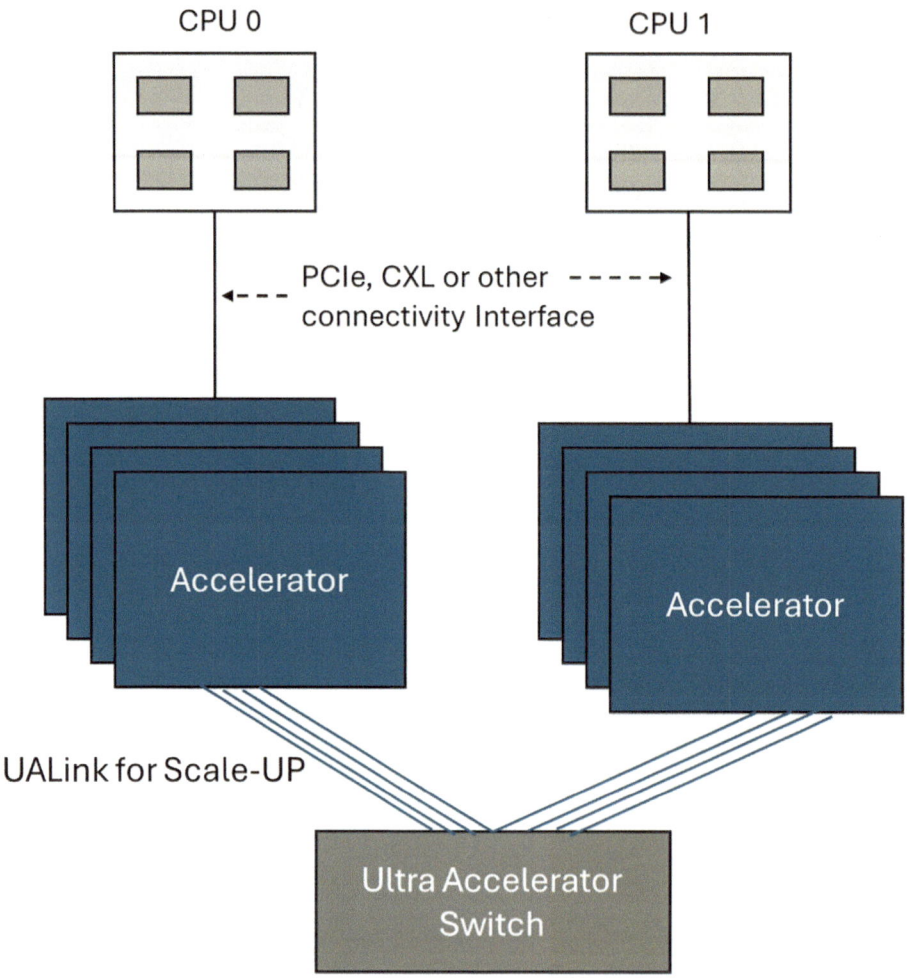

Fig. 8.17 UALink solution for GPU communications

of the Ultra Ethernet Consortium means UALink benefits from broad industry support and adherence to open standards, fostering interoperability and innovation. UALink represents a significant shift toward Ethernet-based interconnect solutions for AI and HPC. As AI workloads grow in complexity and scale, the need for efficient, scalable, and cost-effective interconnect solutions like UALink will become increasingly important. This shift could democratize access to high-performance AI infrastructure by reducing reliance on proprietary technologies.

8.2.2.2 CPUs (Central Processing Units)

CPUs are versatile and widely used in AI infrastructure for general-purpose processing. Although not as specialized as GPUs, they play a crucial role in various AI tasks. CPUs are designed for a broad range of tasks, from simple arithmetic to complex computations, excelling in tasks requiring high single-threaded performance. They handle data preprocessing tasks, such as cleaning and transforming data before feeding it into AI models, and are used for managing data pipelines and orchestrating complex workflows. Many traditional machine learning algorithms, such as decision trees and logistic regression, run efficiently on CPUs. Furthermore, CPUs are often tasked with managing and coordinating the operations of GPUs and other accelerators within a system, handling tasks such as loading data and distributing workloads.

CPUs use technologies like QuickPath Interconnect (QPI) and Ultra Path Interconnect (UPI) to facilitate high-speed communication between processors and other system components. QPI, developed by Intel, provides a point-to-point processor interconnect that significantly reduces latency and increases data transfer rates compared to older front-side bus systems. It enables multiple CPUs to communicate directly with each other and with memory controllers, enhancing overall system performance and scalability. UPI, the successor to QPI, further improves communication efficiency and speed, supporting higher data rates and better power efficiency, which is critical for modern AI and HPC applications.

8.2.2.3 TPUs (Tensor Processing Units)

TPUs, custom-designed by Google, are specialized processors aimed at accelerating machine learning workloads, particularly those involving tensor operations. Optimized for the dense matrix multiplications and additions essential in deep learning, TPUs offer exceptional performance for specific AI tasks. They deliver high performance per watt, making them an energy-efficient option for large-scale AI models.

Available through Google Cloud, TPUs provide organizations access to cloud-based AI infrastructure without the need for significant upfront investment. This integration supports both the training of neural networks and the inference process, enabling faster deployment of AI models. TPUs leverage high-speed interconnects to communicate efficiently with other TPUs and system components, ensuring low latency and high throughput in distributed AI workloads.

Technically, TPUs are built with a large number of arithmetic units to handle tensor operations, allowing for rapid execution of deep learning algorithms. Their architecture is designed to maximize throughput for these operations, which are fundamental in tasks such as image recognition, natural language processing, and generative models. The high-speed interconnects used by TPUs, such as Google's proprietary interconnect technology, facilitate efficient data transfer between TPUs and other components, minimizing latency and enhancing overall system performance. This design makes TPUs particularly well-suited for handling the computational demands of modern AI applications.

8.2.2.4   FPGAs (Field-Programmable Gate Arrays)

FPGAs provide customizable hardware acceleration, enabling tailored performance optimization for specific AI applications. They can be programmed to execute specific algorithms with high efficiency, offering flexibility to optimize performance for particular tasks. FPGAs deliver low-latency processing, making them ideal for real-time AI applications and edge computing scenarios. Additionally, they consume less power compared to traditional processors, enhancing energy efficiency for certain AI workloads. Often used for inferencing in edge devices, FPGAs ensure customized, low-latency, and energy-efficient processing. They typically employ PCIe (Peripheral Component Interconnect Express) for high-speed data transfer, ensuring efficient communication with other system components.

8.2.2.5   Integration and Use in AI Infrastructure

In a typical AI infrastructure, these compute resources are integrated to leverage their unique strengths. For example, GPUs might be used for the heavy lifting during the training phase of deep learning models, while CPUs handle data preprocessing and orchestration tasks. TPUs could be employed for highly optimized training and inference tasks in a cloud environment, and FPGAs might be deployed for real-time inference on edge devices.

By strategically utilizing these different types of compute resources and their interconnect technologies, organizations can build a robust and flexible AI infrastructure that meets diverse computational needs, maximizes performance, and optimizes costs. Understanding the specific applications and advantages of GPUs, CPUs, TPUs, and FPGAs allows for informed decisions in designing and scaling AI systems.

8.2.3   Storage

Storage is a critical component of AI hardware infrastructure, as it manages the vast amounts of data required for training and inference in AI applications. The sheer volume of data generated and consumed by AI models necessitates robust and scalable storage solutions to maintain efficient workflows. Efficient storage solutions ensure quick access to data, enabling high performance and seamless operation of AI workloads. This includes not only the speed and capacity of the storage hardware but also the architecture and technologies used to connect storage to compute resources.

In AI applications, data storage must accommodate the rapid ingestion, processing, and retrieval of large datasets. This necessitates the use of high-performance storage systems that can handle the intensive read/write operations typical of AI workloads. NVMe (Non-Volatile Memory Express) SSDs are commonly used for their ability to deliver high-speed data transfer rates, significantly improving the performance of AI workloads.

These systems are often configured in RAID arrays to enhance data redundancy and performance.

Distributed file systems, such as the Hadoop Distributed File System (HDFS) and Ceph, are essential for managing large-scale datasets across multiple storage nodes. These systems provide scalability, reliability, and high availability, ensuring that data is accessible even in the event of hardware failures. They also facilitate parallel data processing, which is crucial for AI training tasks that require handling large volumes of data. Object storage solutions, like those compatible with Amazon S3, offer scalable and cost-effective storage for unstructured data. This type of storage is particularly useful for storing vast amounts of data generated by AI applications, such as images, videos, and logs. Object storage systems are designed to handle large volumes of data with high durability and availability, making them suitable for long-term storage of AI datasets.

Network connectivity is vital for ensuring seamless data transfer between storage systems and compute resources in AI infrastructure. High-speed interconnects such as InfiniBand and high-bandwidth Ethernet (40Gbps, 100Gbps, or higher) provide the necessary bandwidth and low latency for data-intensive AI workloads. These interconnects enable quick access to data stored in high-performance storage systems and distributed file systems, minimizing bottlenecks and ensuring smooth operation of AI tasks. Storage Area Networks (SAN) and Network-Attached Storage (NAS) are also integral parts of AI storage infrastructure. SAN provides block-level storage that can be accessed by servers over a high-speed network, offering high performance and low latency. NAS, on the other hand, provides file-level storage accessible over a standard Ethernet network, making it easy to share data across multiple compute nodes.

Effective data management strategies, including tiered storage, are essential for optimizing storage infrastructure in AI applications. Tiered storage involves categorizing data based on its frequency of use and importance, storing frequently accessed data on high-performance storage systems, and moving less frequently accessed data to cost-effective storage solutions. This approach ensures that critical data is readily available while optimizing storage costs.

Overall, the storage component of AI hardware infrastructure must be designed to meet the high demands of data-intensive AI applications. This involves integrating advanced storage technologies, scalable architectures, and efficient data management practices to support continuous and high-speed access to large datasets, thereby enabling seamless and effective operation of AI workloads. In the following sections, we will explore the AI/ML data pipeline, the storage challenges associated with it, and how various storage technologies can help resolve these challenges.

8.2.3.1   AI/ML Data Pipeline and Storage Challenges

The AI/ML data pipeline comprises several stages: data ingestion, preprocessing, model training, evaluation, inference, and data management. The following diagram depicts AI/

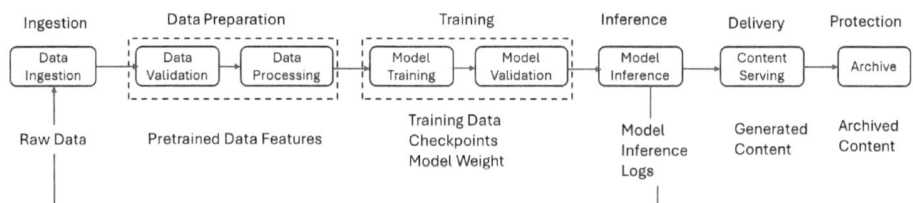

Fig. 8.18 AI/ML data pipeline

ML Data pipeline. Each stage has specific storage requirements and challenges that need to be addressed to ensure efficient and effective AI operations (Fig. 8.18).

Data ingestion involves handling large volumes of data from various sources, necessitating high throughput and ensuring data integrity. High-performance storage systems, such as NVMe SSDs, are essential for rapid data capture and storage. Distributed file systems like HDFS manage large datasets across multiple nodes, providing scalability and reliability.

During preprocessing, data undergoes cleaning, transformation, and normalization, requiring fast access to stored data. High-speed interconnects, such as InfiniBand and high-bandwidth Ethernet, facilitate rapid data transfer, while RAID arrays enhance data redundancy and performance, ensuring efficient preprocessing workflows.

Storing vast amounts of structured, semi-structured, and unstructured data poses challenges in ensuring availability, durability, and accessibility. Object storage solutions, such as MinIO, offer scalable and cost-effective storage for unstructured data. Storage Area Networks (SAN) provide high-performance block-level storage for critical data, offering low latency and high reliability.

Model training is computationally intensive, requiring continuous access to large datasets. NVMe SSDs provide high throughput and low latency, essential for quick data access during training. Tiered storage strategies optimize performance and costs by storing frequently accessed data on high-performance systems and moving less critical data to more economical storage.

Model validation requires similar storage capabilities as training, ensuring quick access to large datasets for evaluating model performance. Distributed file systems ensure data availability and support parallel processing, which is crucial for efficient model validation. High-bandwidth Ethernet provides the necessary bandwidth for rapid data transfer during validation processes.

Inference and real-time processing demand low latency and high throughput. Low-latency storage solutions, such as NVMe SSDs and high-speed interconnects, ensure immediate data access, enabling real-time inference. Network-Attached Storage (NAS) facilitates easy data sharing across multiple nodes, supporting collaborative real-time processing workflows.

Effective data management, including tiered storage, is essential for optimizing performance and costs. Tiered storage stores frequently accessed data on high-performance systems and archives less critical data on more economical storage. Object storage provides a scalable solution for archiving vast amounts of unstructured data, ensuring high durability and availability.

8.2.4   Networks

Recently, there has been significant momentum toward enhancing Ethernet-based data center networks (DCNs) to support AI/ML workloads. Traditionally, DCNs use commodity Ethernet switches to construct spine and leaf architectures, also known as Clos-based Ethernet fabric. These networks operate on an oversubscription model, assuming that not all devices will communicate at maximum bandwidth simultaneously. This approach contrasts with the specialized interconnects and topologies found in high-performance computing (HPC) environments.

AI/ML workloads, more similar to HPC tasks than traditional ones, require extremely high-performance computing nodes distributed across multiple CPUs and GPUs. These nodes need real-time, low-latency, non-blocking, and high-bandwidth communication. AI/ML tasks cannot afford network delays, as GPUs waiting for data from others leads to inefficiencies and increased job completion times. The scalability of AI/ML workloads is achieved by distributing tasks across numerous connected devices, rather than relying on a single, monolithic computer. This concept reinforces the idea that "the network becomes the computer," highlighting the critical role of advanced networking in modern AI infrastructure. To support these AI workloads, specialized data center networks must feature a non-blocking architecture, a 1:1 subscription ratio, ultra-low network latency, high bandwidth availability, and no congestion. Achieving these goals requires real-time, programmatic traffic engineering with intelligence at the data plane level, ensuring minimal latency from control plane traffic. In the following sections, we will present some of the prevailing thoughts around optimized DCN spine and leaf architecture for AI/ML workload environments.

8.2.4.1   Congestion in Ethernet-Based DCN Fabric

Typical Clos-based Ethernet network fabrics in data centers often use oversubscription ratios like 2:1 to 5:1, assuming not all devices will communicate at maximum bandwidth simultaneously. For AI/ML workloads, a non-oversubscribed fabric is crucial, ensuring equal uplink and downlink bandwidth to minimize congestion. Consider a non-blocking design with 128 servers, each with $2 \times 100G$ ports, dual-homed to two-leaf switches. This setup requires $256 \times 100G$ downlink ports. The design involves 10 leaf switches, each with $28 \times 100G$ downlink ports and $8 \times 400G$ uplink ports, connecting to 4 spine

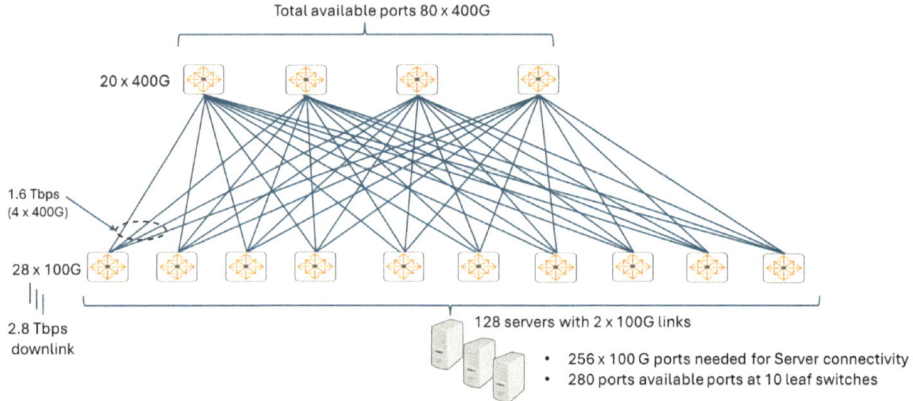

Fig. 8.19 Non-blocking spine-leaf network design for AI/ML workload

switches. Each spine switch has 20 × 400G ports. The following diagram depicts the design (Fig. 8.19).

The leaf switches are connected to four spine switches, each equipped with 20 × 400G ports, resulting in a total of 80 x 400G ports or 32 Tbps capacity. Each leaf switch connects to the spine switches with 4 × 400G (1.6 Tbps) links, forming a non-blocking architecture. Each spine switch uses only 10 × 400G links, leaving another 10 × 400G links available for future expansion without compromising the non-blocking nature of the network.

However, in such a non-blocking design, congestion can still manifest in three primary areas: between leaf and spine switches at the transmit queue of the leaf switch ports toward the spine switches; between spine and leaf switches at the transmit queue of the spine switch ports toward the leaf switches; and between leaf switches and nodes at the transmit queue of the leaf switch ports toward the nodes. Addressing these potential congestion points is crucial to maintaining optimal network performance. Consider the scenario based on our earlier non-blocking design, where GPU nodes at leaf switch 1 (L1) communicate with GPU nodes at leaf switch 3 (L3). There are two paths to L3: one through spine switch 1 (S1) and another through spine switch 2 (S2), each at full bandwidth (400G). When a third set of GPU nodes at L1 also needs to reach L3, the leaf switch uses an Equal-Cost Multipath (ECMP) load-balancing algorithm to decide the path. Please refer to the diagram (Fig. 8.20).

ECMP generates a hash based on attributes such as source/destination IP address, source/destination port, and protocol of incoming packets. Packets with matching hashes are directed to the same uplink toward the spine, creating a "flow." Methods like Hash-Threshold mapping or Modulo-N mapping can determine this flow-to-uplink mapping. ECMP aims to distribute packets evenly across available uplinks while maintaining packet order within each flow. This method works well for typical TCP-based applications

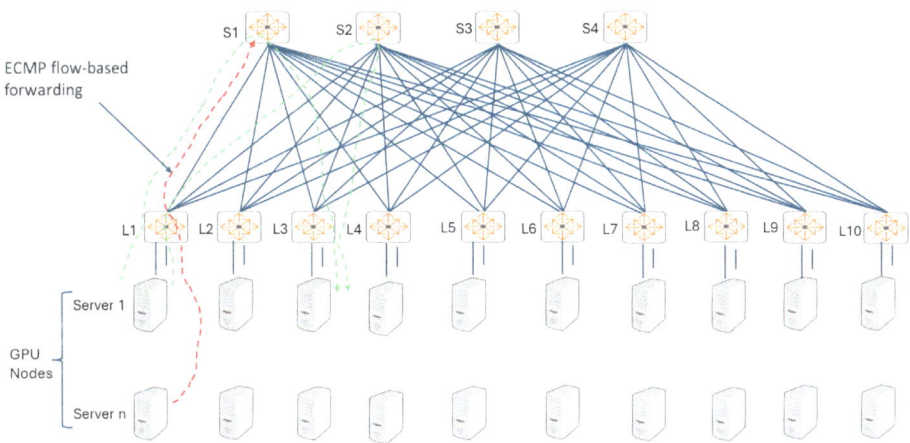

Fig. 8.20 ECMP flow-based forwarding for communication between GPU nodes across multiple leaf switches leading to congestion

with numerous short-lived sessions, generating different hashes and ensuring distribution across various uplinks. However, long-lived flows can present challenges. If multiple high-bandwidth flows are hashed to the same uplink, they may exceed the bandwidth of that link, causing congestion issues. Thus, ECMP flow-based forwarding can lead to suboptimal network utilization and packet loss, adversely affecting RoCEv2 transport for AI/ML workloads. This inefficiency is particularly problematic due to the high bandwidth and long-lived flows typical in these environments, leading to increased congestion and reduced performance.

To address this situation, we can leverage technologies like Global Adaptive Load Balancing, Reactive Path Rebalancing, and Data Center Quantized Congestion Notification (DCQCN).

Global Adaptive Load Balancing dynamically adjusts traffic distribution based on real-time conditions, improving network utilization by rerouting traffic away from congested paths. Reactive Path Rebalancing adjusts traffic flows in response to congestion or failures, shifting traffic to less utilized paths and enhancing performance. DCQCN, used alongside these methods, sends explicit congestion notifications (ECNs) to sources to throttle transmission rates when congestion is detected, preventing packet loss and managing congestion proactively. Integrating these technologies ensures efficient load distribution, adaptive response to network changes, and effective congestion management, all critical for maintaining high performance and reliability in AI/ML environments.

8.2.4.2   Global Adaptive Load Balancing

Global Adaptive Load Balancing emerged from the need to efficiently manage network traffic in large-scale data centers. It is an evolution of traditional load-balancing techniques designed to address the dynamic and unpredictable nature of modern data center traffic. Traditional static load-balancing techniques, which distribute traffic based on predefined rules, proved inadequate for handling dynamic and unpredictable traffic patterns, especially with the rise of cloud computing, big data, and AI/ML workloads. Global Adaptive Load Balancing addresses these challenges by dynamically adjusting traffic distribution across multiple network paths based on real-time network conditions.

Global Adaptive Load Balancing operates by continuously monitoring network performance metrics such as bandwidth utilization, latency, and congestion levels. When a potential bottleneck or imbalance is detected, the system dynamically reroutes traffic to less congested paths. This process involves:

- **Real-Time Monitoring**: Collecting data on network performance metrics from various points in the network.
- **Traffic Analysis**: Analyzing the collected data to identify potential congestion points and underutilized paths.
- **Dynamic Adjustment**: Adjusting traffic flows by rerouting data packets to optimize network utilization and minimize congestion.

The key advantage of Global Adaptive Load Balancing (GALB) is its ability to respond to changing network conditions in real-time, ensuring efficient use of network resources and maintaining high performance even under varying workloads. This technology should not be confused with Broadcom's Global Load Balancing (GLB) that was introduced in the Tomahawk 5 series of SOC (System on Chip), extending the capabilities of traditional dynamic load balancing (DLB). Broadcom's implementation leverages advanced algorithms to enhance traffic distribution and network efficiency.

The following table depicts some of the key differences between GALB and GLB (Table 8.1).

Table 8.1 Comparison between GALB and GLB

Algorithm	Type	Algorithm and implementation	Specificity and optimization
GALB	General concept applicable to all platforms	Global adaptive load balancing can be implemented using different algorithms and techniques depending on the vendor and network infrastructure	It will require vendor-specific optimization
GLB	Proprietary	Broadcom GLB leverages broadcom's proprietary algorithms for traffic management	Broadcom GLB is tailored to optimize broadcom's networking hardware and software, providing potentially deeper integration and optimization compared to GALB

For AI/ML workloads in spine-leaf-based Ethernet data centers, both solutions aim to improve network efficiency and reduce congestion. The choice between Broadcom GLB and a more general Global Adaptive Load Balancing approach may depend on the specific network infrastructure, existing hardware, and performance requirements. Utilizing advanced load-balancing techniques ensures that AI/ML workloads receive the necessary bandwidth and low-latency communication essential for optimal performance.

8.2.4.3   Reactive Path Rebalancing

Reactive Path Rebalancing emerged from the need to dynamically manage network traffic in response to real-time changes in network conditions. Traditional static routing methods were insufficient for modern data centers, where traffic patterns can be highly unpredictable and workloads such as AI/ML can cause sudden surges in data flow. Reactive Path Rebalancing was developed as a solution to address these challenges by adjusting traffic routes based on current network performance and congestion levels.

Reactive Path Rebalancing operates by continuously monitoring the state of the network and making adjustments to traffic flows as needed. Here's how it works:

- **Real-Time Monitoring**: The system continuously collects data on network performance, including metrics such as latency, bandwidth utilization, and packet loss.
- **Congestion Detection**: When congestion is detected on a particular path, the system identifies it as a potential bottleneck. This can be done through various methods, such as monitoring queue lengths or analyzing traffic patterns.
- **Dynamic Adjustment**: Upon detecting congestion, the system dynamically adjusts the routing of traffic. This involves rebalancing the traffic load by redirecting it to less congested paths. The adjustments are made in real-time, ensuring minimal disruption to ongoing data flows.
- **Algorithmic Decision-Making**: The decision to rebalance traffic is typically made using advanced algorithms that consider various factors, such as current traffic load, network topology, and historical performance data. These algorithms aim to optimize network performance and minimize latency.
- **Feedback Loop**: The system continuously monitors the effects of the rebalancing actions and makes further adjustments as needed. This feedback loop ensures that the network remains optimized for performance and can adapt to changing conditions.

For AI/ML workloads, which often involve high-bandwidth, long-lived flows, Reactive Path Rebalancing is particularly beneficial. These workloads can cause sudden spikes in network traffic, leading to congestion and performance degradation. By dynamically adjusting traffic routes in response to real-time network conditions, Reactive Path Rebalancing ensures that the network can handle these spikes efficiently.

This approach is critical for maintaining the high performance and low latency required by AI/ML applications. It helps prevent bottlenecks, reduces packet loss, and ensures that data flows smoothly across the network, thereby optimizing the overall efficiency and reliability of data center operations.

8.2.4.4 Data Center Quantized Congestion Notification (DCQCN)

Data Center Quantized Congestion Notification (DCQCN) is a sophisticated congestion control mechanism designed to enhance RoCEv2 networks, crucial for high-performance computing (HPC) and AI/ML workloads that demand low latency and high throughput. DCQCN integrates Priority Flow Control (PFC) and Explicit Congestion Notification (ECN) to create a lossless network. PFC manages data flow at the interface level by issuing Pause Frames to prevent buffer overflow, while ECN provides end-to-end congestion management by marking packets during congestion. These marked packets prompt receivers to send Congestion Notification Packets (CNPs) to the sender, which then reduces its transmission rate. This integration of PFC and ECN ensures efficient congestion control, maintaining network performance and preventing packet loss.

Priority Flow Control (PFC)

PFC is the most fundamental flow control technology is the Ethernet Pause mechanism defined in IEEE 802.1Qbb standard. It aims to provide a lossless network environment for specific types of traffic. It achieves this by enabling flow control on individual traffic classes, identified by IEEE 802.1p priority values. PFC works by pausing the transmission of frames for a particular priority level when the buffer for that traffic class is about to overflow. This is done using Pause Frames, which are sent from the receiver back to the sender to temporarily halt the transmission of that specific traffic class. Each Pause Frame includes a timer value that specifies how long the sender should stop sending frames (Fig. 8.21).

In the diagram above, eight priority queues on Device A's transmit interface correspond to eight receive buffers on Device B's receive interface. When a receive buffer on Device B, configured for a specific PFC traffic class, becomes congested, Device B sends a PFC Pause Frame (XOFF) to Device A. This frame instructs Device A to halt the transmission of packets for that traffic class while allowing other traffic to continue. PFC manages traffic by controlling individual priority queues instead of the entire interface, enabling different types of traffic to share the same link without interruption. This selective pausing ensures that high-priority traffic proceeds unaffected by congestion in lower-priority queues, optimizing link utilization and maintaining network performance.

Explicit Congestion Notification (ECN)

Explicit Congestion Notification (ECN) enables end-to-end congestion management in TCP/IP networks by allowing communication between ECN-enabled senders and receivers. For ECN to function correctly, it must be enabled on both endpoints and all intermediate devices. ECN works by marking packets instead of dropping them when congestion is detected, signaling the sender to reduce its transmission rate. This process

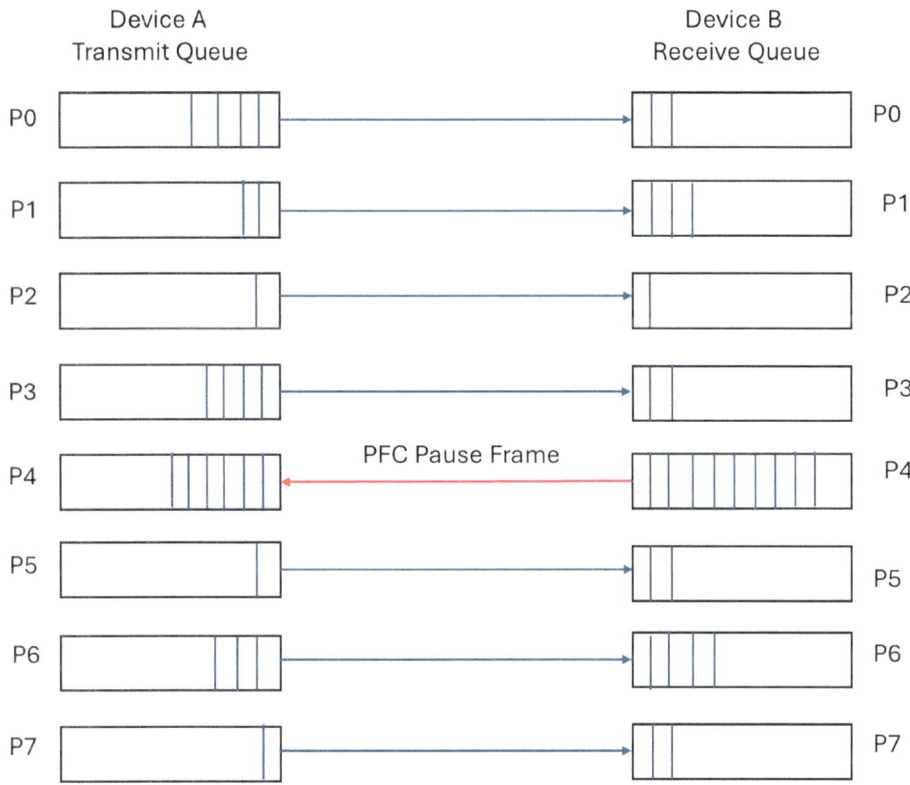

Fig. 8.21 PFC working mechanism

is outlined in RFC 3168, which details the implementation and benefits of ECN in reducing packet loss and delays. In Data Center Quantized Congestion Notification (DCQCN), ECN works with Priority Flow Control (PFC) to achieve lossless Ethernet by addressing PFC's limitations, enhancing overall network performance.

ECN functionality specified in RFC 3168 involves the cooperation of both the network devices (such as routers and switches) and the endpoints (such as servers and clients). ECN marking operates within the IP header and the Transmission Control Protocol (TCP) header. The diagram below illustrates the Explicit Congestion Notification (ECN) process in IP networks, showing the flow of packets and congestion signaling between a Client and a Server through three routers (Fig. 8.22).

Initially, during the TCP connection setup, the Client and Server negotiate ECN capability by setting the ECN-Echo (ECE) and Congestion Window Reduced (CWR) flags in their TCP SYN and SYN-ACK packets. As the Client sends packets to the Server, these packets pass through Router 1 (R1), Router 2 (R2), and Router 3 (R3), with the ECN bits in the IP header set to 00, indicating Non-ECT (Non-ECN-Capable Transport).

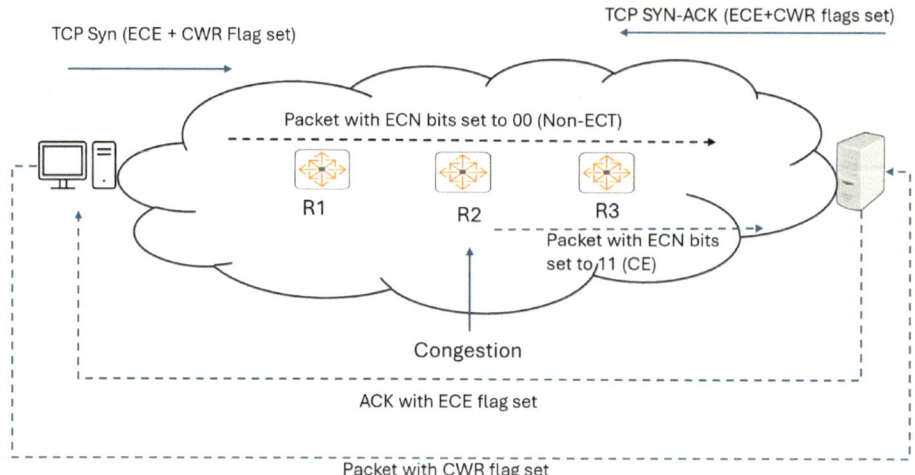

Fig. 8.22 Explicit Congestion Notification (ECN) process in IP networks

When Router 2 experiences congestion, it marks the ECN bits in the packet to 11 (Congestion Experienced or CE). The Server, upon receiving the CE-marked packet, sends an acknowledgment (ACK) back to the Client with the ECE flag set, signaling the detection of congestion.

In response, the Client reduces its congestion window and sends a packet back to the Server with the CWR flag set, acknowledging the congestion signal and indicating that it has taken action to mitigate congestion. The diagram highlights the key components such as the IP header ECN field, TCP header ECE and CWR flags, and the congestion marking at Router 2, providing a clear visualization of the ECN process.

The ECN field in the IP header consists of two bits located in the Differentiated Services Code Point (DSCP) field, previously known as the Type of Service (ToS) byte:

- 00: Not ECN-Capable Transport (Non-ECT)
- 01: ECN Capable Transport (ECT(1))
- 10: ECN Capable Transport (ECT(0))
- 11: Congestion Experienced (CE).

In the TCP header, the ECN-Echo (ECE) and Congestion Window Reduced (CWR) flags are used to facilitate ECN signaling:

- **ECE**: Indicates to the sender that the receiver has received a packet with the CE mark.
- **CWR**: Indicates that the sender has reduced its congestion window in response to receiving an ECE signal.

Explicit Congestion Notification represents a significant advancement in congestion management within IP networks. By enabling proactive signaling of congestion and allowing endpoints to respond without dropping packets, ECN enhances network performance, reduces latency, and improves overall user experience. As more network devices and systems adopt ECN, its benefits will become increasingly widespread, contributing to more efficient and reliable network operations.

8.2.4.5 Rail-Only Topology

In a recent paper[10] a novel network architecture called Rail-only is proposed that is designed to train large language models (LLMs) efficiently and cost-effectively at hyperscale. The key idea is to tailor the network design to the unique communication patterns of LLM training, which generate sparse communication patterns and thus do not require a full-bisection any-to-any network. The Rail-only architecture eliminates the spine layer of switches in traditional GPU clusters, reducing network costs by 38 to 77% and power consumption by 37 to 75% compared to conventional designs, while maintaining similar performance.

Prior to the rail-only design proposal, the prevailing architecture is a "any-to-any" clos fabric that is known as "rail-optimized" and advocated by GPU manufacturers as depicted in the diagram (Fig. 8.23).

The topology uses High Bandwidth Interconnect (HBI) for vendor-specific high-bandwidth GPU platforms, such as Nvidia DGX and AMD Infinity Fabric. These systems provide terabits of bandwidth through HBI technologies like NVLink using NVSwitch. However, they are not scalable unless connected through a rail-optimized topology [10].

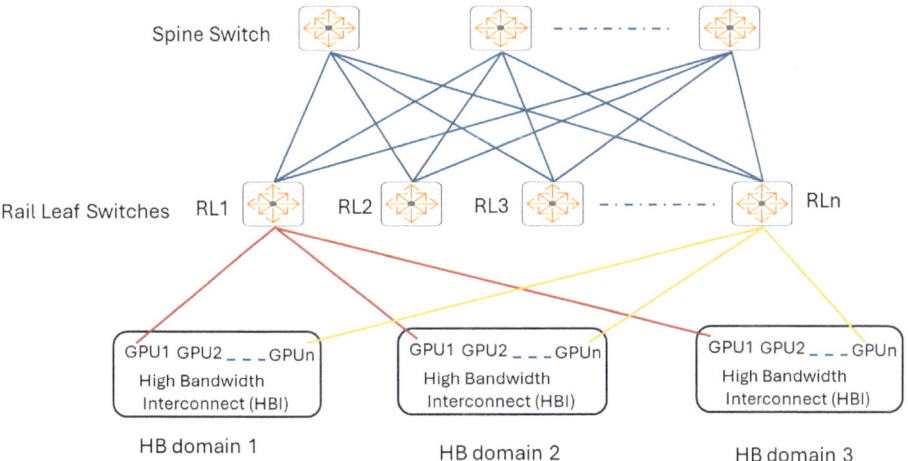

Fig. 8.23 Rail-optimized topology showcasing any-to-any closed networks

---

[10] Wang et al. [10].

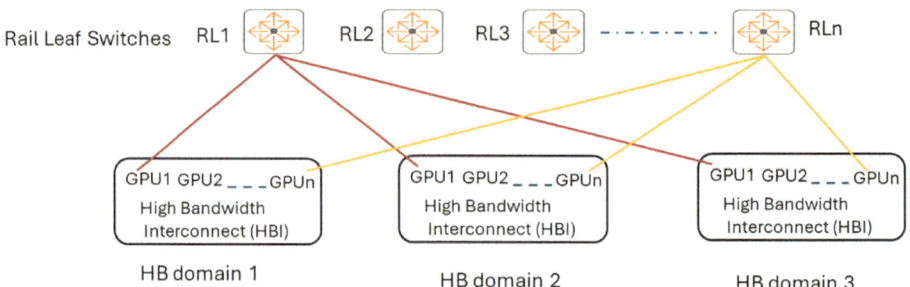

Fig. 8.24 Rail-only architecture for LLM workload

For a GPU platform with an HB domain of size "n," there are "n" total rails, where a rail comprises GPUs with the same local rank that belong to different HB domains. A rail-optimized network places these GPUs under the same set of switches, denoted as "rail leaf switches" (RL1 to RLn). The figure above highlights rail one and rail "n"in red and yellow, respectively. Rail-optimized architectures achieve low latency for communication between GPUs within the same rail. To connect these rails, layers of spine switches are utilized, creating a full-bisection any-to-any Clos network topology. This setup ensures that GPUs in different high-bandwidth domains (HB domains) can communicate efficiently at the network line rate, which can reach several hundreds of Gbps.

The research [10] on MegatronLM traffic pattern analysis for GPT models (ranging from 145.6 billion to 1 trillion parameters) shows that tensor parallelism (TP) traffic dominates within high-bandwidth domains (HB domains). In contrast, data parallelism (DP) and pipeline parallelism (PP) traffic are significantly smaller and occur primarily in the HB domain. Most communication in the HB domain stays within rails, making the spine switches largely unnecessary for efficient LLM training. This finding helps the researchers to propose a rail-only topology that aims to achieve the same performance as in rail-optimized topology without a spine layer thus improving TCO for data center operators. The following diagram depicts a rail-only architecture that excludes spine layer (Fig. 8.24).

An article published on "Outshift by CISCO"[11] discusses an efficient GPU traffic routing mechanism within AI/ML clusters using rail-only connections. This mechanism aligns with the principles of rail-only topology by eliminating spine switches and focusing on direct, high-bandwidth connections within high-bandwidth domains (HB domains). The emphasis on reducing complexity and cost while maintaining performance through simplified routing and direct interconnects indicates agreement with the rail-only topology rather than the rail-optimized topology, which uses spine switches.

---

[11] Ma [11].

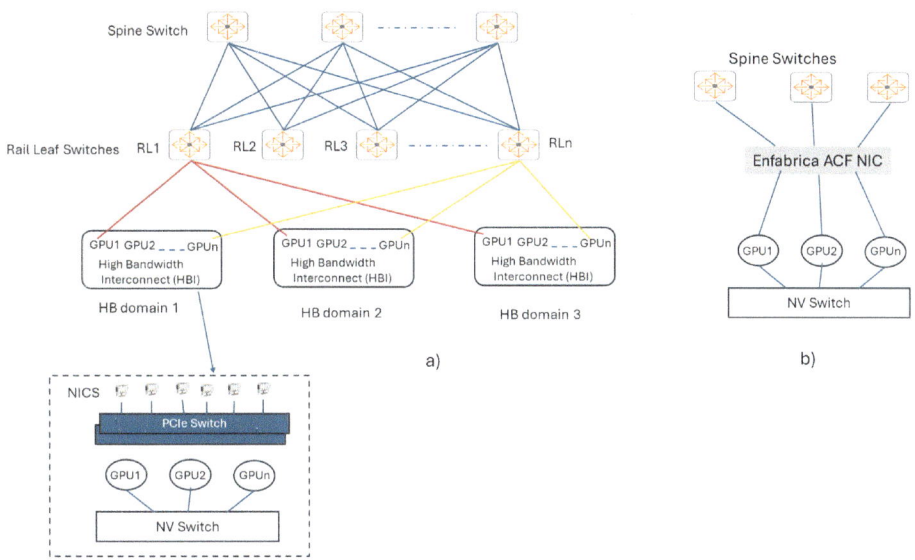

Fig. 8.25 Rail-optimized topology and Enfabrica ACF Network Interface Card (NIC) based network design for Gen AI workload

8.2.4.6   Accelerated Compute Fabric (ACF)

Recently, I came across an article on Forbes that claims a startup called "Enfabrica" has created a solution to improve total cost of ownership (TCO) for data center networks by eliminating rail switches. The company refers to this solution as "accelerated compute fabric" and is positioning the technology for Gen AI workloads. The diagram (8.25a) depicts a rail-optimized topology with HBI, including NV Switch, PCIe Switch, and RDMA NICs. Enfabrica's NIC, termed "Accelerated Compute Fabric (ACF)," replaces PCIe switches, RDMA NICs, and rail switches, connecting directly to spine switches (refer to Fig. 25b). Enfabrica claims its solution provides 8x scale-out RDMA bandwidth and efficient traffic distribution across GPUs[12] (Fig. 8.25).

The solution aims to offer spine connectivity using 100/200/400/800 GbE optical or copper links with high-radix multipathing capabilities, ensuring robust and scalable network performance. This innovative approach replaces traditional components like PCIe switches, RDMA NICs, and rail switches, directly connecting to spine switches, which significantly enhances data center network efficiency and reduces overall costs, making it well-suited for demanding Gen AI workloads.

---

[12] Enfabrica [12].

8.2.5 Virtualization

Virtualization technologies are crucial for optimizing AI infrastructure by enabling efficient resource utilization, scalability, and flexibility. Virtual machines (VMs) are created using hypervisors like VMware ESXi, Microsoft Hyper-V, and KVM, which allow multiple isolated VMs to run on a single physical server. This setup supports resource sharing and diverse operating systems and applications, enhancing management and isolation.

Containers, facilitated by platforms such as Docker and Kubernetes, package applications and their dependencies into lightweight, consistent environments. This approach accelerates deployment, scalability, and resource utilization, making it ideal for AI workloads that require rapid experimentation and deployment.

GPU virtualization, with solutions like NVIDIA vGPU and AMD MxGPU, allows multiple VMs or containers to share GPU resources. This capability is essential for AI workloads that demand high computational power. It enhances performance, reduces costs, and provides flexibility in resource allocation.

Serverless computing, offered by services like AWS Lambda, Azure Functions, and Google Cloud Functions, allows code to run without the need to manage servers. This model is particularly suitable for event-driven workloads and scalable applications, offering simplified management, automatic scaling, and cost efficiency.

Network Function Virtualization (NFV) uses platforms like OpenStack and VMware NSX to virtualize network services, reducing reliance on dedicated hardware and improving flexibility. This technology offers cost savings, easier network management, and rapid deployment of network services.

Overall, virtualization technologies enhance AI infrastructure by maximizing hardware utilization, simplifying management, and providing the scalability and flexibility needed for complex AI workloads. This leads to significant cost savings and accelerates innovation in AI applications.

8.2.6 Data Management and Processing Layer

The data management and processing layer is a fundamental component of AI infrastructure, responsible for handling all aspects of data collection, storage, cleaning, preprocessing, and transformation. This layer ensures that data is adequately prepared and readily available for AI and ML models, providing a solid foundation for accurate and efficient model training and deployment. The key components are:

- **Data Ingestion**: Data ingestion involves collecting data from various sources, such as databases, sensors, web services, and external APIs. Effective data ingestion tools and methods are essential for ensuring that data flows seamlessly into the system without loss or corruption. Examples of data ingestion tools include Apache Kafka, Flume, and

AWS Glue, which enable real-time and batch data ingestion, supporting diverse data types and formats.

- **Data Storage**: Storing large volumes of data reliably and efficiently is crucial for AI and ML applications. Data storage solutions include data lakes, data warehouses, and databases. Data lakes, such as AWS S3 and Azure Data Lake Storage, provide scalable storage for raw, unstructured data. Data warehouses like Amazon Redshift, Google BigQuery, and Azure Synapse Analytics support structured data storage and fast querying capabilities. These solutions ensure that data is easily accessible and manageable, facilitating smooth data processing workflows.
- **Data Cleaning and Preprocessing**: Data cleaning and preprocessing are vital steps to ensure data quality and consistency. This process involves handling missing values, removing outliers, and normalizing data to a standard format. Tools such as Pandas, Apache Spark, and TensorFlow Data API offer functionalities for data cleaning and preprocessing. Data augmentation techniques, like image transformations for computer vision tasks, are also applied to enhance the dataset and improve model robustness.
- **Data Transformation**: Data transformation involves converting data into formats suitable for analysis and model training. This step typically includes Extract, Transform, and Load (ETL) processes. ETL tools, such as Apache NiFi and Talend, automate data extraction from various sources, transformation to a desired format, and loading into storage or processing systems. These tools streamline the data pipeline, ensuring efficient and accurate data transformation, which is critical for downstream AI and ML tasks.
- **Data Workflow Orchestration**: Managing and automating complex data workflows is essential for maintaining consistency and reliability in data processing. Workflow orchestration tools like Apache Airflow, Prefect, and Luigi provide capabilities to define, schedule, and monitor data workflows. These tools help automate repetitive tasks, handle dependencies, and ensure that data processing pipelines run smoothly, from ingestion to final preparation for model training.

The data management and processing layer plays a crucial role in AI infrastructure by ensuring that data is collected, stored, cleaned, preprocessed, and transformed efficiently and effectively. By leveraging robust tools and techniques in each of these components, organizations can establish a solid data foundation that supports accurate and scalable AI and ML model development.

8.2.7   AI/ML Framework Layer

The AI/ML framework layer is the cornerstone of AI infrastructure, providing essential tools and frameworks for developing, training, evaluating, deploying, and managing AI and ML models. This layer supports the entire lifecycle of AI models, ensuring that they are built efficiently, trained effectively, and deployed seamlessly into production environments. Key components are Model Development, Training and Validation, Model Evaluation, Model Deployment, Model Monitoring and Management.

8.2.7.1 Model Development

Model development in the AI/ML framework layer involves utilizing advanced libraries and predefined architectures to create robust models tailored to specific tasks. Libraries such as TensorFlow, PyTorch, and Keras provide extensive tools and APIs that simplify the process of designing and implementing models. These libraries support a range of pre-defined architectures like Convolutional Neural Networks (CNNs) for image processing and Recurrent Neural Networks (RNNs) for sequence-based tasks. Additionally, they offer the flexibility to build custom architectures, allowing developers to innovate and optimize models for unique applications and datasets. This phase is critical for laying the groundwork for subsequent training and evaluation, ensuring that models are well-constructed and capable of achieving high performance in their intended tasks.

8.2.7.2 Training and Validation

Training and validation are crucial stages in the AI/ML model lifecycle, responsible for optimizing model performance and ensuring its generalizability. The training process involves feeding the model with large datasets and adjusting its internal parameters (weights) to minimize the difference between its predictions and the actual outcomes. This is typically achieved using various optimization algorithms, such as Stochastic Gradient Descent (SGD) or Adam. These algorithms iteratively update the model's weights based on the gradients of the loss function, which measures the model's prediction error. During training, the dataset is often divided into training and validation sets. The training set is used to update the model's parameters, while the validation set is used to evaluate the model's performance on unseen data. This separation helps in assessing how well the model generalizes to new data and prevents overfitting, where the model performs well on training data but poorly on new, unseen data.

Distributed training is a technique employed to handle large-scale datasets and complex models. This involves splitting the training workload across multiple GPUs or machines, significantly speeding up the training process. Frameworks like TensorFlow's distributed strategy and Horovod are commonly used to facilitate this process, enabling synchronized training and efficient resource utilization.

Hyperparameter tuning is another critical aspect of the training process. Hyperparameters, such as learning rate, batch size, and network architecture, are parameters that are set before the training begins and directly affect the training process. Techniques like grid search, random search, and more advanced methods like Bayesian optimization are used to find the optimal hyperparameter values that yield the best model performance.

Validation involves using the validation set to evaluate the model at regular intervals during training. This helps in monitoring the model's performance and making adjustments as needed. Metrics such as accuracy, precision, recall, and F1 score are commonly used to assess the model's performance. Additionally, techniques like cross-validation, where the dataset is divided into multiple subsets, and the model is trained and validated on different combinations of these subsets, provide a more robust evaluation of the model's performance.

In summary, training and validation are iterative processes aimed at optimizing the model and ensuring its ability to generalize well to new data. These processes involve the use of advanced optimization algorithms, distributed training techniques, and hyperparameter tuning to achieve the best possible model performance. Regular evaluation using validation data helps in monitoring the model's progress and making necessary adjustments to prevent overfitting and ensure robust performance.

8.2.7.3 Model Evaluation

Model evaluation is a critical phase in the AI/ML pipeline, designed to assess the performance and generalizability of a trained model. This process involves applying various metrics and methodologies to understand how well the model predicts outcomes on unseen data, ensuring that it performs reliably in real-world scenarios. The evaluation begins by using the validation set, which consists of data not seen by the model during training. This helps in providing an unbiased assessment of the model's performance. Key metrics such as accuracy, precision, recall, F1 score, and ROC-AUC are calculated to measure different aspects of model performance. Accuracy indicates the overall correctness of the model's predictions, while precision and recall provide insights into the model's ability to correctly identify positive cases and its performance on imbalanced datasets, respectively. The F1 score, being the harmonic mean of precision and recall, offers a balanced metric when dealing with imbalanced classes. The ROC-AUC metric evaluates the trade-off between true positive and false positive rates, giving a comprehensive view of the model's diagnostic ability.

To ensure the robustness and reliability of the model, cross-validation techniques are employed. Cross-validation involves partitioning the data into multiple subsets and training and evaluating the model on different combinations of these subsets. This method helps in mitigating the variance associated with a single train-test split and provides a more comprehensive understanding of the model's performance. K-fold cross-validation is a popular technique where the data is divided into k subsets, and the model is trained and evaluated k times, each time using a different subset as the validation set and the remaining as the training set.

Another important aspect of model evaluation is the use of confusion matrices. A confusion matrix provides a detailed breakdown of the model's predictions, showing the counts of true positives, true negatives, false positives, and false negatives. This visualization helps in understanding the types of errors the model makes and provides insights into areas where the model needs improvement.

Additionally, evaluation involves analyzing the model's performance across different subsets of the data to ensure it is not biased toward any particular class or feature. This includes examining performance metrics across various demographics, time periods, or other relevant segments to ensure fairness and equity in the model's predictions.

Beyond these quantitative measures, model evaluation also includes qualitative assessment, where domain experts review the model's predictions to ensure they make logical and practical sense. This step is crucial in applications where interpretability and trust in the model's decisions are paramount, such as in healthcare or finance.

Furthermore, evaluation is not a one-time process but an ongoing activity. Models deployed in production are continuously monitored to ensure they maintain their performance over time. This involves setting up automated systems to track key performance indicators (KPIs) and alerting mechanisms to detect any performance degradation. This continuous evaluation is essential for maintaining the reliability and effectiveness of AI/ML models in dynamic environments where data distributions may shift over time.

In summary, model evaluation is a comprehensive process that combines quantitative metrics, cross-validation, confusion matrices, and qualitative assessments to ensure that AI/ML models perform accurately, reliably, and fairly. This rigorous evaluation process is essential for deploying robust and trustworthy AI solutions that can handle real-world complexities and deliver consistent value.

8.2.7.4   Model Deployment

Model deployment is a critical phase in the machine learning lifecycle, transforming a trained model into a functional component of a production system where it can generate predictions on new data. This stage involves several key steps to ensure that the model operates efficiently, reliably, and at scale.

The first step in model deployment is model export, which involves converting the trained model into a deployable format. This can include formats such as TensorFlow's SavedModel, ONNX (Open Neural Network Exchange), or PyTorch's TorchScript. These formats standardize the model, making it compatible with various deployment environments and facilitating interoperability across different platforms.

Once the model is in a deployable format, the next step is setting up the infrastructure for serving the model. Model serving involves creating an environment where the model can accept input data, process it, and return predictions. Systems such as TensorFlow Serving, TorchServe, and Kubernetes-based solutions provide scalable and efficient ways to serve models. These systems manage the lifecycle of the model, handle incoming prediction requests, and ensure low-latency responses, which are critical for real-time applications. In addition to setting up the serving infrastructure, it is essential to consider the integration of the model into the existing production pipeline. This involves connecting the model to data sources, ensuring that it can seamlessly receive the required input data. It also includes integrating the model's predictions into the broader application or service, enabling end-users or downstream systems to consume these predictions effectively.

To maintain the reliability and performance of the deployed model, continuous monitoring and logging are necessary. Monitoring tools track key performance metrics, such as latency, throughput, and error rates, providing insights into the model's operational health. Logging mechanisms record prediction results and input data, which are essential for debugging issues and auditing model performance. Tools like Prometheus, Grafana, and ELK (Elasticsearch, Logstash, Kibana) stack are commonly used for these purposes.

Another critical aspect of model deployment is ensuring scalability. As the demand for predictions grows, the serving infrastructure must scale to handle increased loads. Techniques such as horizontal scaling, where multiple instances of the model server are deployed, and load balancing, which distributes requests across these instances, help achieve this scalability. Cloud platforms like AWS, Google Cloud, and Azure offer auto-scaling features that dynamically adjust resources based on demand.

Security is also a paramount concern in model deployment. This includes securing the model itself, the data it processes, and the predictions it generates. Implementing robust authentication and authorization mechanisms ensures that only authorized users and systems can access the model. Encrypting data in transit and at rest protects sensitive information, while regular security audits and vulnerability assessments help identify and mitigate potential risks.

Finally, model deployment is not a one-time task but an ongoing process. As new data becomes available and the performance of the model is continuously evaluated, it may be necessary to update or retrain the model. Implementing a continuous integration and continuous deployment (CI/CD) pipeline for machine learning models helps automate this process, ensuring that updates are rolled out seamlessly and efficiently.

In summary conclusion, model deployment encompasses a series of steps, from exporting the trained model to setting up serving infrastructure, integrating with production systems, monitoring performance, ensuring scalability, and maintaining security. This comprehensive approach ensures that machine learning models deliver reliable, efficient, and scalable predictions, ultimately driving value in real-world applications.

8.2.7.5 Model Monitoring and Management

Model monitoring and management are critical for maintaining the performance, reliability, and effectiveness of deployed AI/ML models. This phase involves ongoing evaluation, tracking, and updating of models to ensure they continue to perform well over time and adapt to changes in data or operating environments.

Model monitoring begins with setting up comprehensive logging and tracking systems. These systems record various aspects of the model's operations, such as input data, prediction results, and performance metrics. Tools like Prometheus and Grafana are commonly used to monitor system metrics, including latency, throughput, and error rates. The ELK

(Elasticsearch, Logstash, Kibana) stack provides robust logging and visualization capabilities, allowing for detailed analysis of model behavior. This continuous logging helps identify anomalies, detect performance degradation, and diagnose issues quickly. Monitoring also involves tracking key performance indicators (KPIs) that reflect the model's effectiveness. These KPIs can include accuracy, precision, recall, F1 score, and other domain-specific metrics. By regularly evaluating these metrics, organizations can ensure that the model maintains its predictive quality and does not drift away from the expected performance levels. Model drift, which occurs when a model's performance deteriorates due to changes in the underlying data distribution, is a significant concern in production environments. Continuous monitoring helps in the early detection of drift, enabling timely interventions.

In addition to performance metrics, monitoring also encompasses data integrity and security. Ensuring that the data flowing into the model is consistent, clean, and free from corruption is essential. Security monitoring involves protecting the model from unauthorized access and attacks. Implementing encryption, access controls, and regular security audits are vital practices to safeguard the model and the data it processes.

Model management is the process of overseeing the entire lifecycle of AI/ML models, from development to deployment and beyond. This involves version control, lifecycle management, and orchestration of models. Version control systems, such as MLflow and DVC (Data Version Control), track changes to models, datasets, and parameters, ensuring reproducibility and accountability. These tools enable teams to experiment with different model versions and configurations, compare their performance, and manage transitions from development to production.

Lifecycle management includes retraining and updating models as new data becomes available. This can be automated through continuous integration and continuous deployment (CI/CD) pipelines, which streamline the process of deploying updated models. These pipelines ensure that new models are rigorously tested before deployment, reducing the risk of introducing errors or regressions.

Model orchestration involves managing multiple models and their interactions within a production environment. This includes deploying models to different environments, scaling them to handle varying workloads, and routing requests to the appropriate models. Kubernetes and other container orchestration tools play a crucial role in managing the deployment and scaling of containerized models, providing flexibility and resilience in production.

Effective model management also requires robust documentation and governance practices. Documenting model architectures, training processes, and deployment configurations ensures transparency and facilitates collaboration across teams. Governance involves establishing policies and procedures for model development, deployment, and monitoring, ensuring compliance with regulatory requirements and organizational standards.

To summarize, model monitoring and management are essential for sustaining the performance and reliability of AI/ML models in production. By implementing comprehensive monitoring systems, robust version control, automated lifecycle management, and effective orchestration, organizations can ensure their models deliver consistent value and adapt to changing conditions. These practices enable proactive management, early detection of issues, and continuous improvement, ultimately driving the success of AI initiatives.

8.2.8 AI/ML Workload

AI/ML workloads encompass a variety of tasks, each with specific requirements and challenges. These workloads include Natural Language Processing (NLP), Generative AI (Gen AI), Computer Vision, Data Processing, and Machine Learning (ML)/Deep Learning (DL). While data processing is a critical part of both the data management layer and AI/ML workloads, its role in each context is distinct. In the AI/ML framework, data processing focuses on transforming data into a form suitable for model training and inference, ensuring efficient and accurate predictions.

8.2.8.1 Natural Language Processing (NLP)

Natural Language Processing (NLP) is a branch of AI that enables machines to understand, interpret, and generate human language. NLP is pivotal for applications such as text analysis, machine translation, sentiment analysis, and conversational agents. The complexity of human language, with its nuances, ambiguities, and context-dependency, presents significant challenges in NLP.

NLP employs various techniques to process and understand text data. Tokenization breaks down text into manageable pieces like words or phrases. Parsing analyzes the grammatical structure of sentences. Language models like BERT and GPT-3 have revolutionized NLP by providing a deep contextual understanding of the text. Named Entity Recognition (NER) identifies and classifies entities such as names and dates within the text, while sentiment analysis determines the emotional tone behind a body of text. Machine translation technologies, driven by models like Google's Transformer, enable the conversion of text from one language to another with high accuracy.

However, NLP faces challenges such as dealing with the inherent ambiguity and variability of human language. Ensuring data privacy and handling sensitive information in text data are also significant concerns. Moreover, creating models that understand and generate text with cultural and contextual relevance adds another layer of complexity.

8.2.8.2 Generative AI (Gen AI)

Generative AI refers to systems that can create new content, such as text, images, music, or even code, based on learned patterns from existing data. This branch of AI is behind innovations such as deepfakes, automated content creation, and artistic style transfer. Generative Adversarial Networks (GANs) are a cornerstone of Gen AI, comprising a generator that creates data and a discriminator that evaluates its authenticity. Variational Autoencoders (VAEs) encode data into a latent space and then decode it to generate new data samples. Transformer models, like GPT-3, are employed for text generation, leveraging large datasets to produce coherent and contextually relevant text. Techniques like style transfer apply the artistic style of one image to the content of another, enabling new forms of digital art.

Quality control is a significant challenge, as ensuring the generated content is accurate and realistic is critical. Ethical concerns arise with the potential misuse of generative models, such as creating deepfakes or misleading information. Balancing innovation with responsible use is essential for the advancement of generative AI.

8.2.8.3   Computer Vision

Computer Vision involves enabling machines to interpret and understand visual information from the world. Applications include image recognition, object detection, facial recognition, and autonomous driving. Convolutional Neural Networks (CNNs) are fundamental to computer vision, as they can detect spatial hierarchies in images. Object detection techniques, such as YOLO (You Only Look Once) and SSD (Single Shot Multi-Box Detector), identify and locate objects within images. Image segmentation divides an image into segments to simplify its analysis. Facial recognition systems identify individuals by analyzing facial features, which is used in security and authentication systems. Autonomous driving leverages a combination of these techniques to interpret road conditions and navigate vehicles safely.

High-quality, annotated datasets are essential for training accurate models, and obtaining such datasets can be labor-intensive and costly. Ensuring real-time processing capabilities is crucial for applications like autonomous driving, where delays can have serious consequences.

8.2.8.4   Data Processing

Data processing in the context of AI/ML workloads involves preparing data for analysis and model training. This includes extracting, transforming, and loading data (ETL), ensuring data quality, and managing large datasets efficiently.

ETL processes extract data from various sources, transform it into suitable formats, and load it into storage systems. Tools like Apache Spark and Apache Kafka facilitate batch and stream processing, respectively. Data lakes and warehouses, such as AWS S3 and Google BigQuery, store vast amounts of structured and unstructured data. Data cleaning techniques handle missing values, outliers, and inconsistencies to ensure the data used for model training is of high quality.

Managing and processing large volumes of data efficiently is a major challenge. Ensuring data quality and consistency is critical for reliable model training. Handling data privacy and security is also a significant concern, especially when dealing with sensitive information.

8.2.8.5 Machine Learning (ML) and Deep Learning (DL)

ML and DL are subsets of AI focused on developing algorithms that learn from data to make predictions or decisions. ML includes a wide range of algorithms, while DL involves neural networks with many layers, capable of learning complex patterns.

Supervised learning algorithms, such as linear regression, decision trees, and support vector machines, learn from labeled data. Unsupervised learning techniques, like clustering and dimensionality reduction, find patterns in unlabeled data. Reinforcement learning trains models to make sequences of decisions by rewarding desired behaviors. Neural networks, including CNNs, RNNs, and Transformers, are deep learning models that can learn from vast datasets to recognize complex patterns. Training these models involves iterative processes of adjusting parameters to minimize errors, while hyperparameter tuning optimizes settings like learning rates and batch sizes to improve performance.

ML and DL require significant computational power, especially for training large models. Ensuring model interpretability is essential for understanding and trusting the decisions made by complex models. Protecting data privacy and security is also crucial, particularly when using sensitive data for training models.

8.2.9 Management and Monitoring Layer

The management and monitoring layer is critical for ensuring the efficient, reliable, and secure operation of AI infrastructure. This layer encompasses resource management, monitoring and alerting, and security management, all essential for maintaining the health and performance of AI systems.

8.2.9.1 Resource Management

Resource management involves allocating and optimizing computational resources such as CPUs, GPUs, memory, and storage to meet the demands of AI workloads. Advanced scheduling algorithms and orchestration tools like Kubernetes are used to dynamically allocate resources based on workload requirements. Kubernetes, for example, allows for containerized applications to be managed and scaled seamlessly across a cluster of machines, ensuring that resources are utilized efficiently. Auto-scaling capabilities adjust the number of running instances based on the current demand, preventing resource wastage and ensuring that AI applications have the necessary computational power when needed.

In addition to dynamic resource allocation, resource management also includes load balancing to distribute workloads evenly across available resources. This prevents any single node from becoming a bottleneck, enhancing the overall performance and reliability of the AI infrastructure. Tools like Kubernetes' Horizontal Pod Autoscaler and Cluster Autoscaler play a crucial role in managing resource allocation and load balancing, ensuring optimal performance and availability.

8.2.9.2 Monitoring and Alerting

Continuous monitoring of AI infrastructure is essential to ensure that systems are functioning correctly and efficiently. Monitoring involves tracking various performance metrics such as CPU and GPU utilization, memory usage, network bandwidth, and disk I/O. Tools like Prometheus, Grafana, and the ELK stack (Elasticsearch, Logstash, Kibana) provide powerful capabilities for collecting, visualizing, and analyzing these metrics in real-time.

Prometheus, for example, collects time-series data and allows for complex querying to detect anomalies and performance issues. Grafana complements this by providing interactive and customizable dashboards that visualize metrics in real-time, enabling operators to quickly identify and diagnose problems. The ELK stack offers comprehensive logging and search capabilities, making it easy to trace issues back to their source by analyzing logs from different components of the AI infrastructure.

Alerting mechanisms are integrated into monitoring systems to provide immediate notifications when predefined thresholds are breached. These alerts can be configured to trigger various events, such as high CPU usage, low memory availability, or network congestion. Notifications can be sent through multiple channels, including email, SMS, and collaboration tools like Slack, ensuring that the operations team is promptly informed of any issues that require attention. Automated incident response systems can also be set up to execute predefined actions when alerts are triggered, such as restarting services or scaling up resources to handle the increased load.

8.2.9.3 Security Management

Security management is a vital aspect of managing AI infrastructure, as it protects sensitive data, models, and computational resources from unauthorized access and cyber threats. Security measures include implementing robust authentication and authorization mechanisms to control access to resources. Multi-factor authentication (MFA) and role-based access control (RBAC) are commonly used techniques to ensure that only authorized users have access to specific resources and actions within the infrastructure.

Data encryption, both in transit and at rest, is critical to safeguarding sensitive information. Transport Layer Security (TLS) protocols are used to encrypt data moving between clients and servers, while encryption tools like AWS Key Management Service (KMS) and Azure Key Vault are used to protect data stored on disk. Regular security audits and vulnerability assessments help identify and mitigate potential risks, ensuring that the AI infrastructure remains secure against emerging threats.

Moreover, the implementation of network security measures, such as firewalls, intrusion detection systems (IDS), and intrusion prevention systems (IPS), helps protect the infrastructure from external and internal threats. Security management also involves monitoring for suspicious activities and anomalies that could indicate security breaches. Advanced threat detection tools utilize machine learning algorithms to identify patterns indicative of cyberattacks, enabling proactive defense mechanisms to be activated.

In conclusion, the management and monitoring layer of AI infrastructure ensures that resources are efficiently utilized, system performance is continuously monitored, and security measures are robustly enforced. This layer is integral to maintaining the health, performance, and security of AI systems, enabling them to deliver reliable and effective results in various applications. Through advanced resource management, comprehensive monitoring and alerting, and stringent security management, organizations can ensure their AI infrastructure operates optimally and securely.

8.3 Summary

This chapter delves into the critical components of AI infrastructure, emphasizing how advanced networking technologies and high-performance computing (HPC) enable the efficient deployment of AI and machine learning (ML) workloads. The chapter begins by outlining the growing demands of AI/ML applications, which require specialized infrastructure to handle massive data processing, low-latency communication, and high-bandwidth requirements.

The chapter explores the role of AI-native networking, detailing the use of technologies like GPU clustering, NVLink, and NVSwitch that facilitate parallel processing and high-speed data transfer between compute nodes. It discusses how these components integrate within HPC environments to support AI workloads, providing the speed and scalability needed for complex deep learning models.

InfiniBand is highlighted as a key networking technology in AI infrastructure, offering low-latency, high-throughput interconnects that are essential for AI training and inference tasks. The chapter covers InfiniBand's architecture, including Host Channel Adapters (HCAs) and Virtual Lanes (VLs), and explains how these elements contribute to the overall performance of AI clusters. Other critical interconnect technologies like RoCE (RDMA over Converged Ethernet) and adaptive routing methods such as UGAL (Universal Globally Adaptive Load-balanced) Routing are also discussed, illustrating their roles in enhancing data flow and reducing congestion in AI-driven networks.

Storage infrastructure is another focus, with the chapter examining the importance of high-speed storage solutions such as NVMe SSDs, distributed file systems, and object storage to support the vast data requirements of AI workloads. The integration of tiered storage systems and efficient data management strategies are presented as essential for maintaining performance and data availability in AI environments.

The chapter also addresses the infrastructure needed for deploying AI models in both on-premises and cloud-based environments, highlighting the flexibility and scalability of AI infrastructure frameworks. Technologies like ultra-low latency networking and advanced traffic management protocols are described as pivotal in optimizing network performance for AI tasks.

Overall, this chapter provides a comprehensive overview of the components that constitute AI infrastructure, emphasizing the interplay between networking, computing, and storage technologies. By detailing the architecture and technologies that enable AI/ML workloads, the chapter equips readers with a clear understanding of how to build and optimize infrastructure tailored to the needs of modern AI applications.

References

1. Gordon (2020) History and overview of high performance computing. Gordon College. https// math.gordon.edu/courses/cps343/presentations/History_and_Overview_of_HPC.pdf
2. Morgan (2019) How cray makes ethernet suited For HPC And AI with slingshot. The Next Platform. https://www.nextplatform.com/2019/08/16/how-cray-makes-ethernet-suited-for-hpc-and-ai-with-slingshot/
3. Kim J, Dally WJ, Scott S, Abts D (2008) Technology-driven. IEEE Computer Society, Highly-Scalable Dragonfly Topology
4. Valiant GL (1982) A Scheme for fast parallel communication. Soc Ind Appl Math. https://ldh ulipala.github.io/readings/ValiantPermutationRouting.pdf
5. Chang C-S, Lee D-S, Jou Y-S (2001) Load balanced Birkhoff-von Neumann switches, Part I: One-stage buffering. In: Proceedings of IEEE HPSR
6. Keslassy I, Chuang S-T, Yu K, Miller D, Horowitz M, Solgaard O, McKeown N (2003) Scaling Internet routers using optics. In: Proceedings of ACM SIGCOMM '03, computer communication review, vol 33, no 4, pp 189–200
7. Zhang-Shen, R. & McKeown, N., 2008. Designing a Fault-Tolerant Network Using Valiant Load-Balancing. https://www.researchgate.net/publication/312132050_Designing_a_Fault-Tol erant_Network_Using_Valiant_Load-Balancing
8. Chaulagain SR (2024) Enhanced UGAL routing schemes for dragonfly networks. https://www. researchgate.net/publication/381139232
9. Princeton (2024) GPU computing. Princeton University. https://researchcomputing.princeton. edu/support/knowledge-base/gpu-computing
10. Wang W, Ghobadi M, Shakeri K, Zhang Y, Hasani N (2023) Rail-only: A low-cost high-performance network for training LLMs with trillion parameters. https://arxiv.org/pdf/2307. 12169
11. Ma S (2024) Training LLMs: An efficient GPU traffic routing mechanism within AI/ML cluster with rail-only connections. Outshift by CISCO. https://outshift.cisco.com/blog/training-llms-eff icient-gpu-traffic-routing
12. Enfabrica (2024) Elevate networking for the age of Gen AI. Enfabrica. https://enfabrica.net/

Toward Intelligent and Self-Healing Networks

In recent years, there has been a significant shift toward creating network management and orchestration platform that facilitate proactive response to network conditions, moving away from a system construct that build around the notion of a reactive approach to network conditions and issues. In Chap. 1, we discussed the trends in networking technologies, tracing the evolution from open networks to software-defined networks (SDNs), intent-based networks, and the gradual push toward self-organizing and self-healing networks. This evolution has been driven by numerous technical advancements, particularly in network ASICs, SoCs, and software technologies.

As of this writing, there is ongoing work to create neuromorphic network SoCs. In November 2023, Broadcom announced its Trident 5-X12 chip,[1] which integrates neural networking capabilities. The company claims that its Neural Network Inference Engine "NETGNT," integrated into the Trident 5-X12 chip, can provide advanced telemetry, security, and traffic engineering services. For example, the NETGNT can use an ML (Machine Learning) engine to detect specific traffic patterns and provide immediate intervention, such as identifying anycast flows for AI/ML workloads in real-time. This type of hardware will further enhance the network's ability to provide proactive responses.

The use of AI and machine learning is not limited to ASICs or SoCs; it is also prevalent in network management and operations. In previous chapters, I have referred to "AI-Naïve Networking" and "networks for AI/ML workloads" to depict two distinct areas, collectively termed "AI Networking" to align with industry buzz. AI-native networking uses AI/ML technologies to solve network issues and enhance performance, while networks for AI/ML workloads are designed to accommodate the burst demands and latency requirements of such workloads.

---

[1] Broadcom [1].

We have also explored zero-trust architecture to highlight the importance of security in modern networks. Self-healing networks combine AI-native networking and security elements to create a network capable of self-healing. Some industry experts refer to this concept as autonomous networks or self-driving networks. While achieving a fully autonomous network may not be viable in the near term, the goal is clear, and many organizations are working toward it. Regardless of the definitions and terms, the time has come for networks to become more intelligent and capable of proactive response.

My assertion is that networks will become more dynamic and intelligent in the near term, evolving into what I call "Intelligent Networks" within the coming decades. However, the industry's response to creating proactive network management has been piecemeal, sometimes rendering the concept of self-healing networks confusing. Claims of self-driving or self-healing networks by industry giants often lean more toward marketing than reality. Nevertheless, the momentum toward intelligent networks is undeniable, with self-healing as a critical first step.

We certainly need self-healing networks due to the immense demands on user experience. Network downtimes are costly, ranging from $625K to $1 million per minute, depending on the industry segment. Coupled with the increasing threat of security breaches globally, the need for networks that can at least make administrators aware of issues in real-time, if not fix them automatically, is imperative. Self-healing networks can prevent catastrophic failures, optimize performance, predict issues, and automatically intervene when network service degradation is detected or anticipated.

This chapter will delve into the concept of self-healing networks, exploring mechanisms to develop such capabilities. As the solutions require a collective of technologies and lack a standardized definition, my proposed solutions include a comprehensive discussion on this topic. We will begin our journey by understanding AIOps, an integral part of self-healing networks. So, sit tight and let us explore this fascinating topic together.

9.1   AIOps

AIOps, or Artificial Intelligence for IT Operations, represents the convergence of artificial intelligence (AI) and machine learning (ML) technologies with traditional IT operations to enhance the efficiency, accuracy, and agility of managing modern IT environments. This approach is designed to address the growing complexity of IT infrastructures, where the volume of data, the velocity of change, and the diversity of systems have made traditional management techniques increasingly inadequate. The historical evolution of AIOps can be traced back to the origins of IT operations management, which began with simple monitoring and manual processes. In the early days, IT environments were relatively straightforward, consisting mainly of mainframes and a limited number of applications. Monitoring was primarily done through manual checks and basic alerting systems, often

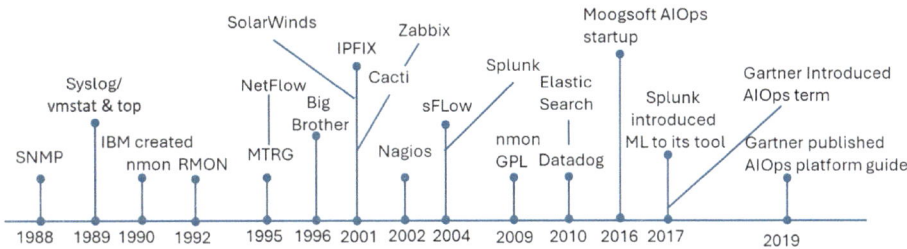

Fig. 9.1 AIOps timeline

resulting in a reactive approach to problem-solving. As IT environments grew in complexity with the advent of distributed computing, the need for more sophisticated tools became evident.

The 1990s and early 2000s saw the rise of network and system management tools that could automate some aspects of monitoring and alerting. These tools, while more advanced than their predecessors, still largely relied on static thresholds and rule-based logic. As IT environments continued to expand, encompassing a broader range of devices, applications, and networks, the limitations of these tools became apparent. They were not equipped to handle the dynamic and increasingly complex nature of modern IT environments, where issues often arose from multiple, interdependent sources. To understand the driving forces behind the transformation in IT operations, it is essential to review the timeline below, which highlights key developments. This timeline should be considered in conjunction with the similar timeline for network visibility presented in Chap. 6 (Fig. 9.1).

The mid-2000s marked a significant turning point with the introduction of big data technologies, which enabled the collection and analysis of vast amounts of data in real-time. This was a crucial development, as it allowed IT operations to move from reactive to proactive management. However, the sheer volume of data generated by modern IT systems posed new challenges, particularly in terms of analyzing and making sense of the data in a timely manner.

The next major leap came with the integration of AI and ML into IT operations. These technologies brought the capability to analyze large datasets at scale, identify patterns, and make predictions about future events. AI algorithms could now learn from historical data, recognize anomalies, and even automate decision-making processes, significantly reducing the time and effort required to manage IT environments. This evolution gave birth to what we now call AIOps.

AIOps platforms combine multiple data sources—such as logs, metrics, events, and traces—and apply AI and ML techniques to correlate, analyze, and provide actionable insights. By doing so, AIOps enables IT teams to detect issues before they impact users, automate routine tasks, and optimize resource allocation. This marks a paradigm shift from traditional operations management, where IT teams were often overwhelmed by the sheer volume of alerts and data, to a more strategic, data-driven approach.

In summary, AIOps has emerged as a critical response to the increasing complexity of IT environments. It represents the culmination of decades of evolution in IT operations, moving from manual processes and basic automation to a sophisticated, AI-driven approach that promises to transform how organizations manage and optimize their IT infrastructures. As AI and ML technologies continue to advance, the role of AIOps in IT management is expected to grow, offering new levels of efficiency, reliability, and innovation.

9.1.1 AIOps Architecture

AIOps (Artificial Intelligence for IT Operations) architecture is designed to manage the vast and complex data generated by modern IT environments, enabling proactive and intelligent management of IT operations. The architecture typically integrates several components that work together to collect, analyze, and act on data in real-time. Below is a diagram that illustrates the AIOps architecture and its key components, followed by a brief discussion of each component (Fig. 9.2).

The key components of the AIOps architecture include the Integration Layer, Data Collection Layer, Data Storage and Management Layer, AI/ML Processing Layer, Correlation and Analytics Layer, Visualization and User Interface Layer, and the Automation and Orchestration Layer. Each of these components plays a crucial role in enabling the efficient and intelligent management of IT operations, working together to collect, process, analyze, and act on data in real-time.

9.1.1.1 Integration Layer

The Integration Layer in AIOps architecture serves as the foundational component that connects the AIOps platform with various ITSM tools, monitoring systems, and other external data sources. It ensures seamless communication and data flow between these systems and the AIOps platform, enabling a comprehensive view of the IT environment. ITSM tools, though primarily focused on managing IT services and workflows, provide valuable data that enriches the AIOps platform's insights.

For instance, incident management data from ITSM tools can be fed into the AIOps platform to correlate incidents with real-time monitoring data, enabling quicker identification of root causes and more effective automated responses. Similarly, change management records from ITSM tools can be integrated to analyze the impact of changes on system performance, helping to predict potential issues before they occur (Fig. 9.3).

The diagram above illustrates the interaction between ITSM tools, the Integration Layer, and the Data Collection Layer within an AIOps framework. The Integration Layer plays a critical role in bridging ITSM tools and other external data sources with the AIOps

Fig. 9.2 AIOps architecture

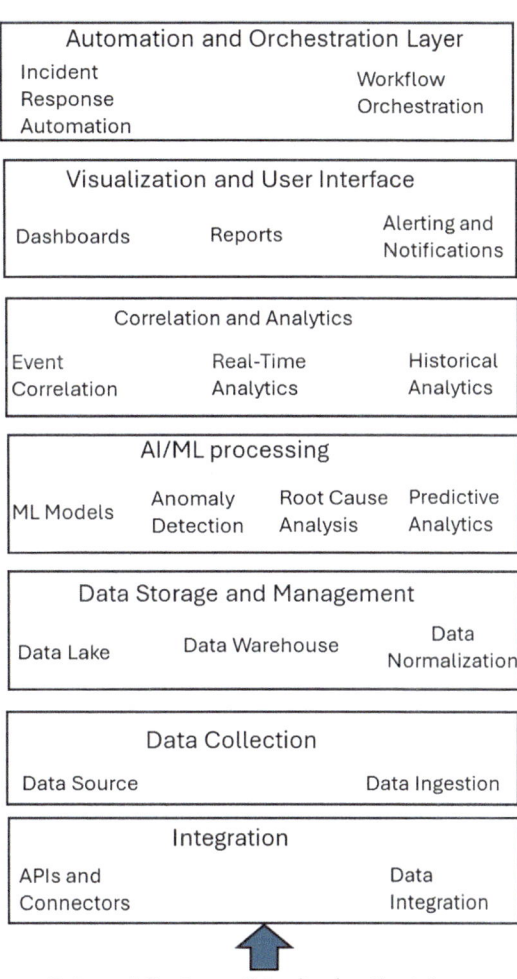

External Systems (Monitoring Tools)

platform, ensuring a seamless flow of data that enhances the overall dataset used for analysis. The Integration Layer is powered by various tools such as Mulesoft, Apache NiFi, Zapier, Boomi, and TIBCO, which are designed to ingest, process, and transmit data from external sources into the AIOps platform. These tools use APIs and connectors to pull data from ITSM tools, monitoring systems, and other external sources. These connectors act as gateways for data to enter the AIOps ecosystem, allowing the platform to receive real-time information about the current state of the IT environment, which is vital for immediate detection and response to issues.

Once the data is ingested, it undergoes normalization to ensure consistency in format and structure. Different ITSM tools and external sources may present data in various

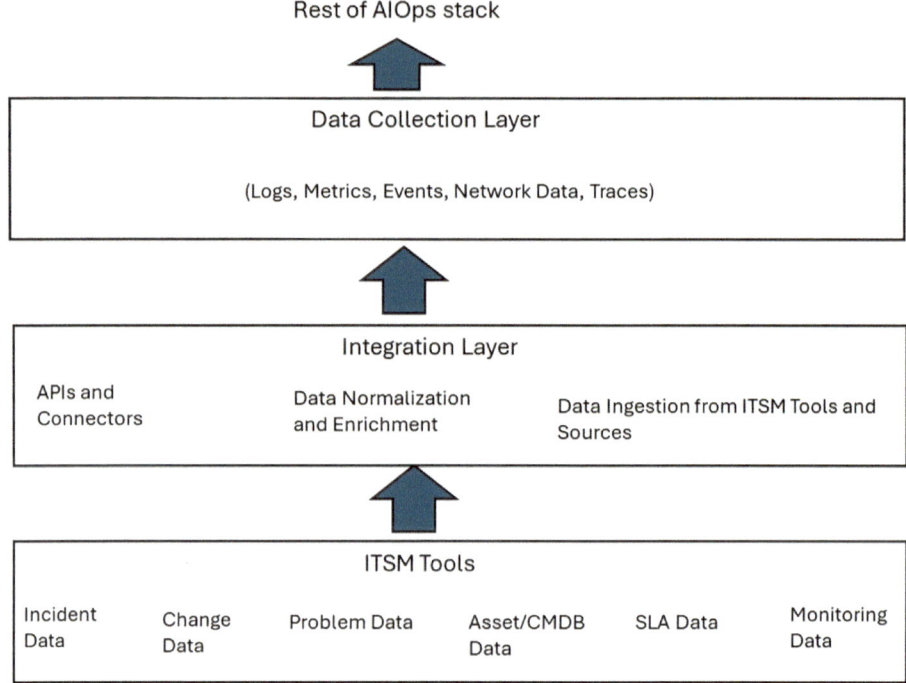

Rest of AIOps stack

Data Collection Layer

(Logs, Metrics, Events, Network Data, Traces)

Integration Layer

APIs and
Connectors

Data Normalization
and Enrichment

Data Ingestion from ITSM Tools and
Sources

ITSM Tools

Incident
Data

Change
Data

Problem Data

Asset/CMDB
Data

SLA Data

Monitoring
Data

Fig. 9.3 Interaction between ITSM tools and integration layer of AIOps framework

formats; the normalization process standardizes this data so it can be effectively processed and analyzed by the AIOps platform. After normalization, the data is enriched
with additional context, such as tagging it with metadata, linking incident data to specific
configuration items in a CMDB, or correlating events with historical trends. This enrichment process adds depth to the data, making it more valuable for predictive analytics and
decision-making.

The Integration Layer, facilitated by tools like Mulesoft or Boomi, supports bidirectional data flow, allowing the AIOps platform not only to pull data in but also to push
data back to ITSM tools and other external systems. For instance, if the AIOps platform identifies and resolves an anomaly, it can update the corresponding incident in the
ITSM tool, ensuring that all systems reflect the latest status. Additionally, the Integration
Layer can trigger automated actions in external tools based on insights generated by the
AIOps platform, such as initiating a change request in an ITSM tool if a predictive model
anticipates a potential failure.

After the data is processed by the Integration Layer, it is passed to the Data Collection
Layer, which serves as the repository for all incoming data. This layer aggregates logs,
metrics, events, and traces from various sources, including the enriched data provided by
the Integration Layer. The Data Collection Layer is vital in the AIOps framework as it

consolidates data from multiple sources, providing a unified view of the IT environment. This aggregated data serves as the primary input for AI/ML models within the AIOps platform, enabling these models to identify patterns, detect anomalies, and predict future issues. The data also supports correlation and analysis, helping to pinpoint the root causes of problems and determine the best course of action.

By seamlessly feeding enriched data into the Data Collection Layer, the Integration Layer ensures that the AIOps platform can leverage both real-time events and historical and contextual information. This integrated approach enhances the platform's predictive capabilities, allowing it to anticipate and mitigate potential issues before they impact the IT environment. Additionally, the ability to automate responses based on these insights reduces the operational burden on IT teams, improving efficiency and minimizing downtime.

In summary, the Integration Layer is a vital component of the AIOps architecture, enabling effective data flow from ITSM tools and other external sources into the AIOps platform. Powered by integration tools like Mulesoft, Apache NiFi, Zapier, Boomi, and TIBCO, this layer normalizes, enriches, and facilitates bidirectional data flow, ensuring that the AIOps platform has comprehensive, high-quality data to deliver actionable insights and automated responses across the IT environment.

9.1.2 Data Collection Layer

The Data Collection Layer in an AIOps (Artificial Intelligence for IT Operations) architecture serves as the core component for aggregating and centralizing vast amounts of data generated across an IT environment. This layer is responsible for collecting logs, metrics, events, network data, traces, and other relevant information from various sources. The data collected here is crucial for enabling AI/ML models and analytics engines within the AIOps platform to perform accurate analysis, anomaly detection, and predictive maintenance. The interaction between the Integration Layer and the Data Collection Layer is critical to the overall functionality of the AIOps framework. The Integration Layer, facilitated by tools such as Mulesoft, Apache NiFi, Dell Boomi, and TIBCO, acts as the intermediary that pulls data from diverse sources, including ITSM tools, monitoring systems, and external data repositories. These tools use APIs and connectors to extract data from these sources and ensure that the data is normalized and enriched before it reaches the Data Collection Layer. For instance, Mulesoft provides comprehensive APIs and connectors that enable the seamless integration of data from various systems, while Apache NiFi offers powerful capabilities for data flow management, including real-time data routing and transformation.

Once the data is ingested by the Integration Layer and processed—meaning it is normalized to ensure consistency and enriched with additional context like metadata or linked configuration items—the data is then fed into the Data Collection Layer. The Data Collection Layer is supported by tools like Elasticsearch and Prometheus, which are designed to handle large volumes of data and provide the necessary infrastructure for real-time and historical data analysis.

Elasticsearch is a search and analytics engine that indexes data in real-time, allowing for fast and efficient querying. It is particularly well-suited for collecting and analyzing logs and metrics from a variety of sources. When data is passed from the Integration Layer to Elasticsearch, it is indexed and stored in a manner that facilitates quick retrieval and analysis. This enables the AIOps platform to perform complex searches and generate insights based on real-time and historical data.

Prometheus, on the other hand, is an open-source monitoring system that excels at collecting time-series data, particularly in dynamic environments like those involving containerized applications. Prometheus collects metrics from different services and stores them in a time-series database. When integrated with the data processed by the Integration Layer, Prometheus ensures that this time-series data is consistently formatted and enriched with contextual information, enabling precise monitoring and alerting.

The seamless interaction between the Integration Layer and the Data Collection Layer ensures that the AIOps platform has access to comprehensive, high-quality data. The Integration Layer continuously streams normalized and enriched data into the Data Collection Layer, where it is aggregated with other incoming data streams to form a holistic view of the IT environment. This process enables the AIOps platform to deliver comprehensive insights, allowing IT teams to optimize performance, improve operational efficiency, and reduce downtime.

In this setup, the Data Collection Layer is not just a passive repository but an active participant in the data processing workflow. Once the data is collected, it is structured and indexed to optimize performance and scalability. This data is then made available to the AI/ML models, correlation engines, and visualization tools that operate in the upper layers of the AIOps architecture (Fig. 9.4).

In the diagram above, the Integration Layer is depicted as the conduit through which external systems and tools interface with the Data Collection Layer. Tools like Mulesoft, Apache NiFi, Dell Boomi, and TIBCO in the Integration Layer handle data ingestion, normalization, and enrichment, which is then passed to Elasticsearch or Prometheus in the Data Collection Layer. These tools within the Data Collection Layer aggregate, structure, and index the data, making it readily available for analysis and insights generation within the AIOps platform.

This integrated approach ensures that the AIOps platform can effectively manage both real-time and historical data, leveraging the enriched data provided by the Integration Layer to deliver actionable insights that enhance IT operations and reduce downtime.

Fig. 9.4 Interaction of integration and data collection layers of AIOps and the tools that can be used for this solution

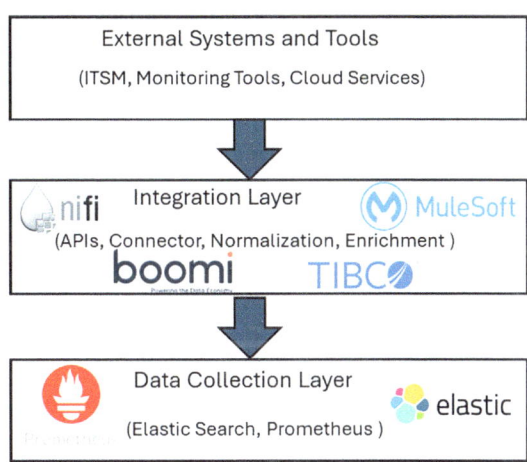

9.1.3   Data Storage and Management Layer

The Data Storage and Management Layer in an AIOps (Artificial Intelligence for IT Operations) architecture is responsible for the long-term storage, organization, and efficient retrieval of the vast amounts of data collected from the IT environment. This layer plays a critical role in ensuring that data is not only stored securely and efficiently but also remains accessible for analysis, correlation, and reporting. The Data Storage and Management Layer typically works closely with the Data Collection Layer to ensure that data flows seamlessly from ingestion to storage, where it can be further processed and analyzed by the AIOps platform.

In an AIOps framework, the Data Collection Layer, supported by tools like Elasticsearch and Prometheus, is responsible for the real-time ingestion and indexing of logs, metrics, events, and traces. Once the data is aggregated and structured in the Data Collection Layer, it is then passed to the Data Storage and Management Layer, where it is organized for long-term storage and efficient retrieval. The interaction between these two layers is crucial for maintaining data integrity, performance, and accessibility.

9.1.3.1   Tools Used in the Data Storage and Management Layer

Hadoop Distributed File System (HDFS) is a widely used tool in the Data Storage and Management Layer. It provides a scalable, distributed storage system that is capable of handling large volumes of data across multiple nodes. HDFS is particularly well-suited for storing unstructured data collected from various sources, and it ensures that this data is replicated across nodes to prevent data loss and ensure high availability.

Amazon S3 is another common choice for data storage in cloud-based AIOps environments. S3 provides scalable object storage that can handle large datasets and offers integration with various data processing and analytics tools. Data from the Data Collection Layer can be stored in S3 buckets, where it is easily accessible for further analysis or long-term archiving.

Apache Cassandra is a distributed NoSQL database that excels in managing large volumes of time-series data, making it an ideal choice for storing metrics and logs collected by tools like Prometheus. Cassandra provides high availability and fault tolerance, ensuring that data remains accessible even in the event of hardware failures.

Apache HBase is another tool often used in the Data Storage and Management Layer. HBase is a distributed, column-oriented database built on top of HDFS, designed to handle large-scale data processing and storage. It is well-suited for storing sparse data and can manage real-time read/write access to large datasets.

9.1.3.2 Functionality of the Data Storage and Management Layer

The Data Storage and Management Layer ensures that data is not only stored efficiently but is also readily available for analysis, reporting, and machine learning. Once data is collected and structured by the Data Collection Layer, it is transmitted to the Data Storage and Management Layer, where it undergoes several processes:

- **Data Organization**: Data is organized into appropriate storage formats, whether as flat files, time-series databases, or columnar storage, depending on the type of data and its intended use.
- **Data Partitioning**: To optimize performance and scalability, data is partitioned across different nodes or storage units. This allows for faster querying and retrieval, especially when dealing with large datasets.
- **Data Replication**: To ensure high availability and fault tolerance, data is often replicated across multiple storage locations. This ensures that even if one node fails, the data remains accessible from another node.
- **Data Compression and Deduplication**: To save storage space and reduce costs, the data may be compressed and deduplicated. This process reduces the overall storage footprint while maintaining data integrity.

The Data Storage and Management Layer works hand-in-hand with the Data Collection Layer to ensure that data flows seamlessly from ingestion to long-term storage. The data stored in this layer can then be accessed by the AIOps platform for in-depth analysis, machine learning, and reporting, providing the foundation for intelligent IT operations management (Fig. 9.5).

In the diagram above, the Data Collection Layer is responsible for the initial aggregation, structuring, and indexing of data as it is ingested from various sources. This data is then passed down to the Data Storage and Management Layer, where tools like HDFS, Amazon S3, Apache Cassandra, and HBase manage the long-term storage and retrieval of this data.

Fig. 9.5 Interaction between data collection layer and data storage and management layer of AIOps

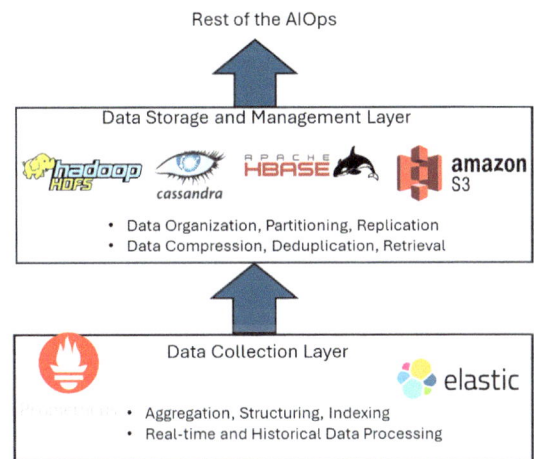

The Data Storage and Management Layer organizes the data into appropriate formats, partitions it for performance, replicates it for fault tolerance, and compresses it to save space. This layer ensures that the data remains accessible and ready for analysis, providing the AIOps platform with the necessary foundation to generate insights, run machine learning models, and deliver intelligent operations management.

In summary, the Data Storage and Management Layer is a critical component of the AIOps architecture, ensuring that data collected and processed by the Data Collection Layer is stored efficiently and remains readily available for further analysis and decision-making. The interaction between these layers, facilitated by tools like Elasticsearch, Prometheus, HDFS, and Amazon S3, ensures that the AIOps platform can effectively manage and utilize the vast amounts of data generated within modern IT environments.

9.1.4 AI/ML Processing Layer

The AI/ML Processing Layer is a critical component of the AIOps (Artificial Intelligence for IT Operations) architecture, responsible for analyzing the vast amounts of data collected and stored by previous layers. This layer leverages artificial intelligence (AI) and machine learning (ML) algorithms to detect patterns, identify anomalies, predict future issues, and automate decision-making processes within the IT environment. The AI/ML Processing Layer transforms raw data into actionable insights, enabling proactive and intelligent IT operations management.

The AI/ML Processing Layer interacts closely with the Data Storage and Management Layer, where it accesses the structured and unstructured data stored over time. The Data Storage and Management Layer, supported by tools like Hadoop Distributed File System (HDFS), Amazon S3, Apache Cassandra, and Apache HBase, provides the necessary data infrastructure that the AI/ML Processing Layer utilizes for training models, performing real-time analysis, and generating predictions.

9.1.4.1 Tools Used in the AI/ML Processing Layer

Apache Spark is a widely used tool for distributed data processing and machine learning. Spark's MLlib library provides a scalable, high-performance machine learning framework that supports a wide range of algorithms, including classification, regression, clustering, and collaborative filtering. Spark can process large datasets in parallel across a cluster, making it an excellent choice for implementing AI/ML models in an AIOps environment.

TensorFlow is an open-source deep learning framework developed by Google. It is designed to handle large-scale machine learning tasks, including image and speech recognition, natural language processing, and predictive analytics. TensorFlow can be integrated with the data stored in the Data Storage and Management Layer to train deep learning models that can identify complex patterns and anomalies in IT operations data.

H2O.ai is another popular AI/ML platform that provides a suite of machine learning algorithms for building predictive models. H_2O supports distributed processing, allowing it to handle large datasets efficiently. It also offers AutoML capabilities, which automate the process of selecting the best model and hyperparameters, making it easier to implement machine learning in AIOps.

Python Scikit-Learn is a versatile machine learning library that offers a range of algorithms for classification, regression, clustering, and dimensionality reduction. Scikit-Learn is often used for smaller-scale AI/ML tasks or as a prototyping tool before scaling up to more powerful frameworks like Spark or TensorFlow.

9.1.4.2 Functionality of the AI/ML Processing Layer

The AI/ML Processing Layer uses the data stored in the Data Storage and Management Layer to train, validate, and deploy machine learning models. These models are then used to analyze real-time and historical data, providing insights that help IT teams make informed decisions. The primary functionalities of this layer include:

- **Anomaly Detection**: AI/ML models are trained to recognize normal patterns of behavior within the IT environment. When deviations from these patterns are detected, the system can raise alerts or automatically initiate corrective actions.
- **Predictive Analytics**: By analyzing historical data, machine learning models can predict future events, such as system failures, performance degradation, or security breaches. This allows IT teams to take proactive measures to prevent issues before they occur.
- **Root Cause Analysis**: When an issue is detected, AI/ML models can correlate various data points to identify the root cause. This helps in reducing the time-to-resolution and minimizing the impact on operations.

- **Automated Decision-Making**: Based on the insights generated by the AI/ML models, the system can automate routine decisions and responses, such as scaling resources, rerouting traffic, or applying patches (Fig. 9.6).

In this diagram, the Data Storage and Management Layer is responsible for organizing, partitioning, replicating, and compressing data to ensure it is stored efficiently and remains accessible for analysis. This layer feeds the AI/ML Processing Layer, where tools like Apache Spark, TensorFlow, H2O.ai, and Scikit-Learn utilize the stored data to train and deploy machine learning models. The AI/ML Processing Layer leverages these models to perform critical functions such as anomaly detection, predictive analytics, root cause analysis, and automated decision-making. By analyzing both real-time and historical data, this layer generates actionable insights that empower IT teams to proactively manage and optimize their operations.

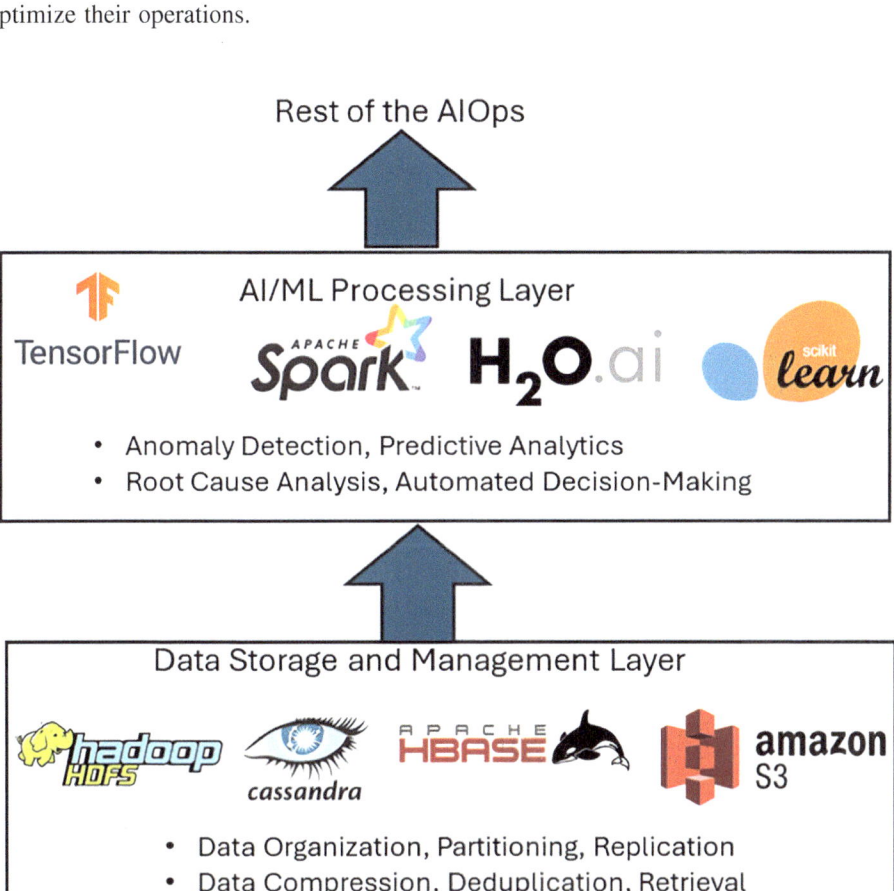

Fig. 9.6 Interaction between data storage and management layer and AI/ML processing layer of AIOps

The AI/ML Processing Layer is the analytical powerhouse of the AIOps architecture, transforming raw data into meaningful insights through the application of machine learning and artificial intelligence. It works closely with the Data Storage and Management Layer to access the data needed for training and deploying models that can detect anomalies, predict future issues, and automate decision-making processes. With the support of tools like Apache Spark, TensorFlow, H2O.ai, and Scikit-Learn, the AI/ML Processing Layer enables proactive and intelligent management of IT operations, reducing downtime and improving overall efficiency.

9.1.5 Correlation and Analytics Layer

The Correlation and Analytics Layer is a pivotal component of the AIOps (Artificial Intelligence for IT Operations) architecture, responsible for synthesizing data insights by correlating different data points across the IT environment. This layer leverages the outputs from the AI/ML Processing Layer to create comprehensive views of system behavior, identify root causes of issues, detect trends, and generate actionable insights. The Correlation and Analytics Layer plays a crucial role in turning raw data and model predictions into coherent, actionable information that can drive operational decisions. The Correlation and Analytics Layer interacts closely with the AI/ML Processing Layer, utilizing the insights generated by machine learning models to perform advanced data correlation and analysis. While the AI/ML Processing Layer focuses on anomaly detection, predictive analytics, and root cause analysis, the Correlation, and Analytics Layer takes these outputs and contextualizes them within the broader IT environment. This involves correlating events, metrics, logs, and traces across different systems and services to identify patterns and relationships that may not be immediately obvious.

9.1.5.1 Tools Used in the Correlation and Analytics Layer

Splunk is a widely used tool in this layer, providing powerful analytics capabilities that allow users to search, monitor, and analyze machine data in real-time. Splunk's ability to correlate data from different sources makes it an essential tool for identifying trends and anomalies across the IT landscape.

Elastic Stack (ELK), comprising Elasticsearch, Logstash, and Kibana, is another powerful suite used in the Correlation and Analytics Layer. Elasticsearch provides fast and scalable search capabilities, Logstash enables data collection and processing, and Kibana offers visualization tools that make it easier to explore and analyze data correlations and trends.

Grafana is often used in conjunction with time-series databases like Prometheus or InfluxDB. It provides visualization and dashboarding capabilities, enabling IT teams to create dynamic, real-time dashboards that correlate various metrics and logs. Grafana's alerting features can also trigger automated responses based on correlations identified in the data.

IBM QRadar is a Security Information and Event Management (SIEM) tool that excels in correlating security events across the IT environment. QRadar collects and analyzes log data, network flows, and other security-related information to detect threats and provide insights into security posture.

SAS Analytics is another tool that provides advanced analytics capabilities, allowing for deep data exploration and the correlation of various datasets to uncover hidden patterns and relationships. SAS is often used in complex environments where detailed statistical analysis and forecasting are required.

9.1.5.2 Functionality of the Correlation and Analytics Layer

The primary function of the Correlation and Analytics Layer is to correlate diverse data points to identify meaningful patterns and relationships within the IT environment. This layer takes the outputs from the AI/ML Processing Layer and enriches them by:

- Event Correlation: Correlating events across systems to identify potential causes and effects. For example, correlating a spike in CPU usage with a specific deployment or configuration change.
- Trend Analysis: Analyzing historical data to identify trends, such as increasing latency in a particular service, that could indicate a developing issue.
- Root Cause Correlation: Building on the root cause analysis provided by the AI/ML Processing Layer, the Correlation and Analytics Layer can correlate additional data to confirm or refine the root cause of an issue.
- Contextual Analysis: Adding context to the insights generated by AI/ML models, such as correlating performance metrics with business KPIs to understand the impact of technical issues on business outcomes.
- Alerting and Visualization: Providing real-time dashboards and alerts that visualize correlations and analytics, enabling IT teams to take timely action based on the insights (Fig. 9.7).

In this diagram (Please refer to Fig. 9.7), the AI/ML Processing Layer is responsible for generating insights through machine learning algorithms, such as detecting anomalies, predicting potential failures, and identifying root causes of issues. These insights are then passed on to the Correlation and Analytics Layer.

The Correlation and Analytics Layer, using tools like Splunk, ELK Stack, Grafana, IBM QRadar, and SAS, take these outputs and correlates them with other data points across the IT environment. For example, it might correlate an anomaly detected by TensorFlow with logs collected by Elasticsearch and visualized in Grafana, identifying a pattern that suggests a developing issue. This layer also provides contextual analysis, alerting, and visualization, turning the raw data and AI/ML insights into actionable information.

Fig. 9.7 Interaction between AI/ML processing layer and correlation and analytics layer of AIOps

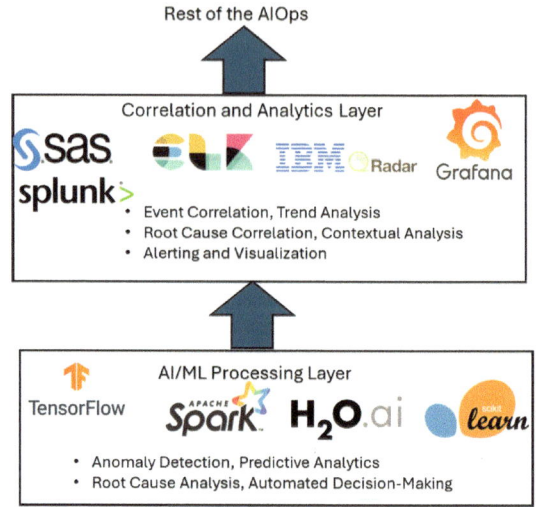

The Correlation and Analytics Layer is a key component of the AIOps architecture that transforms the outputs of the AI/ML Processing Layer into meaningful, actionable insights. It does this by correlating diverse data points, identifying trends, and providing contextual analysis that enhances the understanding of IT operations. With the help of tools like Splunk, the ELK Stack, Grafana, IBM QRadar, and SAS, this layer ensures that IT teams can make informed decisions based on comprehensive data analysis, ultimately improving the efficiency and effectiveness of IT operations.

9.1.6 Visualization and User Interface Layer

The Visualization and User Interface (UI) Layer is a critical component of the AIOps (Artificial Intelligence for IT Operations) architecture, responsible for presenting data insights in a user-friendly and actionable manner. This layer provides IT teams with real-time dashboards, reports, and visualizations that help them monitor system health, identify trends, and respond to issues promptly. The Visualization and UI Layer takes the complex outputs from the Correlation and Analytics Layer and translates them into intuitive visual formats, enabling IT operations to make informed decisions quickly.

The Visualization and UI Layer interacts closely with the Correlation and Analytics Layer, which synthesizes and correlates data from various sources to generate actionable insights. Once the Correlation and Analytics Layer has processed the data, identifying patterns, anomalies, and trends, the Visualization and UI Layer takes these insights and presents them in a way that is accessible and meaningful to end-users. This includes creating dashboards, visual alerts, reports, and other interactive elements that allow IT teams to monitor and manage their environments effectively.

9.1.6.1 Tools Used in the Visualization and User Interface Layer

Grafana is one of the most widely used tools in this layer, offering powerful visualization capabilities and support for a wide range of data sources, including Prometheus, Elasticsearch, and others. Grafana allows users to create dynamic, real-time dashboards that display correlated metrics, logs, and events, making it easier to track system performance and identify issues as they arise.

Kibana, part of the ELK Stack (Elasticsearch, Logstash, Kibana), is another essential tool for visualizing data stored in Elasticsearch. Kibana provides a variety of visualizations, including line charts, bar charts, pie charts, and heat maps, allowing users to explore and analyze data correlations identified by the Correlation and Analytics Layer.

Tableau is a powerful business intelligence and data visualization tool that can be used to create interactive dashboards and reports. Tableau supports a wide range of data sources and offers advanced analytics features, making it suitable for visualizing complex IT operations data and correlating it with business metrics.

Power BI by Microsoft is another versatile tool for creating visual reports and dashboards. Power BI integrates seamlessly with various data sources and offers rich visualization capabilities, including the ability to create interactive reports that provide deep insights into IT operations.

Datadog offers a cloud-based monitoring and analytics platform that includes robust visualization features. Datadog allows users to create custom dashboards that integrate metrics, logs, and traces from different sources, providing a unified view of the IT environment.

9.1.6.2 Functionality of the Visualization and User Interface Layer

The Visualization and UI Layer serves several key functions within the AIOps architecture:

- **Real-Time Dashboards**: The layer provides real-time dashboards that display key metrics, logs, and events in an easily digestible format. These dashboards allow IT teams to monitor system health and performance at a glance.
- **Interactive Visualizations**: The layer offers interactive visualizations that enable users to drill down into specific data points, explore trends over time, and correlate different metrics. This helps in identifying the root cause of issues and understanding the impact of anomalies.
- **Alerting and Notifications**: The Visualization and UI Layer supports alerting and notification features that notify IT teams when certain thresholds are breached or when anomalies are detected. Alerts can be visualized on dashboards or sent via email, SMS, or other channels.
- **Custom Reports**: The layer enables the generation of custom reports that can be tailored to meet the needs of different stakeholders. These reports may include detailed analysis of system performance, historical trends, and predictive insights.

Fig. 9.8 Interaction between correlation and analytics layer and visualization and user interface layer of AIOps

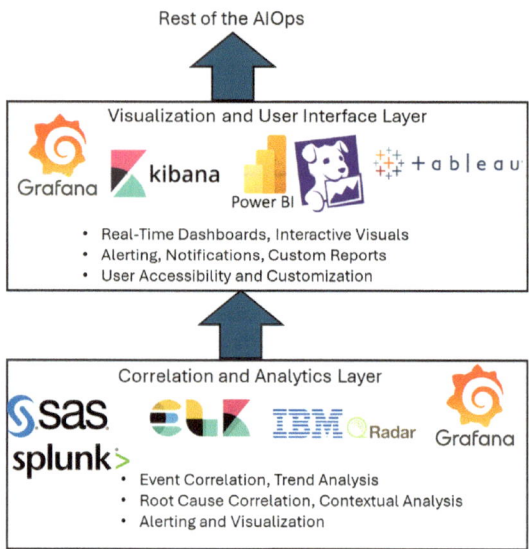

- **User Accessibility**: The Visualization and UI Layer is designed to be user-friendly, ensuring that IT teams of all skill levels can access and interpret the data. This includes the ability to customize dashboards and reports according to individual preferences and requirements (Fig. 9.8).

In this diagram, the Correlation and Analytics Layer is responsible for processing data, correlating events, and generating insights that are then passed on to the Visualization and UI Layer. The Visualization and UI Layer, using tools like Grafana, Kibana, Tableau, Power BI, and Datadog, takes these insights and presents them in real-time dashboards, interactive visualizations, and custom reports. The Visualization and UI Layer also handles alerting and notifications, ensuring that IT teams are promptly informed of any critical issues. By providing a user-friendly interface, this layer makes complex data accessible and actionable, allowing IT teams to monitor system health, identify trends, and respond to issues effectively.

The Visualization and User Interface Layer is a vital component of the AIOps architecture, responsible for transforming the outputs from the Correlation and Analytics Layer into actionable insights that are easy to understand and act upon. With the help of tools like Grafana, Kibana, Tableau, Power BI, and Datadog, this layer provides real-time dashboards, interactive visualizations, and custom reports that enable IT teams to monitor, manage, and optimize their environments efficiently. By ensuring that data is presented in a clear and accessible manner, the Visualization and UI Layer plays a crucial role in the overall success of AIOps.

9.1.7 Automation and Orchestration Layer

The Automation and Orchestration Layer is the final and crucial component of the AIOps (Artificial Intelligence for IT Operations) architecture, responsible for executing automated actions and orchestrating workflows based on the insights derived from the layers below it. This layer ensures that the AIOps platform can not only detect and analyze issues but also take proactive and automated steps to resolve them, thereby minimizing downtime, reducing manual intervention, and improving overall operational efficiency. The Visualization and User Interface Layer provides IT teams with actionable insights through real-time dashboards, alerts, and reports. These insights, once visualized and understood, often necessitate immediate or scheduled actions to address potential issues, optimize performance, or maintain service levels. The Automation and Orchestration Layer interacts with the Visualization and UI Layer by receiving triggers based on the data presented, such as when a certain threshold is breached or an anomaly is detected. Once these triggers are received, the Automation and Orchestration Layer executes predefined workflows or automation scripts to remediate issues, scale resources, deploy updates, or perform other necessary actions.

9.1.7.1 Tools Used in the Automation and Orchestration Layer

Ansible by Red Hat is a widely used automation tool that simplifies the management of IT infrastructure by automating configuration management, application deployment, and other repetitive tasks. Ansible uses a declarative language (YAML) to define automation workflows, making it easy to integrate with AIOps platforms for orchestrating complex operations.

Puppet is another powerful automation tool that automates the provisioning, configuration, and management of IT infrastructure. Puppet's configuration management capabilities allow IT teams to maintain consistency across large-scale environments, making it ideal for automating routine tasks triggered by the AIOps platform.

Chef is a configuration management tool that automates the process of configuring, deploying, and managing servers and applications across an IT environment. Chef uses a domain-specific language (DSL) based on Ruby, allowing for flexible automation and orchestration of complex workflows in response to AIOps insights.

Jenkins is a widely used automation server that enables continuous integration and continuous delivery (CI/CD). Jenkins can automate the process of building, testing, and deploying applications, ensuring that updates and patches are applied quickly and efficiently in response to triggers from the AIOps platform.

ServiceNow Orchestration is an extension of the ServiceNow platform that provides powerful automation capabilities, allowing IT teams to automate IT and business processes across systems and applications. It is often used in conjunction with ServiceNow's ITSM capabilities to automate incident resolution, change management, and other critical tasks.

Terraform by HashiCorp is an infrastructure as code (IaC) tool that automates the provisioning and management of cloud infrastructure. Terraform is particularly useful in dynamic environments where infrastructure needs to be scaled up or down based on real-time data from the AIOps platform.

9.1.7.2 Functionality of the Automation and Orchestration Layer

The Automation and Orchestration Layer provides several key functionalities that enable proactive and efficient IT operations management:

- **Automated Remediation:** This layer automatically takes corrective actions when issues are detected. For example, if the Visualization and UI Layer identifies a server under high load, the Automation Layer might trigger the deployment of additional resources to balance the load.
- **Workflow Orchestration**: The layer coordinates complex workflows that span multiple systems and processes. For example, in response to a security alert, the Automation Layer might orchestrate a workflow that includes isolating the affected system, patching vulnerabilities, and updating firewall rules.
- **Continuous Integration and Delivery (CI/CD)**: Automation tools in this layer facilitate the CI/CD process, ensuring that new code is tested, built, and deployed automatically whenever it is committed to the repository. This helps in maintaining agility and reducing the time to market for new features and updates.
- **Infrastructure as Code (IaC)**: The Automation Layer leverages IaC tools like Terraform to manage and provision infrastructure in a consistent and repeatable manner. This ensures that infrastructure changes are version-controlled and can be rolled back if necessary.
- **Self-Healing**: The layer can implement self-healing mechanisms, where systems automatically detect and repair issues without human intervention. For instance, if a service fails, the Automation Layer might automatically restart the service or roll back to a previous stable version (Fig. 9.9).

In this diagram (Please refer above), the Visualization and User Interface Layer provides IT teams with insights and alerts that may require action. These insights are passed to the Automation and Orchestration Layer, where tools like Ansible, Puppet, Chef, Jenkins, and Terraform execute the necessary actions. Whether it's deploying new infrastructure, rolling out updates, or remediating issues, the Automation and Orchestration Layer ensures that these tasks are completed automatically, reducing the need for manual intervention.

The Automation and Orchestration Layer also enables complex workflow orchestration, where multiple actions are coordinated across different systems and services in response to triggers from the Visualization Layer. For instance, a performance alert visualized in

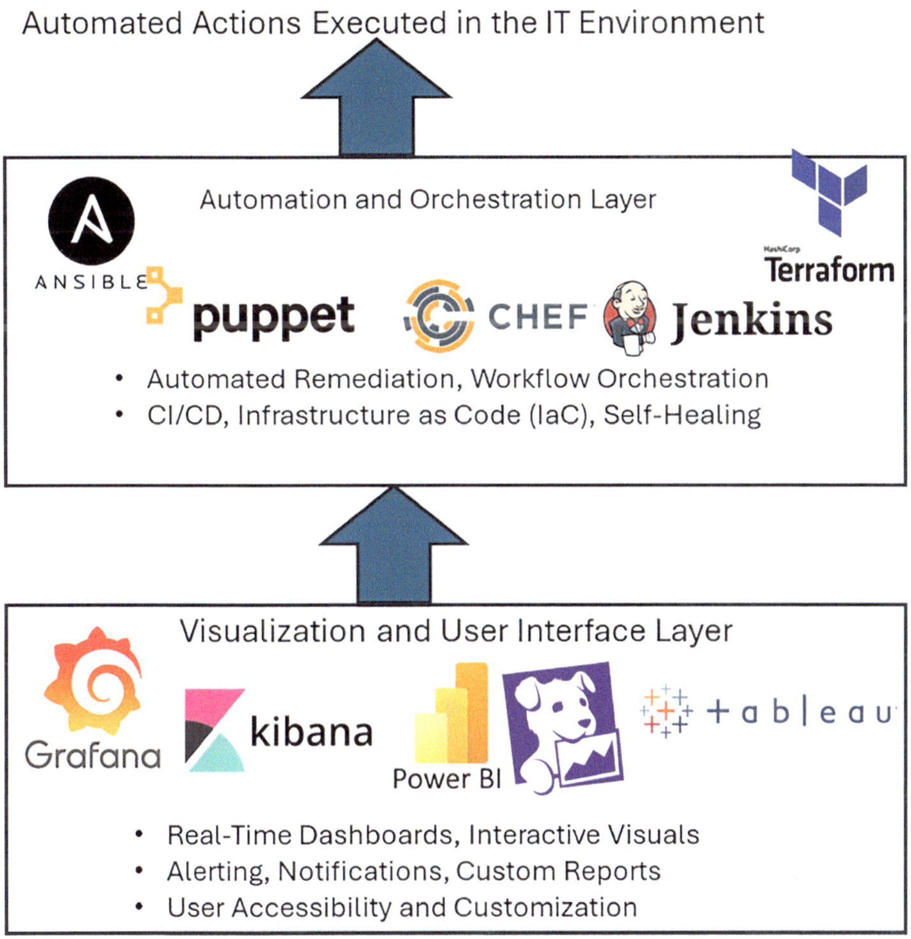

Automated Actions Executed in the IT Environment

Fig. 9.9 Interaction between visualization and user interface layer and automation and orchestration layer of AIOps

Grafana might trigger an automated workflow in Jenkins to deploy additional resources or rollback a problematic deployment.

The Automation and Orchestration Layer is the execution engine of the AIOps architecture, responsible for turning insights into actions. It interacts closely with the Visualization and UI Layer, receiving triggers based on visualized data and executing automated workflows to maintain system health, optimize performance, and reduce downtime. With the support of tools like Ansible, Puppet, Chef, Jenkins, and Terraform, this layer ensures that the AIOps platform can respond to issues in real-time, orchestrate complex processes, and maintain continuous delivery of IT services. The Automation and

Orchestration Layer is essential for achieving the full potential of AIOps, enabling a self-healing, highly efficient IT environment.

9.2   Self-Healing Networks

Self-healing networks represent the next frontier in networking technology, designed to autonomously detect, diagnose, and recover from issues without requiring manual intervention. These networks leverage advanced artificial intelligence, machine learning, and automation to maintain operational continuity, ensuring minimal disruptions and maximizing uptime. The increasing complexity of modern networks—driven by the rise of cloud computing, edge devices, IoT, and distributed applications—demands a more intelligent approach to network management. As a result, the traditional reactive methods of troubleshooting and maintenance are being replaced by proactive and adaptive solutions that can respond in real-time to various network conditions.

The concept of self-healing in networking draws inspiration from biological systems, where resilience and adaptability are key to survival. Similarly, self-healing networks are engineered to continuously monitor their environment, analyze performance data, and implement corrective actions to optimize performance and mitigate risks. These networks operate on a feedback loop where they learn from past events, predict potential future issues, and dynamically adjust configurations to prevent downtime. For businesses, this level of automation is crucial, as network outages can cost $9,000 per minute, $540,000 per hour, or $680,000 per event, significantly impacting operational continuity. Moreover, a 2022 Uptime Institute survey reports that networking issues cause a significant portion of IT outages across industries.

In the previous section, we discussed AIOps (Artificial Intelligence for IT Operations), which plays a foundational role in enabling self-healing networks by providing the intelligence and automation needed to identify and resolve network issues autonomously. Now, let us delve deeper into what exactly constitutes self-healing and how such networks are developed.

With the market for self-healing networks projected to grow from over $600 million in 2022 to more than $12 billion by 2032, at a compound annual growth rate (CAGR) exceeding 27%, it's clear that organizations are increasingly investing in technologies that prevent catastrophic failures. Please refer to Fig. 9.10 for a detailed illustration of the market size forecast. This figure highlights the growing demand for self-healing networks, driven by the need to reduce the risks of downtime and improve network resilience.

Networking issues that lead to IT outages span across industries, with the financial services and healthcare sectors facing 12% of data center network outages in 2022, followed by e-commerce retail and collocation services at 10% each. This underscores the necessity for industries to adopt self-healing networks to ensure rapid detection and mitigation of hardware and software failures.

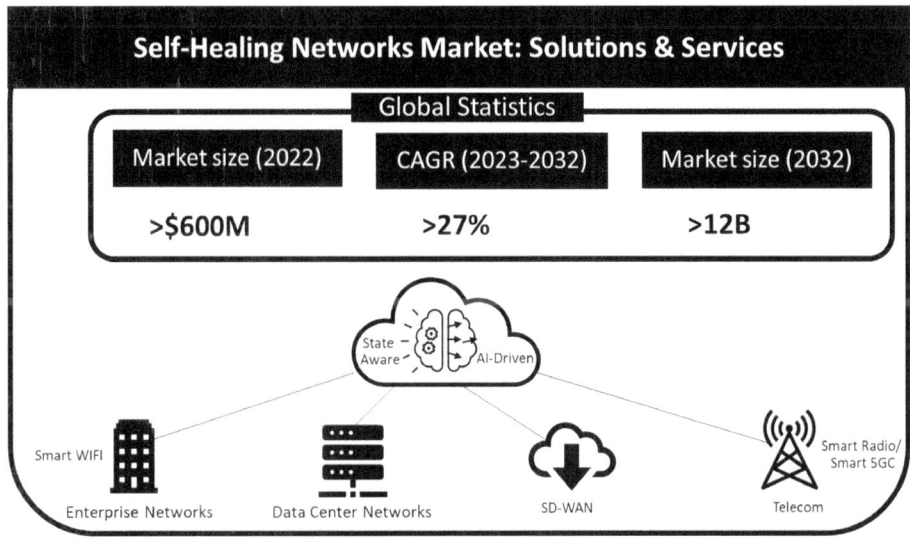

Fig. 9.10 Market demand for self-healing networks, projected to grow from over $600M in 2022 to $12B by 2032

In this section, we will explore the architecture and components of self-healing networks, the role of machine learning and automation, and how these systems evolve through continuous learning and adaptation. The goal of this technological advancement is to create networks that are more efficient, resilient, and capable of self-managing, ultimately transforming how networks are designed and operated.

9.2.1 Addressing Requirements Across Enterprise Networks: Data Centers and Campus Networks

In today's complex IT environments, self-healing networks must address the diverse needs of both data centers and enterprise campus networks, ensuring operational efficiency, resilience, and security. As organizations scale, their networks face increasing demands for visibility, performance optimization, fault tolerance, and security. Let's explore how self-healing networks meet these requirements for data centers and campus networks, based on the depicted architecture and design elements.

9.2.1.1 Self-Healing Networks for Data Centers
Data centers are the backbone of enterprise IT operations, hosting critical applications and services that require high levels of availability and performance. Self-healing networks are

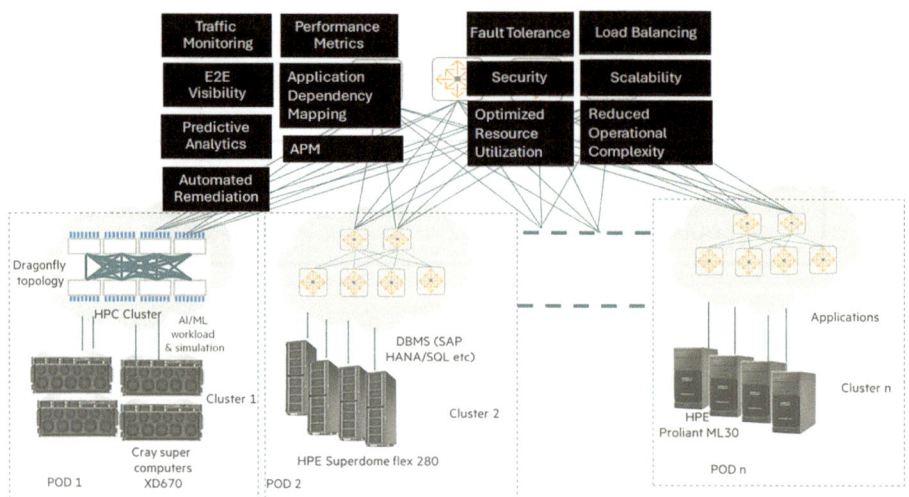

Fig. 9.11 Self-healing networks addressing requirements for data centers

essential to ensure uninterrupted operations within these environments by addressing the following key requirements (refer to Fig. 9.11).

Traffic Monitoring and Performance Metrics: Monitoring network traffic and performance in real-time ensures that applications continue to perform optimally, without congestion or delays. These metrics allow for proactive adjustments and capacity planning, enabling a seamless user experience.

1. **Traffic Monitoring and Performance Metrics**: Monitoring network traffic and performance in real-time ensures that applications continue to perform optimally, without congestion or delays. These metrics allow for proactive adjustments and capacity planning, enabling a seamless user experience.
2. **End-to-End (E2E) Visibility**: Complete visibility across the data center network is critical for identifying potential issues, monitoring resource utilization, and ensuring that the network's health is maintained at all times.
3. **Predictive Analytics and Automated Remediation**: Leveraging AI and machine learning, self-healing networks can predict potential failures before they occur and automatically remediate them. This reduces downtime and operational disruptions, ensuring smooth operations across the data center.
4. **Application Dependency Mapping (APM):** Understanding how different applications rely on each other is critical in preventing cascading failures within a data center. Self-healing networks implement APM to ensure that issues in one part of the network don't inadvertently bring down other critical services.

5. **Fault Tolerance, Load Balancing, and Scalability**: Data centers must be resilient to hardware failures and able to efficiently distribute traffic to avoid bottlenecks. Load balancing ensures that traffic is evenly distributed, while scalability allows the network to grow seamlessly with business demands.

6. **Security, Resource Optimization, and Complexity Reduction**: Security remains a top priority in data center environments. Self-healing networks ensure that the latest security measures are implemented and automatically updated. Additionally, they optimize resource utilization and reduce the operational complexity associated with managing large, distributed environments.

These features of self-healing networks allow data centers to operate with minimal human intervention while maintaining high levels of performance and security (Fig. 9.11).

Traffic Monitoring and Performance Metrics

In a data center environment, self-healing networks continuously monitor traffic across critical infrastructure components such as spine switches, leaf switches, routers, and servers. AI-driven analytics help detect anomalies like congestion, latency spikes, jitter, and packet loss. These systems are often deployed in Spine-Leaf architectures, where Equal-Cost Multi-Path (ECMP) paths are used to balance traffic across multiple paths between leaf and spine switches. While ECMP provides redundancy and balanced traffic distribution, congestion can still occur due to imbalanced traffic or sudden surges.

Network telemetry collects data in real-time, feeding it into the AI system that analyzes performance indicators like bandwidth utilization, packet delay, and error rates. This information enables the self-healing system to take proactive steps before any significant performance degradation occurs. When ECMP is operational, but congestion is detected on specific paths, the self-healing system steps in to optimize traffic flow. For instance, congestion may arise between Leaf 2 and Spine 2 due to high-traffic loads, even though ECMP is distributing traffic across multiple paths. The AI-driven monitoring system identifies this congestion by analyzing key metrics like increased latency or packet loss. As soon as congestion is detected, the AI engine reroutes traffic dynamically.

In this scenario, the AI engine reroutes traffic from the congested path between Leaf 2 and Spine 2 to alternative, underutilized paths, such as Leaf 1 to Leaf 2 via Spine 1 and Leaf 2 to Leaf 3 via Spine 3, for traffic traveling from Server 1 to Server 3. This real-time rerouting alleviates congestion and ensures a smooth, uninterrupted network experience. Please refer to the diagram below, where the green solid lines indicate the AI-driven self-healing mechanism. The dotted green arrows indicate traffic flow from server 1 to server 3.

There may also be additional congestion in other parts of the network. For example, the path between Leaf 1 and Spine 3 could experience high traffic or degraded performance. In response, the AI system detects the congestion and reroutes traffic between Leaf 1 and Leaf 3 through the more efficient path via Spine 1.

In cases where congestion becomes more severe, the self-healing network takes additional steps by adjusting Quality of Service (QoS) policies. These policies ensure that high-priority traffic, such as mission-critical applications or time-sensitive services, is allocated more bandwidth. Meanwhile, lower-priority traffic is deprioritized or throttled to free up resources and maintain the performance of essential applications. For instance, video conferencing traffic could be given more bandwidth, while bulk data transfers are assigned less priority until congestion is resolved.

Beyond immediate congestion relief, the AI engine continuously monitors the network and can predict potential traffic spikes before they happen. By analyzing historical data and trends, the self-healing network can proactively adjust traffic routing even before congestion occurs, ensuring uninterrupted performance. This proactive monitoring allows the network to adapt to changing conditions without manual intervention, ensuring that high-performance levels are consistently maintained across the data center.

End-to-End (E2E) Visibility in Data Centers

End-to-End (E2E) visibility is essential in modern data centers, enabling continuous monitoring of traffic across all network components such as servers, switches, routers, and the links between them. E2E visibility refers to the ability of the network to provide comprehensive insights into traffic flow, performance metrics, and potential issues from one end of the network to the other. Self-healing networks leverage this visibility to detect anomalies like congestion, packet loss, or hardware failures and take corrective action in real-time.

In a Spine-Leaf architecture, telemetry data is collected across every point in the network, from the servers (e.g., Server 1 and Server 3) to the leaf and spine switches. This data is fed into an AI-driven monitoring system that continuously analyzes bandwidth utilization, latency, and error rates. When abnormal patterns emerge, such as congestion between Leaf 2 and Spine 2, the system immediately identifies the issue. The AI engine then traces the source of the problem and dynamically reroutes traffic to less congested paths.

For example (please refer to Fig. 9.12), if congestion is detected between Leaf 2 and Spine 2, the AI system can reroute traffic traveling from Server 1 to Server 3 via Leaf 1 to Leaf 2 through Spine 1 and then Leaf 2 to Leaf 3 via Spine 3. This rerouting minimizes performance degradation and ensures that data continues to flow smoothly. The continuous monitoring and rerouting capabilities are performed automatically, without manual intervention, ensuring high network availability.

E2E visibility allows the self-healing network to be proactive rather than reactive. The system detects issues before they significantly impact end users, reducing downtime and improving performance. This visibility also enhances network security by identifying and mitigating potential threats as they arise, ensuring comprehensive protection across the entire network infrastructure.

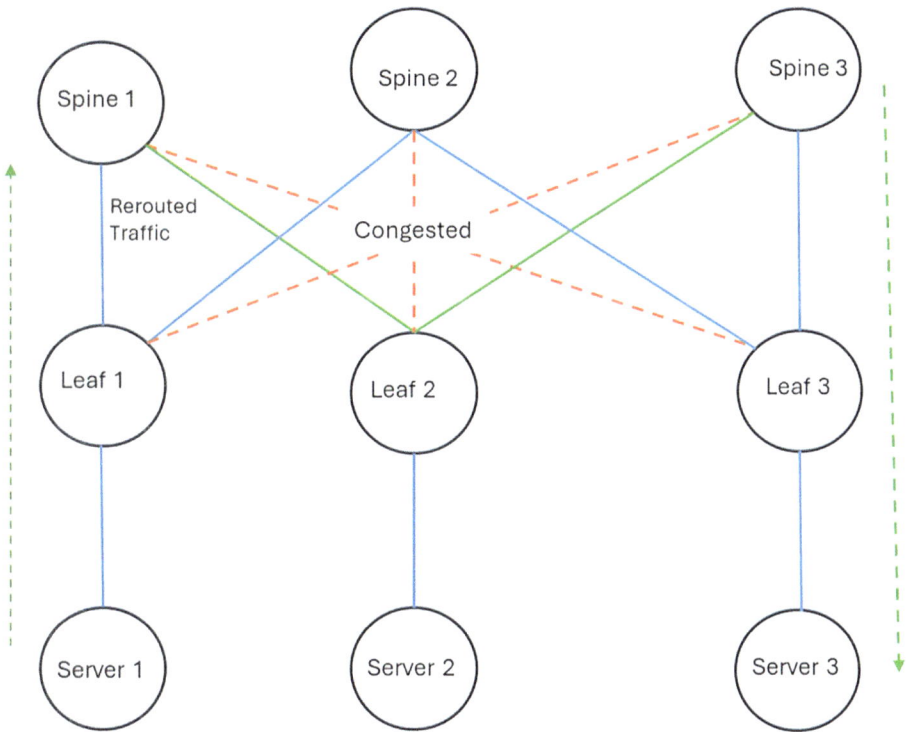

Fig. 9.12 AI-driven traffic monitoring and rerouting in spine-leaf architecture

Predictive Analytics and Automated Remediation in Data Centers

In a modern data center, predictive analytics enables self-healing networks to forecast potential network issues before they escalate, allowing proactive measures to be taken. The AI system uses machine learning models to analyze historical data and real-time telemetry, predicting issues like congestion, hardware failure, or traffic surges. By identifying patterns that suggest an impending problem, predictive analytics allows the network to act in advance.

For example, by analyzing traffic patterns and error rates between Leaf 2 and Spine 2 (as depicted in Fig. 9.12), the system may predict an upcoming hardware failure or traffic overload. Once detected, Automated Remediation immediately kicks in. The AI system reroutes traffic away from the potentially affected path, redirecting data from Server 1 to Server 3 via alternative routes such as Spine 1 and Spine 3. This proactive rerouting ensures that network performance remains unaffected and that no downtime occurs.

Predictive analytics also helps forecast high-traffic volumes. If the system predicts a congestion surge on specific paths, it can increase bandwidth allocation or preemptively reroute traffic to prevent bottlenecks. This automated response minimizes disruptions and enhances the overall reliability of the network, ensuring that even predicted issues are addressed before they cause any performance degradation. By leveraging predictive analytics, self-healing networks shift from reactive to proactive maintenance, reducing downtime and improving efficiency.

Application Dependency Mapping (APM) in Data Centers

Application Dependency Mapping (APM) is a crucial component of self-healing networks, particularly in large data centers where multiple applications interact across a distributed infrastructure. APM refers to the process of identifying and tracking the relationships and dependencies between different applications, services, servers, and network components. This mapping helps the self-healing network understand how various applications rely on one another, ensuring that when an issue arises, it can be quickly isolated, and remediation actions can be taken without causing a cascading failure across interconnected services.

In a typical Spine-Leaf architecture, applications often span multiple servers, switches, and data center components. For instance (please refer to Fig. 9.12), a database application might rely on services running on Server 1 and Server 3, with traffic routed through Leaf and Spine switches. If there's an issue, such as congestion between Leaf 2 and Spine 2, the self-healing system, aided by APM, can quickly identify the applications and services affected by the disruption.

The key function of APM is to create an accurate, real-time map of how these applications communicate. This map allows the AI system to predict the impact of a potential failure and take appropriate action. For instance, if Server 1 hosts a critical application that depends on another service running on Server 3, any disruption in communication between these servers could impact application performance. APM helps identify such dependencies and triggers automatic rerouting or load balancing to ensure continuous availability.

When APM identifies that an application running on Server 1 is dependent on traffic flowing through Leaf 2 and Spine 2, and the self-healing network detects an issue such as congestion or a hardware fault on that path, the AI system can preemptively reroute traffic to avoid disruptions. By rerouting traffic through alternative, less congested paths like Spine 1 or Spine 3, the self-healing network ensures that critical application dependencies remain intact, minimizing performance degradation and preventing cascading failures.

Fault Tolerance, Load Balancing, and Scalability in Data Centers

Fault Tolerance ensures that a data center can maintain operations even when network components fail. In a Spine-Leaf architecture, multiple redundant paths between leaf and spine switches ensure that if one path fails, traffic can be rerouted through another path. For example (please refer Fig. 9.12), if Leaf 2 and Spine 2 experience a failure, traffic between Server 1 and Server 3 is rerouted via Spine 1 or Spine 3. This automatic rerouting, powered by AI, ensures continuous service.

Load Balancing prevents network congestion by distributing traffic evenly across multiple paths. In this architecture, the AI-driven system dynamically adjusts traffic distribution in real-time based on load conditions. If a particular spine switch becomes overloaded, traffic is redirected to other, less congested paths. This real-time load balancing ensures optimal network performance, even under high demand.

Scalability allows the network to grow as more resources are added. The Spine-Leaf architecture is naturally scalable, and with AI-based automation, new resources such as switches or servers can be added seamlessly. For instance, a new server (New Server) and leaf switch (New Leaf) can be connected to the spines, and the AI system will automatically integrate them into the existing network, redistributing traffic and maintaining performance.

These three mechanisms—fault tolerance, load balancing, and scalability—work together to ensure that the data center remains resilient, adaptable, and efficient.

Security, Resource Optimization, and Complexity Reduction in Data Centers

Security, Resource Optimization, and Complexity Reduction are essential pillars of modern data center operations, ensuring not only the protection of data and infrastructure but also efficient utilization of resources and simplified network management. These three elements work together to create a resilient and high-performing self-healing network.

Security: In a Spine-Leaf architecture, security is critical to maintaining the integrity and confidentiality of data as it flows across multiple interconnected devices and paths. Self-healing networks are equipped with AI-driven security mechanisms that continuously monitor traffic for anomalies such as unauthorized access, malware, or Distributed Denial of Service (DDoS) attacks. These systems rely on Network Access Control (NAC) and Segmentation to isolate critical parts of the network, ensuring that sensitive data is protected even when a breach or vulnerability is detected.

For example, if unusual traffic patterns are detected between Server 1 and Server 3, the AI engine identifies this as a potential security risk. It automatically isolates the affected segments and reroutes traffic through secure, monitored paths. Moreover, the AI engine implements microsegmentation, ensuring that compromised parts of the network do not affect other services or applications.

The self-healing network also applies encryption and firewall policies dynamically, adjusting security protocols based on real-time threat detection, and preventing breaches before they cause damage.

Resource Optimization: Resource optimization ensures that the network makes the most efficient use of its available bandwidth, compute power, and storage. In a Spine-Leaf architecture, the AI-driven self-healing system continually monitors resource utilization across the network and optimizes traffic flow to avoid bottlenecks and overuse of specific paths or devices.

For instance, if Spine 2 is becoming overutilized, the AI engine dynamically shifts traffic to less congested spines such as Spine 1 and Spine 3. Similarly, the system can redirect tasks to underutilized servers to balance the workload. This dynamic allocation of resources ensures that no single point in the network is overwhelmed, improving performance and reducing the risk of failures.

In scenarios where traffic is lower, the self-healing system can also deallocate or power down unnecessary resources, such as underutilized servers or switches, to save energy and reduce operational costs.

Complexity Reduction: As data centers grow larger, they become more complex to manage. A Spine-Leaf architecture typically consists of many switches, servers, and links, all of which need to be carefully coordinated to avoid inefficiencies. Self-healing networks simplify this complexity by automating many of the manual tasks traditionally required to manage and configure a network.

The AI engine reduces complexity by automating fault detection, load balancing, security enforcement, and scalability. Instead of network administrators needing to manually configure traffic rerouting, detect performance issues, or implement security measures, the self-healing network takes care of these tasks autonomously.

For example, when a new server or leaf switch is added, the AI system automatically integrates it into the existing network, ensuring balanced traffic distribution, secure access, and optimized resource usage. The automation of these tasks significantly reduces the management burden, freeing up IT teams to focus on strategic initiatives rather than day-to-day troubleshooting.

Self-Healing Action: In a self-healing network, security, resource optimization, and complexity reduction work in tandem. The AI engine continuously monitors the network for potential security risks, dynamically reallocates resources based on real-time conditions, and simplifies management through automation. If a security threat is detected between Leaf 2 and Spine 2, traffic is rerouted via Spine 1 and Spine 3 for increased security. At the same time, the AI system adjusts resource usage, ensuring that the rerouted traffic is optimally distributed across the available paths. Lastly, the entire process is automated, reducing the need for manual intervention.

9.2.1.2 Self-Healing Networks for Enterprise Campus Networks

Enterprise campus networks, which include branch offices, logistics operations, and other distributed environments, face their own set of challenges. As shown in Fig. 9.12 , self-healing networks address the following requirements:

1. **Fault Detection and Correction**: Campus networks must have the ability to automatically detect faults, such as connectivity issues or device failures, and correct them in real-time to ensure uninterrupted service to users.
2. **Load Balancing and Scalability**: Like data centers, campus networks must efficiently distribute traffic to prevent congestion. Scalability is also key to supporting the growth of the enterprise, enabling the network to expand as more users and devices connect.

3. **Wireless Network Optimization**: With the increasing reliance on wireless connectivity, optimizing wireless performance is critical in campus environments. Self-healing networks automatically adjust wireless configurations to minimize interference and maximize coverage.

4. **Security Threat Mitigation**: Campus networks are highly vulnerable to cyber threats due to the large number of connected devices and users. Self-healing networks implement advanced security mechanisms that detect and mitigate threats before they cause harm.

5. **Proactive Maintenance and Quality of Service (QoS) Enforcement**: By continuously monitoring network health, self-healing networks can perform maintenance proactively to prevent outages. QoS enforcement ensures that critical applications receive the necessary bandwidth and priority, enhancing performance.

6. **User Mobility Support and Device Connectivity for IoT**: As users move between different parts of the network, self-healing networks ensure that their connectivity remains uninterrupted. Additionally, with the growing number of IoT devices, the network must provide reliable connectivity and coverage for these devices.

7. **Energy Efficiency, Reliability, and Resilience**: Energy-efficient networking is increasingly important, especially in large campus environments. Self-healing networks optimize energy usage while ensuring the network remains reliable and resilient against failures (Fig. 9.13).

Fault Detection and Correction in Enterprise Networks

In enterprise networks, Fault Detection and Correction are critical processes that ensure continuous availability, reliability, and performance of network services. Given the complexity and scale of modern enterprise environments, where multiple services and applications are interconnected across distributed systems, prompt detection and resolution of faults are essential. A self-healing network powered by AI can autonomously detect, diagnose, and correct faults, minimizing downtime and preventing significant service interruptions.

Fault Detection: Fault Detection involves identifying issues such as hardware failures, misconfigurations, link degradation, or abnormal traffic patterns that can disrupt the network. Traditionally, fault detection required manual monitoring by network administrators, with alerts triggered when network performance metrics fell outside predefined thresholds. However, modern self-healing enterprise networks leverage advanced AI-driven telemetry and real-time monitoring to detect faults proactively.

By continuously collecting and analyzing data from routers, switches, firewalls, and servers, the AI system can identify anomalies such as:

- **Link failures**: Broken or degraded connections between devices, such as a failed link between a switch and a router.
- **Congestion**: Excessive traffic loads causing latency spikes or packet drops.

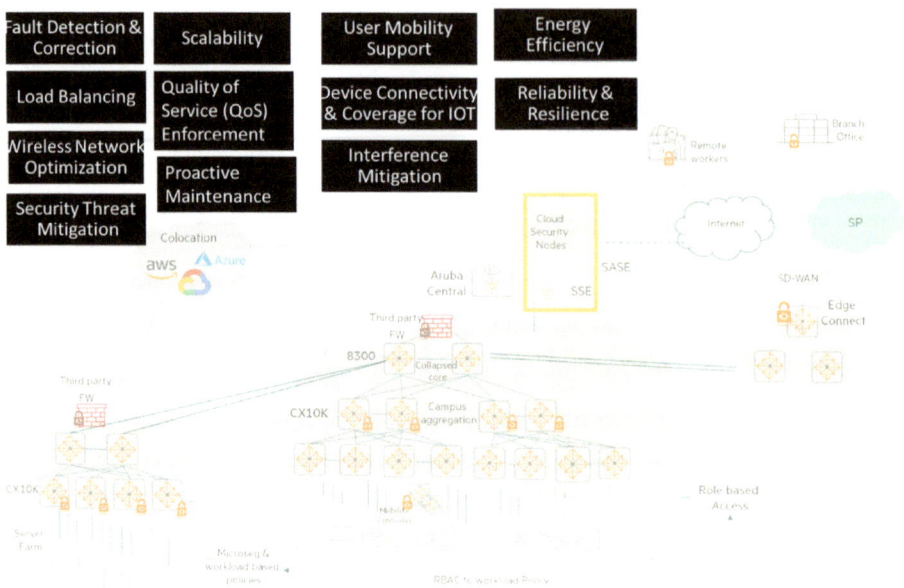

Fig. 9.13 Self-healing networks addressing requirements for enterprise campus networks. This diagram includes references to certain HPE Aruba products as part of the solutions for enterprise campus networks

- **Hardware faults**: Issues in routers, switches, or servers that reduce performance.
- **Security breaches**: Suspicious traffic patterns indicating potential attacks or unauthorized access.

For example, if a switch in an enterprise network begins to experience degraded performance due to hardware issues, the AI engine quickly identifies this as an anomaly, flagging it as a fault. Through predictive analytics, the AI system can also anticipate future faults by analyzing historical patterns and making predictions based on trends in traffic or hardware usage.

Fault Correction: Fault Correction refers to the network's ability to resolve detected issues automatically, minimizing human intervention and reducing downtime. In a self-healing enterprise network, AI systems use predefined rules, machine learning algorithms, and real-time data to take corrective action as soon as a fault is detected. Fault correction mechanisms include:

- Dynamic Rerouting: When a link or device fails, the AI system automatically reroutes traffic through alternate, functioning paths. For example, if a router fails, traffic is dynamically shifted to a backup route, ensuring uninterrupted service.

- Load Balancing: In cases of network congestion, the AI system redistributes traffic across less congested links to avoid bottlenecks and optimize performance.
- Configuration Management: Self-healing networks can automatically adjust configuration settings on network devices to resolve misconfigurations that cause performance degradation.
- Hardware Failover: In cases of severe hardware failure, such as a switch or server going offline, the self-healing system switches over to redundant systems or devices, ensuring minimal disruption.

For example, if a critical server hosting a database application experiences a hardware fault, the AI system quickly identifies the issue and redirects the traffic to a backup server or node in the network. This fault correction process occurs automatically and seamlessly, reducing recovery times and preventing service-level agreement (SLA) violations.

End-to-End Fault Management: In enterprise networks, end-to-end fault management is crucial. Fault detection and correction are not limited to isolated components; instead, they span the entire network, including WAN links, data centers, branch offices, and cloud environments. The AI system has a holistic view of the network and can detect faults across various segments while coordinating corrective actions across the entire infrastructure.

By integrating fault detection and correction, self-healing enterprise networks ensure that performance remains high, even in the face of hardware failures, network congestion, or security threats. Automated fault correction helps maintain service availability, minimizes manual troubleshooting, and improves the overall efficiency of network operations.

In summary, fault detection and correction are fundamental to keeping enterprise networks resilient. With AI-powered self-healing capabilities, enterprise networks can detect potential faults early, apply automatic corrective measures, and prevent downtime, ensuring uninterrupted services and optimal performance.

Load Balancing and Scalability

Load Balancing and Scalability are two critical aspects of enterprise networks that ensure efficiency, reliability, and seamless expansion to meet growing demands. Modern enterprise networks handle a vast amount of traffic across distributed applications, services, and locations. Efficient load balancing ensures that no single component is overwhelmed, while scalability allows the network to grow dynamically without compromising performance. In self-healing networks, AI-driven automation plays a key role in optimizing load distribution and enabling seamless scalability.

Load Balancing: Load Balancing refers to the distribution of network traffic across multiple paths, devices, or servers to prevent congestion and optimize performance. In an enterprise network, traffic can be distributed across routers, switches, data centers, and cloud services. Proper load balancing ensures that no single path or resource becomes overutilized, reducing the risk of performance degradation or downtime.

In a traditional network, load balancing was often handled manually, requiring network administrators to configure routing protocols and manage traffic manually. However, in modern self-healing enterprise networks, AI-driven systems automate load balancing based on real-time traffic analysis. These AI systems continuously monitor traffic patterns, application demands, and resource utilization, making dynamic adjustments to balance the load.

Key benefits of AI-driven load balancing include:

- **Real-Time Traffic Management**: The AI system monitors traffic across the entire network and identifies links or devices that are nearing capacity. It then redistributes traffic to less congested paths, ensuring even load distribution.
- **Application-Aware Balancing**: The AI engine can prioritize traffic for mission-critical applications, ensuring that high-priority workloads, such as enterprise resource planning (ERP) systems or video conferencing, receive adequate resources while less critical traffic is deprioritized.
- **Redundancy and Failover**: In the event of a link or device failure, load balancing ensures that traffic is rerouted to backup paths or devices without manual intervention, ensuring continuous availability and service performance.

For example, if a data center router begins to experience heavy traffic due to multiple high-demand applications running concurrently, the AI system can detect the overload and redirect some of the traffic to other routers or servers with lower utilization, balancing the load and preventing bottlenecks.

Scalability: Scalability in enterprise networks refers to the ability to seamlessly expand network resources, such as servers, switches, and storage, to meet increased demand without causing performance bottlenecks. Scalability is particularly important in today's dynamic enterprise environments, where cloud services, remote work, and IoT devices continuously increase the load on network infrastructure.

In a self-healing network, scalability is enhanced through automation, allowing the network to grow or contract based on real-time demand. This is achieved through:

- **Dynamic Resource Allocation**: The AI system continuously monitors network usage and allocates additional resources when needed. For example, if a surge in traffic occurs due to a major business event or increased use of cloud applications, the AI system dynamically provisions more servers or bandwidth to handle the increased load.
- **Seamless Integration of New Resources**: When new switches, servers, or cloud resources are added to the network, the AI engine automatically integrates them into the network's existing architecture. This ensures that new resources are efficiently utilized without manual configuration or downtime.

- **Horizontal and Vertical Scaling**: The AI system supports both horizontal scaling (adding more nodes or servers) and vertical scaling (upgrading the capacity of existing resources), depending on the needs of the enterprise.

For example, if an enterprise launches a new service that significantly increases traffic on certain parts of the network, the AI system can automatically detect the increased load and scale resources by provisioning additional bandwidth, computing power, or cloud instances. This scaling happens in real-time, ensuring that users experience uninterrupted performance, even during periods of rapid growth.

9.2.1.3 Wireless Network Optimization

Wireless Network Optimization is crucial for ensuring reliable, high-performance connectivity in modern enterprise environments. As wireless networks handle an increasing number of devices and applications—ranging from mobile phones to IoT devices—optimizing these networks is essential to ensure seamless communication, low latency, and robust security. In self-healing networks, AI-driven automation enables wireless networks to adapt dynamically, making real-time adjustments to maintain optimal performance and minimize interference or congestion.

Challenges in Wireless Networks: Wireless networks face unique challenges compared to wired networks due to factors such as signal interference, fluctuating bandwidth demands, and physical obstructions that can degrade signal strength. As enterprise environments become more mobile, with employees, guests, and IoT devices constantly connecting and disconnecting, managing, and optimizing these networks become increasingly complex. Some common challenges include:

- **Interference**: Signals from neighboring wireless devices or other radio frequency (RF) sources can interfere with wireless communication, leading to packet loss or reduced throughput.
- **Congestion**: With many devices accessing the same wireless access points (APs), congestion can occur, resulting in higher latency, slower speeds, and reduced overall performance.
- **Coverage Gaps**: Physical obstructions, such as walls or large equipment, can block or weaken wireless signals, creating coverage gaps in parts of the network.
- **Roaming Issues**: Devices that move across different parts of the network, such as mobile phones or laptops, may experience connectivity issues when transitioning from one access point to another.

AI-Driven Wireless Network Optimization: In self-healing enterprise networks, AI-driven wireless optimization plays a critical role in addressing these challenges. The AI system continuously monitors the performance of the wireless network and automatically adjusts network settings to optimize performance. These adjustments may include altering channel assignments, reconfiguring access points, and redistributing traffic to balance the load across the network. Key aspects of wireless network optimization include:

1. **Dynamic Channel Selection**: Wireless networks operate on multiple channels, and interference from neighboring devices or APs can degrade performance if they operate on overlapping channels. The AI system dynamically selects the best channels for each AP to minimize interference and maximize throughput. This process is continuous, allowing the network to adapt to changes in the RF environment in real-time.
2. **Load Balancing Across Access Points:** As users and devices move throughout the enterprise, some APs may become congested while others remain underutilized. The AI system monitors the traffic load on each AP and automatically redistributes devices to balance the load across the network. For example, if a specific AP becomes congested due to many users connecting in one location, the AI engine can steer new devices to less congested neighboring APs, ensuring optimal performance.
3. **Roaming Optimization**: Mobile devices such as smartphones, laptops, or tablets frequently move across the enterprise network, switching between different APs. AI-driven optimization ensures that devices experience smooth transitions between APs, reducing handover delays or dropped connections. By monitoring signal strength and device movement patterns, the AI system can predict when a device is about to move out of range of one AP and preemptively connect it to another AP with a stronger signal.
4. **Signal Strength and Coverage Management**: AI systems continuously analyze signal strength and identify coverage gaps or areas with weak signal reception. If a coverage gap is detected, the AI engine can adjust the power levels of nearby APs or suggest the deployment of additional APs to improve coverage. In some cases, the AI system can also reconfigure antenna directions to enhance signal distribution across the enterprise.
5. **Automatic Interference Mitigation**: In environments with heavy RF interference, such as manufacturing floors or densely populated office spaces, the AI system can automatically detect interference sources and adjust AP settings accordingly. This may involve changing the frequency band (e.g., from 2.4 GHz to 5 GHz), adjusting transmission power, or moving devices to less crowded channels.

AI-Driven Resource Allocation and QoS (Quality of Service): Wireless networks in enterprise environments must accommodate a diverse range of devices, applications, and traffic types. AI-driven wireless optimization enables the network to prioritize traffic and allocate resources dynamically based on real-time usage patterns and application requirements. The AI system can:

- **Prioritize Critical Applications**: Mission-critical applications, such as video conferencing, voice over IP (VoIP), or cloud-based enterprise applications, can be given higher priority to ensure low latency and minimal packet loss. The AI engine dynamically allocates more bandwidth or reduces latency for these applications, ensuring high-quality performance.

- **Manage Bandwidth for IoT Devices**: IoT devices, such as sensors, cameras, and smart devices, often require consistent but low-bandwidth connections. The AI system ensures that these devices remain connected and function efficiently without consuming excessive network resources.
- **Adapt to Traffic Spikes**: When traffic demand spikes, such as during large meetings or events, the AI system can allocate additional bandwidth or steer traffic to different access points to ensure that performance remains consistent.

Scalability and Flexibility: As enterprises expand, the ability to scale wireless networks while maintaining optimal performance is essential. AI-driven wireless network optimization ensures that the network remains scalable and flexible, allowing new APs and devices to be integrated seamlessly. The AI system automatically configures new APs and adjusts existing configurations to accommodate increased traffic, ensuring consistent performance as the network grows.

For example, if a new office area is opened and new APs are installed, the AI system automatically integrates these APs into the network, balancing traffic, optimizing channels, and ensuring consistent coverage.

Self-Healing Wireless Networks: Self-healing capabilities are particularly important in wireless networks, where issues like interference, coverage gaps, or equipment failures can disrupt connectivity. The AI system continuously monitors the network for faults or performance degradation and automatically takes corrective actions to resolve the issues. For example, if an AP fails, the AI engine reroutes traffic to nearby APs and adjusts power levels to cover the affected area until the issue is resolved.

In summary, **Wireless Network Optimization** in enterprise environments ensures seamless, high-performance connectivity across an expanding range of devices and applications. AI-driven automation enables real-time adjustments to network settings, such as channel selection, load balancing, and interference management, ensuring that the wireless network remains resilient, scalable, and optimized for both performance and security. By continuously monitoring traffic patterns and adapting to changing conditions, self-healing wireless networks provide a reliable, low-latency experience for users across the enterprise.

9.2.1.4 Security Threat Mitigation in Enterprise Networks

Security Threat Mitigation is a crucial component of modern enterprise networks, where protecting data, applications, and services from internal and external threats is paramount. As cyberattacks become more sophisticated and frequent, enterprise networks must be equipped with robust security measures that can identify, respond to, and mitigate security threats in real-time. In self-healing networks, **AI-driven security systems** offer a proactive approach to threat detection and mitigation, automating many of the manual processes traditionally handled by security teams.

Evolving Threat Landscape

Enterprise networks face a broad range of security threats, including:

- **Malware and Ransomware**: Malicious software designed to infiltrate, damage, or take control of network resources.
- **Distributed Denial of Service (DDoS) Attacks**: Large-scale attacks aimed at overwhelming network resources to disrupt services.
- **Phishing and Social Engineering Attacks**: Attempts to trick users into revealing sensitive information, leading to potential data breaches.
- **Insider Threats**: Employees or contractors with access to sensitive data who either maliciously or accidentally compromise security.
- **Advanced Persistent Threats (APTs)**: Sophisticated attacks that often go undetected for long periods, compromising the network's integrity over time.

To address these evolving threats, enterprise networks must be equipped with advanced, intelligent systems that can detect, analyze, and respond to security incidents quickly.

AI-Driven Threat Detection and Mitigation

In self-healing networks, **AI-driven threat detection** plays a pivotal role in monitoring, identifying, and mitigating security threats. The AI engine continuously analyzes traffic patterns, user behavior, and system logs across the network, allowing it to detect anomalies or indicators of compromise (IoCs) in real-time. Once a threat is detected, the AI system immediately initiates **automated threat mitigation**, neutralizing the threat before it causes significant damage.

Key components of AI-driven threat mitigation include:

1. **Anomaly Detection**: The AI system monitors baseline traffic patterns and user behavior to establish what is "normal" in the network. Any deviations from these norms, such as an unexpected spike in traffic to an external IP address or unusual login attempts, are flagged as potential security incidents. For example, if a surge of outbound traffic is detected from a normally low-traffic server, the AI system can flag this as a potential exfiltration of sensitive data and initiate a response.
2. **Behavioral Analytics**: In addition to monitoring traffic, AI-driven systems also analyze user behavior across the network. If a user begins accessing sensitive data or applications they do not normally use, the system can detect this as a potential insider threat or account compromise. In response, the AI can isolate the user's account, restrict access to sensitive resources, and notify security teams.

3. **Threat Intelligence Integration**: The AI system integrates with external threat intelligence feeds to stay updated on the latest cyber threats, vulnerabilities, and attack vectors. By cross-referencing internal data with external threat intelligence, the AI engine can detect known malware signatures, IP addresses involved in cyberattacks, or other IoCs.

4. **Automated Incident Response**: When a security threat is identified, the AI engine can automatically execute **incident response protocols**. These may include:
   - **Isolating compromised devices** from the network to prevent the spread of malware.
   - **Blocking malicious IP addresses** or domains associated with phishing or DDoS attacks.
   - **Quarantining files** suspected of containing malware, preventing them from being executed.
   - **Terminating suspicious user sessions** if they are exhibiting behavior indicative of an account takeover.

For example, if a server within the network begins communicating with an external IP known for distributing ransomware, the AI system can immediately block that communication, isolate the server, and initiate a full security investigation—all without manual intervention.

Network Segmentation and Microsegmentation

Network segmentation is a core strategy in security threat mitigation. By dividing the network into smaller, isolated segments, administrators can limit the movement of potential threats. If an attacker gains access to one segment of the network, segmentation ensures that they cannot easily move laterally to other parts of the network. **Microsegmentation** takes this concept even further, applying fine-grained security policies to individual devices or workloads.

AI-driven self-healing networks enhance this process by automatically implementing network segmentation and microsegmentation policies. If the AI system detects suspicious activity in one part of the network, it can dynamically adjust the segmentation policies, isolating the affected area and preventing the spread of the attack. For instance, if malware is detected on a specific server, the AI system can quarantine that server by restricting its access to other network segments until the issue is resolved.

Real-Time Security Orchestration

AI systems in self-healing networks are capable of orchestrating multiple security tools and technologies simultaneously. This **security orchestration** capability ensures that the right security tools (firewalls, intrusion detection/prevention systems, endpoint protection, etc.) are activated at the right time during an attack. By coordinating these systems, the AI-driven network ensures that:

- **Firewalls dynamically adjust rules** to block malicious traffic.
- **Intrusion detection systems (IDS)** and **intrusion prevention systems (IPS)** scan for suspicious activity and stop potential attacks in real-time.
- **Endpoint protection tools** deploy updates and patches to vulnerable devices automatically.

For example, during a DDoS attack, the AI engine can instruct the firewall to block traffic from specific IP ranges while simultaneously adjusting load balancers to mitigate the impact of the attack.

Threat Mitigation in Hybrid and Cloud Environments

As more enterprises move to hybrid and multi-cloud architectures, securing these environments becomes more complex. Self-healing networks ensure that security threat mitigation extends across both on-premises infrastructure and cloud environments. The AI system continuously monitors cloud-based applications and data flows, applying the same threat detection and mitigation strategies across the entire hybrid network. This unified approach ensures that cloud workloads are protected from threats, whether they originate inside the network or from external sources.

For instance, if a cloud-based application experiences abnormal traffic patterns that suggest a potential attack, the AI system can isolate the cloud environment, adjust security policies, and reroute traffic to protect the enterprise's sensitive data.

Proactive Threat Mitigation with Predictive Analytics

AI systems in self-healing networks also employ **predictive analytics** to anticipate potential security threats. By analyzing historical data and identifying trends, the AI engine can forecast potential attacks, such as increased likelihood of phishing during certain times of the year or DDoS attempts following high-profile events. These insights allow the network to proactively adjust security measures, such as tightening access controls or increasing monitoring, before the attack occurs.

Conclusion

Security Threat Mitigation in enterprise networks is a multi-faceted process that leverages AI-driven automation to detect, analyze, and respond to security incidents in real-time. With continuous monitoring, behavioral analytics, and integration with threat intelligence, self-healing networks are able to identify and neutralize threats before they cause significant damage. By dynamically adjusting security policies, isolating compromised devices, and orchestrating security tools, AI-driven networks provide comprehensive, proactive defense against the ever-evolving landscape of cyber threats.

9.2.1.5 Proactive Maintenance and Quality of Service (QoS) Enforcement in Enterprise Networks

Proactive Maintenance and Quality of Service (QoS) Enforcement are essential to the long-term stability, performance, and user experience of enterprise networks. As network infrastructure grows more complex and traffic demands increase, relying on reactive maintenance is no longer sufficient. Enterprise networks need proactive approaches that anticipate potential issues before they impact performance. Similarly, QoS enforcement ensures that mission-critical applications consistently receive the bandwidth and resources they need, even during times of high-traffic demand. Together, these concepts form the backbone of an enterprise network's ability to remain robust, efficient, and responsive.

Proactive Maintenance in self-healing networks involves continuous monitoring of network components, such as routers, switches, servers, and access points, to detect early signs of wear, misconfigurations, or potential failures. AI-driven systems play a crucial role in this by gathering real-time data from network devices and applying predictive analytics to forecast potential issues before they occur. For instance, if a router's performance metrics begin to degrade—such as increasing latency or packet loss—the AI system can detect this trend and flag the device for proactive maintenance, such as a firmware update or hardware replacement, before it becomes a point of failure. This helps avoid unexpected downtimes and ensures uninterrupted network availability.

Proactive maintenance also extends to routine tasks like software updates, configuration audits, and hardware optimizations. AI-driven networks automate these processes, reducing the need for manual intervention. For example, firmware updates or security patches can be applied across devices during low-traffic periods, ensuring the network remains up-to-date without causing service disruptions. This proactive approach helps reduce the risk of vulnerabilities or misconfigurations that can be exploited by attackers, enhancing overall network security.

The role of Quality of Service (QoS) Enforcement is equally important in enterprise networks. QoS refers to the policies and mechanisms that prioritize certain types of network traffic based on their importance to the organization. For instance, real-time applications like VoIP, video conferencing, and financial transactions must receive priority treatment to ensure they operate smoothly, with minimal latency, jitter, or packet loss. On the other hand, less critical traffic, such as file downloads or non-urgent email, can be deprioritized during peak traffic periods.

In self-healing networks, AI-driven QoS enforcement dynamically adjusts to traffic demands in real-time. By analyzing traffic flows and application usage patterns, the AI engine ensures that high-priority traffic always receives the necessary bandwidth. For example, if the network experiences a surge in demand due to an all-company meeting with video conferencing, the AI system can temporarily allocate more bandwidth to these video streams, ensuring smooth performance while adjusting the bandwidth for less critical applications. This dynamic enforcement ensures that service-level agreements (SLAs) are met, and users enjoy a seamless experience, even during high-traffic loads.

QoS enforcement also plays a critical role in preventing network congestion. In congested environments, AI systems proactively manage traffic by identifying and redirecting bandwidth-heavy but non-urgent activities to avoid overloading specific network segments. For instance, large file transfers or backup operations can be temporarily slowed or rescheduled during off-peak hours, freeing up bandwidth for critical applications. This approach prevents network bottlenecks and ensures that important applications continue to perform at optimal levels. Additionally, QoS enforcement in self-healing networks is adaptive. As enterprise networks expand and incorporate cloud-based services, remote workers, and IoT devices, the traffic profile can change significantly. The AI engine continuously learns from network traffic patterns and dynamically adjusts QoS policies to align with evolving business needs. For example, as IoT devices become more integral to business operations, the network may prioritize its traffic to ensure timely data collection and processing. Similarly, remote workers' access to cloud-based collaboration tools may take priority during peak working hours, while after-hours traffic from non-essential services is deprioritized.

The combination of proactive maintenance and dynamic QoS enforcement helps maintain a high-quality user experience in enterprise networks. Proactive maintenance minimizes the chances of unexpected outages by addressing issues before they escalate, while QoS enforcement ensures that essential applications consistently receive the bandwidth and resources they require to operate effectively. Together, they create a self-healing network that not only addresses issues as they arise but also prevents many problems from occurring in the first place, allowing the network to maintain optimal performance and reliability.

In summary, Proactive Maintenance and QoS Enforcement are key components of enterprise networks, ensuring that potential issues are resolved before they impact performance and that critical applications receive the necessary priority during periods of high demand. By leveraging AI-driven automation and real-time analytics, self-healing networks can maintain high availability, improve network performance, and deliver a seamless user experience.

9.2.1.6 User Mobility Support and Device Connectivity for IoT in Enterprise Networks

As enterprise networks evolve to meet the demands of a mobile workforce and the increasing prevalence of IoT (Internet of Things) devices, ensuring seamless User Mobility Support and Device Connectivity for IoT has become critical. These elements are foundational to modern enterprise networks, providing reliable, secure, and efficient connectivity for users and devices that are constantly on the move or require real-time communication across the network. User Mobility Support in enterprise networks ensures that employees and users can move freely throughout the organization while maintaining consistent, reliable network access. With the rise of wireless devices such as laptops, tablets, and smartphones, it is crucial that users experience seamless connectivity as they roam across different floors, rooms, or even campus locations.

Enterprise networks leverage Wi-Fi and wireless LAN (WLAN) controllers to facilitate user mobility. WLAN controllers manage access points (APs) distributed throughout the network, enabling users to move between different access points without disruptions to their connectivity. The AI-driven systems in self-healing networks further enhance user mobility by dynamically managing handovers between APs, ensuring minimal latency and preventing dropped connections.

When a user moves across the network, their device may switch from one AP to another as signal strength fluctuates. The AI system predicts this movement by continuously monitoring signal strength, user behavior, and network traffic. This allows the system to preemptively establish the next connection to a stronger access point, ensuring seamless mobility without interruptions or degraded performance. Additionally, load balancing across access points ensures that no AP becomes overloaded, allowing for an even distribution of devices. For example, if a user moves from one side of the office to another during a video conference, the AI system ensures that the device transitions smoothly between access points, providing a stable connection without any disruptions. This seamless experience is vital for maintaining productivity in environments where employees frequently move around or access multiple areas within the enterprise.

Device Connectivity for IoT in Enterprise Networks
IoT devices have become an integral part of enterprise networks, enabling real-time data collection, automation, and monitoring across various sectors such as manufacturing, healthcare, and smart buildings. These devices, which include sensors, cameras, smart thermostats, and industrial equipment, often require consistent connectivity to function effectively. Supporting a diverse range of IoT devices in enterprise networks introduces challenges such as ensuring adequate bandwidth, low latency, security, and scalability.

Enterprise networks designed for IoT device connectivity utilize dedicated wireless networks or segmented wired networks to ensure optimal performance. In self-healing enterprise networks, AI-driven systems play a key role in managing IoT device connectivity. These systems continuously monitor device health, performance, and traffic patterns, adjusting network resources dynamically to accommodate the specific needs of IoT devices. Because IoT devices often communicate intermittently but require consistent low-latency connections, the network must prioritize and optimize traffic accordingly. AI systems enforce Quality of Service (QoS) policies that prioritize critical IoT data—such as sensor readings in manufacturing lines or video feeds from security cameras—ensuring that these devices get the bandwidth and low-latency connectivity they need, even during peak traffic hours.

Additionally, IoT devices often communicate with backend services in the cloud or data centers. AI-driven self-healing networks optimize the paths for IoT device communication, ensuring that traffic from these devices reaches the appropriate servers or cloud applications efficiently, without congestion or delay. Security is also paramount when dealing with IoT devices, as many IoT devices have minimal built-in security. Enterprise networks use network segmentation and microsegmentation to isolate IoT traffic from sensitive parts of the network. This limits the potential damage in the event of an IoT device being compromised. The AI engine monitors for any suspicious behavior or anomalies in IoT traffic, such as unusual communication patterns or unauthorized access attempts, and can automatically quarantine affected devices or block malicious traffic to prevent security breaches.

Integrating User Mobility and IoT Connectivity
As enterprises integrate more mobile users and IoT devices, the challenge of managing these diverse endpoints becomes more complex. Self-healing networks leverage AI-driven automation to seamlessly integrate user mobility and IoT device connectivity within the same infrastructure. The AI system ensures that as mobile users move throughout the network, their devices coexist with IoT devices without causing congestion or performance degradation. For instance, an office environment might have mobile employees using laptops and smartphones alongside IoT devices such as smart lighting systems, security cameras, and environmental sensors. The AI-driven system manages the demands of both users and IoT traffic by dynamically allocating bandwidth and prioritizing traffic based on the criticality of the application. It also ensures that mobile users do not experience degraded connectivity when moving between access points or when sharing network resources with IoT devices.

Scalability and Adaptability for Future Devices
Enterprise networks must be designed to scale and adapt to the increasing number of mobile users and IoT devices. AI-driven self-healing networks ensure that as the number of connected devices grows, the network can dynamically scale by adding new access points, switches, or IoT gateways without disrupting existing services. The AI system automatically integrates new devices, optimizes traffic, and adjusts resource allocation to ensure consistent performance across the network. For example, in a smart building, hundreds or thousands of IoT sensors may be added over time to monitor energy usage, temperature, or security. The AI system detects the addition of these devices, segments their traffic for security purposes, and ensures that they communicate efficiently with the cloud or data center, all without requiring manual configuration by network administrators.

User Mobility Support and Device Connectivity for IoT are fundamental components of modern enterprise networks. With AI-driven self-healing capabilities, enterprise networks can seamlessly support mobile users, providing them with uninterrupted connectivity as they move throughout the network. At the same time, these networks manage the growing number of IoT devices, ensuring they remain connected, secure, and optimized for performance. By leveraging AI to dynamically manage resources, enforce QoS, and provide real-time traffic optimization, enterprise networks can scale and adapt to meet the evolving needs of both users and devices.

9.2.1.7 Energy Efficiency, Reliability, and Resilience in Enterprise Networks

As enterprise networks continue to expand in complexity, the need for Energy Efficiency, Reliability, and Resilience has become more important than ever. These elements are crucial not only for maintaining the day-to-day operations of an organization but also for ensuring that the network can handle future growth, respond to disruptions, and operate sustainably. In modern networks, particularly self-healing, AI-driven systems, achieving these goals requires intelligent resource management, predictive maintenance, and automated recovery from failures. Energy efficiency in enterprise networks refers to the ability to minimize power consumption while maintaining optimal performance. With the increasing number of devices, servers, and infrastructure components within networks, energy consumption has become a significant operational cost. Efficient energy management also plays a key role in reducing the environmental impact of large-scale enterprise networks.

Self-healing networks improve energy efficiency by using AI-driven analytics to optimize power usage dynamically. AI systems continuously monitor the network's traffic patterns, resource utilization, and power consumption levels. Based on real-time data, the AI can make decisions to reduce power consumption without impacting performance. For example, during off-peak hours when traffic is low, the AI system may automatically power down or throttle underutilized switches, servers, or access points. Similarly, the AI engine can reallocate traffic to fewer devices, reducing the number of active components and thus saving energy.

Additionally, energy-efficient protocols such as Energy Efficient Ethernet (EEE) can be implemented in self-healing networks, allowing devices to enter low-power idle modes when traffic is minimal. The AI system can monitor network activity and enforce these protocols dynamically, scaling power usage based on actual demand rather than keeping all devices fully powered at all times. As organizations adopt green IT strategies, AI-driven self-healing networks can contribute to sustainability goals by optimizing energy consumption across the entire infrastructure, ensuring that devices only consume the necessary power for current workloads while scaling down during periods of inactivity.

Reliability

Reliability in enterprise networks refers to the ability to provide consistent and dependable service with minimal downtime or disruption. A reliable network ensures that users can access applications, data, and services without interruption and that mission-critical operations are not affected by network failures. As businesses become increasingly dependent on digital platforms and remote access, network reliability has become a cornerstone of operational success. Self-healing networks are designed to enhance reliability by leveraging AI-driven predictive analytics and automation to detect potential issues before they become critical failures. By continuously monitoring network performance, the AI system can identify early warning signs of hardware degradation, traffic congestion, or security vulnerabilities. For example, if a switch begins to show signs of increased packet loss or latency, the AI engine can flag the device for proactive maintenance, reducing the risk of a sudden failure.

Moreover, self-healing networks implement redundancy and automatic failover mechanisms to further enhance reliability. If a critical device, such as a core router or server, fails unexpectedly, the AI system can automatically reroute traffic through backup paths or devices, ensuring that users experience minimal disruption. This automatic failover capability helps maintain high availability even in the face of unexpected hardware failures or network outages. Reliability is also reinforced through real-time diagnostics and automatic configuration recovery. If a device experiences misconfiguration, the AI engine can automatically correct the settings or revert the device to a previously known stable configuration. This eliminates the need for manual intervention and reduces the likelihood of human error impacting network performance.

Resilience

Resilience in enterprise networks is the ability to recover quickly from failures, attacks, or other disruptive events. While reliability focuses on preventing disruptions, resilience emphasizes how well the network can bounce back from adverse conditions, ensuring that services are restored as quickly as possible with minimal impact on users. Self-healing networks are inherently resilient due to their ability to detect, isolate, and recover from failures autonomously. AI-driven fault detection systems continuously analyze network performance, identifying anomalies that could indicate potential issues, such as hardware malfunctions, security breaches, or network congestion. When a fault is detected, the AI system takes immediate corrective action, either by rerouting traffic, isolating compromised segments, or initiating repairs without requiring human intervention.

Resilience is especially critical in the event of cyberattacks. In a self-healing network, AI-driven security mechanisms work in tandem with resilience protocols to respond to threats such as Distributed Denial of Service (DDoS) attacks, malware outbreaks, or insider threats. The AI engine can dynamically adjust firewall policies, isolate affected devices, and reroute traffic to protect critical infrastructure while keeping essential services operational. Another important aspect of resilience is the network's ability to scale resources during periods of increased demand. For example, during a high-traffic event such as a company-wide video conference, the AI system can dynamically allocate additional bandwidth and processing power to maintain performance. Once the event is over, resources can be scaled back to normal levels, ensuring that the network remains responsive and adaptable in various situations. Resilient networks are also built with geographic redundancy, ensuring that if a data center or critical infrastructure component fails, the system can fail over to an alternative location. AI systems in self-healing networks orchestrate these transitions smoothly, ensuring that users experience minimal impact during failover events.

Synergy Between Energy Efficiency, Reliability, and Resilience
In a modern enterprise network, energy efficiency, reliability, and resilience are interconnected. A well-optimized network can achieve energy savings without compromising performance, and a reliable network can maintain performance even in the face of disruptions. AI-driven self-healing systems enable this synergy by constantly monitoring, analyzing, and adjusting the network in real-time. For instance, a self-healing network may achieve energy efficiency by dynamically adjusting power levels based on traffic loads. At the same time, the system ensures reliability through proactive maintenance, minimizing the risk of failures that could disrupt operations. In the event of a network failure or attack, the resilience mechanisms within the network automatically initiate recovery processes, rerouting traffic, and restoring services without significant downtime. By integrating these three elements, self-healing networks create a robust infrastructure that not only supports the day-to-day operations of an enterprise but also provides the flexibility and adaptability needed for long-term sustainability and growth.

Energy Efficiency, Reliability, and Resilience are critical attributes of modern enterprise networks, ensuring that they can operate sustainably, consistently, and adapt to disruptions. Self-healing, AI-driven networks play a key role in achieving these goals by optimizing power usage, proactively maintaining network devices, and ensuring rapid recovery from failures. As enterprises continue to evolve, investing in networks that prioritize these three attributes will be essential for both operational success and sustainability in the future.

9.2.2 Toward a Comprehensive Self-Healing Network Strategy

With a solid understanding of how self-healing networks enhance data center and enterprise networks, it's time to delve into a more comprehensive self-healing network strategy. In recent years, vendors have made various strides toward developing self-healing networks, but these efforts have often been fragmented or narrowly focused. Many solutions address specific network segments, such as data center or campus networks, while others focus on isolated features like automated network models or infrastructure continuity. The result has been a piecemeal approach that lacks the comprehensive strategy needed to fully realize the potential of self-healing networks.

Although these vendor solutions offer critical capabilities like automation, preventive measures, and cognitive control, the need for a unified, holistic framework remains. Modern networks are increasingly complex, and the growing demands for resilience,

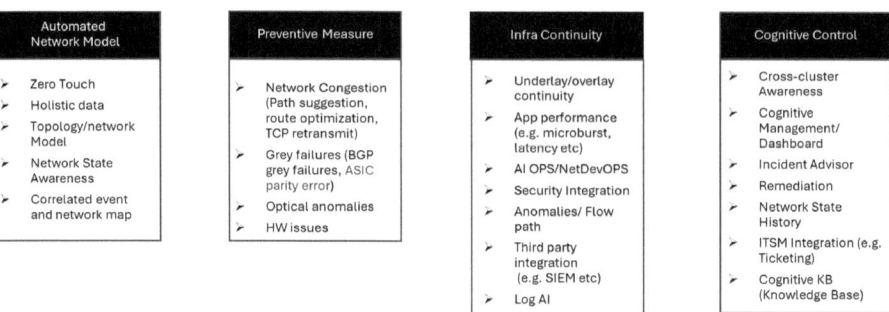

Fig. 9.14 Four pillars of self-healing networks

performance, and security require a strategy that encompasses all aspects of network operations—from automation to continuous learning and adaptation.

To bridge this gap, I propose a four-pillar framework that serves as the foundation for developing and deploying truly self-healing networks. This concept unifies the essential elements needed to create networks capable not only of detecting and repairing issues autonomously but also of optimizing performance in real-time while ensuring continuity and security. The diagram below illustrates the four pillars of self-healing networks, along with examples of key tasks associated with each area of focus. These tasks are examples and not an exhaustive list (Fig. 9.14).

The four pillars are:

1. **Automated Network Model**: A zero-touch approach that leverages holistic data and an adaptive network topology. This pillar focuses on real-time state awareness and correlates network events to create a dynamic, up-to-date network map.
2. **Preventive Measures**: This pillar emphasizes preemptive action by using AI and machine learning to identify potential risks, such as network congestion, hardware issues, and optical anomalies. By implementing predictive analytics, networks can avoid failures before they occur.
3. **Infrastructure Continuity**: Ensuring the network remains operational through both underlay and overlay continuity. It integrates AIOps/NetDevOps to maintain application performance and security, as well as enabling anomaly detection and third-party integrations for deeper monitoring.
4. **Cognitive Control**: The final pillar provides a cognitive management system that allows the network to make informed decisions based on historical data and current network states. This includes cross-cluster awareness, incident management, remediation, and continuous learning through cognitive knowledge bases.

By bringing together these four pillars, the goal is to create a network that can fully manage itself—anticipating issues, addressing them autonomously, and continuously learning from both internal and external data sources. This strategy moves beyond the current fragmented vendor solutions and sets the stage for the next generation of self-healing networks.

The following sections will explore each of these four pillars in more detail, outlining how they contribute to a comprehensive self-healing network and why each is essential for achieving the full potential of network automation and resilience.

9.2.2.1   Automated Network Model

The Automated Network Model is the first pillar in the self-healing network framework and represents the foundation for building a fully autonomous, resilient, and adaptive network. It leverages **zero-touch provisioning**, **holistic data gathering**, and **real-time network state awareness** to create a dynamic, continuously updated view of the network. This model minimizes human intervention by automating routine tasks such as device configuration, topology mapping, and event correlation. The following diagram illustrates how these components interact to form an integrated, AI-driven system (Fig. 9.15).

Key Components of the Automated Network Model

- **Zero-Touch Provisioning**:
  Zero-touch provisioning (ZTP) allows network devices to automatically configure themselves upon connection to the network. When new devices, such as switches or routers, are added, they are detected by the AI system, which then automatically pushes

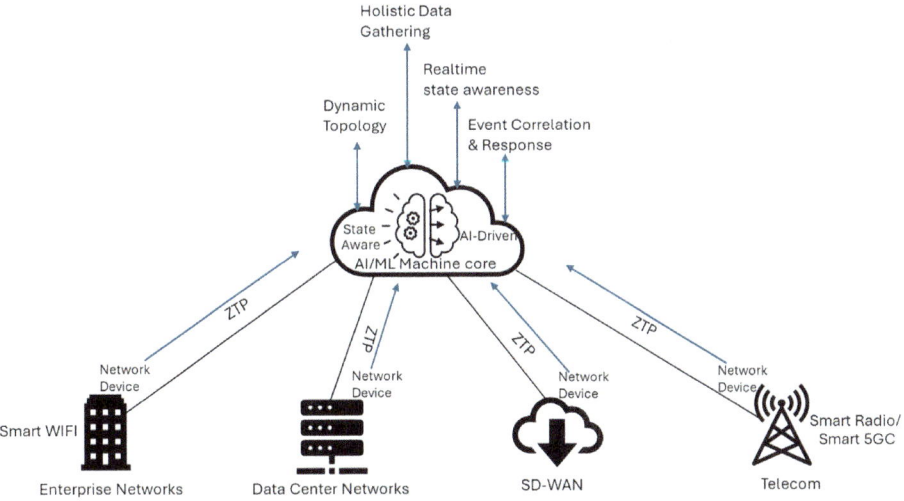

Fig. 9.15   Automated network model in self-healing networks

the necessary configurations based on predefined templates and policies. This eliminates manual intervention, reduces configuration errors, and accelerates deployment times.

- **Holistic Data Gathering**:
  A core feature of the automated network model is its ability to gather data from multiple sources, including routers, switches, firewalls, wireless access points, and even application-layer metrics. This data is collected in real-time and feeds into an AI-driven analytics engine that continuously monitors network health, performance, and security. The system correlates this data to build an accurate and up-to-date network topology that reflects the real-time state of the network.

- **Dynamic Network Topology**:
  The automated network model constantly updates the network topology based on changes in the network environment. As new devices are added, or existing devices are reconfigured, the AI system adjusts the network map to reflect the current state. It also tracks changes in traffic flows, security policies, and device health, providing a holistic view of the network at any given time.

- **Real-Time State Awareness**:
  The AI engine in the automated network model provides continuous monitoring and real-time state awareness of all network components. By understanding the current state of every device and connection, the network can automatically detect anomalies, optimize traffic flows, and respond to configuration changes in real-time. For example, if a link between two switches becomes congested, the automated system can reroute traffic through alternate paths without human intervention.

- **Event Correlation and Response**:
  As part of its real-time monitoring capabilities, the automated network model uses AI and machine learning to correlate events across the network. By analyzing data from multiple sources, the system can identify patterns that indicate potential issues, such as security threats or network congestion. When such patterns are detected, the system can automatically take corrective action, such as updating firewall rules, adjusting bandwidth allocation, or rerouting traffic to prevent failures.

- **Policy-Driven Automation**:
  The automated network model operates on predefined policies set by network administrators. These policies dictate how the network should behave under various conditions, such as traffic spikes, security threats, or device failures. The AI system uses these policies to make autonomous decisions, ensuring the network remains optimized and secure. Over time, the system can refine these policies based on historical data, improving its decision-making capabilities.

As depicted in the diagram, the AI and Machine Learning core is central to all components, dynamically interacting with elements such as zero-touch provisioning, real-time state awareness, and holistic data gathering. This seamless integration allows for autonomous decision-making and continuous optimization of network performance.

The Network Devices at the bottom of the diagram are automatically provisioned and configured by the AI system. The Event Correlation and Response function ensures that any anomalies detected in real-time data are addressed immediately, without the need for human intervention.

The Dynamic Topology adapts the network's structure based on current conditions, ensuring optimal traffic flow and minimizing potential issues. This interconnected, automated framework ensures that the network can evolve, adapt, and heal itself in real-time.

In conclusion, the Automated Network Model provides the foundation for self-healing networks, allowing them to function autonomously, optimize resources, and ensure that performance and security are maintained continuously. By automating key processes and enabling the network to respond to changes dynamically, this model helps create a resilient, adaptable infrastructure.

9.2.2.2 Preventive Measure

The Preventive Measures pillar in a self-healing network is designed to predict and prevent network issues before they lead to failures, ensuring that the network remains highly available and resilient. By leveraging AI-driven analytics, machine learning, and real-time monitoring, preventive measures proactively address potential risks such as network congestion, hardware failures, and security threats. This anticipatory approach helps maintain performance, minimize downtime, and ensure that network operations run smoothly.

The following are key components of preventive measures. This list provides representative examples and is not intended to be exhaustive:

- **Predictive Analytics for Risk Identification**: Predictive analytics is the backbone of preventive measures in self-healing networks. The AI system continuously collects and analyzes data from network devices, traffic flows, and infrastructure components. By analyzing historical data and identifying patterns, machine learning algorithms can forecast potential issues before they escalate into network failures. For example, if the system detects a gradual increase in packet loss or latency on a particular link, it can predict that congestion may occur in the near future and take corrective action.
- **Proactive Congestion Management**: Network congestion is one of the most common causes of performance degradation. AI-driven preventive measures continuously monitor traffic levels and usage patterns across the network, identifying potential bottlenecks before they impact users. When congestion is detected or predicted, the AI system can dynamically adjust traffic routing, allocate additional bandwidth, or prioritize critical

applications to mitigate the risk. For instance, traffic may be rerouted to underutilized links or alternative paths to maintain optimal performance.

- **Detection of Gray Failures**: Gray failures are partial network failures that degrade performance without causing a complete breakdown. These can be particularly challenging to detect, as they often occur intermittently and without clear indicators. Examples of gray failures include BGP route flapping, ASIC parity errors, or minor hardware glitches that may not trigger immediate alerts but still impact the network. AI-driven preventive measures use machine learning algorithms to detect subtle, non-obvious signs of gray failures. Once detected, the system can either perform maintenance or reroute traffic to more stable components, mitigating the impact of such failures.

- **Optical Anomaly Detection**: In modern enterprise and data center networks, optical links are commonly used to connect infrastructure components. Preventive measures include monitoring these optical links for anomalies such as signal degradation, fiber cuts, or component failures. If an optical anomaly is detected, the AI system can reroute traffic to backup links or initiate maintenance procedures to resolve the issue. By catching these anomalies early, the network can avoid service disruptions and costly repairs.

- **Hardware Issue Prediction and Management**: Self-healing networks rely on predictive analytics to monitor the health of hardware components, such as switches, routers, and servers. The AI engine monitors key performance indicators like CPU usage, memory utilization, and temperature, detecting signs of impending hardware failures. For example, if the system detects an increasing temperature on a switch or server, it can predict that the device may soon fail. The system can then schedule proactive maintenance, such as cooling the equipment, adjusting workloads, or replacing hardware, before a failure impacts network performance.

- **Preemptive Security Measures**: Preventive measures also extend to network security. The AI engine continuously analyzes traffic for signs of emerging threats, such as malware, DDoS attacks, or unauthorized access attempts. Using behavioral analytics, the system can identify unusual patterns, such as an increase in suspicious outbound traffic, and take immediate action by blocking the source, adjusting firewall rules, or quarantining affected devices. By detecting and mitigating threats early, the network remains secure and protected from major security incidents.

- **Machine Learning for Continuous Improvement**: Over time, the AI-driven system learns from past incidents, continuously improving its ability to predict and prevent issues. By analyzing historical data from previous network events, such as security breaches, hardware failures, or congestion episodes, the system refines its predictive models. This continuous learning process allows the network to become increasingly adept at preventing failures and maintaining optimal performance.

The Role of AI in Preventive Measures

AI plays a central role in preventive measures, allowing the network to predict and act on potential issues in real-time. The AI system continuously processes vast amounts of data from across the network, identifying subtle indicators of trouble that human operators might miss. By integrating predictive analytics, pattern recognition, and historical data analysis, AI enables a proactive approach to network management, reducing the likelihood of unexpected failures.

For example, if the AI system detects a combination of factors that have historically led to a specific type of failure—such as an increase in traffic and rising CPU temperatures in a switch—it can automatically reroute traffic, schedule maintenance, or increase cooling to avoid a breakdown. This proactive approach prevents costly outages and ensures the network operates smoothly.

The following diagram illustrates how AI-driven systems monitor the network for potential issues, identify risks through predictive analytics, and take preemptive action to avoid network failures. It shows key components such as Predictive Analytics, Congestion Management, Gray Failure Detection, Optical Anomaly Detection, Hardware Issue Prediction, and Preemptive Security, all working together to ensure a resilient network (Fig. 9.16).

As depicted in the diagram, Predictive Analytics acts as the central hub, feeding into various preventive components. These components work in coordination to monitor and respond to potential network risks, ensuring that issues are addressed proactively, before they escalate into larger problems. The AI system plays a critical role in this continuous monitoring and risk mitigation process, enhancing the overall reliability and resilience of the network.

By implementing these preventive measures, self-healing networks can maintain optimal performance and security while reducing the need for manual intervention, ensuring seamless operations and minimizing downtime.

9.2.2.3 Infrastructure Continuity

Infrastructure Continuity is a critical pillar of self-healing networks, ensuring that the network remains operational and resilient even during disruptions. Infrastructure continuity focuses on maintaining **underlay and overlay network integrity**, **seamless application performance**, and **integrating security across the network** to provide a stable and secure environment. By leveraging **AIOps** (Artificial Intelligence for IT Operations), **Net-DevOps** (network development and operations), and other automation tools, this pillar ensures that the network can recover from potential disruptions while maintaining optimal performance.

The following are examples of key components in the infrastructure continuity aspect of self-healing networks.

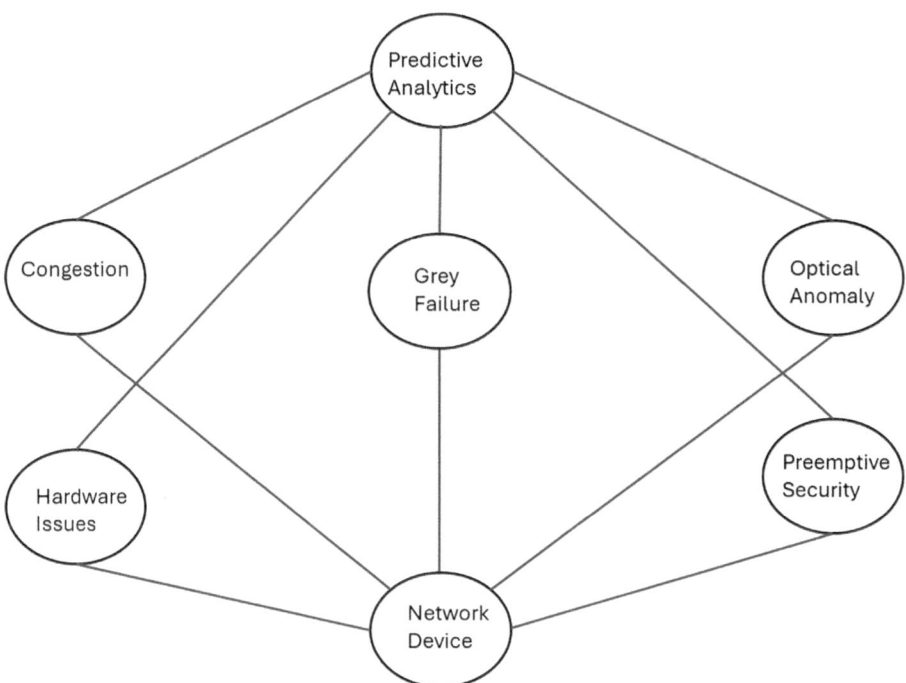

Fig. 9.16 Preventive measures in self-healing networks

- **Underlay and Overlay Continuity**: The underlay refers to the physical network infrastructure, while the overlay encompasses the virtualized network layers that sit on top. Maintaining continuity between these two layers is crucial for ensuring uninterrupted communication. The AI-driven self-healing system continuously monitors both underlay and overlay layers for potential failures or degradations. If a failure occurs in the physical infrastructure (such as a hardware malfunction or link failure), the overlay can quickly adapt, rerouting traffic and leveraging alternative paths to maintain application performance without service interruption.

- **Application Performance Monitoring**: Maintaining infrastructure continuity requires continuous monitoring of application performance. In a self-healing network, AI-driven systems track key performance indicators (KPIs) such as latency, bandwidth utilization, and throughput to ensure applications are running smoothly. If the system detects a drop in performance, it can automatically adjust network resources, reroute traffic, or provision additional capacity to restore application performance. This proactive approach helps prevent downtime and ensures that critical business applications remain available and responsive.

- **Anomaly Detection and Automated Remediation**: AI-powered anomaly detection is a vital component of infrastructure continuity. The system continuously analyzes data from network devices, servers, and applications to identify any deviations from normal behavior. These anomalies might indicate performance degradation, security breaches, or hardware failures. Once an anomaly is detected, the AI system triggers automated remediation procedures. For example, the system may reroute traffic, restart services, or adjust configurations to restore network functionality without requiring manual intervention.

- **Security Integration for Continuity**: Security is a core part of ensuring infrastructure continuity. In self-healing networks, security tools such as firewalls, intrusion detection/prevention systems (IDS/IPS), and threat detection mechanisms are integrated with the AI engine. This enables real-time monitoring of security threats, allowing the network to remain secure even during failures or disruptions. The AI system ensures that any security vulnerabilities are quickly identified and mitigated, keeping the network protected without sacrificing performance.
- **Third-Party Integration for Deeper Monitoring**: Infrastructure continuity is further enhanced by integrating third-party monitoring tools into the self-healing framework. These tools provide deeper insights into specific network components, such as cloud environments, edge networks, or specific applications. The AI system consolidates this data to maintain a comprehensive view of the network, ensuring that any issues affecting infrastructure performance are detected and addressed immediately.

The Role of AIOps and NetDevOps in Infrastructure Continuity

AIOps and NetDevOps play a central role in maintaining infrastructure continuity by automating routine tasks, enhancing monitoring, and enabling proactive remediation. AIOps leverages machine learning and AI to analyze large volumes of network data, identify patterns, and automate troubleshooting processes. NetDevOps focuses on integrating development and operations processes to ensure seamless configuration management, testing, and deployment.

Together, these technologies ensure that the network can recover from failures quickly and efficiently, without compromising application performance or security. By automating remediation tasks, such as rerouting traffic, adjusting configurations, or even spinning up additional network resources, AIOps and NetDevOps reduce the need for manual intervention, allowing the network to self-heal in real-time.

9.2.2.4 Cognitive Control

Cognitive Control is the final pillar of self-healing networks, enabling the network to make informed, autonomous decisions based on real-time data, historical insights, and continuous learning. This pillar leverages **AI-driven decision-making** and **machine learning algorithms** to analyze the network's behavior and make adjustments without human intervention. Cognitive control not only addresses incidents in real-time but also adapts based on past events, ensuring that the network becomes more efficient and resilient over time.

The followings are some of the key components of cognitive control:

- **Cross-Cluster Awareness**: Cognitive control provides cross-cluster awareness by analyzing data across different segments of the network, such as data centers, campus networks, and cloud environments. This ensures that decisions are made with a comprehensive view of the entire network infrastructure. The AI system identifies interdependencies between different clusters and makes decisions that optimize the entire ecosystem rather than just isolated components. For instance, if a failure occurs in one part of the network, the cognitive control system evaluates its potential impact on other segments and adjusts the configurations accordingly.
- **Incident Management and Remediation**: Cognitive control allows for autonomous incident management. When an issue occurs—such as a hardware failure, traffic spike, or security breach—the AI system immediately identifies the root cause and implements corrective actions. For example, if a server becomes overloaded, the cognitive control system automatically redistributes workloads to other servers, preventing a service outage. In the case of security incidents, the AI system can quarantine affected devices, block malicious traffic, and adjust firewall rules to contain the threat, all without manual intervention.
- **Cognitive Knowledge Base and Continuous Learning**: The cognitive control system continually learns from network events and incidents, updating its knowledge base over time. This knowledge base serves as a repository of past incidents, network configurations, performance data, and the outcomes of previous decisions. The AI system leverages this knowledge to improve its decision-making capabilities. By analyzing historical patterns and trends, the system can predict potential future incidents and take proactive measures to avoid them. Continuous learning enables the network to adapt to changing conditions and become more intelligent and efficient as it evolves.
- **Remediation and Automation**: One of the core functions of cognitive control is its ability to automate remediation processes. When an anomaly or failure is detected, the cognitive system initiates automated remediation workflows, such as rerouting traffic, adjusting configurations, or rebooting malfunctioning devices. These automated processes ensure that network issues are resolved quickly, minimizing downtime and preventing further disruptions. The AI system also learns from each remediation process, optimizing future responses to similar issues.
- **Decision-Making Based on Historical Data and Real-Time Insights**: Cognitive control combines real-time monitoring with historical data to make intelligent decisions. The AI engine continuously collects data from network devices, security logs, and performance metrics, comparing current conditions with historical patterns. For example, if the system detects a traffic pattern similar to a past congestion incident, it can preemptively reroute traffic to prevent performance degradation. By considering both real-time conditions and historical knowledge, cognitive control provides more accurate and effective decision-making.

The Role of AI in Cognitive Control

AI is at the heart of cognitive control, enabling the network to perform complex decision-making processes without human intervention. Machine learning algorithms analyze vast amounts of data, identifying patterns, trends, and anomalies. The AI system continuously refines its decision-making capabilities by learning from previous events and adapting to new scenarios. Over time, the cognitive control system becomes more adept at managing incidents, optimizing performance, and ensuring network reliability.

9.3 Summary

This chapter dives into the world of self-healing networks, showcasing the revolutionary shift from traditional network management to intelligent, autonomous systems that can detect, diagnose, and resolve issues on their own. As networks grow increasingly complex, the need for self-healing capabilities has become critical to maintaining performance, security, and reliability. This chapter explores how advancements in AI, machine learning, and automation are driving the evolution of self-healing networks, making them a cornerstone of modern IT infrastructure.

The chapter begins by highlighting the challenges faced by traditional network management, including the high costs of downtime, manual troubleshooting, and the growing threat landscape. It then introduces self-healing networks as the solution, leveraging AI-driven analytics, real-time monitoring, and automated remediation to proactively manage network performance. Key technologies such as AIOps (Artificial Intelligence for IT Operations) and AI-native networking components like neuromorphic SoCs and Broadcom Trident 5-X12 are explained, illustrating how they enable networks to adapt dynamically to changing conditions.

The architecture of self-healing networks is explored in detail, emphasizing their ability to autonomously detect faults, balance loads, optimize wireless connectivity, and mitigate security threats. With end-to-end visibility and continuous learning capabilities, these networks can predict issues before they impact users, reroute traffic in real-time, and adjust configurations on the fly. Practical applications in data centers and enterprise campus networks are discussed, demonstrating how self-healing networks enhance efficiency, support user mobility, and ensure seamless operations.

Throughout the chapter, the integration of AI/ML technologies into network operations is showcased as a game-changer, allowing networks Network routing in containers to self-manage with minimal human intervention. From proactive fault correction and adaptive routing to dynamic resource allocation, self-healing networks represent the next frontier in networking, delivering unmatched resilience, agility, and performance.

Reference

1. Broadcom (2023) The New Trident 5-X12 doubles bandwidth, reduces power By 25%, and adds neural network to enable next-generation telemetry, security and traffic engineering. Broadcom. Available online at https://investors.broadcom.com/news-releases/news-release-details/broadcom-introduces-industrys-first-switch-chip-neural-network

Building the Future Network: A Path Forward 10

In the chapters leading to this point, we've explored the revolutionary trends reshaping the networking landscape. From advancements in network SOCs/ASICs and the cloud-native shift to the rise of containers and automation, each technology plays a vital role in modern networks. We've delved into the power of telemetry and observability for providing deep network insights and examined the security transformations driven by Zero Trust models and adaptive strategies. We also covered how artificial intelligence (AI) and self-healing capabilities, bolstered by AIOps (Artificial Intelligence for IT Operations), are poised to make networks more autonomous, resilient, and responsive to both current and future challenges.

This final chapter is about convergence. It is about understanding that the future of networking is not just an incremental improvement but a transformational leap—where intelligent systems work together seamlessly, self-correcting, optimizing performance, and maintaining security in real-time. AIOps plays a pivotal role here, acting as the brain that processes vast amounts of network data, identifying issues, and enabling proactive solutions before they become critical.

Building such a network requires more than just deploying the latest technologies in isolation; it demands a thoughtful, orchestrated approach that leverages the strengths of automation, AI, and telemetry to create a truly intelligent, self-healing, and secure infrastructure. In this context, AIOps isn't merely an add-on; it's an integral part of the self-healing fabric, enabling predictive and prescriptive actions that minimize downtime and improve operational efficiency.

This chapter serves as a blueprint for integrating these technologies into a cohesive network that not only meets today's demands but is prepared for the evolving complexities of tomorrow. Whether operating at the data center, the cloud, or the network edge, the

© The Author(s), under exclusive license to Springer Nature Switzerland AG 2025   429
D. D. Chowdhury, *Future of Networks*, Synthesis Lectures on Communications,
https://doi.org/10.1007/978-3-031-71440-5_10

following sections will provide insights and practical strategies to construct a network that scales intelligently, secures proactively, and heals autonomously.

By the end of this chapter, you will have the tools and knowledge to not only understand but build a state-of-the-art network that supports innovation, enhances performance, and ultimately, shapes the future of digital infrastructure.

10.1   Charting the Course to the Network of Tomorrow

The future of networking is far more than the sum of its individual technologies—it is the critical foundation upon which modern enterprises will innovate, grow, and thrive in the digital age. With the exponential increase in data, the expansion of cloud services, the rise of AI-driven applications, and the ever-growing importance of cybersecurity, networking is no longer just about connectivity. It has become the backbone for digital transformation, enabling businesses to deliver new services, enhance customer experiences, and remain competitive in an increasingly digital world.

As we step into this new era, the role of the network is transforming. It is no longer a static infrastructure that simply supports business operations; instead, it must become a dynamic, intelligent system capable of adapting to the rapidly evolving demands of the digital enterprise. This transformation requires a holistic approach, where network architects, business leaders, and IT professionals work together to build networks that are not only scalable and high-performing but also agile, secure, and resilient. The networks of tomorrow must be built to seamlessly integrate emerging technologies while remaining flexible enough to accommodate future innovations that have yet to be conceived. One of the key shifts driving this change is the growing need for adaptability. Businesses are under pressure to respond quickly to market shifts, evolving customer expectations, and disruptive technologies. A static, rigid network cannot keep up with these demands. Networks need to be designed with the ability to scale up or down rapidly, reconfigure in real-time, and support new applications and services without requiring significant manual intervention. This adaptability will be powered by automation, AI, and self-healing capabilities, which will allow networks to respond to changes autonomously and ensure continuous, uninterrupted service.

At the same time, networks must become more intelligent. The future network will not simply move data from point A to point B; it will analyze, interpret, and act on that data in real-time. Leveraging advanced telemetry, observability, and artificial intelligence (AI) tools, tomorrow's networks will provide deep insights into performance, security, and operational health. This intelligence will be crucial for proactive decision-making, enabling businesses to anticipate and resolve issues before they impact end-users or business operations. As network traffic becomes increasingly complex, with more devices, users, and services vying for bandwidth and resources, this intelligence will be the key to ensuring optimal performance and maintaining a high-quality user experience.

Security, too, must be reimagined as a core function of the network, rather than an afterthought. The traditional perimeter-based security model is becoming obsolete as businesses embrace cloud computing, remote work, and decentralized architectures. The future network will need to embed security at every layer, leveraging Zero Trust principles and adaptive security measures to protect against increasingly sophisticated threats. As cyber-attacks become more frequent and advanced, networks must be capable of detecting and responding to anomalies in real-time, mitigating risks before they escalate into major breaches.

This section will explore how the network of tomorrow is being shaped by current trends, such as AI-driven automation, edge computing, cloud-native applications, and Zero Trust security, among others. More importantly, it will demonstrate how these technologies, when woven together into a unified and cohesive strategy, can create a network that is not only ready for today's challenges but also built to evolve alongside the demands of the future. Through a blend of strategic foresight and technical innovation, businesses can build a network that will enable them to scale rapidly, innovate continuously, and protect their most valuable assets in a rapidly changing digital landscape.

In this new paradigm, the future of networking is not just about implementing the latest technologies; it's about creating a foundation that empowers businesses to thrive, innovate, and transform in the digital age. This section will serve as a guide to understanding how to build this future-ready network, aligning technology choices with business objectives, and preparing for the next wave of advancements that will reshape how networks operate. By thinking holistically and acting strategically, organizations can ensure that their networks become enablers of growth and innovation, rather than bottlenecks to progress.

10.1.1 Aligning Network Strategy with Business Goals

To build a future-proof network, enterprises must start by aligning their network strategy with overall business objectives. The network is no longer a supporting player in business operations—it is an enabler of business innovation. The goal should be to create a network that is not only efficient and high-performing but also flexible enough to support new business models and services as they emerge.

10.1.1.1 Assess the Current State and Build a Shared Vision

The first step in aligning your network strategy is to thoroughly assess the current state of your network infrastructure. This involves taking a detailed inventory of existing hardware, software, cloud deployments, and edge devices, as well as evaluating network performance metrics such as bandwidth, latency, and throughput. Identify legacy systems and outdated technologies that may be bottlenecks to scalability, performance, or security. Utilize telemetry and observability tools to gain data-driven insights into traffic patterns, bottlenecks, and areas of underutilization. This assessment will highlight the strengths,

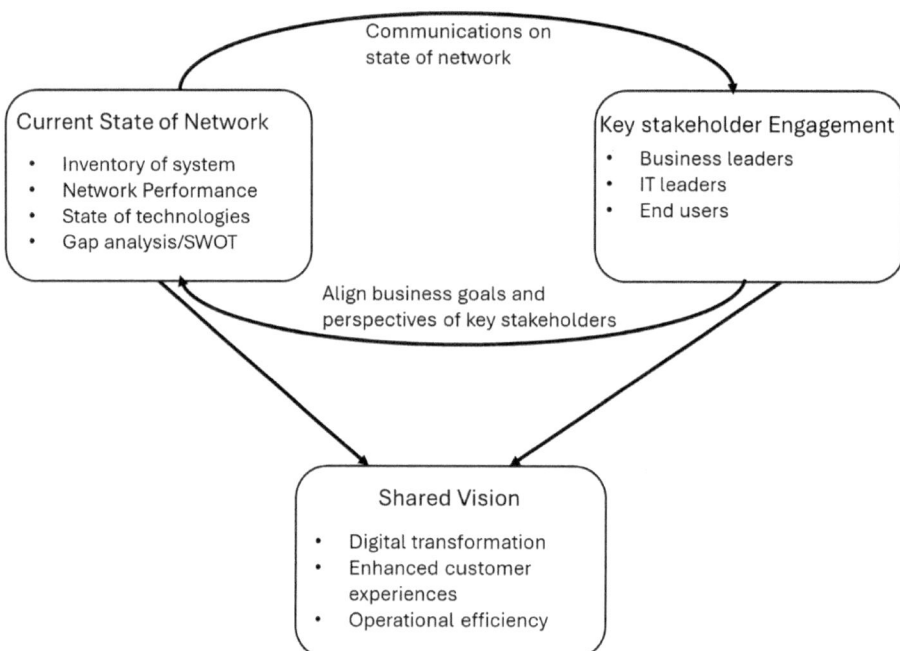

Fig. 10.1 Assessing the current state and building a shared vision for network strategy alignment

weaknesses, and gaps within your current network environment, setting a clear baseline for future improvements (Fig. 10.1).

Engaging with key stakeholders—including business leaders, IT managers, and end-users—is crucial in understanding the current performance of the network and its alignment with business needs. Gathering feedback on pain points and aligning these insights with broader business strategies such as digital transformation, enhanced customer experiences, and operational efficiency will help build a shared vision. This shared vision ensures that all parties are aligned and invested in the strategic direction of the network.

10.1.1.2　Evaluate and Define the Desired State

After assessing the current state of your network, the next crucial step is to define the desired future state, creating a clear vision of what the network needs to become to support business goals effectively. This process involves envisioning a modern, agile, and secure network capable of accommodating emerging technologies such as AI, automation, multi-cloud architectures, and real-time data processing. The attached diagram visually represents the process of evaluating and defining this desired state. It captures the key decision points and actions necessary for transitioning from the current state to the future

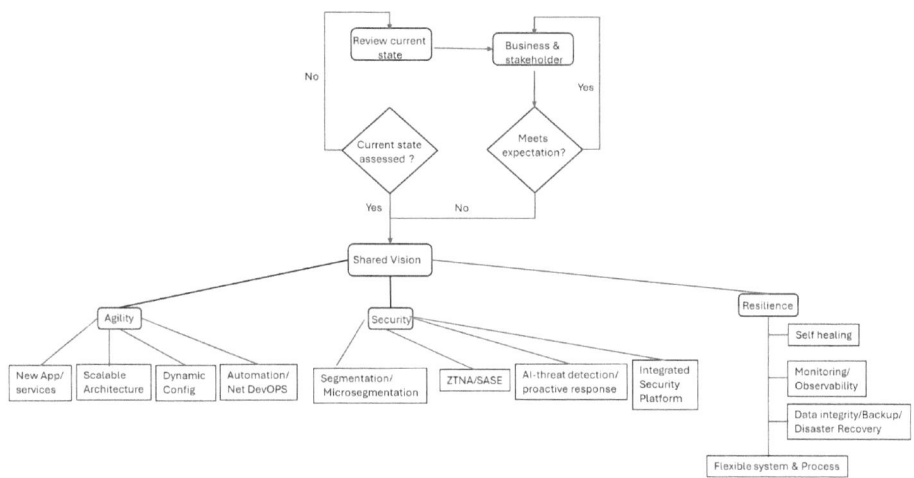

Fig. 10.2 Defining the desired future network state: A structured evaluation and implementation approach

network architecture, ensuring a systematic approach from evaluation through to execution (Fig. 10.2).

The diagram illustrates the process of defining the desired future state of the network, emphasizing the development of key criteria derived from a shared vision. These criteria—Agility, Security, and Resilience—represent essential attributes that align the network with evolving business needs, though they are not exhaustive. Each criterion encompasses specific elements that guide network modernization and strategic alignment.

Agility reflects the network's capability to quickly adapt to new demands, such as deploying new applications and services, supporting scalable architectures, and enabling dynamic configuration and automation. Agility also integrates NetDevOps and DevOps practices, which streamline the development and operational processes, enhancing the overall responsiveness and flexibility of the network.

Security focuses on establishing a comprehensive and integrated approach to protecting the network. This includes advanced strategies like segmentation and microsegmentation to control access and minimize attack surfaces, Zero Trust Network Access (ZTNA), and Secure Access Service Edge (SASE) frameworks that enhance security at the edge. AI-driven threat detection and proactive response mechanisms are crucial, ensuring the network can identify and respond to potential threats in real-time. An integrated security platform consolidates these measures, providing a unified approach to managing security across the entire network.

Resilience is about creating a robust network that can withstand disruptions and maintain continuous operations. Key aspects include self-healing networks that can automatically detect and correct issues, and comprehensive monitoring and observability tools, such as Application Performance Monitoring (APM), that provide real-time insights into network health. Ensuring data integrity, implementing backup and disaster recovery plans, and maintaining flexible systems and processes are all vital to resilience, enabling the network to recover quickly from any disruptions.

These criteria serve as a framework for defining the desired state of the network, guiding organizations through the transition from their current capabilities to a future-ready network that is adaptable, secure, and resilient. By focusing on these critical areas, the network strategy can effectively support broader business objectives, drive innovation, and ensure that the network remains aligned with the evolving needs of the organization.

10.1.1.3 Create and Communicate the Action Plan
Developing an actionable plan is essential for transitioning from the current state to the desired future state. The action plan should include specific steps for upgrading infrastructure, adopting new technologies, implementing automation, and enhancing security protocols. It should also outline the roles and responsibilities of various teams, timelines, resource allocation, and measurable goals. Effective communication of this plan across the organization is vital to ensure alignment and buy-in from all stakeholders.

Regular communication and updates keep everyone on the same page, fostering a culture of trust and collaboration. It's important to articulate how each action item contributes to the overall business strategy and to provide visibility into progress through dashboards and performance metrics.

10.1.1.4 Execute, Review, and Improve
Executing the action plan requires robust change management practices to navigate the complexities of modernizing network infrastructure. Start with small-scale pilot projects to test new technologies and approaches in a controlled environment, allowing teams to learn and refine their strategies before broader implementation. This iterative approach reduces risks and provides valuable insights into optimizing the network transformation process. Continuous improvement is critical. Establish a feedback loop that includes monitoring key performance indicators, stakeholder input, and periodic reviews to adjust the strategy as needed. As the business environment evolves, so too must the network strategy. Ongoing evaluation helps ensure that the network remains aligned with business objectives, embraces new opportunities, and maintains a competitive edge.

10.1.1.5 Integrate Data and AIOps for Operational Excellence
Leveraging data is essential to maintaining a responsive and proactive network. Advanced analytics and AIOps enable predictive maintenance, automate routine tasks, and provide insights that drive better decision-making. Data integration across the network ensures that performance, security, and capacity are continually optimized, supporting business resilience and innovation.

By thoroughly assessing the current state, defining a shared vision, evaluating future options, creating a detailed action plan, and maintaining a focus on continuous improvement, organizations can effectively align their network strategy with business goals. This comprehensive approach ensures that the network not only supports the business today but also propels it toward a more agile, innovative, and secure future.

10.2 Strategic Considerations for Brownfield and Greenfield Networks

Designing or evolving a network strategy involves making critical decisions about whether to work within a brownfield or greenfield environment. Each scenario presents unique challenges and opportunities that significantly impact how the network can support current and future business objectives. Understanding these distinctions helps organizations tailor their network modernization efforts to best fit their specific context, ensuring optimal integration of advanced technologies while maintaining operational stability. The brownfield approach focuses on evolving existing network infrastructures, leveraging legacy systems while integrating new technologies. This strategy is often seen as a practical and less disruptive option, especially for organizations with established processes and significant investments in their current systems. Brownfield implementation allows companies to maintain business continuity by incrementally upgrading their network without completely overhauling existing configurations.

A strategic approach to brownfield implementation involves several key steps. First, organizations need to conduct a detailed assessment of their current state, identifying critical areas for improvement, such as outdated protocols or security vulnerabilities. The next step involves planning the integration of new technologies like automation, AI-driven monitoring, and security enhancements in a phased manner, allowing for adjustments without disrupting ongoing operations. This gradual modernization can include adding microservices, software-defined networking (SDN), and network function virtualization (NFV) to enhance scalability and flexibility. Additionally, organizations should prioritize maintaining robust change management processes to address integration challenges while safeguarding the stability of legacy systems.

One significant challenge with brownfield strategies is the accumulation of technical debt and the limitations imposed by legacy constraints. Careful planning is essential to balance modernization efforts with the need to maintain legacy system integrity, ensuring that each upgrade aligns with long-term strategic goals without adding unnecessary complexity or risk. To define a comprehensive brownfield network strategy that aligns with long-term goals, it is crucial to consider multiple advanced strategies beyond those typically discussed. Incorporating elements like change management, hardware strategy, software strategy, design strategy, network management and orchestration, and optimization strategy can create a well-rounded approach that ensures the network not only meets current needs but is also positioned for future evolution.

10.2.1 Change Management

Change management is a cornerstone of brownfield network strategies, providing a structured approach to implementing updates and modernizations without disrupting current operations. It facilitates smooth transitions by carefully planning, communicating, and managing each phase of network upgrades. This strategy minimizes downtime, aligns stakeholder expectations, and ensures that network changes are integrated seamlessly with business processes. By proactively managing risks and maintaining clear communication channels, change management helps ensure that the upgrades contribute positively to long-term organizational goals, fostering an adaptive and resilient network environment. The following table depicts some of the key attributes to implement proper change management for brownfield network upgrades (Table 10.1).

These strategies, along with other hardware, software, design, management, and optimization strategies, provide a robust framework for evolving brownfield networks. By carefully considering and integrating these elements, organizations can maintain operational continuity while aligning their network modernization efforts with long-term business goals, creating a resilient, adaptable, and future-ready network infrastructure.

10.2.2 Hardware Strategy

A well-structured hardware strategy is essential for evolving brownfield network infrastructures while ensuring long-term alignment with business goals. This strategy emphasizes upgrading key components incrementally rather than overhauling the entire system, allowing organizations to enhance network performance, scalability, and security without disrupting ongoing operations. The hardware strategy focuses on selecting compatible new hardware that integrates seamlessly with existing systems, supporting modern capabilities such as multi-cloud integration, edge computing, and increased automation.

A comprehensive hardware strategy involves careful planning, selection, and deployment of network hardware that meets current and future demands. This approach ensures that the network can handle growing data volumes, support advanced technologies, and maintain high levels of performance and security. The following table outlines the key attributes and tools involved in a robust hardware strategy for brownfield network upgrades (Table 10.2).

Table 10.1 Change management strategy for brownfield network upgrades

Key attributes	Description	Tool
Planning	Effective planning helps minimize risk and mitigates unwanted situations	Service mapping tools
Communication	Create feedback loops with team members and power users for each change applied to the network architecture	Collaboration platforms like confluence may help in such works
Network configuration and change tracking	Constantly monitor configuration changes in real-time with change management software	NCCM (Network Change and Configuration Management) tools
Reduce the risk of unauthorized access	Role-based access control (RBAC) manages unauthorized access at the network and object level	Many network vendors provide solutions for this
Compare configuration	Maintain a baseline configuration for stable network operations and compare with new configurations to mitigate issues immediately	
Version control	Label each configuration version clearly to state the applied changes. Helps mitigate mismatched conditions against stable configurations	Change management software tools
Document changes	Document every change, including network topology, applications, dependencies, and configuration changes. Use slides for better visuals	
Backup and rollback	Create backup and rollback plans for unexpected issues, keeping track of every change for easier rollback	Cisco prime infrastructure, solarwinds NCM and blackbox

These hardware strategy components align with long-term goals by ensuring that each upgrade enhances overall network performance, scalability, and security without disrupting existing operations. Incremental improvements allow the network to evolve gradually, reducing costs and maintaining stability while supporting future growth. By selecting scalable, energy-efficient, and secure hardware that integrates seamlessly with current systems, organizations can optimize their infrastructure to handle increased demands, support multi-cloud and edge computing, and meet modern cybersecurity standards. This approach

Table 10.2 Hardware strategy for brownfield network upgrades

Key attributes	Description
Assessment and compatibility check	Evaluate existing hardware to identify bottlenecks and determine compatibility with new technologies. Ensure new components are compatible with legacy systems to avoid disruptions
Incremental upgrades	Replace outdated components like switches, routers, and firewalls with higher bandwidth, more secure models that support modern protocols. Upgrades should align with current needs and future scalability
Scalability planning	Select hardware that supports future growth, including additional ports, modular designs, and stackable configurations. This helps accommodate increased data flow without frequent replacements
Integration with multi-cloud and edge computing	Choose hardware that supports multi-cloud connectivity and edge computing. This includes edge routers, SD-WAN appliances, and cloud gateways that enhance data distribution and processing closer to end-users
Enhanced security features	Upgrade to hardware that offers built-in security features such as encryption, intrusion detection, and secure boot. This strengthens the network's security posture and aligns with modern security frameworks
Energy efficiency	Opt for energy-efficient hardware that reduces operational costs and aligns with sustainability goals. Features like lower power consumption and intelligent cooling systems help minimize the environmental impact
Vendor support and longevity	Select hardware from vendors with strong support ecosystems, long product lifecycles, and a clear upgrade path. This reduces the risk of obsolescence and ensures ongoing support and updates
Monitoring and management integration	Integrate new hardware with existing network management systems to ensure seamless monitoring, configuration, and maintenance. Tools that support telemetry and provide visibility into hardware performance

(continued)

Table 10.2 (continued)

Key attributes	Description
Redundancy and reliability	Implement redundant hardware components such as dual power supplies, redundant links, and failover systems to enhance network reliability and uptime
Lifecycle and maintenance planning	Plan for hardware maintenance, including regular updates, firmware upgrades, and end-of-life replacements. Maintaining a lifecycle strategy ensures continuous performance and avoids unexpected failures

not only enhances operational efficiency but also ensures that hardware investments are sustainable, adaptable, and aligned with the organization's strategic objectives, providing a resilient foundation for ongoing technological advancements.

A thorough **assessment and compatibility check** of existing hardware ensures that new components integrate smoothly, enhancing overall network performance rather than disrupting it. Incremental upgrades to critical hardware, such as switches and routers, allow organizations to enhance their network gradually without the expense and disruption of a full overhaul, supporting continuous improvement in line with evolving needs. **Scalability planning** is essential, as choosing hardware that supports future growth ensures that the network can accommodate new applications, increased traffic, and additional users, maintaining operational stability over the long term.

Selecting hardware that supports **integration with multi-cloud and edge computing** allows organizations to optimize data flow and enhance user experiences by efficiently handling distributed workloads, which aligns with strategic business initiatives. Upgrading to hardware with **enhanced security features** mitigates the risks associated with legacy systems and ensures the network remains resilient against evolving cyber threats, safeguarding data and assets. **Energy efficiency** is also a key component, as opting for energy-efficient hardware reduces operational costs and aligns with sustainability goals, providing financial benefits and supporting broader CSR initiatives.

Choosing hardware from reputable vendors with strong **support and longevity** ensures the network remains adaptable and operationally sound, mitigating risks related to obsolescence and protecting hardware investments over time. **Monitoring and management integration** with existing tools helps maintain network visibility and control, optimizing performance, quickly identifying issues, and streamlining maintenance processes. Implementing **redundancy and reliability** measures, such as redundant power supplies and failover systems, ensures continuous network availability, supporting business continuity even in the face of component failures.

A clear **lifecycle and maintenance planning** strategy for hardware upgrades and replacements helps avoid unexpected failures and maintain optimal performance, reducing the total cost of ownership and supporting long-term network stability. By implementing these hardware strategies, brownfield networks can effectively evolve to meet modern demands while maintaining a focus on long-term goals. This comprehensive approach allows organizations to enhance performance, security, and scalability, ensuring that their network infrastructure supports ongoing growth and technological advancement.

10.2.3 Software Strategy

A well-defined software strategy is crucial for modernizing brownfield networks and aligning them with the demands of current and future technologies. As networks evolve, software plays a pivotal role in enhancing automation, scalability, and security, enabling the network to adapt to dynamic business requirements. The goal of a software strategy in a brownfield context is to update existing software components, integrate new technologies, and ensure that the network remains agile and resilient. This approach not only improves operational efficiency but also prepares the network to handle increased traffic, diverse workloads, and emerging cyber threats.

A comprehensive software strategy involves upgrading and integrating various software solutions that can enhance the network's functionality, performance, and security. Below are the key elements of a robust software strategy for brownfield network upgrades, with an emphasis on how each component aligns with long-term strategic goals (Table 10.3).

Implementing a software strategy that integrates SDN provides the network with the capability to automate management, streamline provisioning, and dynamically adjust traffic flows, aligning with the needs for scalability, agility, and adaptability in modern enterprise environments. Automation and orchestration platforms complement SDN by further reducing manual processes and enabling consistent, rapid deployment of complex workflows. This combination fosters proactive network management that anticipates and resolves issues before they impact operations, maintaining resilience and supporting continuous growth.

AI-driven analytics and monitoring (AIOps) enhance this proactive approach by leveraging AI and machine learning to analyze network performance, predict potential issues, and automate responses. This results in a more intelligent and self-healing network that aligns with the organization's need to minimize downtime and optimize performance, thus supporting long-term operational stability.

Enhanced security software, including SASE, ZTNA, and microsegmentation, fortifies the network by enforcing Zero Trust principles, dynamically adapting to evolving threats, and preventing lateral movement within the network. These advanced security measures ensure that the network is resilient against cyber threats, safeguarding organizational data and assets, and aligning security practices with modern standards to reduce risks and support business continuity.

Table 10.3 Software Strategy for Brownfield Network Upgrades

Key attributes	Description	Example tools/Software
SDN integration	Modern SDN automates network management, provisioning, and control through orchestration platforms, enhancing scalability, reducing manual errors, and improving security and agility	Cisco ACI, VMware NSX, OpenDaylight, HPE Aruba Central
Automation and orchestration platforms	These platforms provide extensive automation of network tasks beyond SDN, allowing for granular control, coding, and orchestration of complex workflows to optimize operations	Ansible, Terraform, Cisco NSO, HPE Aruba Central, HPE GreenLake
AI-Driven Analytics and Monitoring (AIOps)	Utilizes AI and machine learning to analyze network performance, predict issues, and automate responses, supporting proactive network management	IBM Watson AIOps, Splunk, Moogsoft, HPE Aruba Central
Enhanced security software	Incorporates advanced security solutions, including SASE and ZTNA, which provide Zero Trust access, AI-powered threat detection, and secure connectivity across all environments	Palo Alto Cortex XDR, Cisco SecureX, HPE Aruba SASE, ZTNA
Microservices and containerization	Supports modular, flexible application deployment and cloud-native approaches, enhancing scalability, operational efficiency, and accelerating development cycles	Kubernetes, docker
API integration and customization	Enables seamless integration, customization, and enhanced interoperability across software and network components, providing tailored solutions	RESTful APIs, GraphQL

(continued)

Table 10.3 (continued)

Key attributes	Description	Example tools/Software
Zero-Touch Provisioning (ZTP)	Automates device configuration during deployment, eliminating manual setup and significantly reducing provisioning time, streamlining network operations	Most of the network equipment vendors support ZTP in their products, e.g., Juniper ZTP, HPE ZTP and Cisco ZTP
Version control and patch management	Maintains software integrity through strict version control and automated updates, enhancing security, compliance, and operational efficiency	Git, Jenkins, Chef
User training and documentation	Provides training and documentation to facilitate effective adoption and management of new software tools, enhancing network management capabilities	Internal training programs, vendor-provided documentation

Microservices and containerization enable a modular approach to application deployment, improving scalability, operational efficiency, and the ability to adapt quickly to changing market demands. By isolating applications and reducing dependencies, these technologies facilitate faster updates and more flexible development cycles, essential for staying competitive in today's fast-paced environment.

Version control and patch management practices ensure that software components remain secure, updated, and compliant with regulatory standards, reducing vulnerabilities and minimizing downtime. This proactive management approach supports a stable and reliable network environment. Additionally, user training and documentation equip IT teams with the skills to manage and optimize these advanced software tools effectively, reducing errors and enhancing overall network performance.

Together, these components create a comprehensive software strategy that not only meets current needs but also positions the network for future success by enhancing automation, security, and operational efficiency in alignment with long-term organizational goals.

10.2.4 Design Strategy for Brownfield Networks

Designing a network strategy for brownfield environments involves reconfiguring existing network architectures to support scalability, enhance performance, and incorporate emerging technologies while minimizing disruptions to current operations. Implementing an effective design strategy is crucial for ensuring that networks can meet future demands without requiring complete overhauls. Below are best practices and key components to consider when developing a robust network design strategy.

- **Adopt a Modular Network Architecture**: A modular architecture allows you to build your network like Lego blocks—adding, removing, or upgrading components without disrupting the entire system. This approach simplifies upgrades and expansions, making it easier to integrate new technologies like SD-WAN, edge computing, or additional security measures. Modular designs also improve scalability, helping the network grow alongside the business without significant downtime or cost-intensive reconfigurations.
- **Utilize Hierarchical Network Design**: Implementing a hierarchical network design with core, distribution, and access layers provides a clear structure that optimizes data flow and performance. This design improves manageability and fault isolation, allowing organizations to better implement network policies at different layers. For instance, the core layer manages high-throughput traffic across the organization, the distribution layer controls data flows between different network segments, and the access layer manages endpoint devices like computers and printers. This structured approach helps maintain a high level of performance and security while simplifying maintenance and scaling efforts.
- **Focus on Future-Proofing with Scalable Solutions**: Future-proofing is essential in modern network design. This involves selecting hardware and software that can scale easily with growing demands, such as integrating IPv6 for expanded address space and using cloud-based solutions to provide flexibility and scalability. Implementing high-performance wireless access points and optimizing for increased mobile and IoT connectivity are also crucial as organizations continue to expand their digital footprints.
- **Integrate Robust Security Measures Early**: Security should be embedded into the network design from the outset rather than being an afterthought. By adopting practices like microsegmentation, Zero Trust models, and integrating security policies at every layer of the network, you can significantly reduce vulnerabilities and prevent lateral movement of threats. These security measures protect data flows between segments and ensure compliance with regulatory standards, making the network more resilient against cyber threats.
- **Incorporate Hybrid Cloud Design**: Adopting a hybrid cloud design allows you to blend on-premises data centers with public and private cloud environments, enhancing data management flexibility and scalability. This approach ensures that workloads can

be efficiently managed and distributed based on performance, security, and cost considerations. For example, sensitive data might remain on-premises while non-critical workloads are processed in the cloud, optimizing resources and maintaining control over data privacy.

- **Embrace Automation and AI-Driven Network Management**: Automation tools and AI-driven analytics significantly improve network efficiency by automating repetitive tasks such as configuration management, monitoring, and fault detection. Implementing these technologies allows the network to respond dynamically to changes and optimize performance in real-time, minimizing manual interventions and reducing the potential for human error.
- **Plan for Redundancy and Resilience**: Network availability is critical for business continuity. Incorporating redundancy through N + 1, 2N, or 2N + 1 configurations ensures that if one component fails, another can immediately take over, maintaining network operations without interruption. Balancing redundancy against budget constraints is crucial to achieving high availability without excessive costs.
- **Implement Comprehensive Documentation and Training**: Creating detailed documentation of network designs, configurations, and policies is essential for ongoing maintenance and future upgrades. This documentation should be kept current to reflect changes and provide clear guidance for troubleshooting. Additionally, training IT staff on the latest best practices ensures that your team can effectively manage the network, reducing downtime and enhancing overall performance.
- **Use Top-Down Design Approach When Feasible**: In complex network environments, a top-down design approach, which starts with business requirements and then addresses network needs, is often more effective. This approach ensures that the network design aligns closely with business goals, supporting strategic initiatives and providing the necessary infrastructure for future growth.

One practical example of this strategy is the implementation of a hybrid cloud design, where an enterprise combines its existing on-premises infrastructure with cloud services. This setup allows for more agile data management, optimized resource usage, and improved disaster recovery capabilities. By integrating with existing systems through modular upgrades and maintaining security through microsegmentation, hybrid cloud environments can significantly enhance the overall performance and resilience of brownfield networks.

By incorporating these best practices into your design strategy, you can ensure that your brownfield network is not only capable of supporting current operations but also well-prepared to adapt to future challenges and technological advancements. This comprehensive approach to network design enhances scalability, security, and performance, aligning the network closely with long-term organizational goals.

10.2.5 Network Management and Orchestration Strategy

Adopting advanced network management and orchestration strategies is essential for transforming how brownfield networks operate. This approach involves leveraging centralized platforms that provide visibility, automation, and control across all network segments, thereby enhancing performance, security, and operational efficiency. Here are the key aspects and best practices for implementing an effective network management and orchestration strategy:

Centralized Management and Visibility: Implementing a centralized network management system allows organizations to gain complete visibility into all network components, including multi-cloud, on-premises, and edge environments. Centralized platforms, such as Cisco DNA Center or HPE Aruba Central, integrate various network functions into a single dashboard, enabling real-time monitoring, performance analytics, and seamless policy enforcement. This comprehensive view helps network operators quickly identify issues, optimize traffic flows, and maintain consistent compliance across both legacy and new systems.

Automation and Policy-Driven Orchestration: Modern orchestration tools automate routine tasks such as configuration changes, software updates, and security policy enforcement, reducing manual intervention and the potential for human errors. Policy-driven management enables networks to adapt automatically to changes in demand, such as adjusting bandwidth allocation or rerouting traffic during peak times. Automation platforms like Itential and Terraform allow for high-code customization, enabling network teams to define and enforce complex workflows that align with business objectives.

Integration of AI and Machine Learning: AI and machine learning are becoming integral to network management and orchestration. These technologies analyze network behavior, predict potential issues, and recommend optimizations, significantly improving network reliability and responsiveness. AI-driven orchestration tools, such as IBM Watson AIOps and Splunk, enable proactive management by continuously learning from network data and automating responses to performance anomalies and security threats.

Validation and Intent-Based Networking: Network validation is crucial in ensuring that automated changes produce the intended outcomes without causing disruptions. Integrating validation into the orchestration process helps organizations verify that configurations align with network design and business intent before they are implemented. This approach supports intent-based networking, where the network is managed based on high-level business objectives rather than individual device configurations, enhancing consistency and reliability.

Shift from Node-Based to Design-Based Management: Historically, network management focused on individual nodes or devices. However, the trend is shifting toward managing the entire network as a unified design entity. This approach treats the network holistically, from physical infrastructure to configuration automation, ensuring that all components work together seamlessly to achieve desired outcomes. By shifting focus to the overall design, organizations can reduce discrepancies between intended and realized configurations, ensuring that the network operates as expected.

Integration of Open Source and High-Code Automation Tools: The use of high-code, open-source tools like Ansible and Python has grown, providing network teams with customizable solutions to tailor automation to their specific needs. These tools allow organizations to integrate various automation and orchestration capabilities, facilitating the coordination of network changes across multiple domains and technologies. Open-source automation also supports a collaborative community approach, where best practices and innovations are continually shared and improved.

Emphasizing Redundancy and Scalability: An effective management and orchestration strategy must also prioritize redundancy and scalability. By designing automation workflows that account for failover and backup scenarios, organizations can ensure continuous network availability, even during maintenance or unexpected outages. Scalable orchestration tools enable networks to expand seamlessly, supporting additional devices, users, and services without significant reconfiguration efforts.

Adopting SaaS-Based Network Management Solutions: SaaS-based network management solutions are becoming increasingly popular due to their scalability, ease of deployment, and reduced need for on-premises infrastructure. These platforms offer automated updates, built-in security, and integration with cloud services, making them ideal for organizations looking to modernize their network management without substantial upfront investments. Examples include Cisco Meraki and HPE GreenLake.

Example:Cisco DNA Center for Policy-Driven Orchestration

Cisco DNA Center exemplifies the power of network orchestration by integrating automation, security, and analytics into a single platform. It enables automated provisioning, policy enforcement, and continuous monitoring, adapting network behavior to meet dynamic business needs. By centralizing management, Cisco DNA Center helps streamline operations, reduce costs, and ensure compliance, making it a key tool in modern network management strategies.

By implementing these advanced network management and orchestration strategies, organizations can optimize their brownfield networks, ensuring they are resilient, scalable, and aligned with long-term business goals. This approach not only enhances operational efficiency but also positions the network to support future technological advancements and evolving business requirements.

10.2.6 Optimization Strategy

Optimizing brownfield networks is crucial for maintaining high performance, efficiency, and scalability while leveraging existing infrastructure. A systematic approach to optimization, which includes continuous monitoring, performance testing, and targeted adjustments, ensures that networks remain responsive to evolving business demands. Here are key best practices for implementing an effective optimization strategy for brownfield networks:

- **Data-Driven Network Optimization**: Data-driven optimization relies on real-time analytics from monitoring tools such as Application Performance Monitoring (APM) and AI-driven telemetry. These tools provide valuable insights into traffic patterns, latency, and bottlenecks, enabling proactive adjustments to network configurations. By using such data, organizations can continually refine network operations to prevent performance degradation and maintain service quality.
- **Scalability and Load Handling**: Scalability is a critical aspect of network optimization. It involves both vertical scaling (increasing resources like CPU, memory, or storage) and horizontal scaling (distributing the workload across multiple instances or servers). Techniques such as load balancing and distributed processing help manage increased traffic and prevent any single point of failure, ensuring that the network can handle growing user demands seamlessly. Regular load and performance testing are essential to identify bottlenecks and validate the effectiveness of scalability efforts.
- **Continuous Monitoring and Iterative Optimization**: Continuous monitoring helps maintain optimal performance by providing insights into resource utilization, response times, and error rates. Tools that offer real-time analytics allow network administrators to detect inefficiencies and make immediate adjustments. Regular performance testing, including load and stress tests, simulates real-world conditions, allowing for the identification of performance limitations and enabling iterative optimizations based on feedback loops.
- **Quality of Service (QoS) and Traffic Prioritization**: Implementing QoS policies allows organizations to prioritize critical applications and services, ensuring that essential traffic like voice and video is not disrupted by lower-priority data. This strategic allocation of bandwidth improves the user experience, particularly during peak traffic times, and ensures that the most important services receive the resources they need to operate effectively.
- **Resource Utilization and Efficiency**: Analyzing resource utilization—such as CPU, memory, and bandwidth usage—helps identify inefficiencies and optimize performance. Monitoring tools can highlight high CPU usage, memory leaks, or excessive disk I/O, prompting targeted optimizations to reduce resource consumption. Techniques like caching, data compression, and asynchronous processing can significantly enhance network responsiveness by reducing the need for repetitive computations and minimizing latency.
- **Automation and AI-Driven Adjustments**: Automation tools and AI-driven solutions enable dynamic optimization by automatically adjusting network configurations based on current conditions. For instance, AI can reroute traffic to avoid congestion, allocate bandwidth where it's needed most, and predict potential failures before they impact operations. This level of automation reduces the burden on network administrators

and ensures that performance optimizations are continually applied without manual intervention.

- **Centralized Management and Change Control**: Centralizing management of network assets and change control processes is essential in brownfield environments where systems and documentation are often fragmented. Implementing a centralized platform that integrates asset data and operational processes helps streamline optimizations and ensures that all changes are well-documented and controlled. This approach also facilitates better coordination between engineering, maintenance, and IT teams, ensuring that optimization efforts are aligned with overall business objectives.

By adopting these optimization strategies, organizations can enhance the performance and scalability of their brownfield networks, ensuring that they remain agile, efficient, and capable of meeting current and future demands. This proactive approach aligns network operations with business needs, supporting long-term growth and competitiveness.

10.2.7 Security Strategy

Implementing a comprehensive security strategy for brownfield networks involves integrating modern security practices while managing the challenges posed by existing infrastructure. Here are key best practices and strategies to enhance the security posture of brownfield networks:

- **Adopt a Zero Trust Architecture (ZTA)**: Zero Trust Architecture (ZTA) has become a foundational approach for securing modern networks, moving away from traditional perimeter-based defenses. Zero Trust operates under the principle of "never trust, always verify," ensuring that no user or device is implicitly trusted, regardless of location within the network. This approach includes microsegmentation, strict access controls, and continuous monitoring of all traffic. ZTA is particularly beneficial for brownfield environments, as it helps reduce the attack surface and prevents lateral movement of threats within the network, even when integrating legacy systems with new technologies.
- **Conduct Regular Security Audits and Assessments**: Conducting regular security audits and assessments is essential to identify vulnerabilities within existing infrastructure. This process involves reviewing configurations, detecting unpatched vulnerabilities, and assessing compliance with security best practices. A comprehensive security audit helps to uncover shadow IT, unnecessary open ports, and other security gaps that could be exploited by attackers. Implementing a security audit framework, such as NIST CSF or ISO 27001, provides a structured approach to assessing and improving the security posture of a brownfield network.

- **Implement Advanced Threat Detection and Response**: Deploy advanced threat detection and response tools, such as Extended Detection and Response (XDR) and Security Information and Event Management (SIEM) systems, which provide visibility into network activities and enable rapid response to security incidents. These tools leverage AI and machine learning to detect anomalies, flag potential threats, and automate response actions, significantly reducing the time to detect and respond to security incidents. Integrating threat intelligence feeds further enhances the capability to anticipate and mitigate evolving cyber threats.

- **Harden Network Devices and Segment Network Access**: Hardening network devices involves securing configurations, disabling unnecessary services, and regularly updating firmware to protect against known vulnerabilities. Network segmentation goes beyond firewalls by creating internal boundaries within the network, isolating sensitive systems, and reducing the scope of potential breaches. Implementing VLANs, SDN, or firewall rules allows organizations to control access between segments, protecting critical assets from unauthorized access and limiting the spread of threats within the network.

- **Secure Endpoints and Enforce Least Privilege Access**: Endpoints are often the weakest link in network security. To mitigate risks, deploy Endpoint Detection and Response (EDR) solutions that monitor and block malicious activities at the device level. Enforcing the principle of least privilege ensures that users and applications only have access to the resources necessary for their roles, minimizing the potential damage caused by compromised credentials or malware infections. Regularly reviewing access controls and implementing risk-based conditional access policies further enhance endpoint security.

- **Encrypt Data in Transit and at Rest**: Encryption is a critical defense mechanism for protecting sensitive data both in transit and at rest. Utilizing strong encryption protocols ensures that intercepted data remains unreadable without the appropriate decryption keys. Organizations should also adopt robust key management practices to secure encryption keys against unauthorized access. When encryption is not feasible, data masking and tokenization can be employed to protect sensitive information from exposure during processing.

- **Develop a Proactive Incident Response Plan**: A well-defined incident response plan is essential for minimizing the impact of security breaches. This plan should outline clear roles and responsibilities, communication protocols, and containment measures for various types of incidents. Regularly conducting tabletop exercises and red-blue team drills helps ensure that the incident response team is prepared to act swiftly and effectively in the event of a cyberattack.

- **Foster a Security First Culture**: Cultivating a security-first mindset across the organization involves continuous education and engagement with employees about the importance of cybersecurity. Regular training sessions on phishing, secure password practices, and safe Internet behavior can significantly reduce the risk of human errors that lead to security breaches. Empowering employees to recognize and report suspicious activities strengthens the overall security posture of the network.

By integrating these best practices, brownfield networks can enhance their security strategy to address current threats while preparing for future challenges. A proactive and layered approach ensures that security measures evolve alongside technological advancements and emerging threat landscapes.

10.2.8 Greenfield Network Strategy

Greenfield network strategies involve designing and deploying a new network infrastructure from the ground up, without the constraints of legacy systems. This approach allows organizations to fully leverage the latest technologies, methodologies, and best practices, creating an optimized and future-proof network environment. Below are key components and best practices for an effective greenfield network strategy, incorporating some relevant elements from brownfield strategies that can be adapted to a greenfield context.

- **Start with Clear Objectives and Business Alignment**: A successful greenfield network strategy begins with clearly defined goals that align with business objectives. This step involves understanding the unique needs of the organization, such as scalability requirements, performance targets, and specific security considerations. By aligning the network design with these objectives, organizations can ensure that their investment directly supports business outcomes.
- **Embrace Cloud-Native and Software-Defined Technologies**: Greenfield environments are ideally suited for adopting cloud-native and software-defined networking (SDN) technologies. Cloud-native solutions, such as Kubernetes for container orchestration, and SDN platforms like Cisco ACI or VMware NSX, enable automated, policy-driven network management that enhances agility and scalability. Leveraging these technologies allows for dynamic, on-demand resource allocation and better integration with multi-cloud and hybrid cloud architectures, which are essential for modern businesses.
- **Design for Security from the Ground Up**: Security should be integrated into the greenfield network design from the outset, rather than being added later. This can be achieved by incorporating Zero Trust principles, microsegmentation, and advanced threat detection solutions like SASE (Secure Access Service Edge). Designing the network with security in mind helps protect against threats while enabling secure remote access, a critical feature in today's distributed work environments.
- **Implement Automation and Orchestration**: Automation is a cornerstone of greenfield network strategies, enabling rapid provisioning, configuration, and scaling of network resources. Orchestration platforms such as Ansible, Terraform, and cloud-native orchestration tools automate routine tasks, reduce the likelihood of human error, and

accelerate deployment timelines. This approach mirrors successful brownfield strategies but is especially impactful in greenfield environments, where automation can be integrated seamlessly from the beginning.

- **Focus on Scalability and Future-Proofing**: Designing for scalability ensures that the network can grow alongside the business without needing major overhauls. This includes selecting scalable hardware like modular switches and routers, and implementing high-performance wireless networks that support increasing device density and IoT deployments. Future-proofing also involves planning for emerging technologies such as AI-driven network management, which can dynamically optimize performance as business needs evolve.
- **Optimize with Direct Peering and Low-Latency Connections**: Direct peering connections, where the network directly links to major cloud providers and content delivery networks, can significantly reduce latency and improve performance for cloud applications. This setup is particularly beneficial in greenfield deployments where there are no legacy routing constraints, allowing for optimized data flows that enhance user experience and application responsiveness.
- **Develop a Comprehensive Testing and Validation Plan**: Before going live, rigorous testing and validation are essential to ensure that the greenfield network meets performance, security, and reliability standards. This phase involves stress-testing the network under various load conditions, simulating potential failure scenarios, and validating that security controls function as intended. By addressing potential issues upfront, organizations can avoid costly disruptions after deployment.
- **Leverage Modular and Microservices-Based Architectures**: Greenfield projects benefit from modular network designs and microservices-based application architectures, which allow for independent scaling and maintenance of different network components. This approach simplifies upgrades, enables rapid innovation, and reduces the risk of network-wide outages due to localized failures.

Greenfield network strategies offer a unique opportunity to build a highly optimized, secure, and scalable network environment that supports modern business needs. By incorporating elements such as automation, cloud-native technologies, security-by-design, and direct peering, organizations can maximize the value of their investment while ensuring that the network can adapt to future technological advancements. This strategic approach provides a robust foundation for innovation, operational efficiency, and long-term growth.

10.2.8.1 Hybrid Strategy

Hybrid approaches combine elements of both brownfield and greenfield strategies, enabling organizations to modernize their networks incrementally while maintaining critical customizations and existing systems. This approach is especially beneficial for large organizations with complex environments, where a complete overhaul (greenfield) or a simple upgrade (brownfield) might not be feasible on their own. Here's an expanded look at hybrid strategies, including best practices and key considerations:

- **Selective Modernization and Incremental Upgrades**: The hybrid approach allows organizations to selectively redesign portions of their network that need modernization while retaining and optimizing areas that are still effective. For instance, critical business processes or applications that are stable and functional can be maintained while newer technologies, such as cloud-native applications or AI-driven analytics, are introduced where they will have the most impact. This selective modernization enables businesses to gradually improve their networks without the disruption of a complete overhaul.
- **Balancing Innovation and Stability**: Hybrid models provide a balanced way to adopt cutting-edge technologies while retaining the stability of proven systems. For example, organizations can leverage greenfield methods to introduce new, innovative solutions in specific departments or geographic locations while using brownfield approaches to maintain existing infrastructure that continues to support business operations. This dual approach ensures that critical legacy systems remain operational while modernizing elements that drive innovation.
- **Phased Implementation to Minimize Disruption**: One of the primary advantages of a hybrid approach is the ability to phase in changes, reducing operational disruptions. Instead of a single large-scale deployment, updates can be rolled out gradually, allowing the organization to test new components, integrate them with existing systems, and adjust processes as necessary. This incremental rollout helps maintain business continuity and minimizes risks associated with large-scale transformations.
- **Enhanced Data Management and Integration**: Hybrid strategies also support improved data management by allowing organizations to keep critical data on-premises while utilizing cloud solutions for less sensitive workloads. For instance, businesses can deploy cloud-native analytics to optimize performance insights while keeping sensitive financial or customer data within their own controlled environments. This split approach helps achieve a balance between data accessibility, security, and compliance.
- **Leveraging Existing Investments**: A key benefit of hybrid approaches is the ability to leverage existing investments in technology and infrastructure. By integrating new technologies with current systems, organizations can maximize their return on investment without having to discard or completely replace valuable assets. This approach allows for a more cost-effective modernization process, optimizing resources and maintaining the value of past investments.
- **Integrating Automation and Orchestration**: Automation and orchestration are essential components of hybrid strategies, allowing for seamless coordination between old and new systems. By using tools like Ansible or Terraform, organizations can automate the integration of new technologies, streamline operations, and reduce the complexity of managing mixed environments. These platforms enable unified management across

diverse technologies, ensuring that network performance and security are maintained throughout the transition.

- **Comprehensive Testing and Validation**: Given the complexities involved in hybrid deployments, thorough testing and validation are critical. Each phase of the modernization should include comprehensive testing to ensure that new integrations do not disrupt existing systems. Simulation and stress-testing tools can help identify potential conflicts between legacy and new technologies, allowing for adjustments before full implementation. Regular validation ensures that the network continues to meet performance, security, and compliance standards.

A hybrid network strategy offers the flexibility to gradually modernize while preserving essential aspects of the existing infrastructure. By integrating greenfield innovations with brownfield reliability, organizations can optimize their networks for current needs and future growth without the high risks and costs associated with a complete overhaul. This approach ensures a smoother transition, maintains business continuity, and aligns with long-term strategic goals.

10.3   Selecting the Right Technologies and Products

Choosing the right technologies and products for a modern network involves a deep understanding of current requirements while anticipating future needs. It is crucial to stay informed about emerging technologies while ensuring that these technologies have reached a level of maturity suitable for integration into your network architecture. While some organizations can afford to be pioneers in adopting cutting-edge technologies, experimenting with early-stage solutions may not be feasible for others due to resource constraints, skillset requirements, and the risks associated with untested technologies.

Network environments have become increasingly complex, and leveraging advanced hardware, cloud-native solutions, AI, and automation platforms is critical to building a robust, future-proof network. As a result, adopting the right technology stack is essential to enhance performance, scalability, and security. Below, we outline key best practices, examples, and strategies to help guide the selection of technologies that will shape the network of the future.

10.3.1 Evaluating Network SOC/ASICs

In Chap. 2, we introduced advances in Network SOCs (System on Chips) and ASICs (Application-Specific Integrated Circuits), providing the fundamental knowledge needed to select the right chipset when choosing network hardware. SOCs and ASICs form the backbone of network hardware, directly influencing performance and scalability. Modern network processors are designed for high throughput, low latency, and complex packet processing, supporting advanced features such as deep packet inspection, encryption, and AI-driven analytics.

For instance, processors from companies like Broadcom and Intel are optimized for data-intensive workloads, making them particularly suited for cloud-native and AI-integrated networks. Selecting the right SOC/ASIC based on factors such as performance benchmarks, scalability, and power efficiency is essential to meet both current and future requirements.

Broadcom and Marvell are industry leaders in networking SOCs, while vendors like Cisco, Juniper, and HPE also offer their own ASICs, each providing distinct capabilities tailored to their products. Broadcom®, a key player in this space, offers two primary series of SOCs: StrataXGS® and StrataDNX®, each tailored to specific networking needs such as data center switches and service provider routers. The following diagram illustrates examples of SOCs/ASICs; these examples serve as suggestions rather than definitive guidelines (Fig. 10.3).

Understanding the specific protocols, routing needs, and functions required in various segments of the network is crucial when selecting the appropriate SOCs/ASICs for deployment. Matching these network requirements with the capabilities of specific chipsets ensures that the selected networking equipment will deliver optimal performance and meet the desired operational goals.

Moreover, the advent of Data Processing Units (DPUs) and their integration into networking hardware, as discussed in Chap. 2, further enhances the ability to design next-generation networks with a security-first mindset. DPUs offload and accelerate critical network functions, including security enforcement, telemetry, and traffic management, enabling more agile and secure network designs. This approach allows for more efficient processing of data-intensive tasks, reducing the burden on traditional CPUs and enhancing the overall performance and security of the network.

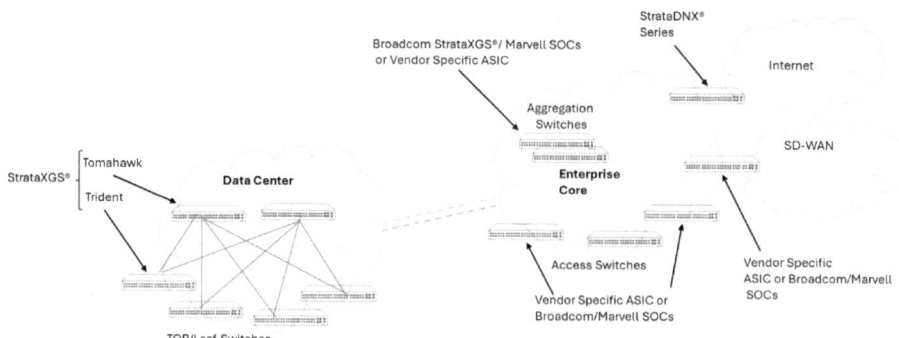

Fig. 10.3 Examples of network SOCs and ASICs that can be used in different segments of enterprise networks

By strategically selecting and aligning the functionalities of SOCs/ASICs with network requirements, organizations can build resilient, scalable, and high-performing networks that are well-equipped to handle future technological advancements.

10.3.2 Cloud-Native, Container-Based, and Hybrid Networks

Cloud-native networks are designed to optimize the use of microservices, containers, and dynamic scaling, making them ideal for modern, agile application development. The key to selecting the right technologies for cloud-native environments includes focusing on tools that support automation, resilience, and scalability:

- **Microservices and Containers**: Cloud-native applications consist of microservices—independent components that work together to form a complete application. Containers, like Docker and Kubernetes, are essential technologies that package microservices with their dependencies, allowing them to run consistently across different environments (development, testing, production). Containers provide flexibility, enabling rapid scaling and reducing infrastructure conflicts between development and operations teams.
- **Service Meshes and APIs**: Service meshes (e.g., Istio) manage the communication between microservices, offering load balancing, observability, and security features without altering application code. APIs facilitate the integration of these microservices, allowing them to communicate efficiently and securely. This setup helps manage the increasing complexity of applications in cloud-native environments.
- **Immutable Infrastructure and Continuous Delivery (CD)**: Adopting immutable infrastructure means that servers are replaced rather than modified, providing consistency and minimizing errors during updates. Continuous delivery tools, like Jenkins and GitLab CI/CD, automate the deployment pipeline, ensuring rapid and reliable software releases without manual intervention, a key advantage in cloud-native operations.

Hybrid networks blend on-premises infrastructure with public or private cloud resources, allowing organizations to maintain control over critical data while leveraging the scalability of the cloud. Key considerations when selecting technologies for hybrid networks include:

- **Integration Capabilities**: Technologies that facilitate seamless integration between on-premises and cloud environments are essential. Solutions like VMware NSX, Cisco ACI, and AWS Outposts enable consistent network management, security policies, and operational efficiency across both cloud and on-premises components.

- **Scalability and Flexibility**: Hybrid architectures should be capable of scaling workloads dynamically, based on current demand. This includes selecting software-defined networking (SDN) tools that support automated network provisioning and scaling, enabling consistent performance regardless of workload location.
- **Security and Compliance**: Hybrid networks need robust security solutions to manage data sovereignty, compliance, and secure access across environments. Implementing Zero Trust models and tools such as Secure Access Service Edge (SASE) ensures that data remains secure as it moves between cloud and on-premises infrastructures.

The right choice of technologies and products for cloud-native, container-based, and hybrid networks hinges on aligning solutions with business needs, understanding operational requirements, and evaluating the maturity of technologies. Organizations should consider:

- **Performance and Scalability**: Evaluate the ability of the technology to handle peak loads, support high-availability requirements, and scale according to business growth.
- **Operational Fit**: Ensure the technologies integrate seamlessly with existing workflows and operational practices, minimizing disruptions during implementation.
- **Security**: Choose solutions that embed security at every layer, from network access controls to application-level defenses, ensuring a holistic approach to protecting assets.
- **Cost-Efficiency**: Analyze the total cost of ownership (TCO) and consider managed services that can offload maintenance tasks, allowing teams to focus on strategic activities.

These guidelines help ensure that the network technology stack not only meets current operational needs but also provides a robust platform for future innovation and growth in cloud-native and hybrid environments.

10.3.3 Automation and AIOps Platforms for Self-Healing and Operational Efficiency

Selecting the right AIOps (Artificial Intelligence for IT Operations) and automation platforms is crucial for building modern, self-healing networks that enhance operational efficiency and minimize manual intervention. As network environments become more complex and dynamic, leveraging AIOps allows organizations to transform traditional IT operations into highly automated, intelligent systems capable of predicting, diagnosing, and resolving issues proactively. Choosing the best platform involves aligning technology capabilities with specific network needs, understanding the integration requirements, and assessing the potential for scalability and adaptability.

The rapid evolution of network infrastructures, including the adoption of cloud-native applications, SD-WAN, and hybrid environments, has increased the need for intelligent automation. Effective AIOps platforms offer a combination of real-time monitoring, advanced analytics, and automation that work together to reduce downtime and optimize performance. They go beyond simple alerting, using machine learning to detect patterns, identify anomalies, and predict potential failures before they impact network operations. This predictive power enables networks to transition from reactive troubleshooting to proactive, preventative management.

Key considerations when selecting AIOps platforms include the depth of integration with existing network systems, the ability to handle multi-vendor environments, and the strength of the platform's AI-driven insights and automation capabilities. A successful AIOps platform should offer robust support for closed-loop automation, where detected issues trigger automated remediation actions without human input. This approach significantly reduces Mean Time to Resolution (MTTR) and ensures that networks can self-heal and adapt quickly to changes, maintaining optimal performance with minimal manual oversight. By prioritizing these factors, businesses can leverage AIOps to drive significant improvements in network reliability, security, and overall operational efficiency, making them well-equipped to meet the challenges of modern digital landscapes.

10.3.3.1 Proactive Monitoring and Incident Management

Proactive monitoring and incident management are crucial for enhancing operational resilience and achieving self-healing capabilities in modern networks. AIOps platforms equipped with advanced monitoring capabilities use AI-driven analytics to detect anomalies, reduce alarm noise, and correlate events in real-time, enabling rapid triage and resolution of incidents. These platforms help network teams move beyond reactive management, instead adopting a proactive stance that minimizes downtime and optimizes resource utilization.

Selecting the right AIOps platform involves understanding how each solution can address the unique needs of network operations. Here is a comparison of leading AIOps platforms and their strengths in proactive monitoring and incident management (Table 10.4).

Augtera Network AI is purpose-built for network-specific applications, offering extensive multi-vendor and multi-layer support that makes it an excellent choice for managing complex and diverse network environments. Its advanced AI/ML algorithms effectively reduce alarm fatigue by filtering out noise and delivering actionable insights, allowing network teams to focus on resolving critical issues quickly and efficiently.

Cisco Crosswork complements Cisco's broader ecosystem while maintaining flexibility for integration with multi-vendor networks, making it versatile for large-scale, dynamic operations. By automating network changes and enhancing service quality through real-time fault correlation and proactive monitoring, Crosswork supports seamless network management and optimization.

Table 10.4 Comparison of leading AIOps platforms for proactive monitoring and incident management

Vendor	Platform	Core feature	Strength
Augtera	Augtera network AI	Purpose-built AI/ML for networks, anomaly detection, auto-correlation	Multi-vendor, multi-layer capabilities; widely adopted in hyperscale and enterprise networks
Cisco	Cisco crosswork	Automated analysis, fault correlation, service monitoring	Best for large scale, multi-vendor networks; strong service provider focus
HPE	Aruba central	AI-driven network management, proactive insights	Well-suited for networking and provides advanced capabilities for HPE Aruba Products
Juniper	Mist AI	AI-driven insights, proactive anomaly detection, network automation	Ideal for Juniper wired/wireless and SD-WAN environments; user experience focus
Arista	CloudVision	Network-wide telemetry, AI-based traffic analysis, anomaly management	Excellent for data center and cloud-integrated environments
IBM	Watson AIOps	Incident correlation, intelligent recommendations, full-stack observability	Advanced ML models, broad applicability in IT environments
Dynatrace	Dynatrace AIOps	Full-stack observability, anomaly detection, auto-remediation	Strong integration with application performance monitoring; best for comprehensive observability
Moogsoft	Moogsoft AIOps	Incident correlation, workflow automation, noise reduction	Effective in reducing alert fatigue in complex IT environments

Aruba Central, designed specifically for network operations, provides AI-driven insights that proactively manage network health, optimize performance, and predict potential issues. This tailored focus on networking allows for efficient management across a variety of scenarios, ensuring that operational goals are met consistently.

Juniper Mist AI further enhances network management by focusing on user experience across wired, wireless, and SD-WAN environments. Its AI-driven insights provide proactive monitoring and rapid anomaly detection, enabling swift issue resolution and improved network reliability.

Together, these AIOps platforms leverage machine learning to identify emerging patterns, detect anomalies, and minimize alarm noise, creating a proactive monitoring environment that allows network teams to preemptively address issues. This integrated approach moves operations closer to a self-healing network, minimizing manual interventions, reducing operational overheads, and ensuring consistent performance and reliability across the entire network landscape.

Integrated Automation within AIOps Platforms: Integrated automation is a critical component of AIOps platforms, enhancing their capabilities beyond just monitoring and incident detection by enabling closed-loop processes where detected issues automatically trigger remediation actions without human intervention. This functionality significantly reduces Mean Time to Resolution (MTTR) and ensures that network issues are addressed promptly and efficiently. The AIOps platforms we previously discussed, including Cisco Crosswork, Juniper Mist AI, Aruba Central, Augtera Networks, Moogsoft, and Dynatrace, all integrate these automation capabilities into their core functions, creating a seamless blend of monitoring, incident management, and automated response.

Cisco Crosswork, for instance, combines real-time fault correlation with intent-based networking, enabling automated adjustments to network configurations that align with business policies and service quality requirements. Similarly, Juniper Mist AI leverages AI-driven insights to automate adjustments in Wi-Fi settings or SD-WAN policies based on observed performance metrics, enhancing network reliability without manual oversight. Aruba Central provides AI-powered management that includes workflows for routine tasks such as firmware updates and fault remediations, directly acting on detected issues to maintain optimal network performance.

Augtera Networks also integrates the automation into its platform, using machine learning to detect anomalies and trigger tailored responses for specific network challenges, such as traffic congestion or configuration errors. Moogsoft and Dynatrace offer auto-remediation features that address incidents by restarting services, adjusting application settings, or reallocating resources, driven by real-time analytics and performance data.

These platforms demonstrate that integrated automation is not a separate function but an inherent part of AIOps, linking proactive monitoring with automated responses to create a cohesive, self-healing network environment. The automation capabilities extend to a range of actions, from simple auto-remediation and configuration adjustments to complex system optimizations, ensuring that networks can adapt dynamically to emerging conditions. This comprehensive approach reduces operational overheads, enhances reliability, and moves network management closer to a fully autonomous model, where manual intervention is minimized, and operational efficiency is maximized.

10.3.3.2 Integrated Automation within AIOps Platforms

Integrated automation is a core feature of AIOps platforms, significantly enhancing their capabilities beyond basic monitoring and incident detection. This automation allows the platforms to automatically trigger remediation actions in response to detected issues, effectively closing the loop between detection and resolution without requiring human intervention. By reducing Mean Time to Resolution (MTTR), integrated automation ensures operational continuity and increases the overall efficiency of network management.

Platforms like Cisco Crosswork provide robust closed-loop automation, automatically adjusting network configurations and initiating corrective actions based on real-time telemetry data, fault correlations, and service policies. This approach aligns changes with business objectives, making the platform highly effective in large-scale, service provider, and enterprise environments. Juniper Mist AI also integrates the automation into its operations, using AI-driven insights not only for anomaly detection but also for automated adjustments, such as modifying Wi-Fi settings or SD-WAN policies based on performance metrics, enhancing user experience without the need for manual input.

Aruba Central's AI-powered network management platform includes automation workflows for routine tasks such as firmware updates, network configuration changes, and fault remediation. Its AI Insights module detects issues and can automatically suggest or execute remediation steps, reducing the need for human oversight. Augtera Networks extends these capabilities by offering automated incident responses tailored for multi-vendor environments. It uses machine learning to detect anomalies and, when combined with automation scripts, resolve specific network issues like traffic congestion, configuration errors, and performance degradation.

Other platforms like Moogsoft and Dynatrace incorporate automation into their AIOps solutions, triggering workflows to address common incidents or initiate service restarts. Dynatrace, in particular, offers auto-remediation features that adjust application settings or resource allocations based on real-time performance analytics, optimizing system operations dynamically.

These integrated automation capabilities, including auto-remediation, automated configuration adjustments, and system optimizations, directly support the role of AIOps platforms as comprehensive solutions for self-healing, operational efficiency, and network optimization. The ability to blend proactive monitoring with automated responses creates a cohesive system that enhances network performance, reduces manual intervention, and ensures continuous, optimized operations across diverse environments.

10.3.4 Telemetry and Observability Tools

Telemetry and observability tools are critical for managing and optimizing modern networks. Telemetry collects data on network performance, traffic, and events, while observability platforms analyze this data to provide actionable insights, detect anomalies, and enable troubleshooting. Selecting the right tools requires a careful evaluation of various attributes, including data collection capabilities, scalability, ease of integration, real-time analysis, and security.

The followings are a few key attributes for selecting Telemetry and Observability Tools:

- **Data Collection Capabilities**: Ability to collect data from various sources (network devices, applications, cloud, etc.) using protocols like IPFIX, NetFlow, sFlow, and SNMP.
- **Scalability**: Support for large-scale deployments with the ability to handle high data volumes without compromising performance.
- **Real-Time Analysis**: Capability to provide real-time insights and alerts for faster incident response.
- **Ease of Integration**: Seamless integration with existing network infrastructure, other monitoring tools, and APIs.
- **Visualization and Reporting**: Advanced dashboards, visualizations, and customizable reports for better data interpretation.
- **Security**: Secure data transmission, compliance features, and role-based access controls.
- **Cost**: Total cost of ownership, including licensing, maintenance, and operational expenses.

Considering these attributes, we have listed some of the leading vendor-specific and open-source telemetry and observability tools in the Table 10.5.

Selecting the right telemetry and observability tools depends on your specific network needs, including scale, integration capabilities, real-time analysis requirements, and security. Vendor solutions like Cisco ThousandEyes and Kentik are powerful but come at a higher cost, while open-source alternatives like Prometheus and OpenTelemetry offer flexibility and cost-effectiveness, albeit with a steeper learning curve. Evaluating these tools against your key attributes will help ensure you choose a platform that provides the right balance of visibility, insights, and operational efficiency for your future network.

10.4 Security First: Embedding Zero Trust and Adaptive Defenses

In Chap. 7, we explored the foundational concepts, technologies, and strategies essential for protecting modern networks. Building on that foundation, this section outlines a practical approach to implementing a Security First Network, emphasizing the integration of Zero Trust principles, adaptive defenses, and advanced security measures. Unlike traditional perimeter-based defenses, a Zero Trust approach assumes no implicit trust within or outside the network. It requires continuous verification of every user, device, and application attempting to access resources, enforcing strict access controls to ensure that only authorized entities have the minimum necessary privileges. This strategy mitigates risks by limiting exposure and reducing the impact of potential breaches.

The accompanying diagram is provided as a guideline, not as an exhaustive list of all components needed for a Security First Network. Instead, it illustrates the layered Zero

Table 10.5 Comparison of telemetry and observability tools

Tool	Vendor	Data collection capabilities	Scalability	Real-time analysis	Ease of integration	Visualization and reporting	Security	Cost
Cisco ThousandEyes	Cisco	IPFIX, NetFlow, SNMP, HTTP, BGP monitoring	High	Yes	Easy integration with cisco devices	Advanced visualizations with network path analysis	Secure data encryption	High
Kentik	Kentik	sFlow, NetFlow, IPFIX, BGP, DNS, HTTP, cloud data	Very High	Yes	Multi-vendor support, cloud integration	Customizable dashboards, advanced analytics	Strong compliance features	Moderate
Datadog network monitoring	Datadog	SNMP, IPFIX, cloud-native data sources	High	Yes	API-driven integration with popular platforms	Detailed dashboards, anomaly detection	RBAC	Moderate to high
Splunk observability cloud	Splunk	SNMP, Syslog, IPFIX, cloud, kubernetes metrics	Very high	Yes	Integrates well with IT and security tools	Highly customizable, AI-driven insights	Strong security compliance	High
Prometheus	Open source	SNMP, HTTP, kubernetes, custom exporters	High	No (real-time via Grafana)	Open APIs, integrates with grafana, kubernetes	Basic but customizable with grafana	Community-driven, basic security	Low (free, with operational costs)

(continued)

Table 10.5 (continued)

Tool	Vendor	Data collection capabilities	Scalability	Real-time analysis	Ease of integration	Visualization and reporting	Security	Cost
Grafana Loki	Open source	Logs collection from cloud and on-premises	High	No	Strong support for kubernetes, prometheus	Advanced visualization via grafana	Basic security via grafana	Low
Opentelemetry	Open source	Traces, metrics, logs from any platform	Very high	Yes (with backend integration)	Vendor-neutral, supports most observability backends	Depends on backend (grafana, etc.)	Secure data handling features	Low
InfluxData telegraf	InfluxData	SNMP, HTTP, Syslog, IoT sensors, custom plugins	High	Limited	Easy integration with influxDB, grafana	Simple, time-series focused dashboards	Limited, open security	Low to moderate

(continued)

Table 10.5 (continued)

Tool	Vendor	Data collection capabilities	Scalability	Real-time analysis	Ease of integration	Visualization and reporting	Security	Cost
SolarWinds NPM	SolarWinds	SNMP, NetFlow, sFlow, IPFIX	Moderate to high	Yes	Best with solarWinds suite, limited open integrations	Intuitive UI with drill-down capabilities	Robust security controls	Moderate
Elastic observability	Elastic	Logs, metrics, APM, security data from elastic stack	High	Yes	Strong integration within elastic stack	Powerful search and visualization with Kibana	Advanced security features	Moderate to high

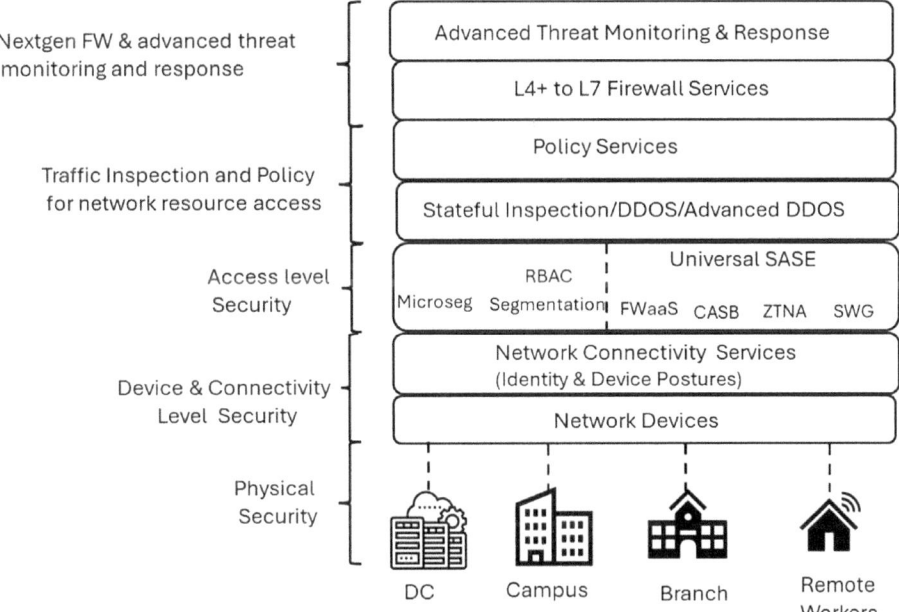

Fig. 10.4 A framework for security first networks

Trust architecture, showing how each component contributes to creating a secure, adaptive, and resilient network environment. The layered concept allows readers to understand the framework and formulate the appropriate tools and strategies needed for each area, tailoring their security approach to the unique requirements of their networks (Fig. 10.4).

The diagram depicts a multi-layered approach to securing networks using Zero Trust principles, showcasing how data centers, campuses, branch offices, and remote users connect securely through a series of security measures that address physical, device, network, access, and advanced threat levels.

1. **Layer 1: Data Center, Campus, Branch Office, and Remote Workers/Home Connectivity**
   - **Security Focus**: Physical security and secure connectivity from various locations.
   - **Design Alignment**: Zero Trust begins with securing all endpoints and access points within the network, whether they are data centers, campuses, branch offices, or remote workers. Physical security controls are essential for protecting data centers and other physical assets from unauthorized access, while secure configurations of network devices ensure a strong foundation.
   - **Enhancements:** Incorporate identity and device posture assessments using solutions like Microsoft Intune and Cisco Secure Endpoint. This ensures that only compliant

devices can connect, protecting the network from potential threats introduced by unauthorized or compromised devices.

2. **Layer 2: Network Devices**
   - **Security Focus**: Device-level security, including firmware updates, secure boot, and hardware-based protections.
   - **Design Alignment**: Securing network devices such as routers, switches, and firewalls is crucial for maintaining the integrity of the entire Zero Trust architecture. Device security measures prevent unauthorized access and ensure that network devices cannot be easily compromised.
   - **Enhancements**: Implement continuous monitoring of device behavior to detect anomalies and unauthorized changes, leveraging telemetry solutions that provide real-time visibility into device performance and security status.

3. **Layer 3: Network Connectivity Services**
   - **Security Focus**: Device and connectivity security, ensuring that data is securely transmitted across the network.
   - **Design Alignment**: Zero Trust emphasizes securing all communication channels. This layer focuses on protecting data in transit using encrypted connections (e.g., TLS, IPSec) and secure tunneling, ensuring that communication paths are authenticated and monitored.
   - **Enhancements**: Implement segmentation at the connectivity level, using technologies like SD-WAN with integrated security to isolate traffic and prevent unauthorized lateral movement within the network.

4. **Layer 4: Universal SASE with FWaaS, CASB, ZTNA, and SWG**
   - **Security Focus**: Secure Access Service Edge (SASE) providing comprehensive security controls across locations and devices.
   - **Design Alignment**: SASE integrates Firewall as a Service (FWaaS), Cloud Access Security Broker (CASB), Zero Trust Network Access (ZTNA), and Secure Web Gateway (SWG) to enforce consistent security policies across the network. This layer supports the Zero Trust model by ensuring secure, policy-driven access for all users, regardless of location.
   - **Microsegmentation and Network Segmentation with RBAC**: These techniques further secure the network by creating isolated zones and enforcing role-based access control (RBAC), limiting access strictly to necessary resources and minimizing the attack surface.
   - **Enhancements**: Utilize adaptive segmentation policies that automatically adjust based on user context, device posture, and threat intelligence. Tools like VMware NSX and Aruba Dynamic Segmentation provide dynamic segmentation capabilities, enhancing security across data centers, cloud, and campus environments.

5. **Layer 5: Stateful Inspection, DDoS, and Advanced DDoS Protections**
   - **Security Focus**: Stateful inspection of traffic and protection against DDoS and advanced DDoS attacks.

- **Design Alignment**: This layer enhances Zero Trust by continuously monitoring and filtering traffic based on context and state, blocking malicious activity before it impacts the network.
- **Enhancements**: Integrate AI-driven analytics for real-time adaptation to evolving attack patterns. This approach not only stops known threats but also learns from new ones, continuously enhancing the network's defensive capabilities.

6. **Layer 6: Policy Services (Network Resources Access and Workgroup Load Policy)**
    - **Security Focus**: Enforcing policies that govern access to network resources and workloads.
    - **Design Alignment**: Policies define who can access what resources, under what conditions, and with what privileges. They are central to Zero Trust, ensuring that all access is explicitly granted based on defined rules and continuously validated.
    - **Enhancements**: Implement continuous policy reviews and updates based on real-time threat intelligence and changes in business needs. Integrating policy management solutions that automatically adjust access based on current risk assessments will further strengthen your Zero Trust architecture.

7. **Layer 7: L4 + to L7 Firewall Services**
    - **Security Focus**: Advanced firewall services that inspect traffic from the transport layer to the application layer.
    - **Design Alignment**: Layer 4 to Layer 7 firewalls provide deep packet inspection, application awareness, and granular control over traffic, crucial for enforcing Zero Trust policies at a deeper level.
    - **Enhancements**: Ensure that firewalls are identity-aware and integrated with broader security policies. This integration helps refine access decisions based on user and device context, enhancing security at the application level.

8. **Layer 8: Advanced Threat Monitoring and Response**
    - **Security Focus**: Continuous monitoring and adaptive response to advanced threats.
    - **Design Alignment**: Advanced threat monitoring and response capabilities are essential for maintaining a proactive Zero Trust posture. This layer integrates AI and machine learning-driven analytics to detect, analyze, and respond to threats in real-time, ensuring that the network is always protected against the latest attack vectors.
    - **Enhancements**: Deploy platforms like CrowdStrike Falcon, Cisco SecureX, or Microsoft Sentinel to provide comprehensive threat intelligence, automated response actions, and seamless integration with the overall security architecture. These tools enhance your network's ability to adapt to new threats dynamically, providing a continuous feedback loop that refines security policies and responses.

The Zero Trust design should effectively integrate security at every level of the network, from physical access controls to advanced threat monitoring. By embedding Zero Trust principles throughout, the network is positioned to adapt and respond to the evolving threat landscape. Continuous monitoring, adaptive segmentation, and AI-driven insights further enhance the architecture, ensuring that security is not just a static set of rules but an evolving framework that grows alongside your network's needs.

The layered approach depicted in the diagram provides a comprehensive and robust security posture, ensuring that every access point, device, and communication channel is secured and continuously validated. This Security First Network strategy aligns perfectly with the Zero Trust model, offering a resilient and forward-thinking approach to modern network security.

10.5 Architecting the Self-Healing Network: A Framework for Implementation

In Chap. 9, we explored self-healing networks, focusing on the role of AIOps in enabling automation and intelligence within network operations. Chapter 10 extended this discussion to the selection of the right tools and products, emphasizing AIOps and telemetry as critical components for building future-ready networks. This section shifts the focus toward architecting a self-healing network, providing a structured approach that integrates these technologies into a cohesive, self-sustaining system. The aim is to outline the architectural elements and strategies essential for developing a self-healing network without reiterating tool-specific discussions previously covered.

One of the first steps in designing a self-healing network architecture is to ensure that network devices can communicate effectively with the orchestration platform, providing information about their condition and the state of the network they are connected to. Some vendors implement lightweight agents within their devices to collect data on issues occurring at the device level. This information is then shared with the network management and orchestration platform through logs, protocols (such as IPFIX, NetFlow, SNMP), and APIs. The diagram below illustrates this concept (Fig. 10.5).

The architecture of a self-healing network must ensure that network devices and tools are capable not only of providing proactive responses to detected issues but also of protecting against catastrophic failures, including major security breaches. In Sect. 10.14, we discussed implementing the Zero Trust model within the network architecture; this security envelope must be capable of interacting seamlessly with the self-healing network framework presented here. Given these considerations, the proposed framework for self-healing network architecture relies on an orchestration platform that enforces automated, intelligent responses across the network. This platform integrates with network devices, analyzes data in real-time, executes automated decisions, and continuously learns from past incidents. Below, we detail how each part of the framework contributes to the overall self-healing architecture:

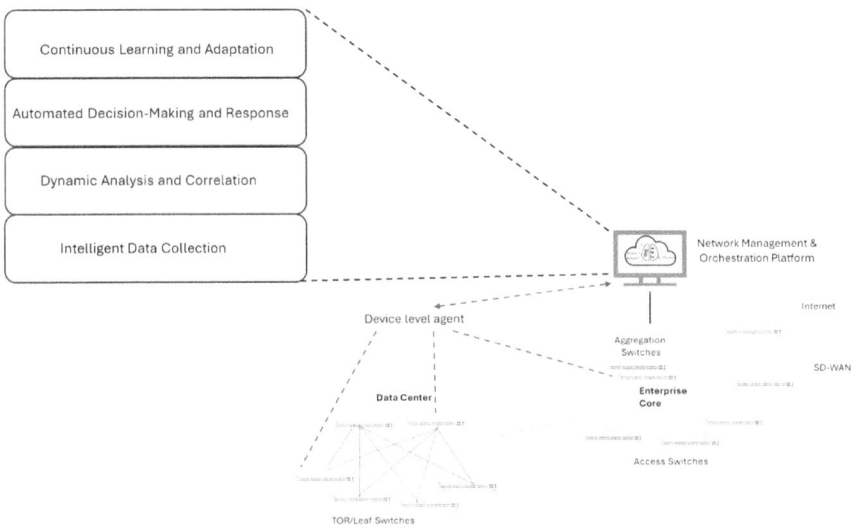

Fig. 10.5 Typical self-healing network with an orchestrating platform that supports self-healing framework

1. **Intelligent Data Collection Layer**: Network devices provide continuous telemetry data using standardized protocols, which is crucial for real-time visibility and proactive management.
2. **Dynamic Analysis and Correlation Layer**: This layer processes the collected data using advanced analytics and machine learning models to identify issues and predict potential failures.
3. **Automated Decision-Making and Response Layer**: Insights from the analysis are used to trigger automated responses through the orchestration platform, such as rerouting traffic, applying patches, or adjusting configurations.
4. **Continuous Learning and Adaptation Layer**: Feedback loops refine AI/ML models and automation policies, allowing the network to evolve and improve its self-healing capabilities over time.

Each layer works in tandem to support a resilient and adaptive self-healing network that aligns with the Zero Trust principles and provides robust protection against evolving threats and operational challenges.

10.5.1 Intelligent Data Collection Layer

The Intelligent Data Collection Layer forms the foundation of a self-healing network, responsible for gathering telemetry data, logs, metrics, and other relevant information from network elements such as routers, switches, firewalls, endpoints, and applications. This layer provides the critical insights required to monitor the network's health, identify performance issues, detect anomalies, and ensure compliance. The key functions of the Intelligent Data Collection Layer include:

- **Real-Time Data Collection**: Continuous data collection from network devices allows for real-time monitoring of the network's state, capturing traffic flows, device health, configuration changes, security logs, and more to provide a comprehensive view of network performance.
- **Data Aggregation and Normalization**: Collected data is often unstructured and raw, requiring aggregation and normalization. This process filters out noise and formats the data in a standardized way that the orchestration platform can easily analyze and interpret.
- **Protocol Integration**: Protocols like IPFIX, NetFlow, sFlow, SNMP, and syslog are critical in collecting telemetry data. These protocols enable devices to communicate information about traffic patterns, performance metrics, and device status, which are essential for proactive network management.

Implementation Strategies

- **Deploy Distributed Telemetry**: Today many network devices are capable of providing streaming telemetry, ensure those network devices are strategically deployed across the network, including at data centers, branch offices, edge locations, and cloud environments, to ensure comprehensive visibility.
- **Use Lightweight Agents**: Implement lightweight agents on network devices to monitor internal state and external interactions, collecting and transmitting data without significantly impacting device performance.
- **Integrate APIs for Data Sharing**: Ensure network devices can communicate with the orchestration platform using APIs, facilitating seamless data exchange and providing the platform with up-to-date information directly from the devices.

Benefits

- Enhanced visibility into network operations.
- Early detection of issues before they escalate into larger problems.
- Improved decision-making based on real-time, accurate data.

10.5.2 Dynamic Analysis and Correlation Layer

The Dynamic Analysis and Correlation Layer serves as the analytical core of the self-healing network. This layer processes the data collected from the Intelligent Data Collection Layer to identify patterns, detect anomalies, and predict potential issues. By using event correlation, predictive analytics, and machine learning models, it enables the network to understand and respond to dynamic conditions proactively. Key functions of the Dynamic Analysis and Correlation Layer include:

- **Event Correlation**: This function links related events across different network components to identify the root causes of issues, reducing noise and focusing on actionable insights that improve fault diagnosis accuracy.
- **Predictive Analytics:** Machine learning models analyze both historical and real-time data to forecast future failures, performance degradation, and security threats, allowing the network to address issues proactively.
- **Anomaly Detection**: Establishes baselines for normal network behavior to detect deviations, indicating potential malfunctions, security breaches, or performance bottlenecks.

Implementation Strategies

- **Deploy Event Correlation Engines**: Implement engines capable of processing vast data sets and identifying patterns that reveal underlying problems, enhancing root cause analysis.
- **Leverage Machine Learning Models**: Use predictive models that continuously train on new data, adapting to evolving network conditions to maintain predictive accuracy.
- **Integrate Real-Time Analytics Tools**: Utilize tools capable of processing and analyzing data in real-time, allowing immediate detection of anomalies that require prompt intervention.

Benefits

- Faster identification of root causes and resolution of issues.
- Proactive management of network health, reducing downtime.
- Enhanced accuracy in diagnosing complex problems through comprehensive data analysis.

10.5.3 Automated Decision-Making and Response Layer

The Automated Decision-Making and Response Layer translates insights from the Dynamic Analysis and Correlation Layer into real-time actions that resolve identified issues. This layer enables the network to autonomously adjust configurations, reroute traffic, apply security patches, or restart services, ensuring continuous performance and security. Key functions of the Automated Decision-Making and Response Layer include:

- **Policy-Based Automation:** Defines automated rules and policies dictating how the network should respond to specific conditions such as performance anomalies, security threats, or device failures, ensuring consistent and repeatable actions.
- **Orchestration and Workflow Automation:** Orchestration tools execute complex workflows across multiple network components, coordinating actions like adjusting bandwidth, updating firewall rules, or rerouting traffic.
- **Fail-Safe Mechanisms:** Include safeguards to monitor the impact of automated actions, ensuring that responses do not introduce new issues and allowing rollbacks if needed.

Implementation Strategies

- **Develop Automation Playbooks:** Create playbooks that define steps for handling various network scenarios, guiding the automation engine in executing appropriate responses.
- **Utilize Orchestration Platforms:** Deploy orchestration platforms that interface with network devices, executing workflows automatically based on predefined triggers and conditions.
- **Integrate Closed-Loop Automation**: Establish closed-loop automation, continuously monitoring the effectiveness of actions and adjusting responses to ensure optimal results.

Benefits

- Reduced Mean Time to Resolution (MTTR) due to quick, automated responses.
- Consistency and reliability in network operations through repeatable actions.
- Dynamic adaptation to network changes, maintaining service quality without manual intervention.

10.5.4 Continuous Learning and Adaptation Layer

The Continuous Learning and Adaptation Layer are critical for ensuring that the self-healing network evolves and improves over time. This layer incorporates AI and machine learning techniques to refine predictive models, update automation policies, and adapt responses to new challenges. By learning from past incidents and feedback, it enhances the network's predictive and corrective capabilities. Key functions of the Continuous Learning and Adaptation Layer include:

- **AI/ML Model Retraining**: Regular retraining of machine learning models with new data maintains their accuracy and relevance, allowing the system to adapt to evolving network conditions and emerging threats.
- **Adaptive Policy Management**: Automation policies are continuously refined based on insights gained from recent incidents and operational feedback, enabling the network to adapt effectively to new challenges.
- **Human-AI Collaboration**: Operators provide feedback on AI-driven actions, validating or refining decisions, which helps integrate human expertise into the automation process.

Implementation Strategies

- **Establish Retraining Pipelines**: Develop automated pipelines for updating machine learning models with the latest data, ensuring the system remains agile and responsive to new conditions.
- **Incorporate Operator Feedback Loops**: Create feedback mechanisms that allow network operators to review and refine AI-driven actions, improving the system's decision-making capabilities.
- **Deploy Explainable AI (XAI) Tools**: Utilize XAI tools to make AI-driven decisions transparent, fostering trust in automated actions and enhancing collaboration between humans and AI.

Benefits

- Continuous improvement in predictive model accuracy and effectiveness.
- Enhanced adaptability to new network conditions and threat landscapes.
- Increased operator confidence and effectiveness through transparent and explainable AI actions.

The proposed framework integrates four critical layers—Intelligent Data Collection, Dynamic Analysis and Correlation, Automated Decision-Making and Response, and Continuous Learning and Adaptation—into a cohesive self-healing network architecture. This comprehensive approach ensures that the network can autonomously detect, diagnose, and resolve issues, maintaining high performance, security, and adaptability in an ever-evolving technological landscape.

Index

© The Editor(s) (if applicable) and The Author(s), under exclusive license
to Springer Nature Switzerland AG 2025
D. D. Chowdhury, *Future of Networks*, Synthesis Lectures on Communications,
https://doi.org/10.1007/978-3-031-71440-5